Creative Problem Solving and Engineering Design

TEACHING MANUAL

A teaching manual will be available from **www.engineering-creativity.com** by Fall 1999. It will provide teaching hints and (when printed out) hardcopy for overhead transparencies and examples of handouts. Directions will be given on how instructors can obtain the manual files containing discussions of class activities and possible outcomes of homework assignments suitable for different levels of students. We also envision that the web site will be used for networking with questions and answers, directions to sites showcasing student projects, sharing of creative ideas, hands-on activities, innovative assignments, effective teaching strategies, etc.

Creative Problem Solving and Engineering Design

Edward Lumsdaine, Ph.D.

Professor of Mechanical Engineering
Michigan Technological University

Monika Lumsdaine

President and Management Consultant for Corporate Behavior
E&M Lumsdaine Solar Consultants, Inc.

J. William (Bill) Shelnutt, P.E.

Professor of Engineering Technology
University of North Carolina at Charlotte

McGraw-Hill, Inc.
College Custom Series
New York St. Louis San Francisco Auckland Bogotá Caracas Lisbon London Madrid
Mexico City Milan Montreal New Delhi San Juan Singapore Sidney Tokyo Toronto

Creative Problem Solving and Engineering Design

ISBN-13: 978-0-07-235909-1
ISBN-10: 0-07-235909-9
Part of
ISBN-13: 978-0-07-236058-5
ISBN-10: 0-07-236058-5

Word processing, desktop layout, and general editing by Monika Lumsdaine (using Microsoft Word 5.1 and Adobe PageMaker 6.5 on an Apple Macintosh Quadra 700 and Laser Writer Pro).

Permissions and Copyrights

The five cartoons designed for this book are by Don Kilpela, Jr.. Copyright ©1999, Don Kilpela, Jr.

The Ned Herrmann materials presented in Chapter 3 and the HBDI forms and worksheets in Chapters 4 and 5 are used by permission of the inventor, Ned Herrmann. Copyright ©1998, 1986 by Ned Herrmann. Figure 3.17 shows the Lumsdaine creative problem-solving model and associated mindsets (with ranges indicated by color) superimposed on the Ned Herrmann four-quadrant model of thinking preferences. These mindset metaphors are "ideals" and are independent of the occupational norms published by Ned Herrmann.

The proposed learning structure for engineers (Figure 1.1) and the cartoon for systems thinking (Figure 4.4) are used with permission from the Boeing Company.

The two BLONDIE comic strips in Chapter 5 are reprinted with special permission of King Features Syndicate; we found them originally in the *Toledo Blade* on September 14 and 15, 1991.

The "what if" story in Chapter 6 by Roger Von Oech is used by permission of the author; it is condensed from the original told in *A Whack on the Side of the Head*, Warner Books, New York, 1983.

The old legend retold by Iron Eyes Cody in Chapter 10 was originally entitled "Words to Grow On" and is reprinted with permission from *Guideposts* Magazine. Copyright © 1988 by Guideposts, Carmel, New York, NY 10512. It first appeared in the July 1988 issue.

The engineering ethics case studies in Chapter 10 are used by permission of *Engineering Times*. They are from the February 1989 and April 1989 issues.

Material on the Pugh method (Chapter 11), FMEA (Appendix D), and FTA (Appendix E) were obtained from the American Supplier Institute, Dearborn, Michigan, and are used by permission.

The example of the design concept drawing (Figure 17.12) is used with permission of C. Bruce Morser. It appeared in the September 1994 issue of *Scientific American* and has been reproduced from there with permission..

The tractor sales drawing in Figure 17.22 is used by permision of Case Corporation (taken from the web site).

The computer program, COMPARE 1.0, and the project planning templates included in the CD-ROM at the back of the book were developed by Bill Shelnutt, copyright ©1999 by Bill Shelnutt.

To our children Andrew, Anne, Alfred, and Arnold,
their spouses Wendy, Jim, Becky, and Sarah,
and our grandchildren Benjamin, Emily, David, Bethany, and Stephen.
E.L. and M.L.

To my wife Joy and our children Greg and Michelle, their spouses Ellen and Joe,
and our grandchildren Emily, Amanda, and Colby.
J.W.S.

Contents

Note:
Resources for further learning (including references, exercises, and a review summary with key concepts and action checklist) are given at the end of Chapters 1 through 13.

List of Activities in the Text

List of Figures

LIST OF TABLES

ABOUT THE AUTHORS

Edward Lumsdaine is currently Professor of Mechanical Engineering at Michigan Technological University and Management Consultant at Ford Motor Company. He previously worked as research engineer at Boeing and held faculty positions at South Dakota State University, the University of Tennessee, and New Mexico State University. He directed the New Mexico Solar Energy Institute and the Energy, Environment and Resources Center at the University of Tennessee, and he was a visiting professor in Egypt, Qatar, and Taiwan. His research projects have spanned many fields from heat transfer, fluid mechanics, turbomachinery, aeroacoustics, solar energy, and robust design (Taguchi methods) to teaching with microcomputers. For thirteen years, he was Dean of Engineering—at the University of Toledo, the University of Michigan-Dearborn, and Michigan Tech. He has pioneered the contextual approach to teaching engineering courses, and for many years he has taught math review using integrated software to engineers in industry. He served as on-site reviewer for the National Science Foundation's engineering education coalition program. Dr. Lumsdaine is a fellow of the American Society of Mechanical Engineers and an associate fellow of the American Institute for Aeronautics and Astronautics. He received the 1994 Chester F. Carlson Award from the American Society for Engineering Education for "designing and implementing significant innovation in a changing technological environment." He has been instrumental in developing the high-tech C3P education and training program at Ford Motor Company, and his current focus is on how innovation can be enhanced in the technical workplace. Ed grew up in Shanghai and made his way to the United States by working for nearly two years on a Danish freighter. He joined the U.S. Air Force and four years later entered junior college in California where he met Monika.

Monika Lumsdaine came to the U.S. from Switzerland in 1958, received a B.S. degree in mathematics with highest honors from New Mexico State University, and became involved in solar energy work through her husband. She founded her own consulting company, and she designed the visitors/operations center of the photovoltaic facility in Lovington, New Mexico, as well as a number of private residences. Her second design won a national award from DOE/HUD. She has extensive technical writing experience in energy conservation, passive solar design, product quality, and engineering. Edward and Monika Lumsdaine developed the math/science Saturday academy for secondary school students in Ohio, and they team-teach creative problem-solving workshops in the U.S. and abroad. Monika is certified in the administration and interpretation of the Herrmann Brain Dominance Instrument (HBDI) and has conducted longitudinal research into the thinking preferences of engineering students. As a visiting scientist, she has team-taught creative problem solving courses at the University of Toledo and Michigan Tech. Her current work as management consultant for corporate behavior is in team building in industry, hospitals, and universities (for students, faculty, staff, engineers, managers, and physicians). Through Monika's HBDI consulting project at the University of North Carolina at Charlotte, the Lumsdaines became acquainted with the work of Bill Shelnutt and realized he would make an ideal co-author for strengthening the engineering design focus in their creative problem solving book. Monika is the main author of the teaching manual that accompanies Parts 1 and 2 this book. It will be available for downloading from the web—the new paradigm of publishing answer books.

James William (Bill) Shelnutt, P.E. is currently Professor of Engineering Technology at the William States Lee College of Engineering, University of North Carolina at Charlotte. In early 1998 he was appointed to serve in the office of the Provost at UNC Charlotte as a faculty associate for teaching/learning/technology, distance education, and program assessment. He earned a B.S. in mechanical engineering from General Motors Institute and an M.S. in systems engineering from the Air Force Institute of Technology. Professor Shelnutt has taught capstone design courses for over 20 years. While at the University of Cincinnati's OMI College of Applied Science, he worked with a faculty team to develop the senior design project course sequence, and he served as Head of the Department of Mechanical Engineering Technology. At UNC Charlotte, he developed the senior design project courses for Mechanical and Manufacturing Engineering Technology, and he served as Chair of the Department of Engineering Technology. He also developed courses and workshops in statistical process control, total quality systems, and designed experimentation. Most recently, he led a faculty team to develop a successful new sequence of introductory courses stressing conceptual design and team skills for all students entering the College of Engineering, and he headed a team of faculty from five universities in developing and presenting a multimedia course in total quality systems. He is certified in the administration and interpretation of the HBDI and certified to facilitate training in the Seven Habits of Highly Effective People for UNC Charlotte personnel. He is also experienced in creative problem solving and team building at universities and in industry. He and his wife Joy live in the beautiful Blue Ridge foothills.

PREFACE

The purpose of this book is to enable engineers and technologists to be more innovative in conceptual design. The integration of creative problem solving with engineering design incorporates a unique double focus: (1) Visualization, cognitive models, teamwork, communications and creative problem solving respond to the needs of industry for employees who have these foundational thinking skills and to the ABET Criteria 2000 (which require that engineering and technology students are able to work on multidisciplinary teams and understand the global context of their work). (2) Application to the twelve steps to quality by design, including "how to" guidelines, planning and economic analysis tools (attached on a DC-ROM) and a library of design documentation formats which enable its users to concentrate on opimizing their design projects and solutions and prevent dysfunctional teams.

The book can be used for three different types of courses depending on the degree of emphasis plased on process (creative problem solving) or product (a rigorous yet innovative design project outcome):

- First-year courses, such as Introduction to the Engineering Profession, CAD, and Conceptual Design—to begin developing the skills that will form the foundation for everything that follows.
- Creative problem solving courses (including design competitions or other multidisciplinary student team projects) for sophomore and junior level students—with topics delivered in a just-in-time format.
- Capstone courses, such as Senior Design Projects.

The book is also a useful resource for engineers and design professionals just starting to work in environments where teamwork is emphasized or where rapid technological change is occurring. Key topics can be taught in on-site seminars or workshops.

To instructors and students alike, the book is challenging, user-friendly, and very practical. At each step, we tried to answer the question of what tools and techniques we could provide to make learning and engineering design easier, more effective, and of higher quality. Although the three parts can be studied sequentially in the conventional manner, the parallel tracks of Part 2 (creative problem solving) and Part 3 (conceptual engineering design) shown in Figure 1.2 on page 8 offer unique flexibility for addressing a variety of needs and learner levels. Many learning activities reinforce theoretical knowledge with immediate application and practice of a broad range of thinking skills.

This book is an ideal companion to the software manuals that teach students a particular design tool. It supplements the more traditional design methodologies with a global, future-oriented outlook and an emphasis on thinking. Here is a brief summary of the chapter content:

Part 1—Fundamental Skills and Mental Models. Chapter 1 provides the big picture—what thinking skills are needed for succeeding in the rapidly changing, global world of the twenty-first century? It also gives hints for effective learning. Chapter 2 enhances memory, visualization, and sketching. Chapter 3 presents three interconnected mental models: Herrmann brain dominance, knowledge creation, and creative problem solving—these frameworks are powerful tools for optimizing learning, teamwork, communication, and innovation. Chapter 4 discusses team development and how to form and manage whole-brain project teams, and Chapter 5 focuses on verbal communication, negotiating a win-win outcome, and technical design communication. Chapter 6 recapitulates Part 1 by showing how to overcome mental blocks to creative thinking.

Part 2—The Creative Problem Solving Process. Three knowledge creation cycles are represented by the creative problem solving process. Chapter 7 teaches how to explore the context and analyze the causes as part of defining the real problem (first cycle). In Chapter 8, students learn the principles of brainstorming, in Chapter 9 the process of idea synthesis, and in Chapter 10 idea judgment. Chapter 11 discusses the Pugh method of creative design concept evaluation and optimization (completion of second cycle). Chapter 12 constitutes a third cycle, as ideas are "sold" and implemented and as the process is monitored and evaluated. Each chapter includes directions for individual and team exercises to practice the creative problem solving process.

Part 3—Applications to Engineering Design. Here, the techniques of creative problem solving are applied to engineering design processes. This part can also serve as a curriculum guide and source of assignments for various types of design courses. Chapter 13 defines "engineering design" as communication in a way that leads to implications for all stages and aspects of the design process: customers, products, processes, systems, ethics, and stewardship. Chapter 14 gives the twelve steps to quality by design, including the concept, parameter, and tolerance design stages; identification of constraints, quantitative design objectives, planning, economic analysis, optimization, evaluation, and presentations.

Chapter 15 presents templates based on Microsoft Project 98 to help designers and students plan their team design projects and stay on track. Chapter 16 introduces economic decision making principles that need to be applied during the design process; their application is made easy through a new program, COMPARE, based on Microsoft Excel. Chapter 17 is a compilation of the entire set of design documentation formats needed in the twelve steps of quality by design. In Chapter 18 students will learn how to spot creativity in organizations and how to function in a creative way (whether or not the workplace environment is supportive of innovation). A technical Appendix provides an awareness of analysis and quality tools used in industry.

Our basic belief is that students can be taught to think more creatively when using the creative problem-solving framework with the design constraints in the optimal sequence of divergent and convergent thinking. In a recent seminar in Singapore, one student asked, "What if brainstorming results in something that is against government policy—what would you do?" Creative problem solving requires that we apply good judgment consistent with the values of the group and understanding the benefits and consequences of the decisions that are being made. Students also learn negotiation skills that can help in getting ideas and continuous improvement accepted. Weak or wrong solutions can be prevented when none of the steps in creative problem solving are omitted or interchanged in the design process.

The grander vision for the benefits of this book (which go beyond engineering design) can be summarized in the words of Paul MacCready, the inventor of such low-energy aircraft as the Gossamer Condor and the Solar Challenger: "No single technological advance will be the key to a safe and comfortable long-term future for civilization. Rather, the key, if any exists, will lie in getting large numbers of human minds to cooperate creatively and from a broad, open-minded perspective to cope with the new challenges." We trust that this book and what it teaches will become a valuable resource for students as they progress through the engineering or technology curriculum and then move on to the industrial workplace, to positions of organizational management and leadership, or to being entrepreneurs in their own businesses.

ACKNOWLEDGMENT

We are deeply indebted to so many people—known and unknown—for what we have learned about creative thinking and for many valuable and intriguing ideas that have found their way into the different versions of our books focused on creative problem solving.

We began with *Creative Problem Solving/Brainstorming*—a grey workshop manual for engineers and managers in industry. Next came training manuals for workshop instructors, teachers, managers, and engineers. The fifth version was a manual for an engineering orientation course at the University of Toledo, followed by a manual for Dana Corporation managers, engineers, and trainers. It underwent a major revision to emerge as the first edition (white cover, 1990) of our textbook published in the College Custom Series by McGraw-Hill: *Creative Problem Solving: An Introductory Course for Engineering Students.* A major revision, *Creative Problem Solving: Thinking Skills for a Changing World,* was published by McGraw-Hill in its College Custom Series (black cover, 1993) and incorporated our effort to reach a broader audence. With thorough editing and updating, this became the green-cover edition published by McGraw-Hill in 1995 under the same title.

With the changes being mandated in engineering education through ABET, we saw the need for a new edition that would include a strong emphasis on engineering design, teamwork, communication, and innovation. Thus Bill Shelnutt joined us as co-author—his extensive experience with teaching engineering design and with quality teams in industry has been extremely valuable in making the book more practical for engineering and technology students. He brought the twelve steps to quality by design to the book, including the design documentation and the planning and economic analysis tools on the attached CD-ROM.

We are grateful to many students, workshop participants, readers, reviewers, and faculty members who have used our books—their feedback has enabled us to continuously improve our material. Also, many ideas and interesting quotes from other authors have found their way into our lectures. When we were ready to publish, we wanted to give credit to all these contributors. Alas, we were unable to identify the source of many of these items. For this we apologize. Where possible, we have added brief comments to the references listed at the end of each chapter, identifying key concepts, ideas, and special vocabulary that we have incorporated into our text. Ken Hardy, an elementary teacher in

Toledo with a sense of humor, sketched many of the original illustrations and deserves a special thank-you. We are still using his line drawings of the creative problem solving mindsets in the overheads (Teaching Manual). The mindset drawings for the 1995 edition were made by Geoffrey Ahlers, an artist from Copper City, Michigan. Don Kilpela, Jr., of Copper Harbor, Michigan—yes, he is the captain of the *Isle Royale Queen III*—developed and drew the cartoons for the present edition. It was fun to observe his creative mind at work as we brainstormed ideas, and his contributions (the drawings as well as the messages) are very much appreciated.

The partnership with Bill Shelnutt added a new dimension to our teamwork and synergy, aided by phone, fax, a face-to-face meeting and work session for the final "design review" and frequent file exchanges by e-mail for feedback, discussion, and integration during the writing and revision process for continuous improvement. We had to employ much creative thinking to cope with balky or incompatible equipment and many detours in our schedules along the way.

Three individuals stand out in our own journey to increased creativity. At the stimulating Creativity Institute at the University of Wisconsin in Whitewater in the summer of 1987, Roger Von Oech really did give us "a whack on the side of the head." Paul MacCready's designs of low-energy vehicles are wonderful examples of his creative spirit and concern for a sustainable future. Ned Herrmann, the creator of brain dominance technology, is a tremendous inspiration to us for his enthusiasm and work in all aspects of creativity and whole-brain thinking. We want to thank him for unstintingly sharing his wisdom and materials with us.

We appreciate the people at McGraw-Hill who kept us organized and watched over the details, particularly Margaret Hollander, B.J. Clark, Margery Luehrs, and Ann Craig on the earlier editions, and Shirley Grall and Pat Dausener this time around.

Above all, the more we learn about thinking, design, and innovation, the more we stand in awe before the Mind of God, the Great Designer and Source of all Creativity.

April 1999 *Edward and Monika Lumsdaine*

Part 1

Foundational Skills and Mental Models

I WAS JUST THINKING YOU GUYS, IF WE COMBINE OUR INDIVIDUAL TALENTS INTO ONE CREATIVE ENTITY WHO KNOWS WHAT GALAXIES WE COULD DISCOVER.
D.K.

1 Introduction

What you can learn from this chapter:
- Why study this book? The benefits of creative problem solving skills.
- Vision and overview—the road map for choosing the best path for your circumstances.
- Definition of key concepts: creativity, problem solving, paradigm shift.
- Hints on how to make learning easier.
- Further learning: references, exercises, review, and action checklist.

Why study this book—the bottom line

We live in a world that is changing rapidly, and in times of change, creative thinking is the key that lets us adapt and succeed. During such times of change, the usual approaches and routine methods are no longer adequate for optimum problem solving, product design, and innovation. We need a framework that will encourage exploration, flexibility, play with ideas, idea synthesis, and constructive evaluation, all within the context of teamwork, good communication, and constraints such as cost and schedules. Creative problem solving is such a framework which employs many different tools and thinking skills. In creative problem solving, we use the whole brain and the capabilities of many different people. Ultimately, this is crucial for the success of any enterprise—be it a relationship, a business, a single task, or a complex design.

Industry is at the vanguard of this change, but the structure of our educational systems has sheltered most administrators, faculty, and teachers from recognizing the critical need for change. For example, the Boeing Company has been working with engineering schools to provide leadership on what is needed in engineering education for the twenty-first century. Figure 1.1 illustrates this vision. Take a moment to compare this model with the way you have been or are being educated. Which of the foundational skills and basic enabling tools have you been taught?

Rivers and mountains are more easily changed than a person's nature.

Old Chinese proverb

Well-prepared students entering an engineering program in the past were expected to have a solid background in math and physics. Period. Most if not all foundational skills and basic enabling tools were missing

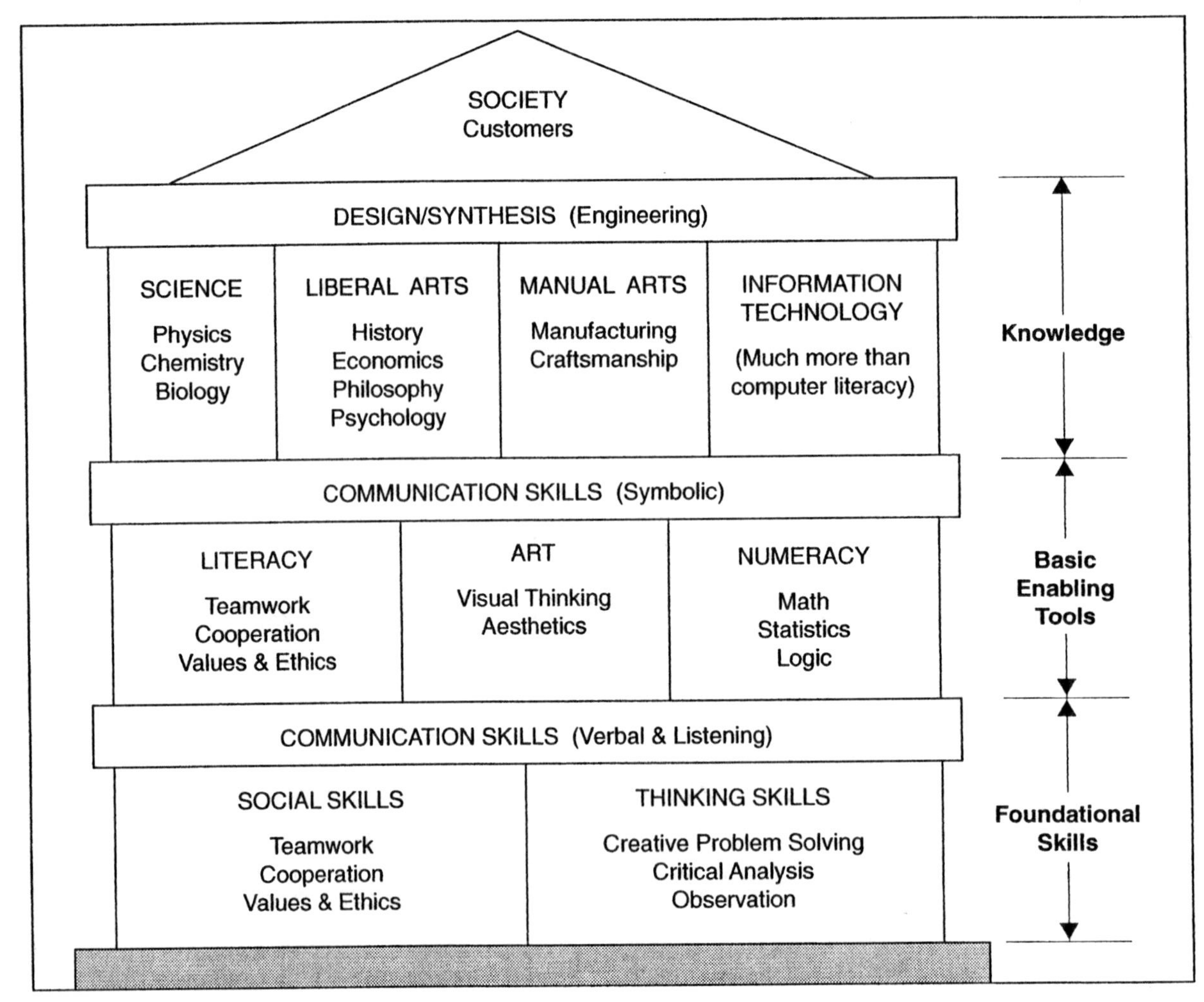

Figure 1.1 A proposed learning structure for engineers.
Boeing draft report (10-24-94) on the second Boeing-university workshop on an industry role in enhancing engineering education, prepared by J.H. McMasters and B.J. White.

because they are not explicitly taught in our secondary schools. Sadly, many of them will still be missing by the time the students graduate and start working in industry. They will have learned some science, some liberal arts, and much procedural engineering analysis, and they may have been introduced to teamwork in an engineering capstone design course. This book is designed to fill the gaps in both learning the foundational skills and in applying them to engineering design:

- Instructors are enabled to integrate these skills into their curricula and their teaching.
- Students can learn these skills either in formal courses or by studying on their own.
- Engineers and technical staff working in high-tech environments can develop and apply these skills on the job—thus this innovative book is above all a resource for lifelong learning in a changing world.

In November 1997, the Accreditation Board for Engineering and Technology approved the ABET Criteria 2000 for accrediting engineering programs. These new criteria represent a major and very important shift from a prescriptive "bean-counting" method based on what courses students have taken to a much broader approach driven by outcome assessments of what students know and how they are succeeding in the workplace. The programs must demonstrate that their graduates have all the abilities and knowledge listed in Table 1.1.

Table 1.1 ABET Requirements for Engineering Graduates

1. **Ability to apply knowledge of mathematics, science, and engineering.**
2. **Ability to design and conduct experiments, as well as to analyze and interpret data.**
3. **Ability to design a system, component, or process to meet desired needs.**
4. **Ability to function on multidisciplinary teams.**
5. **Ability to identify, formulate, and solve engineering problems.**
6. **Understanding of professional and ethical responsibilities.**
7. **Ability to communicate effectively.**
8. **The broad education necessary to understand the impact of engineering solutions in a global societal context.**
9. **A recognition of the need for and an ability to engage in life-long learning.**
10. **Knowledge of contemporary issues.**
11. **Ability to use the techniques, skills, and state-of-the-art engineering tools necessary for engineering practice.**

This book encompasses several learning cycles as it introduces creative thinking (Part 1) and creative problem solving (Part 2) and as it applies the principles to engineering design and product development (Part 3). Effective learning cycles demand your active involvement in doing the exercises and studying the examples so you can "experience" the principles in action—this will require mental effort and interaction with other people's minds and ideas. Thus bring along an open mind, a good dose of curiosity, and a willingness to work hard and communicate—then be prepared to enjoy yourself!

Once you start using your creative problem solving skills, you will reap important and unexpected benefits. Here are a few possibilities:

➧ With creative problem solving, you will be able to overcome many of the shortcomings in your education. Our school systems, all the way from first grade through college, have tended to emphasize the use of our minds for storing information. We have been taught the mechanics or "cookbook" methods of problem solving, also known as the "plug-and-chug" approach. Now you will be able to develop your brain's marvelous power for producing new ideas and turning these into reality and thus enhance your continuous (life-long) learning efforts.

■ Creative problem-solving skills will enrich your life because you can use these thinking skills at home, in recreational pursuits, and in all your interaction with others. You will be energized with a sense of adventure, surprise, and enjoyment. The new ideas and innovative solutions that you will generate will be unique and often of much higher quality than those obtained "the way it has always been done."

You will be able to change an "If it ain't broke don't fix it" attitude to "If it ain't broke make it even better."

■ You will be able to offer more value to your employer beyond mere technical competence since you will have added creative thinking, teamwork, and communication skills. Working together with others, you will identify customer needs, set goals, and find the best solutions to problems. With creative problem solving, you have the power to invigorate the workplace and to direct change.

■ Because calculators and computers are now widely available, many problems can be solved routinely, without much thinking. This leads to the belief that computers allow us to solve complex problems with the same problem-solving skills that we are accustomed to using. However, to properly take advantage of the powerful capabilities of computers, we must substantially expand our creative problem-solving abilities. Our productivity will be enhanced when we use our brain to question, explore, invent, discover, and create—in other words, when we use the brain for creative thinking, a task that computers cannot do.

■ You will be able to help your company, your business, and your service organizations to innovate and find better solutions to complex technical and social problems within the context of the entire community and the global marketplace. You should be able to invent products and ways of doing things that will truly satisfy your customers all over the world—products that are technologically appropriate to culture and the environment while making life better for individuals and communities.

Three-Minute Activity 1-1: Problems

With two other people, brainstorm and jot down problems in the news right now that could use a creative approach because the old ways of dealing with these problems are just not working. Also think about some personal problems that would make good targets for creative problem solving.

The big picture: vision and overview

We believe that just about everyone is born creative, but through early experiences, many of us learned to hide our creativity. For most individuals and corporations, the potential to be creative far exceeds its expression and accomplishments. This book wants to show you how to unlock, develop, nurture, and apply this creativity. Metaphorically speaking, this book is like a trunk full of maps and tools—you will have to use them if you want to find treasure.

This book is divided into three parts. Part 1 presents foundational models and thinking tools used in creative problem solving and engineering design. Part 2 details the creative problem solving process in the proper sequence, including solution optimization. Part 3 focuses on applications in engineering design, and you can emphasize those topics that best meet your present needs in your career development. Figure 1.2 shows major connections between the chapters and topics of the book.

I value this approach of teaching teamwork and creative problem solving to complement the 3 R's. I would love to be able to hire enginers with these "whole-brain approach" skills. We as a nation would be well served if all of our educated youths came to industry packing the skills taught in this book.

John Faust, engineering manager in a Fortune 500 company in Toledo, Ohio. His daughter learned creative problem solving in a precollege program developed for students and their parents.

Part 1—Fundamental skills and mental models. After introducing the context of the book, its benefits, and key definitions, we will focus on visualization: you will quickly learn some skills that serve to enhance memory, as well as observational and spatial skills required in design and solid modeling. We will examine three mental models: (1) the Herrmann four-quadrant model of brain dominance or ways of knowing; (2) the knowledge creation model, and (3) the creative problem solving model. You will see how these models are related and how they can enhance teamwork, communication, and organizational behavior. Separate chapters present applications to creating and managing successful engineering design teams and to the communication skills they need to develop. Also, you will learn to overcome mental blocks to creativity.

Part 2—The creative problem-solving process. The stages of creative problem solving are associated with specific thinking skills or mindset metaphors. You will learn how to find and define the real problem by adopting the mindset of the "explorer" and "detective." You will generate many ideas through brainstorming, using the mindset of the "artist." You will make these ideas better and more practical through synthesis in the mindset of the "engineer," and you will determine the optimum solution in the mindset of the "judge." The related topics of values, ethics, and critical thinking will be discussed, and the iterative Pugh method for creative design concept evaluation and optimization will be introduced at this point. Finally, you will adopt the mindset of the "producer" to put the best solution into action.

Part 3—Application to engineering design. First, engineering design is defined to include all stages and aspects of the design process at the freshman or sophomore level, at the level of industrial or senior capstone design projects, and in engineering practice. The twelve steps leading to quality by design are examined. Templates are provided (based on Microsoft Project 98) that can help students and designers plan their team projects and stay on track. A separate chapter shows how to make economic decisions during the design process, using COMPARE—a new analysis program in a simple spreadsheet format based on Microsoft Excel. The book ends with a look forward to innovation in the workplace. The nature of corporate creativity is explored, and ideas are presented on how to implement change successfully.

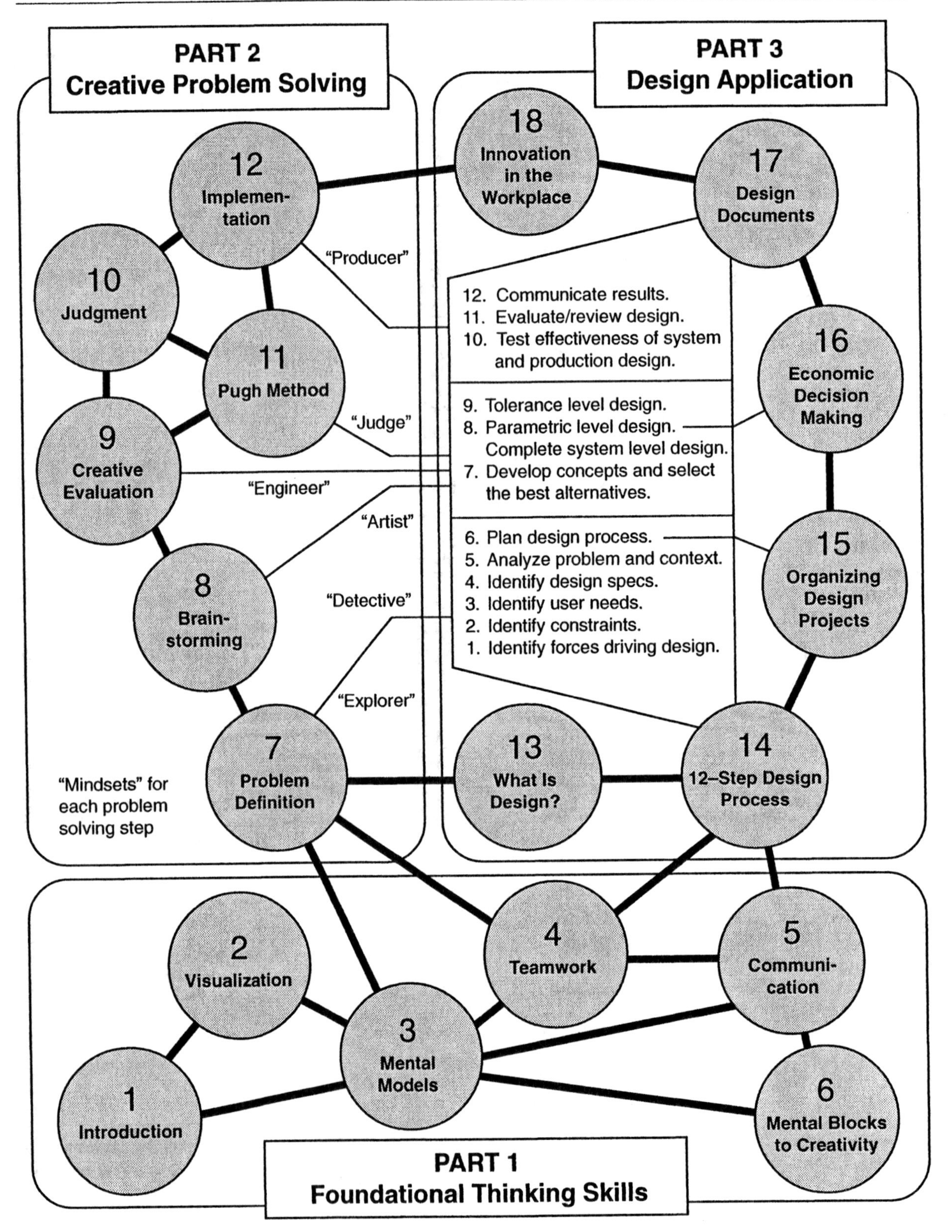

Figure 1.2 Major links between thinking skills, the creative problem solving process, and engineering design.

Definitions of important concepts

Creativity is playing with imagination and possibilities while interacting with ideas, people, and the environment, thus leading to new and meaningful connections and outcomes.

Before we go any further, we need to have a clear and common understanding of key words that we use in this book. So let's look at the definitions of creativity, problem solving, and paradigm change.

What is creativity?

We can think about creativity in many ways. Is it something external, something in the environment, that encourages creativity? Is it primarily internal mental processing that makes up creativity? Does creativity require a tangible output—a product or application—to be valid? We believe that creativity involves all three of these aspects. It is fun to describe aspects of creativity as slogans—you will discover examples throughout the book. As you become more familiar with the subject, you can develop your own definition of creativity.

Ned Herrmann, author of *The Creative Brain,* sees creativity is a dynamic activity that involves conscious and subconscious mental processing—it involves the whole brain. He defines creativity this way:

> My own thinking is that creativity in its fullest sense involves both generating an idea and manifesting it—making something happen as a result. To strengthen creative ability, you need to apply the idea in some form that enables both the experience itself and your own reaction and others' to reinforce your performance. As you and others applaud your creative endeavors, you are likely to become more creative.

When the implementation of a creative idea results in permanent change, we can say that innovation has occurred. In studying the development of innovation in technology over the last 1000 years, it is fascinating to note the many instances when just hearing about an invention or advance in a faraway place has lead to a blossoming of creativity and innovation in another culture (Ref. 1.6). Creativity rarely happens in isolation—it needs other people's minds, ideas, and inventions. Thus in the broadest perspective, creativity is expressed in the quality of the solutions we develop in problem solving.

We can look at creative problem solving as a tool for changing and improving an unsatisfactory situation by using new ideas.

However, if we want to make a permanent change—in essence if we want to innovate—we must overcome resistance to change. This is very difficult to do, and the book is written with this context in mind.

Businesses often use the terms creativity and innovation interchangeably, because many managers feel more comfortable with the word innovation. One key difference between the two processes is originality—which is part of the domain of creativity. Innovation can build on a creative idea, or it can combine creative ideas in novel ways. As we shall see, the mindsets required for each are different: creativity primarily belongs to the "artist," innovation belongs to the "engineer." Creative problem solving provides training and opportunities for both.

Table 1.2 Problem-Solving Schemes of Various Fields

Scientific Method Science	Creative Thinking Psychology	Polya's Method Math	Analytical Thinking Engineering	Team Problem Solving Industry	Creative Problem Solving Many Problems
Inductive data analysis and hypothesis.	Exploration of resources.	What is the problem?	Define and sketch the system. Identify unknowns.	1. Use a team approach. 2. Collect data; define the problem.	1. Problem definition: exploration of trends and context; data collection/analysis.
Deduct possible solutions.	Incubation—possibilities.	Plan the solution.	Model the problem.	3. Deal with the emergency. 4. Find the root causes.	2. Idea generation → many ideas. 3. Creative idea evaluation → better ideas.
Test alternate solutions.	Illumination—definite decision on solution.	Look at alternatives.	Conduct analysis and experiments.	5. Test corrective action and devise best action plan.	4. Idea judgment and decision making → best solution.
Implement best solution.	Verification and modifications.	Carry out the plan. Check the results.	Evaluate the final results.	6. Implement plan. 7. Prevent problem recurrence. 8. Congratulate team.	5. Solution implementation and follow-up. What was learned?

Five-Minute Activity 1-2: Problem Solving

In a brief paragraph, describe the method that you use most often to solve problems. If you are in a class or group, share your answer with one or two people sitting next to you. Then compare your approach with the schemes outlined in Table 1.2.

Problem solving

A problem is not only something that is not working right or an assignment teachers give to students—a problem is anything that could be improved through some change. A problem is finding the best birthday gift ever for the most important person in your life; a problem could be designing, building, or inventing something that fills a specific need; or a problem could be finding a better way of managing an organization or providing a service. As we shall see in Chapter 7, a problem has two aspects, although one may be more apparent: difficulty (or danger), and opportunity (or challenge). It is easy to overlook the opportunity when in the midst of an emergency. When we have dealt with the crisis, we have a chance to introduce a policy of continuous improvement or creatively make a fundamental change leading to true innovation.

Think about the problem-solving approaches you learned in school. Math or science courses usually provide some training in analytical thinking. Perhaps you had an exceptional English teacher who taught you creative thinking and brainstorming to improve your writing. It has been estimated that about eighty percent of all problems in life need to be approached with creative thinking. The creative problem solving process involves all three types of thinking: analytical, creative, and critical, and it employs them in the most appropriate sequence for solving problems well. Table 1.2 compares problem-solving approaches that are taught in various contexts. In addition, some people may use unguided experimentation, trial and error, or guessing—these commonly have unreliable outcomes and are not included here.

The **scientific method** uses inductive data analysis to arrive at a hypothesis. For example, let us say that we are in the business of manufacturing brakes for trucks. We are having a problem: some brakes fail after a relatively short time. We examine the data and hypothesize that heat build-up in the brake rotor disk causes the problem. We design a number of different disk brake configurations with fins and holes to allow the brake to cool faster. We run a series of tests on these prototypes and pick the brake design that seems to solve the problem best. An interesting feature of the scientific method is that it must be on the lookout for data that will *disprove* the hypothesis. The results of problem solving with the scientific method are then reported sequentially, whereas the actual process included many detours—intuition and idea synthesis—that are rarely recognized and acknowledged explicitly.

Psychologists regard **creative thinking** as a process where the available resources and information are explored first, according to researcher Graham Wallas. The mind then subconsciously incubates ideas and possibilities until—quite suddenly—a definite decision on the solution emerges. This is the "aha" phenomenon. The conscious mind verifies this solution and makes minor modifications to make it practical. Since the first idea that comes to mind may not necessarily be a superior idea, a method that invites many different ideas before making a judgment may result in a higher-quality solution.

Mater artium necessitas. Necessity is the mother of invention.

Ancient saying

George Polya devised a set of steps for **solving mathematical problems.** First, we ask: What is the problem? Then we plan the solution and look for alternate ways on how we may be able to get there. Finally, we carry out the plan and check the results. We have all been taught this method in some way in school; our difficulties arise when we use it for other types of problems where such an analytical approach does not work well because it discourages contextual, holistic, and intuitive thinking.

In **engineering problem solving,** we define and sketch the system and identify the key elements before applying the appropriate physical laws. In electrical engineering, we draw a circuit diagram, in mechanics, a free-body diagram. In thermodynamics, the system is defined in terms of a control volume. Next, the known and unknown quantities are listed separately, and the problem is modeled mathematically. In computer-aided engineering, this process is done graphically. Then the model is analyzed. Tests may be needed to determine the accuracy of the model and the assumptions made in modeling. Other items that need checking are the units and the "reasonableness" of the answer. It helps to learn to make quick estimates on the order of magnitude the answer is expected to have. To solve the problem of the overheating disk brake rotor with engineering analysis, the heating, cooling, and internal stresses in the disk rotor are modeled with mathematical equations and then confirmed with tests. Industry now uses tools such as the Taguchi method of designed experiments to optimize product design and testing.

Creative problem solving provides a framework for creative thinking.

Large companies have developed their own problem solving method based on a **team approach.** These methods usually focus on data collection to find the root causes of the problem and then on devising corrective actions. Because few people on these teams have training in creative thinking, analytical thinking predominates. Analytical methods are useful for problems such as finding and fixing a "clank" in an engine. They do not generate innovative design concepts or contextual solutions to problems in the field, because these require creative thinking. For example, analytical methods came up with an alternator design that was able to dump heat more efficiently to prevent damage to the rectifier—but it was creative thinking that conceived a placement for the heat-sensitive rectifier outside the alternator, where it could be cooled by ambient air flow. Creative problem solving might also find a different way for converting alternating current to direct current.

Creative problem solving has five steps that are related to different mindsets. It is a sequence of successive phases of divergent thinking followed by convergent thinking. As "explorers" we brainstorm the context of a problem, as "detectives" we collect as much information about a problem area as possible, then analyze the data and condense it to its major causes or factors, culminating in a problem definition statement expressed as a positive goal. As divergent-thinking "artists" we use brainstorming to get many "wild and crazy" ideas. As "engineers" we first use divergent thinking as we elaborate on ideas but then shift to idea synthesis and convergence to obtain better, more practical solutions. As "judges" we use divergent thinking to explore criteria and constructively improve the final ideas to overcome flaws. We follow with convergent thinking that results in decisions on the best idea for implementation. Implementation itself is a new problem that requires another round of creative problem solving. Thus, as "producers" we repeat the creative problem-solving cycle and again use alternate periods of divergent and convergent thinking. Creative problem solving may employ aspects of all five methods in Table 1.2. Each phase in the process is like an open tool box, with many different techniques available to enhance the process and achieve an optimum result, depending on the type, goals and context of the problem, the time and resources available, the experiences and training of the team members, and the organizational culture.

Divergent thinking is an effort to search, to stretch our thinking, and to consider many possibilities and directions.

Convergent thinking is an effort to screen, select, or choose the most important or promising possibilities, closing in on one or a few items.

Scott G. Isaksen and Donald J. Treffinger

Change and paradigm shift

If change is not important in your life, you may not see the need for having creative thinking skills—yet. Let's think about change. Change is a natural part of living. We have experienced infancy, childhood, adolescence, maturing adulthood with major career moves, marriage, a growing family, an empty nest, the death of loved ones, planning for retirement—all tremendous changes. However, in recent years, the rate of social and technological change has accelerated greatly. Table 1.3 is a list of changes that have affected our world and life in the United States during the last thirty years or so—you can probably think of others.

Table 1.3 Important Changes in the World Since 1970

- We truly are "Spaceship Earth." Ecological concerns are: acid rain, the ozone level, global warming, rain forest preservation, water quality (including the oceans), endangered species, recycling of materials, and waste disposal.
- Energy choices are difficult: oil use has environmental, economic, and political costs; nuclear energy has problems with safe operation of aging plants, radioactive waste disposal, and high costs due to regulation, research, and decommissioning of plants. Coal impacts health and the environment.
- U.S. manufacturing declines while Europe and Pacific Rim countries (notably Japan and Korea) have become leading manufacturers for a global marketplace. The world's economies have become strongly linked.
- Total quality management represents a revolution in manufacturing; zero defects is now the standard; continuous improvement is an attitude leading to innovation.
- Personal computers and Internet access are widely available at reasonable cost. Satellites allow real-time communication worldwide; information is a key resource.
- Women have become important in the work force and in public life. Union power is declining. Minority rights are becoming widely guaranteed.
- Nationalism is rising; communism is disintegrating. The developing world has global influence (politically and economically). Terrorism knows no national borders. Massive population migrations have transformed many countries besides the U.S., notably England, Germany, and the Middle East. In many countries, population flows from rural to urban areas (and the influx of a multitude of war refugees) have created ecological and social problems.
- A loss of values is seen in the decline of marriages and in TV programming, a decreasing respect for authority (police, parents, courts, teachers, government) and increasing drug use, child abuse, and violent crime.
- High-tech health care is available, but access to health care and insurance coverage is not. AIDS, homosexuality, and abortion rights are difficult social issues. Exercise, low-fat foods, and a smoke-free environment have become important.
- Video games, videotapes, TV, air conditioning, and convenience foods have changed family life, recreation, and the sense of neighborhood. CNN and affordable air travel have shrunk the world dramatically.
- A high school (or even a college) education is no guarantee of a job. Downsizing is a threat to employee security even in successful businesses.

We are entering an economic environment that will reward those who can adapt to change and punish those who can't—or won't!

Cordell Reed, senior vice president, Commonwealth Edison Company, Chicago

Let's think about the opposite angle—what are some things that have had little or no change? As a society we have decided that some things are so valuable in their original form that they should not be changed—such as the U.S. Constitution. But even here changes have been made through the amendments, and the interpretation by the Supreme Court has undergone fundamental changes as the values undergirding our culture are changing. Some things change very little, such as the way baseball is played—although in the early days, it took "seven balls for a walk." Some inventions needed little further improvement, such as the zipper. What about education? School systems and universities are examples of institutions that are very resistant to change, even when they are no longer working well in a changing world. U.S. industry as a whole has been quite slow during the 1980's in recognizing the need for change toward higher-quality products and meeting customer needs. Also, because creative ideas demand change, many people face quite a battle to get these ideas accepted and implemented. Why is there often such resistance to change—unless a major crisis makes change imperative?

A classic illustration of this resistance is told by Joel Barker in his videotape on *Discovering the Future: The Business of Paradigms.* Here is a condensed version:

> The Swiss watch manufacturing industry in the 1970's had about 65 percent of the world market in watches and over 80 percent of the profits. Yet from 1979 to 1982, they had to cut employment from 65,000 to 15,000, and their market share fell to less than ten percent. What caused such a rapid decline? The answer of course is the invention of the quartz watch. Do you know who invented the quartz watch? A team of researchers at the Federal Watch Research Center in Neuchâtel, Switzerland, created the first prototype in 1967, but when they presented the model to the Swiss manufacturers, the idea was rejected, instead of patented. When the model was exhibited the following year, Seiko people saw its possibilities. Which nation has the largest share of the watch market now?

A paradigm is a set of rules and regulations that defines boundaries and helps us be successful within those boundaries, where success is measured by the problems solved using these rules and regulations.

Joel Barker, futurist

Why did the Swiss watch manufacturers not recognize the potential of the invention? In its January 14, 1980 issue, *Fortune* magazine wrote:

> The manufacturers simply refused to adjust to one of the biggest technological changes in the history of time-keeping, the development of an electronic watch. Swiss companies were so tied to traditional technology that they couldn't—or wouldn't—see the opportunities offered by the electronic revolution. It was a classic case of vested interests blocking innovation.

This story has a sequel. When Joel Barker returned to Switzerland years later and asked one of these manufacturers about this loss of jobs, he received the astonishing comment that losing all these jobs was not important. Here is the bigger picture: because the early opportunity to get workers trained in small electronics was missed, Switzerland lost out on the next development which was much more significant, namely the manufacture of small electronics components for computers and instruments—a much larger employment market.

Joel Barker explains what happened to the Swiss in terms of paradigm shift. Their old, successful watchmaking paradigm simply blocked them from being able to recognize a different way of keeping time. Paradigms tell you what the game is and how to play it successfully according to the "rules" (even though the rules are not usually spelled out). When a paradigm shifts, past success can be a barrier to future success because it can blind you to visions of the future and possible alternatives. When a paradigm shifts, everyone goes "back to zero."

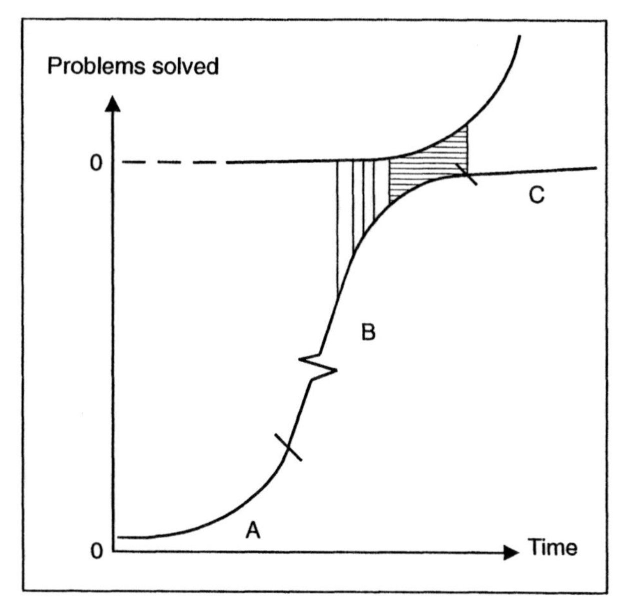

Figure 1.3 The paradigm life-cycle curve.

Paradigms as tools for problem solving have a life cycle in the shape of a typical S-curve as shown in Figure 1.3. In the early phase (Segment A), problem solving is slow because of the learning curve and because only a few pioneers are beginning to use the paradigm. During the main phase (Segment B), problem solving with the paradigm is quite successful and is getting well established, although some "impossible" problems are set aside in the hopes that further development with increased experience, refinement, and precision will help solve these cases. In the last phase (Segment C), problem solving becomes more costly, more time-consuming, and less satisfactory, not only because the problems solved in this stage are the more difficult problems but also because the solutions no longer fit the larger context because of changes elsewhere.

Unsolved, intractable problems create a feeling of uneasiness and uncertainty—a climate that encourages outsiders to look for a new paradigm, even though the current paradigm is still very useful and doing well in solving most problems in its field. This stage of creative thinking by the outsider is shown by the thin vertical lines in Figure 1.3. Once these so-called paradigm shifters are beginning to be successful in solving problems the new way, they are joined by the paradigm pioneers, the people who are adopting the new paradigm and change. This shift may happen over time, as indicated by the hatched area. Note that problem solving now has shifted to a new S-curve. The longer the delay of jumping to a new paradigm, the higher are the costs of making the shift and the lower the probability of being able to compete. In today's rapidly changing world, the window of opportunity for making a timely and profitable paradigm shift can be quite narrow.

A new scientific truth does not triumph by convincing its opponents and making them see the light, but rather because its opponents eventually die, and a new generation grows up that is familiar with it.

Max Planck, originator of quantum theory

Examples

When students turn in writing assignments, some use the old paradigm of doing it by hand. They are competing with students who are using a new report-writing paradigm—a word processor. These students can go through ten or more rewrites; they can use spell and grammar checking tools to put out a higher-quality product. Instructors cannot help but give such reports a higher grade, everything being equal, just because they are much easier to read and thus take less time to grade. Even good-quality hand-written reports can be improved with the new paradigm because when word-processed, they will be easier to proof-read and check for logic. When switching from hand writing to word processing, efficiency is lost at first because learning the new skill requires time. Later, the gain may not be an advantage in time, but definitely an advantage in quality (which is no longer a fixed standard but has expanding boundaries)—writing now requires desktop publishing skills!

Some examples of paradigm shift are very striking—one multi-stage development is illustrated in Figure 1.4. In real-time communication, paradigm shifts have made it possible to send increasingly detailed and accurate messages over greater and greater distances as people have progressed from shouting, smoke, fire, drum, and flag signals to electrically transmitted impulses such as the telegraph, telephone, fax, and live video by wire, optical fiber, and communications satellite. Note that paradigm shift is different in scope from continuous improvement. Two hundred horses hitched together, no matter how powerful and fast, cannot get a carriage to go from zero to 50 miles per hour in five seconds, although they represent the same horsepower as a modern automobile engine. We often overlook the benefit of an attitude of continuous improvement: it prepares the mind to recognize good ideas and to become a paradigm shifter or paradigm pioneer. Innovation represents a paradigm shift.

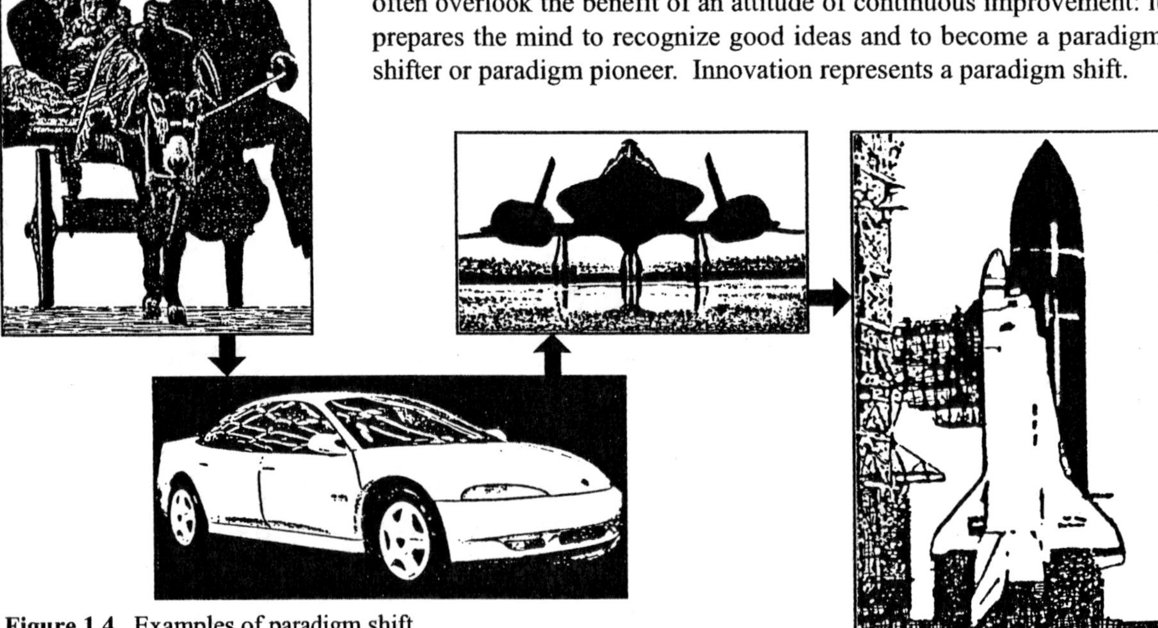

Figure 1.4 Examples of paradigm shift.

Five-Minute Activity 1-3: Paradigm Shift

Depending on your interest, select one of the following:

a. With a teacher or student, discuss where along the paradigm curve you would place the educational systems in your community. Have some paradigms already been discovered through tinkering or breakthrough thinking by people who did not follow the "rules? Are these new paradigms being adopted by paradigm pioneers?

b. With a colleague in your line of work, identify a paradigm shift that has occurred in your organization or in your industry within the last ten years. Discuss what happened. Describe the factors that were instrumental in making the change.

c. With another person, select an area of technology and innovation that is personally impacting you (i.e., home fax machine, high-speed Internet link). Sketch a paradigm shift curve and mark the position of your chosen subject. If it is located in Segment A, discuss the specific advantages and disadvantages of having made the jump. If it is located in Segment B, discuss the possibilities of new paradigms and how they may be discovered. If it is located in Segment C, describe what changes must be made to shift to a new paradigm already being pioneered.

Creative thinking is the key in all phases of paradigm shift. Creativity is exhibited by paradigm shifters and their ideas. To be a paradigm pioneer able to take advantage of the new rules also requires creative thinking, because taking risks, changing directions, and following a vision take flexibility and feeling comfortable with change. Creative thinking will let us recognize good ideas in others, so we can support them and seize the opportunities they represent. Joel Barker gives an example of creative thinking and paradigm shift in a hand-grenade company (Ref. 1.1). The company's president noticed that air bags in cars "go off on impact" and "blow up." He perceived a potential application for his company's expertise. His engineers in less than a year developed a trigger for an automobile air bag that would cost less than $50. The trigger is by far the most expensive component in the air bag system, which in 1995 cost around $600. Can you guess the reaction the hand-grenade people received when they presented their innovative trigger idea to engineers at one of the Detroit auto makers? They were sent away to look for interest elsewhere—at Toyota in Japan and Jaguar in Great Britain.

To prevent us from having an inflexible mind that is incapable of recognizing a coming paradigm shift, we must develop a habit of frequently asking ourselves the paradigm shift question posed by Joel Barker:

What is impossible to do in my field or organization today, but if it could be done, would fundamentally change what I do?

Look at your education or your career this way. Trends and change will not surprise you or pass you by. You will be prepared to become a paradigm pioneer and take advantage of opportunities to innovate.

Chapter organization and other hints to make learning easier

Now that you have seen the overall approach to the book and learned key ideas, we want to give you hints on how each chapter is organized to make learning easier.

Each chapter represents roughly a three-to-five hour study unit or learning cycle. The chapter begins with a listing of learning objectives or goals and a story or question at the personal level for motivation. Then key concepts are introduced, if possible with a metaphor or analogy, followed by information about the subject. Examples and activities are interspersed to enhance tacit understanding. To encourage you to apply the new learning and gain hands-on experience, additional exercises, as well as a list of resources and action items are provided. The knowledge-creation model presented in Chapter 3 will help you understand why we are using this particular sequence of learning modes.

The learning challenges for teenagers and executives are the same: learning to do a job well; facing and enduring hardships; learning from role models; learning from mistakes; and learning in the classroom.

Michael M. Lombardo, Center for Creative Leadership, Greensboro, North Carolina

If you are studying this book on your own, assign yourself as a minimum the exercises and activities marked with a check mark (✓). Advanced exercises are bracketed with a pair of stars (★ ... ★) and will require extra time, but they offer an opportunity for more in-depth thinking and extension of learning.

We use three techniques to encourage your mind to switch from routine reading to an active thinking mode:

1. We will pose questions in the text. Pause to think how you might answer these questions before continuing on with the material.
2. The illustration, comments, highlights, and quotes in the sidebars invite you to make connections between these supplements and the text for better learning and recall. Use your imagination to add your own comments, questions, and sketches in the margins as you read.

Asking questions—a superb learning tool:

- **Think up questions as you study or read about a topic. Jot them downin the margin or in a notebook. You will be surprised how this will prepare your mind to "hear" the answers when you unexpectedly come across them in the next few days.**
- **If the answer to an intriguing question does not appear, explore resources: other people, books and publications, or the Internet.**
- **To sharpen your attention and thereby increase learning and retention, do get into the habit of rehearsing two questions at the end of a lecture or chapter you are reading:**
  1. **What is the most important concept I have just learned?**
  2. **What is an important question I still have?**

3. Watch out for brief assignments given in a rounded box and marked with a timer . We designed these activities or discussions to immediately reinforce your learning—thus take the time to do them. Many assignments in this book integrate academic and social aspects because learning should be integrated with the rest of your life, not compartmentalized. If possible, study with one or two friends, family members, or colleagues. You will learn more about teamwork in Chapter 4 of this book.

Learning is the new form of labor.

Shoshana Zuboff, professor of organizational behavior and human resource management, Harvard Business School

WARNING!
If you want your life to go on as usual, don't read and learn from this book—because learning to think creatively will change your life.

Resources for further learning

Developing a habit of life-long learning will be an important key to maintaining professional skills, success, and employment in the twenty-first century. In high school, students still have the perception that teachers are responsible for making sure the students learn the subject. In college, effective instructors and professors present the subject in a way that makes it interesting and challenging; they are responsible for knowing their subjects well and for discovering new knowledge and new applications. However, it is primarily the responsibility of you as a student to learn, to find your way around, to buy textbooks and use them, and to seek out resources like computer labs, software, seminars, special speakers, and the more challenging (not the easiest) courses. If you do not have good study habits, remedy the situation immediately! Find the learning, study, or writing centers on your campus. Check out guidebooks or videotapes on study skills. Develop good habits!

A textbook is a tool to be used, underlined, highlighted, annotated—not to be preserved in pristine condition.

This "value added" will transform it into a treasure trove and good friend.

For learning more about the topics presented in each chapter, we recommend the references (given with brief "clues") at the end of each chapter. If you are not a good reader, check out audio or videotapes from libraries and bookstores. As soon as you can, invest in a reading course.

References

1.1 ✓ Joel A. Barker, *Future Edge: Discovering the New Paradigms of Success,* Morrow, New York, 1992. The book includes powerful messages about overcoming the resistance to change and creating an innovative environment from his two *Discovering the Future* videotapes: *The Business of Paradigms* and *The Power of Vision,* Charthouse International Learning Corporation, 221 River Ridge Circle, Burnsville, Minnesota 55337. These tapes might be borrowed from a University Extension Service or from a large corporation.

1.2 John Fabian, *Creative Thinking & Problem Solving,* Lewis Publishers, Chelsea, Michigan, 1990. This book targets scientists, engineers, and project leaders. Its breakthrough discovery process has four phases: define the target, search for options, check for fit, and take action, thus putting creative problem solving into a different framework/ vocabulary than we are using in our book.

1.3 Scott G. Isaksen and Donald J. Treffinger, *Creative Problem Solving: The Basic Course,* Bearly Limited, Buffalo, New York, 1985. This softcover workbook emphasizes the cycles of divergent and convergent thinking in the five steps of creative problem solving: mess and data (fact)-finding, problem-finding, idea-finding, solution-finding, and acceptance-finding. This book includes a brief history of research and development of the area of creativity and problem solving.

1.4 ✓ Raymond B. Landis, *Studying Engineering: A Road Map to a Rewarding Career,* Discovery Press, Burbank, California, 1995. This paperback teaches students academic success strategies needed to excel in math, science, and engineering courses, as well as development of behaviors and attitudes that will help them become successful engineers.

1.5 Don Koberg and Jim Bagnall, *All New Universal Traveler,* Kaufmann, Los Altos, California, 1981. This softcover book is a veritable "horn of plenty" for creative ideas, approaches to solving problems, and processes of reaching goals, all presented within a travel metaphor.

1.6 ✓ Arnold Pacey, *Technology in World Civilization—A Thousand-Year History,* MIT Press, Cambridge, Massachusetts, 1990. This interesting book surveys the development of technology in many cultures. It discusses the cross-cultural flow of ideas that can lead to creative thinking and innovations, in both large-scale industrial and appropriate "survival" technologies.

One of the ingredients of survival will be flexibility, tolerance of ambiguity, and creativity in facing issues that will unfold, gain in complexity, and mutate as we grapple with them.

Hunter Lovins, president and executive director, Rocky Mountain Institute, Snowmass, Colorado

Exercises

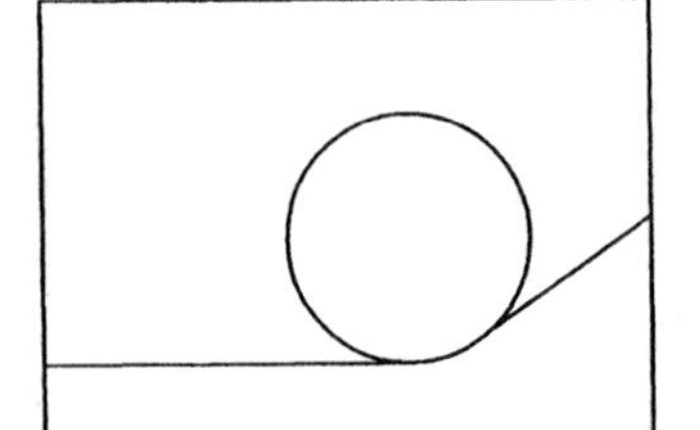

1.1 ✓ Diagnostic Quiz: How Creative Are You?

a. Briefly describe the processes or approaches you use most frequently to solve (1) math problems, and (2) "life" problems.

b. Briefly summarize your previous training in creative thinking and brainstorming.

c. What do you see in the figure on the left? You can give more than one answer.

1.2 ✓ Setting Goals

What do you expect to learn from this book? Make a list of personal benefits that you would like to gain by learning creative problem-solving skills. Also make a list of short-term goals (one year or less) and long-term goals (five to ten years) in your life. Be sure to include these aspects of a balanced life: spiritual, family, career, social, self, health, leisure, money. Now consider your weekly schedule. Have you set aside sufficient time to study this book and do practice problems? Are you incorporating into your weekly schedule activities that are related to your accomplishing your short-term as well as your long-term goals? Hint: See the section on time management in Chapter 12.

1.3 Name Tag and Logo

Design a creative name tag or logo for yourself, with a design or symbol that expresses something meaningful about yourself and your interests.

1.4 ★ Creativity Bulletin Board ★

Over a few weeks or months, make up a bulletin board with comics, cartoons, jokes, and puns that illustrate creative thinking and give a positive message about learning.

1.5 Outside Materials Related to Creativity and Learning

During an entire week, pay attention to anything that relates to the topics presented in this chapter. Do you hear a TV news report mention some government action regarding the quality of education in this country? Do you notice a newspaper or magazine article about a creative learning project being done in an inner-city school? Do you participate in a discussion on the influence of computers on learning? Is your campus involved in curriculum restructuring? Make a folder with clippings and notes. You may also find it fascinating to search the Internet.

No longer is it true that having a skill will guarantee a successful career. Markets, technology and priorities are so quick to change that the only security or assurance of success people have is their productive capabilities. The implicit rule is, "What value can you add today?"

James S. Pepitone, management consultant and business builder

1.6 ★ Tinker Toy™ Invention (Group Project) ★

Preparation: Gather a sack full of Tinker Toy pieces. Form groups of three (or four) people each. Hand out a generous handful of Tinker Toy pieces selected at random for each group.

Assignment: Each group is to invent a model of a new and useful product with two moving parts, come up with a name for the product, and prepare a "sales" presentation—all within 20 minutes. Then enjoy the results as each group does the skit! Each group then writes up a brief summary about the entire experience (5 minutes).

Chapter 1 — review of key concepts and action checklist

The purpose of this book: People need creative thinking skills to succeed in a rapidly changing world. Engineers need creative thinking for innovation and for solving design and manufacturing problems. This is a practical book—a treasure chest of tools to enhance thinking and engineering design—with many applications for acquiring hands-on knowledge. The only prerequisites are an open mind and a willingness to work hard. Creative and innovative thinking can be learned or "unlocked."

Benefits of creative problem-solving skills: Creative problem solving is needed to make up for shortcomings in our education. Personally, the benefits are an enriched life and a successful career, as we learn enhanced communication and teamwork skills. Creative problem solving lets us find optimum solutions to many serious economic, social, and environmental problems for the global community.

Jeder Tag muss das Denken verändern. Each day must change your thinking.

Hans Erni, Swiss artist, in an interview on his 85th birthday

Creativity is playing with imagination and possibilities, then making new and meaningful connections while interacting with ideas, people, and the environment. This process results in a product or application that will encourage more creativity. It can happen within an individual, an interactive team, or as idea transfer between separate cultures.

Problem solving: Math, science, industry, and engineering methods are analytical; creative thinking (from psychology) is a subconscious process. Creative problem solving combines aspects of all of these and purposefully alternates divergent and convergent thinking for defining the problem, generating many ideas, synthesizing better, more practical concepts, finding the best solution, and putting the solution into action.

Paradigm shift: "A paradigm is a set of rules and regulations that defines boundaries and helps us be successful within those boundaries, where success is measured by the problems solved using these rules and regulations" (Joel Barker). Successful people often resist change or are unable to recognize a paradigm shift. The best time to seek new paradigms is while the current paradigm is still useful. Through creative thinking, paradigm shifters come up with new ideas to solve "impossible" problems. Through creative thinking, paradigm pioneers recognize the value of these new ideas, take the risk, and adopt them.

Action checklist

- ☐ Keep learning! Learning is a lifelong occupation, duty, joy, and adventure. Discover, practice, and apply new knowledge and skills.
- ☐ Once a month, alone or with others, brainstorm answers to the paradigm shift question: "What is impossible to do in my field or organization today, but if it could be done, would fundamentally change what I do?" Jot down the ideas in a notebook.
- ☐ Also in your notebook, jot down your short-term and long-term goals. Evaluate your progress once a month. Remember to regularly schedule items from the action checklists in this book in your calendar.
- ☐ Look at James Pepitone's definition of learning given in the quote on the left. List the two most recent occurrences when you learned at this level (in regular studies, at work, in your hobbies or religious activities). Determine to learn something daily at this level for at least 15 minutes!

Learning is an individual self-directed activity associated with developing, changing, and growing which goes beyond acquired knowledge to determine its meaning, significance, and limitations, thus creating new knowledge.

James S. Pepitone
Future Training, *1995*

2

Visualization

What you can learn from this chapter:
- Memory as a complex function of the brain; traditional approaches for improving memory.
- Four mental "languages": verbal, mathematical, visual, and sensory.
- Visual images for large quantities and for changing behavior. The impact of television.
- Visualization techniques to enhance memory: association, substitute word, story link, and phonetic alphabet. Remembering lists of items, names, and numbers.
- Sketching—a tool for visualization and thinking.
- Visualization for design and three-dimensional solid modeling.
- Further learning: references, exercises, review, and action checklist.

Think about getting to the moon!

What thoughts came into your mind first when you read the sentence above? Were you thinking about numbers, all the mathematics, science, engineering, and technology that would be involved? Did you zero in on NASA's Apollo program and its step-by-step problem solving and procedures, or did you recall Neil Armstrong's "giant leap for mankind"? Did your imagination soar to have you instantly walk on the lunar landscape and see the Earth from that perspective (and all this without donning a space suit)? Or did you experience—in your mind—barefoot climbing a sand dune on a balmy evening, hand-in-hand with your beloved to watch the full moon rise over a silvery, softly murmuring, tangy sea? Perhaps you noticed the ambiguity and wondered: "Do I write a science fiction short story, a research report, a poem, or what?"

Visualization or "seeing with the mind's eye" can expand creative thinking and problem solving.

Dorothy A. Sisk, chaired professor, Lamar University

These alternative responses illustrate different but complementary mental languages that we use in thinking. We will focus on visual thinking because schools rarely teach this mode, yet it is a key to creativity and one of the basic enabling tools shown in Figure 1.1 of the Introduction. We will demonstrate powerful memory techniques based on visualization. You will be introduced to exercises that can enhance your sketching and observational skills as well as spatial thinking for solid modeling and design.

Conscious effort is necessary to pursue new directions. Perspiration is, in fact, an excellent investment.

Perhaps the most common inhibition to creativity is our usual reliance upon traditional problem-solving routines and the fantasy that creative problem solving should be easier, rather than more difficult, than producing answers to routine problems.

James L. Adams

Memory and the brain

It is tempting to draw analogies between computer memory and the human brain. The short-term or "scratch-pad" memory is like the work on a computer screen. Unless it is saved into the hard-disk or long-term memory, it can disappear at the touch of a button. When you look up a telephone number and then dial it, you are using short-term memory. If the phone is busy and you have to redial, you will most likely find that you have forgotten the number and will have to look it up again. Human minds have many properties that are very different and far more complex than computers. Human memory may be less reliable than computer memory, but it is much vaster. But most importantly, the human mind is not just an information processor; it is able to associate things in many different ways, and these connections can be very unexpected and creative. In essence, the brain is experience-based, not a logic machine. When our brain learns paradigms and "scripts," it can use these not only to help navigate life efficiently, it is also able to adapt, change, and move these scripts around, either in response to changing circumstances or in response to imagination. Unfortunately, many schools primarily teach passing tests and plugging into the formulas, not exploring different ways of thinking, changing the scripts, and being flexible in the creative use of knowledge and different mental languages.

The first step in learning involves memorizing information—we have to make deposits into our memory banks to enable us to do higher levels of thinking. The main approach used in schools to get students to memorize is through rote learning and repetition, a strictly verbal approach. The mental activity of visualization is commonly ignored, yet this is a powerful tool for remembering because of the way the brain functions.

Scientists in the past thought that people used only about ten percent of the brain's capacity. However, researchers at the University of California at Los Angeles have estimated that we use less than two percent. An adult human brain is roughly the size of a grapefruit and weighs three pounds. To build a computer with the memory capacity of a human brain—if that were possible—would cost many trillions of dollars; it would be the size of a skyscraper and require a huge cooling system. Although computation is one specialized task that computers can do much faster and more accurately than the human brain, we have an amazing array of unbelievably complex thinking abilities.

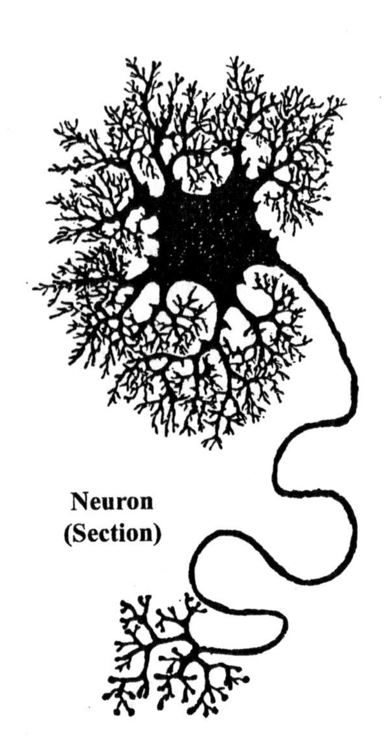

Neuron (Section)

The active cells in the brain and nervous system are called neurons. Neuropsychologists see the neuron as an independent, unique cell not physically connected to other neurons. Think of each neuron as an information processing system. Unlike other cells in the body, neurons do not replicate themselves. It is estimated that a human brain consists of as many as 150 billion neurons. A neuron has a very large number of tentacle-like protrusions called dendrites that make it possible for each neuron to receive synapses (signals) from as many as 1,000 to 15,000 neighboring neurons. Scientists think that the number of possible connections between neurons in a human brain exceeds the number of atoms in the known universe. Some of the synapses are determined by genes, but most are made by experience. In the first five years of life, these connections occur most easily. Outside of some diseases like Alzheimer's, we retain the capability of making new connections into old age, especially if we keep on learning.

The brain is the last and greatest biological frontier; it is the most complex thing we have yet discovered in our universe.

James Watson, codiscoverer of the double helix in DNA

When people say that they are too old to learn a new skill, it is because they are no longer willing to spend the time and effort. How long did it take you to learn math in school? If we are to learn a similar amount of new material later in life, it would take just as long. The good news is that our brain is surprisingly changeable—we can intentionally change the way it functions. Learning new habits, new thinking, and new problem-solving skills will take a considerable effort, since we have to establish new connections in our neural networks that will override our old habitual patterns. Most adults when polled will say that they have a bad memory, yet their brains could function much better if they learned to use the best techniques that work with the brain and its design. As a first step, which of the techniques summarized in Table 2.1 could help you improve your memory?

Mental languages

Imagine the following scenario: It is your sister's birthday, and you are giving a party. A friend arrives, carrying a small gym bag. He announces that your sister must guess what the gift is by sketching it from your description. But you are to be blindfolded, and then you are invited to touch the gift by inserting one hand into the bag. As you explore the mystery object (which seems to have an irregular geometric shape), you are asked to give a running commentary about its attributes, without naming the object. Although this sounds like an easy exercise, you will find it in most cases surprisingly difficult. It is not a simple task to identify shape by feel, and it is even more difficult to describe this type of an object verbally. And making a sketch of the mystery object is a very baffling task for most people, because we are not taught sketching skills.

A well-armed problem solver is fluent in many mental languages (verbal, mathematical, visual, and sensory) and is able to use them interchangeably.

James L. Adams

In our culture, great emphasis is placed on verbal thinking which is constrained by syntax (the sequential word order in sentences). This linear, logical structure exerts a powerful influence on thinking and problem solving toward a single end. However, verbal thinking is not the only thinking mode suitable for problem solving. When we are dealing with quantities, a verbal approach can become very complicated, whereas

Table 2.1 Non-Visual Techniques to Improve Memory

1. **Practice new techniques:** Repetition strengthens a particular new path or structure in the brain if you need exact recall, but understanding is enhanced when new learning is connected in different ways: thus apply it, experiment with it, teach it to someone else.

2. **Time repetitions effectively:** Is cramming intensely the night before an exam the best way to study? A repeated, timed exposure will fix the material in your mind much more solidly than a single, stressed time of hard study. As a minimum, review important new information within 10 minutes (or the same day), followed by reviews one, two, and seven days later. Most people will stop there (and most students will stop after taking an exam), but for long-term retrieval, review in six-month intervals.

3. **Make purposeful connections with what you already know:** The more we know, the easier it is to learn and connect new knowledge. Yet if the new information is very similar to something already in our memory, or if we know very little about the topic, the mind will not pay much attention unless we look for something "odd" in the new input, note these differences, and draw comparisons. Thus ask yourself questions: "What's new about this? How is this different? Does this make sense—in what way"?

4. **Pay attention:** Interest, motivation, and a highly emotional or sensory context in the learning situation sharpen our attention and increase the amount of information transferred to long-term memory. A classic example is the John F. Kennedy assassination—people clearly remember what they were doing when they heard this shocking news. To help recall important information, attach it to sensory "tags" by making an unusual change in your routine learning environment (fragrances are especially effective).

5. **Observe carefully:** Sharpen your observational skills. Scan your environment. Look for things that are odd, different, interesting—things that do not belong in the particular context. Observation differs from passive seeing; it gets you actively involved. Observation is the first step in enabling you to think up useful questions. When you are curious and can ask questions on phenomena that you have observed, you will remember the answers when they come.

6. **Use memory aids:** Who says that you have to remember everything? Make lists; use alarm clocks and timers, appointment books and daily lists of things to do. Take notes. Use videos and computers as teaching and memory assistants. Use maps, charts, and other visual aids. Organize your desk; keep things you use frequently in their assigned place. Develop a filing system and maintain it. Keep a journal or diary. Post notes to yourself in strategic places. Put your name on umbrellas, clothing, books, and pencils. Make use of mnemonics.

7. **Support your brain with a healthy life style:** Exercise regularly; get enough sleep. Choose nutritious foods and eat in moderation. Avoid stimulants like alcohol and drugs; avoid white flour, refined sugar, artificial sweeteners, and saturated fats. Eat protein for alertness; eat larger proportions of carbohydrates after work and study, since these tend to soothe the brain. Fish, soy, oatmeal, rice, and peanuts boost choline (which is a chemical precursor of the neurotransmitter acetyl-choline essential to memory). The folic acid in green, leafy vegetables helps improve brain function and learning. Judicious, non-routine use of caffeine can increase alertness. The "heart-smart" and cancer-prevention diets are also good for the brain.

Although visual thinking can occur primarily in the context of seeing, or only in imagination, or largely with pencil and paper, expert visual thinkers flexibly utilize all three kinds of imagery. They find that seeing, imagining, and drawing are interactive.

Learning to think visually is vital to integrated mental activity.

Robert H. McKim

a mathematical approach using symbols can easily solve the problem. Yet abstract, mathematical thinking also follows structured patterns and conventions and leads to a predetermined outcome. When perceptions are verbally (or mathematically) labeled before they are fully savored, stereotyped thinking is often the result.

Sensory thinking is important to memory and to creative problem solving. Input from our senses of smell, touch, taste, and hearing, as well as from kinesthetic sensations in our body, can have a direct bearing on the problem that needs to be solved, whether you are inventing a prize-winning recipe, devising a marketing strategy for a new toothpaste, designing a steering wheel for a futuristic car, or investigating why a baby is crying. And a physical environment that includes pleasant textures, sounds, and smells indirectly stimulates and enhances brain function for creative thinking because these stimuli are processed primarily in the right hemisphere of the brain. Watch a few television commercials and note how much sensory information is conveyed visually, verbally, and with sound effects and rhythm to help you remember the product being advertised.

Visual imagery—visualization—is a key thinking mode that is involved in many different activities, yet it is only now beginning to receive increased attention. We need to learn to really "see" things around us; we need to practice imagination (or making mental pictures), and we need to develop our skills in graphically representing our ideas. When we sketch images and make diagrams about data and relationships, we make them more concrete and help our thinking processes. Sketching also helps in communicating ideas and information to others. Because visual thinking is holistic, spatial, and not bound by rules, it can lead to synergy, discovery, and surpassing creativity.

⌛ ✓ Ten-Minute Activity 2-1: Ducks and Lambs

Use different mental languages to solve this problem in four or five different ways. A farmer's child received a present of 8 animals (ducks and lambs) with a total of 22 legs. Determine the number of ducks.

Visualizing quantities

All thinking modes can be enhanced through visualization. Let's illustrate this in relation to mathematical thinking. We can develop a feel for size and quantities and how they are related to each other. For example, what is the difference between one million and one billion, or one billion and one trillion? If we use time as a measure of comparison, one million seconds is approximately eleven and a half days; one billion seconds is almost 32 years, and one trillion seconds is over 30,000 years. How do the sizes of viruses, atoms, and protons compare to each other? A virus

Which can you remember better, directions to a certain location conveyed verbally or directions given with a roughly sketched map?

compares to a person's size in the same proportion as a person relates to the size of the Earth. An atom is to a person as a person is to the Earth's orbit around the sun. A proton is to a person as a person is to the distance from Earth to its nearest star, Alpha Centauri. Which is the smallest? Are you "seeing" the scale of comparative smallness?

It is useful to develop a mental image of some common, large quantities. Get two rolls of pennies or dimes and experience what "100 of something" looks and feels like. Examine a brick wall in your neighborhood—how large a wall area contains 1000 bricks? What is easier to compare: a quarter-inch segment on a line that is nine yards long, or the area of a small, dark fingerprint on a standard-size white sheet of paper, or one cubic centimeter (about one teaspoon) in one liter (or one quart) of milk? Each of these represents roughly one part of one thousand.

⌛ ✓ **Five-Minute Activity 2-2: Visualizing Large Quantities**

With two people, brainstorm different ideas on how to visualize 1,000, 10,000, 100,000, and one million. Then select the best ideas and write them down here:

1,000 can be visualized as ______________________________

10,000 can be visualized as ______________________________

100,000 can be visualized as ______________________________

1,000,000 can be visualized as ______________________________

Visualization is a very powerful thinking tool. We will examine applications in memory techniques and sketching in the following sections. Right now we want to show how you can use it to change a habit (a positive effect). Conversely, we want to give you a cautionary message about the effect of unguarded mental images on your brain.

Changing a habit

We can help our mind solve problems when we encode information about the problem in visual form. The subconscious mind will work with this information while we are sleeping or busy with other tasks. This is one reason why having explicit goals and visualizing them is so important. If you frequently picture your goals in your mind in detail, your mind will help you do things that will move you toward achieving these goals. Eventually, you will become the kind of person you imagine yourself to be. Having positive role models works! Visualization can improve your interaction with people, as you mentally rehearse positive behavior in various situations. Visualization is now being used in the training of athletes, where they visualize peak performance and mentally rehearse the skills, routines, and behaviors needed to achieve their goals.

We want to give you a demonstration of how you can use the power of visualization to change an undesirable habit. Let's say you want to improve your health by changing your habit of eating a junk-food breakfast. Let's try an experiment. Tell yourself three times:

Don't eat donuts for breakfast!

What is happening in your mind? What will you be thinking of all day? The mind does not "hear" the "don't"—it sees a vivid image of a donut instead, with the result that you will be thinking of donuts and will give in to your craving, especially if one comes within sight of you. So—what can you do that will help your subconscious mind establish a new habit? Give yourself a positive command; you can tell your mind:

Eat a healthy breakfast of oatmeal and fruit!

Picture in your mind the positive command and the result! Visualize yourself preparing and enjoying a steaming bowl of cinnamon oatmeal together with a baked apple on a cold winter morning or with fresh peaches or berries in the summer. Use sensory thinking—imagine the smells, texture, and taste! This technique of strong visualization works for many situations like giving up smoking, restraining a bad temper, or strengthening the immune system when fighting an illness.

We need to keep this ability of the brain in mind when we interact with small children. If we tell them, "Don't touch the vase," we are giving them an image of touching the vase—no wonder most toddlers will touch the vase within minutes of our warning. Thus, help the child by giving positive directions and opportunities for exploration, play, learning, and creative thinking. Show the child a safe object and ask, "How does this feel when you touch it? How does it smell when you rub it? How does it sound when you tap it?" Identify the object by name. Keep treasures out of reach or better yet out of sight until the child is older.

Whatever you practice, you will perform.

Jerry Lucas, memory expert and gold-medal winning basketball star

What we spend our time thinking about is important because the way we think will affect our behavior. We should not let uncontrolled images influence us—instead, we should use the power of visual thinking to make us more effective thinkers. Some people fear visualization as a tool of philosophies or political programs which they oppose. But each person is the final judge who decides what to think and how to use the marvelous capabilities of the human mind. An informed, thinking mind is the best defense against unwanted subconscious influences.

Watching television or playing video games

The subconscious mind cannot distinguish between real situations and make-believe images. What are we feeding our subconscious minds when we indiscriminately watch television without pausing to do some evaluative thinking? In mid-1989, a *Time* magazine article stated that by the time a typical American child is sixteen, he or she has watched 200,000 episodes of mostly glorified violence on television. The situation has hardly improved since then. Can this be a contributing factor to the alarming increase in the crime rate among young people?

Television can give a distorted, incomplete view of the world. Most of the news is packaged into 30-second sound bites which are accompanied by graphic and often negative visual images seen through the eye and "filters" of the reporter and editors, thus manipulating the audience. Dramas rarely portrait good problem solving; the use of violence predominates to a much larger degree than happens in the real world. We have to make an effort to search out those programs that encourage thought and inform us about the marvels of our world, the lives of good role models, and the background of important issues.

Passively watching television or playing video games for hours can harm the brains of our children. Research is showing that this constant, intense visual input may be neurologically addictive by changing the frequency of electric impulses which can block normal mental processing, including the capacity for creative thinking. Frequent visual and auditory changes force the brain to pay attention in ways that overpower its natural defense mechanisms. Reading, in contrast, develops the language and reasoning skills needed for problem solving, and it encourages imagination. Experts now recommend that children's television viewing be stopped completely or severely curtailed (and then be accompanied by discussion and reading to encourage conscious thinking). In Chapter 6, we will discuss other barriers, besides television, that can keep us from thinking creatively.

Whatever is true,
whatever is noble,
whatever is right,
whatever is pure,
whatever is lovely,
whatever is admirable—
if anything is excellent
or praiseworthy—
think about such things.
Whatever you have
learned or received
or heard from me,
or seen in me—
put it into practice.

Paul's Letter to the Philippians, Chapter 4, Verses 8-9, NIV Bible

Four visualization techniques to enhance memory

Visualization—or thinking in images—is very important to good memory. This mode works because it is based on how the subconscious mind processes information. Scratch-pad memory can only remember about seven unrelated items, whereas in one visual image, the mind is able to link and store thousands of bits of information. Since the mind remembers unusual images best, we also must construct "memorable" images and "weird" linkages for best effect. Some of these techniques were already known in ancient times, when knowledge was transmitted from person to person and generation to generation through memory, not books. In this section, we will demonstrate four methods: association, word substitution, the story link, and the phonetic alphabet. A more advanced method—the memory pegs—can be investigated as a special assignment but cannot be covered within the scope of this book.

Association

In this technique we link the material we want to remember with something we already know well. Speakers in ancient Greece and Rome memorized the topics of their speeches by imagining a walk through their own homes. As they mentally walked from room to room, they associated in their minds the different points of the speech with different places in their home. To be effective, these connections must be unusual and unexpected. This particular technique is known as "loci" or "places."

Memory is the vital source of all aspects of human intelligence, imagination, and accomplishment.

Jack Maguire

Example 1—remembering the outline for a speech: Three different concepts are involved in good memory: understanding the material, filing, and retrieval. We want to use the technique of "places" to memorize the three concepts in sequence. The house we are thinking of has a front porch with a roof supported by a post. This roof—in your mind—transforms into a heavy slab, and on the side facing the street, the word MATERIAL is chiseled in large letters. Now picture yourself standing under the slab of material. Do you have this image of "standing under the material" firmly in mind? If your house does not have a front porch that can be mentally modified into a similar image, you must construct a different image, one that will have meaning for your situation. Now step into the entry. You want to take off your coat. But instead of hanging it in the closet as you usually do, imagine a huge filing cabinet in place of the closet. You pull out an enormous drawer and file away your coat. The third place in your walk through the house is the guest bathroom. Imagine an immense wash basin filled with rubber ducks. You are reaching for a "red sieve" to *retrieve all* these ducks. The red sieve as well as the action should remind you of "retrieval."

If you paid attention when you made the image links, you will be able to "see" this association again at any time. When you are in front of an audience to give your speech or your mind goes blank during a test, think of walking through your house. The associated concepts will pop right into your conscious mind in the correct order. If you want to remember long lists of items, use a system where you will remember five items per room. Visualization is a powerful technique for retrieving information because the visual image is like the code that unlocks the "safe" to a particular area of knowledge. If you understand some topic well, yet do not file the information effectively, you may have difficulty retrieving it later. Visualization will remind you of key words, but it will be your understanding that will fill in and let you use the information.

Example 2—remembering an intention: Association works well when you want to remember an intention—something you want to do at a later time. What if you wake up in the middle of the night and want to remember to take a certain book with you in the morning but don't want to jot down a note? Make a strong visual image of the book percolating in your coffee pot, then go back to sleep. In the morning, when you are ready to pour the coffee from the pot, this visual image will pop into your mind. You can now place the book in your briefcase or backpack. Make sure that you build the "crazy" association with an activity that you will always do, even in the case of oversleeping.

What if during the day at work or school, you suddenly think that you must buy orange juice on your way home? This time you have to remember two things—to stop at the store and to buy a specific item. In your mind, picture the road you normally take and the critical turnoff that will get you to the store. Now imagine an event that would prevent you from continuing straight home: a huge snake rearing up to hiss at you, a large wall of water sweeping you in the right direction, a bonfire in the middle of the road. Follow this image by visualizing a giant bottle of orange juice perched above the entrance to the store, drenching people with juice as they enter. After you have made these images, continue with your tasks of the day. Then, on your way home, your mind will automatically get you to the store where you will remember what you are supposed to buy—the images will pop into your mind on cue.

A word of caution: As you read through the examples in this chapter, do not try to memorize the particular images or stories. They are merely illustrations of techniques you can use. Visualization works best when you make up your own images that are meaningful to you!

Word substitution

Memory expert Jerry Lucas begins his presentation on learning and memory by saying that we first learn as young children by associating the name of an object with seeing the object. The image is linked with the name in our memory because we have "photographic" minds—not a photographic memory which is an ability that only a few retain into adulthood. When you hear the name of the object again, you instantly see its picture in the mind. Let's demonstrate. You are forbidden in the next moment to think of a zebra. So, what mental image popped into your mind as soon as you read the word "zebra"? Of course, you saw a zebra. You cannot NOT think of a zebra. The difficulties come when we are asked to memorize abstract, intangible things. Here we can use a technique called word substitution; we substitute a tangible object for the intangible word or concept. We can't pick just any word; we must select a word that will remind us of the intangible word—a sound-alike word or phrase. Jerry Lucas gives the example of visualizing the word "pronoun" by imagining a nun playing golf—a "pro" nun. This technique is useful for learning vocabulary, and Table 2.2 shows examples. People who are good at remembering names have a big advantage and create much goodwill for themselves. To remember names, we combine word substitution with the association technique, as shown in Table 2.3.

The average man is more interested in his own name than he is in all the other names on earth put together. Remember that a man's name is to him the sweetest and most important sound in any language.

Dale Carnegie

Table 2.2 Vocabulary Examples of the Word Substitution Method

English: *Actinoid* sounds like "act annoyed." Imagine a five-pointed star on a stage being buzzed by an insect. It is acting very annoyed. The picture gives the clue to the meaning of the word ➛ star-shaped.

Italian: *Prezzo ridotto* sounds like "pretzel, rid a toe." Imagine using a pretzel to rid a toe of a huge price tag attached to it. This activity causes the tag to shrink, giving the meaning of the word ➛ reduced price.

Dutch: *Rok* sounds like "rock." Imagine a lady wearing a rock instead of a skirt. This crazy subject-substitution gives the meaning of the word ➛ skirt.

Japanese: *Ahiru* sounds like "Ah hear you." Imagine a duck putting a wing up to its ears and saying the phrase to you in an accented voice. The fowl gives the meaning of the word ➛ duck.

Mandarin: *Wan fan* sounds like "one fun," with the vowel sound in "fun" prolonged a bit. Imagine that the one fun thing to do in China is eating dinner, because there is not much evening entertainment. Thus dinner is synonymous with "number one fun." This gives you the meaning of the word ➛ dinner!

Arabic: The word for book in Arabic sounds like "key tab." Imagine a large key with a tab attached. When you pull the tab, a book emerges from the key, giving you the meaning of the word ➛ book.

Table 2.3 Procedure for Remembering People's Names

1. You must hear and understand the name. If the name is mumbled, ask the person to repeat it slowly or spell it out. People will be flattered that you are interested enough to want to know their name.
2. Next, repeat the name slowly. Pay attention to the way it sounds. Select substitute words that will remind you of the name. For example, the name Traynum sounds like "train of M's."
3. Now look at the person's face and select a prominent feature (excluding eye glasses).
4. Link the image to the feature, preferably associated with some action. If Ms. Traynum has a very high forehead, picture the train of M's chugging across her forehead. When you see her again, one look at her face will bring the image of the train of M's to your mind, and you will instantly remember her name.
5. Review. You will fix people's names in your memory much better, if you review what you have learned about them. Within a few minutes, try to speak to the person, using the name. If you can, jot down the name, the feature, your image, and the related story on a note card (or make a sketch). Review the name and person mentally in the evening, and again one week later.

Example: How would you remember a Mr. Hoppendorfer? A substitute word for the name could be "hopping dwarfs." If this man happens to have very spiky hair, you can visualize several dwarfs hopping across these hair spikes. Make it a vivid, crazy image—it's your private memory aid! The next time you see him (and his hair), the image—and the name—will immediately come to your mind.

⧖ Three-Minute Activity 2-3: Remembering Names

If you are in a class, form a team of three, preferably with people that you do not know yet. Make up a word substitution, story, and associated image to remember each other's names. If necessary, link the first name to the picture also.

The memory link or story chain

What if you had to remember a long list of unrelated items, such as a shopping list? For a history quiz, you might need to remember the list of all the presidents of the United States or a list of technological achievements of the twentieth century.

✓ Five-Minute Activity 2-4: Remembering Lists

Read through the list of twenty technological achievements below and try to memorize the items by going over them for a minute or two. Then cover up the list and try to reconstruct it from memory, in the correct sequence.

Technological Achievements of the Twentieth Century

Henry Ford's assembly line	Nuclear energy	Movies	Freeze drying
Alternating current	Antibiotic	Pacemaker	Communications satellite
Airplane	Transistor radio	Television	Lasers
Nylon	Household appliances	Credit cards	Walk on the moon
Plastics	Telephone	Computers	Solar cells

How many of the items were you able to recall? Without visualization, the average person is usually able to remember, in the scratch pad memory, about seven bits of unrelated information. You will probably remember items at the beginning of the list and a few at the end. The ones in the middle are the most difficult to recall, especially if they have no personal meanings attached to them.

✓ Ten-Minute Activity 2-5: Linking Images

Use the link method and visualization to write or sketch a wild story about the twenty technological achievements. Use your own imagination. If you have trouble doing this, read through the example while making a strong effort to visualize the image and links.

Picture in your mind an automobile assembly line of Model T's because Henry Ford is famous for these cars. Next, imagine wires extending from this assembly line, with light pulsating rapidly in one direction, then in the opposite direction, to indicate alternating current. These wires are attached to an airplane. Now imagine a gigantic nylon stocking dangling from the plane. The nylon stocking is filled with all kinds of plastic objects like spoons and toys. The nylon stocking bursts, and the plastic items spill out over a nuclear plant. This makes the power plant sick; it needs some treatment with antibiotics. You hear the news of this strange treatment from a huge transistor radio rolling along on wheels and passing by the nuclear plant. The radio stops in front of you, and you see that it is full of appliances: toasters, mixers—the whole works. You look through the pile and you find a shiny, purple telephone. With the telephone, you call to reserve tickets for an exciting movie. But the show is too much excitement—you faint. When you wake up, a doctor tells you that you have just been given a pacemaker. You decide to watch television from your hospital bed. This strange TV set can only be turned on by inserting two credit cards. Something goes wrong when you insert the credit cards—they turn into computer equipment programmed to produce freeze-dried products. These freeze-dried packets are put together to form a communications satellite being launched into space. A laser show originates from the satellite—it lights up the whole sky, replaying the first moon walk by Astronaut Neil Armstrong who is opening up a bag and scattering solar cells all over the moon.

Close your book, take a piece of paper, and jot down the list of items from memory.

How did you do this time? We think you will be surprised at the results if you have never used the story link before. Repeat your experiment in a day or even a week. Most likely, you will remember the entire string of items forward and backward—this technique is very effective. If you have a problem recalling some of the items, it will most likely be because you did not have a good image or link. When you make up your own striking links and imagine (and perhaps even sketch) your own interesting story, the associations and thus the memory will be strong.

Association, word substitution, and the story link are very useful techniques for memorizing lists of words and names. But what do you do if you want to memorize numbers? This requires some additional tools—the phonetic alphabet and the peg system. We will introduce the phonetic alphabet in the next section. The peg system can be investigated as part of an advanced exercise at the end of this chapter.

The phonetic alphabet

To remember dates and numbers, a simple phonetic alphabet is used. Each numeral from 0 to 9 is assigned to a distinct sound in the English alphabet. The consonants making these sounds are given the respective numerical value. As a help for memorizing these pairings, you can use visualization. As you read through the list in Table 2.4, try to make a sketch of each explanation.

Table 2.4 Associating Phonetic Alphabet Sounds with Numerals

1 = T, D	Think of an umpire signaling one touchdown at a football game. Also, the letter T (or t) has one downstroke.
2 = N	A letter N, when tipped over to lay on its side, looks like the numeral 2. Also, the letter N (or n) has two downstrokes.
3 = M	A letter M (or m), when tipped over to lay on its side, looks like the numeral 3. Also, the letter M (or m) has three downstrokes.
4 = R	Four is a four-letter word ending in r — emphasize the "r" sound in the word.
5 = L	Five fingers on the *left* hand, when held up with thumb out, form the letter L.
6 = J, SH, CH, soft G	A capital G resembles a 6. "Shell to jewel, a giant change" gives you the representative sounds.
7 = K, Q, hard C, hard G	The letter K looks like it is made up of two 7s, back to back and laying horizontally. "Queen, go kick a cow" will remind you of the sounds.
8 = F, V, PH	Visualize the number 8 eating "phony fruits and vegetables" (or drinking V-8 juice). Also, the script letter f resembles the number 8.
9 = P, B	A reversed P looks like a number 9. Imagine a 9 scooping up some peanut butter.
0 = S, Z, X, soft C	These are all "hissing" sounds. The image of a snake rolled into a zero and hissing at a cent perched on an x will remind you of the sounds that go with the number 0.

Silent letters make no sounds; thus they have no numerical value. Vowel sounds (A, E, I, O, U, W, Y) and the letter H also have no value. Repeated consonants and combinations of consonants count as one letter if they make only one distinct sound. Thus, batter = 914; elephants = 58210; recharge = 4646; wheat = 1; muck = 37. The relationship of numerals to consonant sounds is very useful for remembering all kinds of numbers. As a first step, assign letter sounds to each numeral. By trial and error and imagination, make words and phrases out of the string of sounds. Look for words and images that can relate the meaning of the word to the event or person associated with the number. With this technique linked to the story chain, adults and middle-school students have memorized the number pi to a hundred digits or more.

Substituting numbers with words: Here are three examples of words that have been developed to help remember the encoded numbers:

- 43610 is a zip code in Toledo, Ohio. The sounds of R, M, J, T, S can be made into RAMJETS.
- Phone number 363-8744. The sounds of M, J, M, F, K, R, R can be made into MUSHY MOVIE CRIER. Picture your sentimental friend crying when he watches a romantic movie.
- Frequent flyer number 074-684-724. The sounds of S, K, R, J, F, R, K, N, R can be made into SCREECH, FREE CANARY (appropriate for a person who exuberantly loves to fly). For a person who is very quiet, this could be remembered as SCREECH-FREE CANARY.

Remembering historical events and years: For important events in history, you will already know to which millennium the events belong. Thus you will not usually have to remember the 1 in front of the years from 1000 to 1999.

- To remember that Napoleon's final defeat at Waterloo happened in 1815, associate the sounds for 815—namely F, T, L—with Waterloo. You could make a sentence that says: Waterloo was FATAL to Napoleon's career. The numerical value of the word "fatal" will give you the year.
- When and where was the traffic light invented? Picture a headline that reads: "Traffic jams BEATEN in Salt Lake City!" You have the place and the odd word in the sentence is "beaten"; it is the clue for the year: 1912.
- What year was ether first used as an anesthetic for surgery? If you remember that a big FERN plant was the first patient, you will know the year: 1842.

⌛ Ten-Minute Activity 2-6: Remembering Numbers

With another person, practice the conversion of numbers into a memorable word or sentence, either with a historical event or a telephone number. Try different combinations, then select the best one. Relate the image to the event (or the organization or person to the phone number).

Sketching—a tool for conceptual thinking and visualization

Are you thinking that visualization and sketching are all well and good for someone who is an artist, but are not for you because you just can't draw? If this is the case, we strongly recommend that you take a break at this point to do the following exercise.

⌛ ✓ One-Hour Activity 2-7: Learning How to See and Draw

Preparation (ahead of time): Do this exercise with a group of people—it is fun to observe the results, especially if some of them are convinced that they can't draw. Set a time and place. Find a line drawing of a person in an art book (for example a portrait done by Henry Matisse or Pablo Picasso) or a line-drawn cartoon of a person's face in a newspaper. Make a copy for each group member. Also, for each person, have five sheets of blank paper, pencils, and a piece of masking tape. Have an assistant with a watch give the instructions. Note: This exercise is for adults and students high-school age on up; it does not have the same kind of results for younger students.

Instructions: The pencil-in-hand symbol ✍ indicates a task needs to be executed before continuing.

Sheet 1: Draw (don't trace) your hand—you have about five minutes. ✍ When finished, sign your name and label it Drawing #1. ✍

Sheet 2: Draw a simple profile of a human face (see the figure on page 165 in this book). ✍ Go over the line again, naming the different body parts you are tracing: forehead, nose, lips, chin, etc. This is left-brain, analytical drawing. ✍ Next, draw the horizontal line at the top and bottom to turn the profile into a vase. ✍ Now, complete the vase by drawing the mirror image of the profile. This is right-brain, holistic, spatial drawing: you are concentrating on the spacing of the line, not on the body part it represents. ✍ What differences did you notice in the types of thinking you needed to do this drawing?

Sheet 3: Turn the line drawing (portrait or cartoon) upside down and copy it—your drawing will also be upside down. You will have 10 to 15 minutes. ✍

Sheet 4: Fasten a blank sheet of paper to your table with a piece of masking tape. Sit sideways, so you will not look at the paper. Instead, look intently at the palm of your hand. Imagine a small insect slowly following the lines of your hand. With your writing hand, copy these paths and lines onto the paper. Keep negative thoughts out of your mind. DO NOT TALK (since this will engages the left brain)! You will be given a signal to stop in 10 minutes. Repeat lines if you have drawn all the tiniest lines you think you see. The purpose of this exercise is to bore your left brain so it will "go to sleep." ✍

Sheet 5: Now we will draw the hand again. This time, form it into an interesting shape with bent fingers and hollow spaces between. The left brain does not like to deal with complexity, so it will leave this drawing task to the right brain. Look for negative space around the hand, rather than at the hand itself—negative space shares edges with the object you are to draw. Do not name the parts of the hand; look for the intersection of lines instead. Add fine details, such as shading and lines, if you wish. Closing one eye may help flatten the image. You will have 10 to 15 minutes. ✍ Sign your work and label it Drawing #2. ✍

Evaluation: Now compare Drawing #2 with #1. Share the results with the group. Does the outcome for each person in this exercise surprise you? You have discovered how to see! Practice this new-found thinking skill. See Betty Edward's books (Ref. 2.4) for other exercise ideas.

There is another fun activity that can help you overcome the "fear of sketching." This is Pictionary™—a marvelous game where a minimal sketch has a better chance of being understood quickly than a detailed, carefully executed drawing.

Strictly speaking, sketching and drawing are not the same thing, and they do not have the same purpose. Sketching is above all an aid to your own thinking. Sketching is a help for developing visual ideas worth communicating. Drawing comes after this thinking and playing stage. Drawing is for communicating a well-formed idea to a knowledgeable audience. With computer-aided design, engineers, architects, and designers have a tool that can manipulate data and produce well-executed drawings. But these drawings cannot usually be understood by an untrained person. Because of the encoded symbols, reading blueprints depends on analysis with the left hemisphere of the brain, whereas freehand sketching, visualization, and three-dimensional modeling involve spatial thinking and imagination in the right hemisphere.

Most people can learn to sketch well by following these steps:

1. *Learn to see.*
2. *Learn to handle the tools (paper and pencil) and what they can do.*
3. *Learn specific techniques (contour drawing, shading, perspective).*
4. *Practice to develop the necessary eye-hand coordination.*

If you follow the steps outlined above or are using Betty Edward's book *Drawing on the Right Side of the Brain* for detailed instructions, you can become proficient at sketching physical objects you see around you. Begin by practicing your observation skills while waiting in a crowd, taking a walk, sitting on a park bench, or relaxing on a shore. Notice the texture, colors, and details of the things you see; become aware of shapes and contours; ponder relationships in perceived size by imagining yourself being a camera and visually recording the world as a two-dimensional image. Use your hands to frame vistas and picture this view projected onto a flat canvas.

Still—in a way—nobody sees a flower—really—it is so small—we haven't time—and to see takes time, like to have a friend takes time.

Georgia O'Keeffe

Being able to sketch objects and landscapes directly from your observation is one thing, but you also will need to practice another kind of sketching—one that is more closely related to playing Pictionary™. This kind of sketching deals with visualizing and communicating ideas; it involves "seeing" the object in your mind's eye. How would you sketch the concept of a cow without actually looking at one? Certainly, you can sketch a head with horns and a body with four legs and perhaps a tail. But there is one other aspect of a cow that distinguishes it from most other four-legged, horned animals—and that is its milk delivery system. If you emphasize that anatomical part, your sketch will be immediately understood without any verbal explanation.

During the conceptual design phase in product development, sketching enables engineers to concentrate on essentials and leave out distracting details, allowing right-brain intuitive and creative thinking and idea synthesis. Thus this type of communication is useful for brainstorming as well as for clarifying ideas. But sketches do not only visualize objects and ideas; they can represent processes and relationships through flow charts and other types of diagrams. In many instances, making a sketch helps the mind to answer a question or "see" a solution. Thus sketching is an essential thinking tool, not just for engineers but for everyone. When you study, summarize and connect the material by making charts and diagrams, such as the horizontal flow diagram below.

The creative problem-solving process

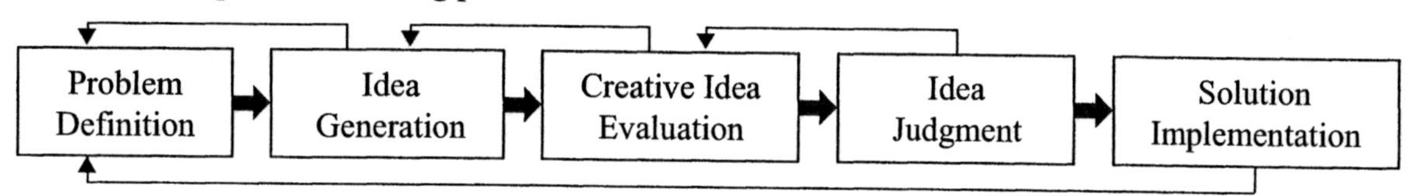

This diagram shows at a glance that creative problem solving is a sequential process with iterative loops or flexibility built in. This means that to achieve satisfactory results at each step, it is often necessary to return to an earlier thinking phase to get additional ideas. Also, implementation is a new problem in itself that requires another cycle of creative problem solving.

3-D visualization in solid modeling

Skills in visual thinking are essential for designers and engineers, since design is central to engineering. How do we develop these skills? Children who frequently play with construction toys such as building blocks, Erector Sets™, Tinker Toys™, Lego™ tiles, or Lincoln Logs™ seem to develop spatial visualization skills. Play with certain "spatial" video and computer games also appears to be beneficial.

How do you know you have spatial visualization ability? A number of tests for low-level, two-dimensional visualization exist, such as the Minnesota Paper Form Board Test (see Figure 2.1) and the Group Embedded Figures Test (see Figure 2.2). Higher-level spatial ability tests, such as the Spatial Relations Subtest of the Differential Aptitude Test (see Figure 2.3) or the Purdue Spatial Visualization Test: Rotations (see Figure 2.4) involve "seeing" and "manipulating" three-dimensional, solid figures. Some engineering schools offer remedial courses to help students develop spatial visualization skills (see Ref. 2.15 for an example). Research conducted at the University of California at Berkeley has shown that even one-day workshops in spatial visualization can substantially increase students' success rate in engineering.

Visualization = perception + imagination + communications.

We see, we imagine, we draw.

Walter Rodriguez

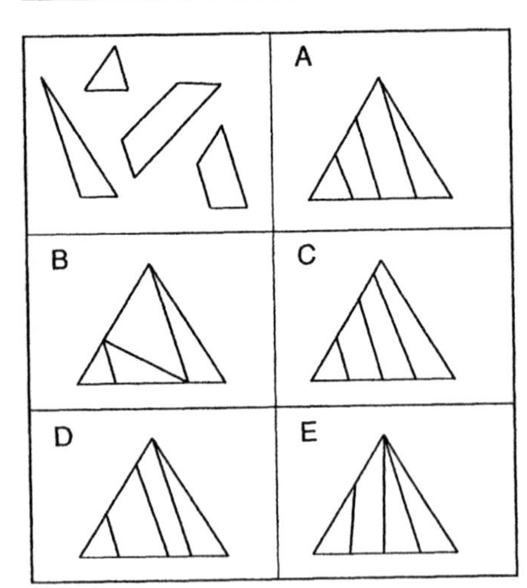

Figure 2.1 Select the option that shows how the four separate parts fit together.

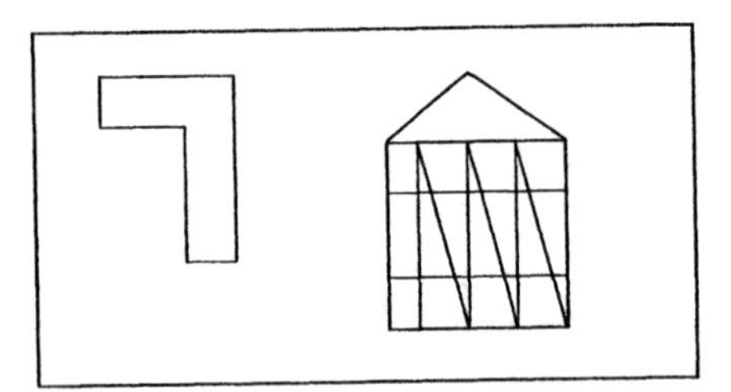

Figure 2.2 Find the simple shape in the complex drawing (without any rotation), then shade the shape.

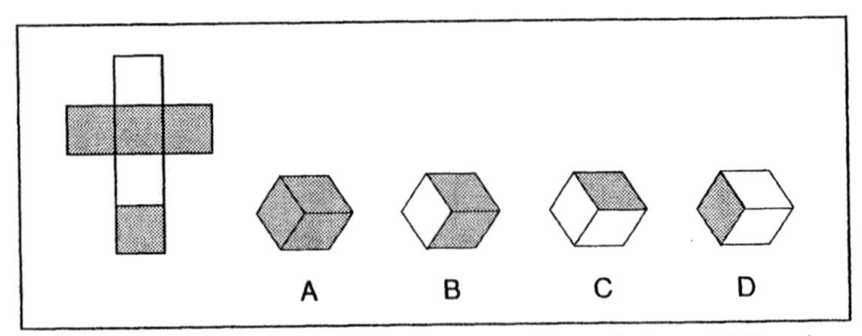

Figure 2.3 Choose the 3-D cube that would result from folding the 2-D pattern shown to the left of the cubes.

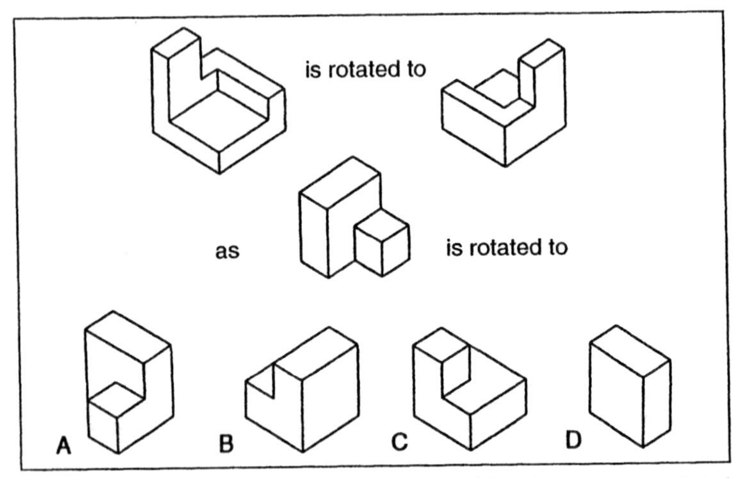

Figure 2.4 Select the option that shows the object in the center rotated in exactly the same way as the object shown at the top of the figure.

Here is a personal illustration by Bill Shelnutt of how visualization was practiced in the days before powerful workstations were available.

> Studying the meaning of the different two-dimensional "views" of a mechanical drawing was more complicated than it might appear at first, because different lay-out conventions exist in different parts of the world. We used a Plexiglas box with hinged sides in which we placed a simple object. With a grease pencil we drew on each side what we "saw," then folded out the sides in various ways to show alternative "views" on one flat surface. Thus we were continuously aware all along that we were translating views of a three-dimensional object into a two-dimensional representation.

Until the 1960's when computer-aided design (CAD) was invented, engineers had to be able to visualize three-dimensional objects from the two-dimensional drawings they drew with instruments and T-squares on drafting boards. CAD was faster and more accurate, but it used the same design process and the same 2-D layout for the drawings. The new paradigm of solid modeling was being developed in the early 1970's and blossomed by the mid-eighties with many 3-D solid modeling software

Industry has vaults filled with old design drawings. With the changes in design technology, we no longer need to learn how to construct such drawings, but we still need to know how to read the blueprints.

packages available. With these visualization tools, both technical and non-technical people could see the object as it was being designed and could give immediate input. This communication capability is a key in concurrent engineering. A mere ten years later, increasingly sophisticated solid modeling tools allow for engineering analysis, thus enabling companies to use just one integrated software system for all design, engineering, manufacturing, and information management activities.

Industry is under tremendous pressure to use state-of-the-art design tools to remain competitive. Educational institutions have been insulated for the most part from such pressures, creating a time lag before advances in engineering design technology find their way into a majority of classrooms. By the late 1980's, CAD was being widely adopted but had not yet completely replaced "graphite graphics." In the late 1990's, solid modeling (Figure 2.6) is being more widely adopted but has not yet replaced 2-D CAD and wireframe techniques (Figure 2.5). This is creating some unexpected problems for designers. Experience in industry is showing that people who are well versed in designing in the

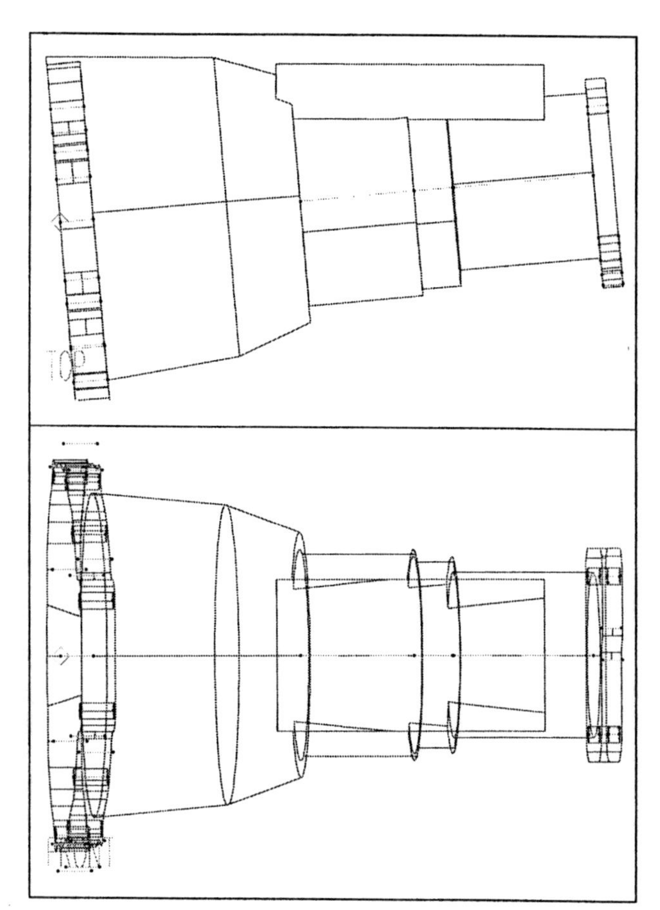

Figure 2.5 Example of a wireframe drawing of a part (top and front views) drawn by Prof. William Shapton.

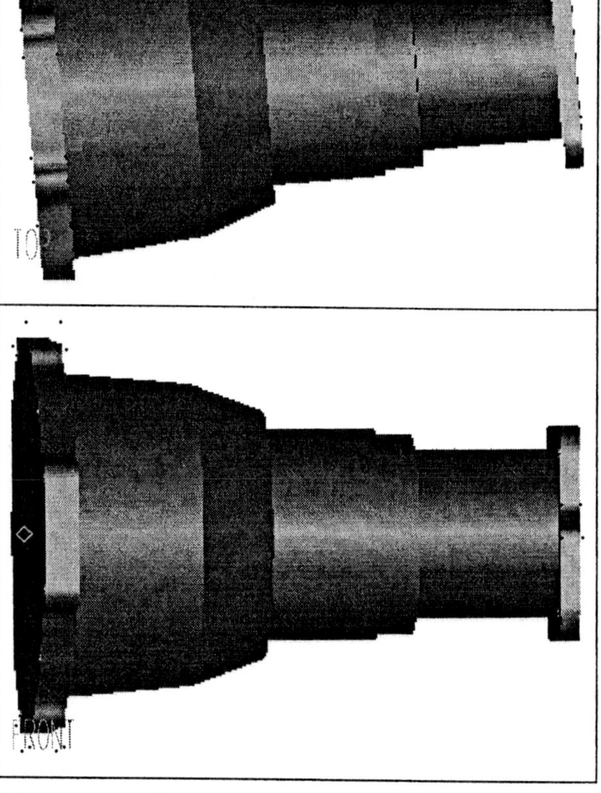

Figure 2.6 Solid model of the part in Figure 2.5 drawn by Prof. William Shapton, Michigan Technological University.

If you, as a student, are given a choice of which software to use to learn design, we urge you to pick the most advanced, most sophisticated tool, even though it may appear to require more effort to pass the course.

wireframe configuration have a much harder time learning solid modeling than people without previous wireframe experience because solid modeling requires different thinking skills.

Developments in industry are occurring at a very rapid rate, thus students are at risk of being obsolete before they graduate. With competence in the latest tools, on the other hand, they can look forward to very attractive job offers and great opportunities, not just in the U.S., but in the global marketplace. Thus check out what your university is offering; insist that state-of-the-art software be taught, and check that it is being upgraded periodically to keep pace with developments in industry. For example, at Ford Motor Company, I-DEAS™ training is being implemented world-wide, and major upgrades are incorporated every few months. If you are dreaming of a job in a particular company or industry, find out early in your education which design tools are being used and then do your best to be prepared, either by choosing the right courses and instructors or by participating in co-op programs in industry.

A major difference between two-dimensional representation and three-dimensional computer graphics is the information in the computer that a line represents. In 2-D it is simply a line on a flat surface—the drawing—and this line may be part of more than one surface edge. In 3-D, it is a feature of a particular surface, such as an edge or an intersection between two surfaces. These features are the important things for a designer. The power of good 3-D programs to produce a section or slice through an object at any angle or position for engineering analysis and scrutiny now represents a very different working environment—students will no longer need classes in descriptive geometry. The computer is now the "glass box" that allows students to generate two-dimensional drawings if desired; it also allows them to view the object from any angle. What students now need to develop is good judgment on how to select the "best"—most useful—position for working on the design and for communicating the information to others. This is just one example to illustrate why the thinking requirements have changed with the new tools.

Hewlett-Packard has found that solid modeling has many advantages (from Ref. 2.6, Ch. 31):

- The solid model is complete and unambiguous; drawings for manufacturing are easily generated.
- Immediate understanding of the design by all people on the product development team.
- Improved quality, teamwork, communication throughout the company; reduced warranty costs.
- Faster product development; fewer physical prototypes; easily generated technical illustrations.
- Solid modeling simulation predicts ease of assembly and identifies interference problems.
- Integration with analysis tools allows evaluation of product performance: fluid dynamics, tolerance analysis, structural analysis and shape optimization, cost and manufacturability analysis.
- More creativity and innovation; engineers can easily experiment and try out new ideas.
- Solid modeling provides documentation history; traditional 2-D drawings can be generated easily.

Resources for further learning

2.1 ✓ James L. Adams, *Conceptual Blockbusting—A Guide to Better Ideas,* third edition, Addison-Wesley, Reading, Massachusetts, 1986. Chapter 6 gives a thought-provoking discussion of mental languages, especially visualization and other sensory modes; it includes good examples and illustrations.

2.2 ✓ Tony Buzan, *Use Both Sides of Your Brain,* revised edition, Dutton, New York, 1983. This book includes technique and examples of mind-mapping—a way to take notes, brainstorm, and connect ideas visually. The memory pegs used differ slightly from those listed by Jerry Lucas.

2.3 Jeremy Campbell, *The Improbable Machine: What the Upheavals in Artificial Intelligence Research Reveal about How the Mind Really Works,* Simon & Schuster, New York, 1989. This book describes the discoveries made by researchers trying to create a "thinking" machine. It shows that experience, not logic, is the governing characteristic of the human mind.

2.4 ✓ Betty Edwards, *Drawing on the Right Side of the Brain,* J. P. Tarcher, Los Angeles, 1979. This classic on drawing, right-brain thinking, and creativity contains many facts, examples, and exercises. It lets the reader experience the difference between "verbal" drawing and right-brain processing of visual information. Also check out *Drawing on the Artist Within* by the same author.

2.5 Jane M. Healy, *Endangered Minds: Why Our Children Don't Think and What We Can Do About It,* Simon & Schuster, New York, 1991. Dr. Healy has investigated the influence of television on language development and thinking skills of children. When TV replaces reading, the ability to process language on a level needed for academic success will not develop.

2.6 Donald E. LaCourse, *Handbook of Solid Modeling,* McGraw-Hill, New York, 1995. Included are contributions from more than 60 industry professionals on solid modeling concepts, methodology, and applications.

2.7 Harry Lorayne and Jerry Lucas, *The Memory Book,* Ballantine, New York, 1985. This book describes different techniques and schemes (such as the memory pegs) for improving memory. Jerry Lucas has also created a learning system based on visualization with audio tapes, video tapes, and workbooks called *How to Learn: Learning That Lasts,* Lucas Learning, Mansfield, Texas.

2.8 Jack Maguire, *Care and Feeding of the Brain: A Guide to Your Gray Matter,* Doubleday, New York, 1990. This book discusses the functions of the mind, the myths, and the discoveries on the frontiers of brain science.

2.9 ✓ Robert H. McKim, *Experiences in Visual Thinking,* second edition, PWS Publishers, Boston, Massachusetts, 1980. This softcover book contains many exercises for flexible thinking, moving from visual thinking and how to see and draw to the use of imagination and sketching of brainstormed ideas.

2.10 Philip Morrison and Phylis Morrison, *Powers of Ten: About the Relative Size of Things in the Universe,* Scientific American Library, Redding, Connecticut, 1982. Stunning photographs illustrate this tour on magnitudes from the atom's interior to the far reaches of the universe.

When I started teaching and did demonstrations, I found that I could either talk or draw, but I couldn't do both at once.

Betty Edwards

2.11 ✓ John Allen Paulos, *Innumeracy: Mathematical Illiteracy and Its Consequences,* Hill & Wang, New York, 1988. This small and easy-to-read book shows how we must and can become more comfortable with numbers, quantities, and probability—how we can overcome the mathematical ignorance so pervasive in our society.

2.12 Walter Rodriguez, *The Modeling of Design Ideas,* McGraw-Hill, New York, 1992. This textbook on computer graphics and modeling includes some free-hand sketching. However, the main focus is on visualization as expressed in 2-D and 3-D computer-aided drawings in a structured, analytical (non-software specific) approach to design. A discussion of solid modeling is included.

2.13 Moshe Rubinstein, *Tools for Thinking and Problem Solving,* Prentice-Hall, Englewood Cliffs, New Jersey, 1987. This book offers interesting and useful tools for representations.

2.14 Roger Schank (with Peter Childers), *The Creative Attitude: Learning to Ask and Answer the Right Questions,* Macmillan, New York, 1988. This book looks at various aspects of creativity; it discusses memory as a phenomenon of "reminding" and has an interesting chapter on script-based thinking.

2.15 Sheryl A. Sorby, Kim J. Manner, and Beverly J. Baartmans, *3-D Visualization for Engineering Graphics,* Prentice Hall, Englewood Cliffs, NJ, 1998. This book helps develop 3-D spatial skills. The authors have found that many women engineering students particularly benefit from a 3-D visualization course.

The nations that lead the world in the decades to come will be those that encourage creative people to become engineers.

Gary Tooker, CEO, Motorola, 1997

Exercises

2.1 Graphs

Find five different visual representations of data (graph, diagram, table, histogram, etc.) or invent your own. You are allowed to copy these "charts" and make any additions or modifications that you desire to improve the presentation. The five charts should all have a different visual form and contain data about different subjects. Check that the purpose of the data and chart is clearly represented.

2.2 ✓ Mountain Path

Read through the following problem. The primary objective here is not getting the answer—your assignment is to be aware of the thinking strategies that you are employing in your attempts to solve the problem. Please jot down some notes on the different ways you are thinking about the problem and on the different mental languages that you are using to arrive at an answer.

A certain mountain in Nepal has a shrine at its peak and only one narrow path to reach it. A monk leaves his monastery at the base of the mountain at 6 a.m. one morning and ascends the mountain at a steady pace. After some hours, he tires and takes a long rest. Then he resumes his climb, albeit more slowly, and he pauses often to meditate or enjoy the view. He also takes a couple of breaks to refresh himself at a spring and to enjoy the meal he has carried along.

Finally, at sunset, he reaches the shrine where he spends the night. At sunrise, he begins his descent, quickly at first, and then more slowly as his knees begin to ache. After a couple of rest stops, he accelerates his pace again—he does not want to miss dinner at the monastery. Prove that there is a point in the path that the monk reached at exactly the same time of day on his ascent and descent.

2.3 ✓ Pop Song

Imagine the following situation. You are in a taxi in a city in China. Your friend, who can speak Chinese, has gone into a store to do some shopping. You are tired and choose to wait in the cab for her return. The cab driver, who does not speak English, turns on a tape of cheerful Chinese pop songs. Suddenly, he becomes aware of your presence and switches to a Beethoven symphony. You want him to switch back to his songs. You sing, you gesture, but he does not understand. In desperation, you grab a pencil and note pad and draw a sketch. His face lights up in sudden understanding, and he restores the song tape. Draw a couple of sketches you think would have this kind of result.

2.4 ✓ Airplane Seating

Read through the following problem and devise a seating scheme that makes the maximum number of people happy, taking the stated facts into account. Note down the steps in your thinking that help you solve this problem. Here is the scenario:

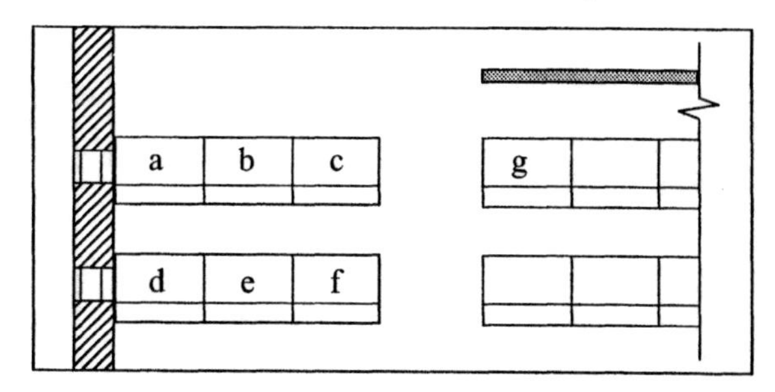

Seven passengers have just boarded a Boeing 747 aircraft for a transpacific flight. They find their assigned seats (see sketch on the left) and sit down. For this 14-hour flight, the people, their seat assignments, and their needs and wishes are:

a. A Korean man who speaks some English; he has the bulkhead window seat.

b. His wife who appears to be ill; she is in the bulkhead middle seat.

c. A big English-speaking Filipino carrying a large bag which he refuses to stow in an overhead luggage bin. He is in the bulkhead aisle seat but demands a seat farther back.

d. A Korean lady who does not speak English; she has the window seat in the row behind the bulkhead. She carried on a large package which does not fit under the seat in front of her. She stows it in the leg space and covers it with a blanket. Of necessity, her legs extend into the space of the middle seat.

e. A hunky U.S. serviceman; he squeezes into the middle seat.

f. A middle-aged American woman on crutches with a broken foot; she has the adjoining aisle seat. She finds that it will be impossible for her to elevate her foot from this seat using the small folding camping stool she has brought along for this purpose.

g. The woman's son, a six-foot-four-inch skinny guy with very long legs. He has the bulkhead seat across the aisle (behind the lavatory partition). He trades seats with his mother to give her more leg room. However, this is not sufficient to allow her to prop up her foot.

The stewardess has found a seat in the back of the crowded plane for the Filipino. The Korean couple is delighted at first, but then they find that the armrests in the bulkhead seats cannot be raised; thus the ill wife cannot lie down. How would you help out these six remaining passengers to achieve win-win trades (where each person ends up with an improved situation that meets their needs as well as the safety regulations on board the aircraft)? Which person do you think was the most desperate and had to think up these trades? If you are working on this problem in a group, make up a role-playing skit to illustrate the interrelated problems and the solution process. What principles can be applied to design?

2.5 Money

Solve the following problem and again pay attention to the mental languages that you are using: Becky and Cory together have three times as much money as Arnold. Dotty has twice as much money as Ernie. Arnold has one-and-a-half times as much money as Dotty. Cory and Dotty together have as much money as Becky plus twice the amount that Ernie has. Ernie, Dotty, Arnold, Becky, and Cory together have $60. How much money does each child have? What thinking modes would you use to solve this problem? Can you think of possible ways of solving this problem without algebra?

2.6 Observation (Classroom Activity)

In preparation, the leader changes five to ten items in the room (or brings in some objects that are not usually found in the classroom or conference room). Then after the group or class members have entered, they must identify the "odd" items or changes.

2.7 Sketching a House

For this exercise, you will need a timer or a stop watch.

1. Take 30 seconds to quickly sketch a house.
2. Take 2 minutes to sketch a house that you see from your window.
3. Take 3 minutes to sketch your dream house.
4. Take 3 minutes to discuss with another person how each of the three sketching exercises differed in the type of thinking you had to do to carry out the assignment. Which one was the easiest for you to do?

2.8 Pictionary™

This excellent game combines visual thinking, sketching, and free association. Play the game according to the rules—or modify some of the rules if you are playing with friends from other countries.

Practice is the best of all instructors.

Publilius Syrus, Maxim 469, First Century BC

2.9 Vocabulary Sketches

Use word substitution and sketches to visualize five foreign or difficult English words. Select your best one and teach the word to your class, group, or a friend. One week later, check to see if the word is remembered. Has teaching (and sketching) helped *you* remember the words?

2.10 ★ Memorizing all the Presidents of the United States ★

Using word substitution and the story link, make sketches for remembering all U.S. presidents in sequence. Then use flip chart paper to develop your sketches into a teaching aid. Find a group of people and teach them the technique in 30 minutes or less. Go through the story slowly; repeat once or twice; then ask for a pair of volunteers to repeat the presidents without looking at the chart. If possible, get the group together the next day for a checkup and reinforcement.

Is learning equal to "remembering forever"? What can be deliberately forgotten or stored elsewhere (and how would you remember the keys to retrieval)?

Susan M. Brookhart,
National Forum,
Vol. 78, No. 4, page 4

2.11 ★ Chemical Elements—Team Project ★

This will require a library search for a memory book containing the peg method (see Ref. 2.2 and Ref. 2.7). Using this method, develop memorable images for learning the number, abbreviation, and name of chemical elements in the periodic chart. Then sketch your images. Your group may want to put together a booklet with the sketches of the most important elements encountered in a chemistry class.

Hints: To do this effectively, use word substitution combined with mnemonics and the peg method to create an action image to visualize:

1. Find a tangible substitute word or phrase to identify the element.
2. Recall or make up a peg word for the element's atomic number.
3. Connect the substitute name and peg word with one or two words whose first letter consists of the letters in the element's symbol, resulting in a phrase that brings to mind an unusual image that can be sketched, as for example:

Mercury: Visualize this "weird" headline about an aristocrat: Marquis **H**ugs **G**reen fox! This will remind you of the symbol for mercury, Hg, and its atomic number, 80. Fox is a peg word for 80.

Tungsten: Imagine the picture (and the phrase): Tongue **W**hacks car. Tongue will remind you of tungsten; you will know that the symbol is W, and that the atomic number is 74.

Chapter 2 — review of key concepts and action checklist

Memory and the brain: The human brain is the most complex arrangement of matter in the known universe. Different types of memory are located in different parts of the brain, depending on the type of learning involved. The brain is experience-based, not logic-based (computers are logic machines). Many techniques can be employed to improve memory.

Mental languages: Verbal thinking is linear and sequential; it is not suitable for solving certain kinds of problems, although it is heavily emphasized in Western school systems. Mathematical thinking is used to solve quantitative problems. Visualization and sensory thinking help in memorization, and both enhance creativity.

Visualization, a tool for improving memory: Graphs clarify relationships for better understanding. Special visualization techniques help in filing and retrieval: *Association*—Link the item to be learned to something you already know (i.e., your home) in a "wild" image. *Substitute word*—A sound-alike tangible word is substituted for an intangible word; the tangible word is then visualized. *Story link*—A list of unrelated items is memorized by making up an image for each item and then linking the images in a "weird" story. Remembering just one of the items will bring the entire chain into the conscious mind. *Phonetic alphabet*—numbers can be transformed into tangible words and then visualized.

Sketching can be learned in four steps: 1. Learn to see. 2. Learn to use the tools. 3. Learn specific techniques for representation. 4. Practice to develop eye-hand coordination. Use sketching for brainstorming, for clarifying ideas, and to visualize processes and relationships with diagrams. Sketching is an essential thinking tool for everyone!

3-D visualization: Skill in spatial visualization seems to be acquired through early life experiences. Special courses and workshops can help students develop this skill needed for success in engineering. Powerful solid modeling design tools have undergone rapid advances and now allow concurrent engineering and global communication; learning how to use this state-of-the-art technology will be an advantage for students.

Action checklist

☐ Practice remembering names while waiting somewhere—with the people or photos you see around you. Make up a name if needed.

☐ Among the items in Table 2.1, select the one you think could really help you improve your memory. Then make a plan on how you would implement one change in study habit or life-style to reach your goal. Make a pact with a friend to help you put your plan into action during the next month. Remember that it takes a minimum of three weeks of steady practice to adopt a new habit.

Use association, word substitution, the story link, and the phonetic alphabet to practice visualization and enhance memory.

☐ If you did not learn to sketch in school, look for a sketching course at a museum or in adult education. Or teach yourself by following the instructions in a book—see Reference 2.4 in this chapter.

☐ This week, when watching television, be on the lookout for visual and sensory information. Also note how viewing influences your mood. Do the values in the show agree with your own?

☐ Get into the habit of visualizing problems. When faced with a problem, shut your eyes and take a "look" at the problem from imaginary, unusual angles—this can lead to creative solutions. Or try to come up with three different metaphors or sketches to visualize the problem.

3

Mental Models

What you can learn from this chapter:

- Mental models undergird learning, communication, and teamwork.
- The Herrmann brain dominance model has four distinct thinking styles: analytical, sequential, interpersonal, and imaginative. Engineers need to think in all four modes. Each person has a unique HBDI profile.
- The knowledge creation model cycles through the four quadrants of sympathized, conceptual, systemic, and operational knowledge.
- Ceative problem solving builds on the other two models for individual, team, and organizational problem solving and innovation.
- Resources for further learning: references; thinking assessments and exercises; review, and action checklist.

Overview and purpose

Engineering success today requires more than up-to-the-minute technical capability; it requires the ability to communicate, work in teams, think creatively, learn quickly, and value diversity.

George D. Peterson, executive director of ABET, 1997

Mental models are powerful thinking tools or metaphors. Like the piers of a bridge, the three mental models that we will discuss—brain dominance, knowledge creation, and creative problem solving—undergird all aspects of engineering, as shown in Figure 3.1. When the mental models are shared and understood in an organization, they enhance communication and teamwork, accelerate knowledge creation and innovation, improve learning and information management, and lead to better engineering design and problem solving—all crucial components to providing successful products and services for a globally competitive, rapidly changing world. Figure 3.2 is an exploded view of the bridge pier of Figure 3.1 and depicts how the three mental models relate to each other. This chapter will explain the four-quadrant model of brain dominance in some detail, since the other two models will build on it. The knowledge creation model and its application to learning and organizational change will be surveyed next, followed by a summary of the creative problem solving model. Applications of the models to teamwork will be shown in Chapter 4 and to communication in Chapter 5. Many problems in learning, communication, teamwork, and organizational functioning can be traced back to a lack of understanding and appreciation of particular thinking styles, skipped steps in the knowledge creation cycle, and an absence of creative problem solving skills.

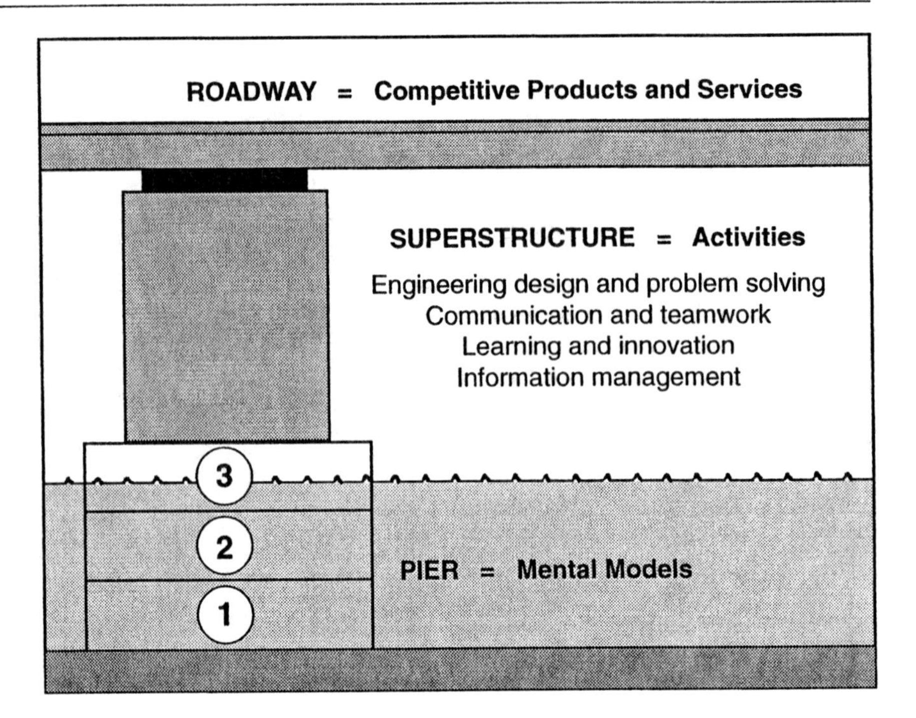

Figure 3.1 Key role of mental models to support team and organizational functioning.

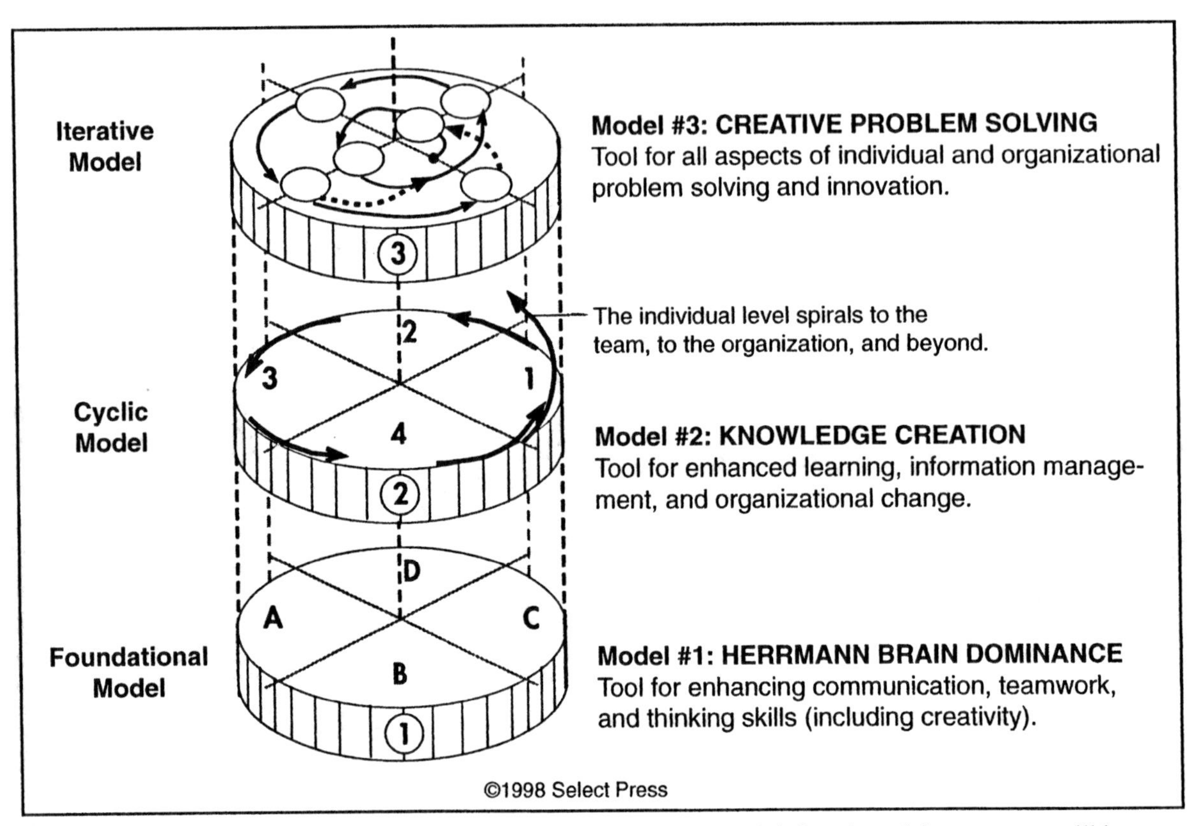

Figure 3.2 Relationship between the three mental models. The sequential direction of the processes will be explained in the discussion of the respective models.

The Herrmann brain dominance model

As you compared mathematical and verbal thinking with visual and sensory thinking in the preceding chapter, did you notice that the different mental languages required distinct and perhaps unfamiliar thinking abilities? Our thinking preferences characterize our approaches to problem solving, creativity, and communicating with others. For example, one person may carefully analyze a situation before making a rational, logical decision based on the available data; another may see the same situation in a broader context and will look for alternatives. One person will use a very detailed, cautious, step-by-step procedure; another has a need to talk the problem over with people and will solve the problem intuitively. All use their particular approaches based on successful experiences. We will now explore a model of thinking preferences that will help you learn to become a more effective thinker and problem solver.

Most of us assume that we are seeing the world the way it really is.

Ned Herrmann

Brain dominance

Ned Herrmann earned a degree in physics and was hired as the first member of GE's physics program. In his first rotating assignment, an attempt was made to convert him into an engineer. He resisted this throughout his career. In years of research into the creativity of the human brain, he came to recognize that the brain is specialized in the way it functions. These specialized modes can be metaphorically organized into four distinct quadrants, each with its own language, values, and "ways of knowing." Each person is a unique mix of these modes of thinking preferences and has one or more strong dominances. Dominance has advantages: quick response time and higher skill level, and we use our dominant mode for learning and problem solving. The stronger our preference for one way of thinking, the stronger is our discomfort for the opposite mode. "Opposite" people have great difficulty communicating and understanding each other because they see the world through very different "filters." Is there a best way? Ned Herrmann found that each brain mode is best for the tasks it was designed to perform. Because our school systems concentrate heavily on sequential reasoning skills, creative abilities are often discouraged by teachers, well-meaning family members, and employers. What is sorely needed is a better balance and an appreciation for all thinking abilities. We must learn how to use and integrate these abilities for whole-brain thinking and problem solving.

Why are some people so smart and dull at the same time? How can they be so capable of certain mental activities and at the same time be so incapable of others?

Henry Mintzberg, management professor at McGill University, Harvard Business Review, *July 1976*

To understand the origin of the four-quadrant brain dominance model, we need to visualize the physical brain. Most people are familiar with the main hemispherical division into the left brain and right brain. Strictly speaking, these are the *cerebral hemispheres* and contain about 80 percent of the brain. Each cerebral hemisphere has a separate structure nestled into it, the corresponding half of the limbic system. The

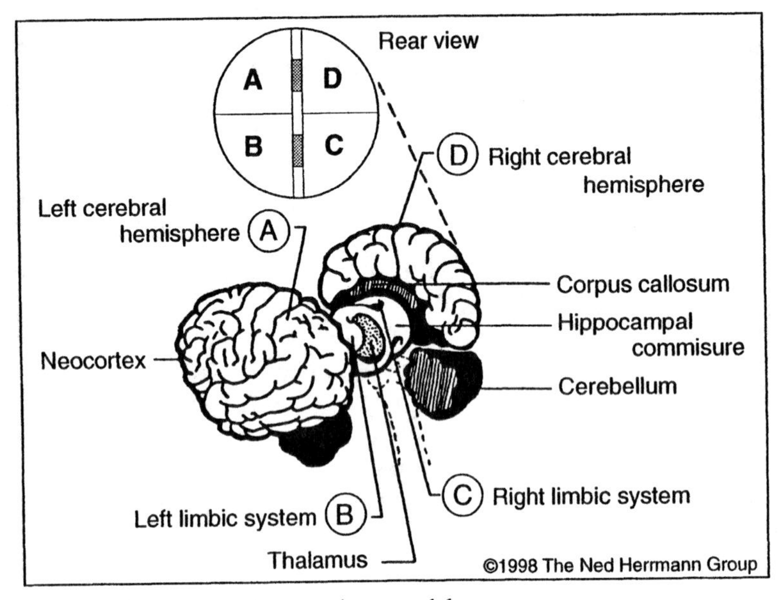

Figure 3.3 How the four-quadrant model relates to the physical brain.

limbic system plays a crucial role in learning by transferring incoming information to memory. Figure 3.3 shows a sectioned view of the human brain together with its relationship to Ned Herrmann's four-quadrant concept.

The hemispheres are connected with fibers that carry communication within and between the hemispheres. *Association fibers* form a complex network connecting the different specialized areas within each hemisphere. The two limbic lobes are linked through the *hippocampal commissure*, and the two cerebral hemispheres are connected by the *corpus callosum* that contains from 200 to 300 million axonic fibers. When one part of the brain is actively thinking, the other parts are more in "idle" mode so they do not interfere with the specialized thinking task. However, when solving a complex problem, more than one thinking skill is involved, and the brain is able to switch signals back and forth very rapidly between different specialized areas within and across the hemispheres.

Each person thinks and behaves in preferred ways that are unique to that individual. These dominant thinking styles are the result of the native personality interacting with family, education, work, and social environments.

No part of the brain works as fully or creatively on its own as it does when stimulated or supported by input from the other parts.

Ned Herrmann

When Ned Herrmann looked around for a method to diagnose thinking preferences based on brain specialization, he could not find any existing tools suitable for his purposes. So he developed his own assessment, now known as the Herrmann Brain Dominance Instrument (HBDI™). When the answers to its 120 questions are evaluated by a computer at Herrmann International headquarters in North Carolina, the results are numerical scores together with a graphical profile. Recent advances in brain research support the validity of the "descriptive" metaphorical model that divides the brain into left and right halves and into the cerebral and limbic hemispheres, resulting in four distinct quadrants, as shown in Figures 3.3 and 3.4. While millions of HBDI's have been administered, Ned Herrmann now has an active data base of over 200,000 individual profiles. His daughter, Ann Herrmann Nehdi, personally trains and certifies people in the administration, evaluation and interpretation of the HBDI to ensure the quality and reliability of the technology.

Although the four-quadrant model was organized based on the divisions in the physical brain, it is a metaphorical model—the newest imaging techniques show the brain's complexity, subtlety, and versatility involved in even the simplest thinking tasks. Yet, the model is useful for clarifying how we think, and it allows for multidominance.

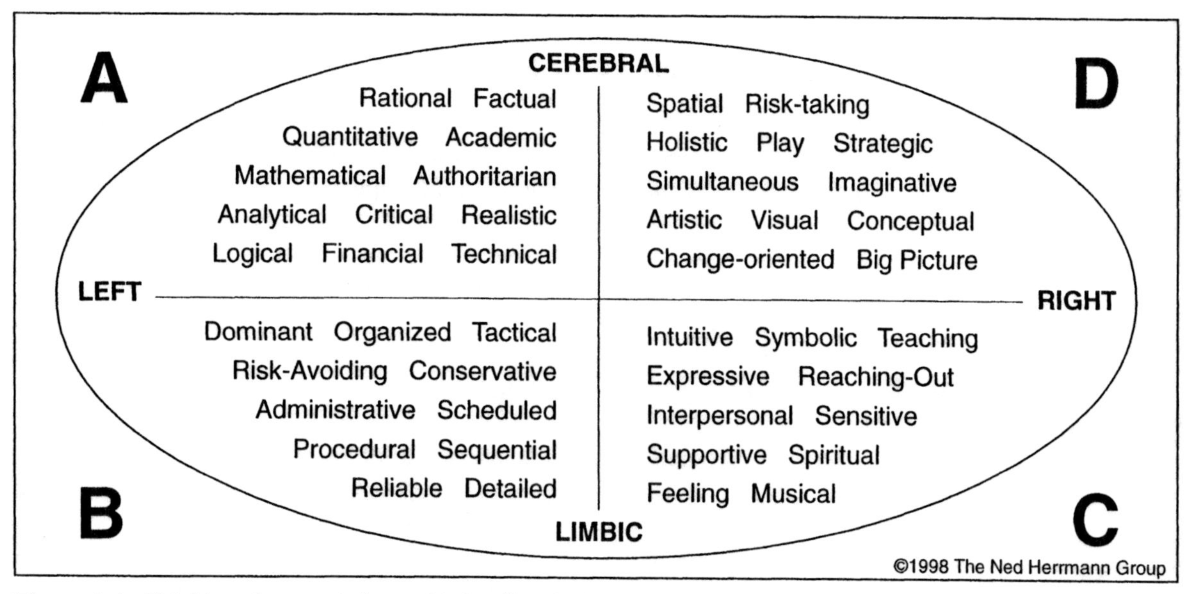

Figure 3.4 Thinking characteristics and "clues" of the Herrmann brain dominance model.

The typical average Herrmann brain dominance profile for engineering faculty is shown in Figure 3.5. A score in a particular quadrant that falls in the innermost region (Circle 3) denotes a low preference (possibly an avoidance); one in the next band (Circle 2) indicates comfortable usage, and scores extending further out indicate strong preferences for those thinking modes. Thus the HBDI profile shows at one glance the intensity of preference for an individual or a group. Having a low score in a quadrant does not mean we cannot think this way; it means when given a choice, we prefer to use other modes. Students can earn top grades in subjects that require thinking in modes they tend to avoid, *if* they make a strong effort. Conversely, having a strong preference does not mean we know how to be good thinkers, but it is easier to develop competencies in areas of strong dominance. The dashed line in Figure 3.5 shows the proforma profile of an engineering curriculum taught by a certain group of faculty at a particular time (dotted line). In a *proforma* study, the thinking required to perform tasks, the contents of documents, the lifework of a person, or a management philosophy or culture are analyzed in terms of the four-quadrant Herrmann model.

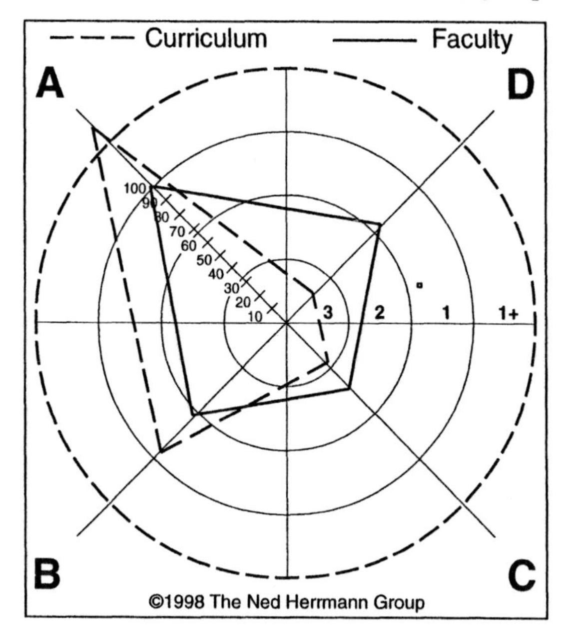

Figure 3.5 Typical average HBDI profile for engineering faculty together with a proforma profile of a mechanical engineering curriculum.

The following sections will illustrate the characteristics of each quadrant from the viewpoint of how people learn and how they can strengthen these abilities. Strong preferences as well as avoidances are expressed in "clues" that can be observed in a person's behavior. A separate section discusses the implications for engineers.

Characteristics of analytical quadrant A thinking

Quadrant A thinking is factual, analytical, quantitative, technical, logical, rational, and critical. It deals with data analysis, risk assessment, statistics, financial budgets and computation, as well as with technical hardware, analytical problem solving, and making decisions based on logic and reasoning. Quadrant A cultures are materialistic, academic, and authoritarian. They are achievement-oriented and performance-driven. An example of a quadrant A thinker is Star Trek's Mr. Spock; others are George Gallup, the pollster, and Marilyn Vos Savant, known as a person with one of the highest IQ scores in the world.

What I want is Facts. Teach these boys and girls nothing but Facts. Facts alone are wanted in life. Plant nothing else, and root out everything else.

Mr. Gradgrind in Charles Dickens, Hard Times, *1854*

People who prefer quadrant A thinking also have preferences for certain subjects in school and for certain careers—algebra, calculus, accounting, as well as science, engineering, and technology. Most textbooks are written in the quadrant A mode. Lawyers, engineers, computer scientists, analysts, technicians, bankers, and surgeons generally show strong preferences for quadrant A thinking. People with quadrant A thinking talk about "the bottom line" or "getting the facts" or "critical analysis." They are talked about as "number crunchers" or "human machines" or "eggheads" (and when accompanied by avoidance in quadrant C as "cold fish"). Learning activities preferred by quadrant A thinkers are listed in Table 3.1.

Table 3.1 "Quadrant A" Learning Activities and Behaviors

- Looking for data and information; doing library searches.
- Organizing information logically in a framework (but not down to the last detail).
- Listening to informational lectures.
- Reading textbooks—most textbooks are written for quadrant A thinkers.
- Analyzing (studying) example problems and solutions.
- Thinking through ideas (rationally or critically).
- Doing research using the scientific method.
- Making up a hypothesis, then testing it to find out if it is true.
- Judging ideas based on facts, criteria, and logical reasoning.
- Doing technical and financial case studies.
- Knowing how computers work; using them for math and information processing.
- Dealing with hardware and things, rather than people and social issues.
- Dealing with reality and the present, rather than with future possibilities.
- Knowing how much things cost.
- Traveling to other cultures to study technological artifacts (bridges, machines, etc.).

⌛ **One-Minute Activity 3-1: Quadrant A Learning**

Circle the dots in Table 3.1 for those items that are easy for you *and* that you enjoy doing.

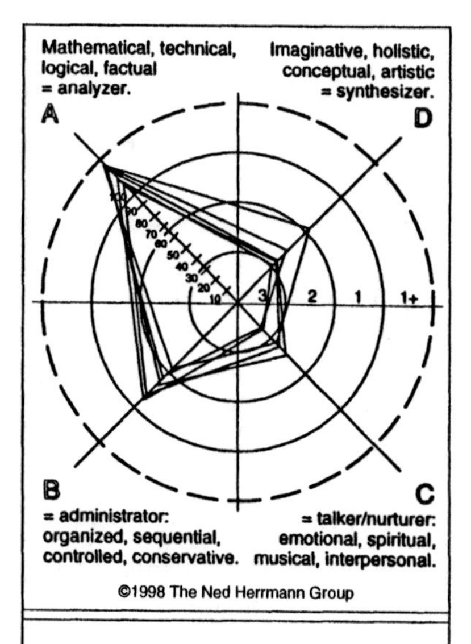

Engineers:
ARE Not Necessarily a Train Drivers.
ARE Creative Person in a Technical Way.
See and understand how things work.
Use factual info to solve problems.
Make big bucks.

Figure 3.6 Team profile and quadrant A "definition of engineer."

Example: To demonstrate that thinking preferences are expressed in vocabulary, we used a brief exercise in one of our classes with first-year engineering students. We grouped students according to their learning preferences, with each group having five members. The students at the time of this exercise did not yet have any knowledge of the Herrmann model. The thinking profiles shown with the group results are the actual HBDI profiles obtained later into the course. The assignment was to come up with a definition of "What is an engineer?" in five minutes, with the answer to be written on a flip chart. The results for the A group are shown in Figure 3.6. Note the quadrant A words and phrases such as "technical, understanding how things work, factual info, and making big bucks." This group wrote down facts; they were not concerned with the details of correct grammar or a nice layout.

Practice: Many engineers have strong preferences in quadrant A. What if you are not a quadrant A thinker but would like to develop this thinking mode? Mathematics, science, and especially engineering analysis courses (and their homework problems) develop quadrant A thinking. We have found in our research that many students increase their quadrant A preference as they go through the engineering curriculum. Exercises in Part 2 of this book for the "detective" and the "judge" will develop quadrant A thinking skills. Table 3.2 lists a variety of quadrant A activities that do not specifically involve math. Many of the exercises in Tables 3.2, 3.4, 3.6, and 3.8 are recommended by Ned Herrmann in his Personal Growth and Development Plan (©1989 Ned Herrmann); others we have added to the list.

Table 3.2 Activities for Practicing Quadrant A Thinking

- Define work, learning, or study goals.
- Get data and information about a subject you do not yet know anything about.
- Find out how a frequently used machine actually works by reading about it.
- Take a broken small appliance apart and diagnose the problem.
- Take a current technical problem situation and analyze it into its main parts.
- Review a recent impulse decision and identify its rational, logical aspects.
- Analyze some politicians running for office—where do they stand on the issues?
- Join an investment club or financially plan your retirement.
- Do logic puzzles or games; play chess.
- Learn how to use an analytical software tool or spreadsheet on your computer.
- Play "devil's advocate" in a group decision process.
- Write a critical review based on logical reasoning of your favorite TV program, movie, poem, play, book, song, or work of art.

Characteristics of sequential quadrant B thinking

Quadrant B thinking is organized, sequential, controlled, planned, conservative, structured, detailed, disciplined, and persistent. It deals with administration, tactical planning, procedures, organizational form, safekeeping, solution implementation, maintaining the status quo, and the "tried-and-true." The culture is traditional, bureaucratic, and reliable. It is production-oriented and task-driven. Edgar Hoover, former Director of the Federal Bureau of Investigation (FBI), Prince Otto von Bismarck, Prussian Chancellor of Germany (1871-1900), and the tactical American Indian Chief Geronimo exemplify quadrant B thinkers.

Order and simplification are the first steps toward the mastery of a subject— the actual enemy is the unknown.

Thomas Mann,
The Magic Mountain, *1924*

People who prefer quadrant B thinking want their courses to be very structured and sequentially organized. Planners, bureaucrats, administrators, and bookkeepers usually exhibit preferences for quadrant B thinking. People with dominant quadrant B thinking modes talk about "we have always done it this way" or "law and order" or "self-discipline" or "play it safe." They are talked about as "pedants" or "picky" or "nose to the grindstone." The learning activities preferred by quadrant B thinkers are listed in Table 3.3. Quadrant B behavior is easy to notice in the area of time—these thinkers stick to schedules, and they are annoyed when others do not have the same kind of discipline! This can be a source of conflict with D-quadrant people who have no sense of time at all.

Table 3.3 "Quadrant B" Learning Activities and Behaviors

- Following directions carefully, instead of improvising.
- Doing detailed homework problems neatly and conscientiously.
- Testing theories and procedures to find flaws and shortcomings.
- Doing lab work, step by step.
- Writing a sequential report on the results of lab experiments.
- Using computers with tutorial software.
- Finding practical uses for knowledge learned—theory is not enough.
- Planning projects and schedules, then executing them according to plan and on time.
- Listening to detailed lectures.
- Taking detailed, comprehensive notes.
- Studying according to a fixed schedule in an orderly environment.
- Making up a detailed budget.
- Practicing new skills through frequent repetition.
- Taking a field trip to learn about organizations and procedures.
- Writing a "how-to" manual about a project.

⌛ One-Minute Activity 3-2: Quadrant B Learning

Circle the dots in Table 3.3 for those items that are easy for you *and* that you enjoy doing.

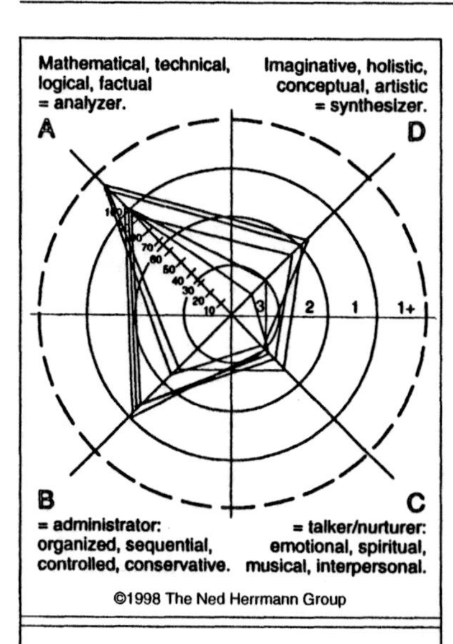

A WHOLE-BRAINED THINKER;
A COMMUNICATOR;
A CREATIVE PROBLEM SOLVER;
A DESIGNER & INVENTOR OF NEW & INNOVATIVE THINGS;
A PERSON WITH GOOD JUDGMENT;
A PERSON WHO BREAKS THE RULES;
A LEADER;
A CORRELATOR OF ABSTRACT IDEAS.

Figure 3.7 Team profile and quadrant B "definition of engineer."

Example: Figure 3.7 shows the exercise results for the more B tilted group of engineering students. Note the obvious structure and consistency, including punctuation. This group was the only one whose members organized themselves before doing the job. They elected a leader and a scribe. Also, they reviewed their work and corrected the sequence of what they had written. They paid careful attention to the details of correct grammar and spelling. Here, the identifying words are "breaking the rules" and "leader." Although this group had strong quadrant A preferences, the quadrant B behavior took precedence in this unfamiliar problem-solving situation. It is a trait of quadrant B teams that they come to closure and finish their tasks quickly.

Practice: Many engineers have strong preferences in quadrant B thinking, although on the average, these are not as pronounced as their dominance in quadrant A. The traditional engineering curriculum, with its emphasis on plug-and-chug problem solving, increases the B quadrant preferences of many students. Also, quadrant B is the preferred teaching style of many instructors. Highly talented people who lack quadrant B skills may not be successful simply because they are not taking care of the details needed to get their good ideas implemented. Developing a comfortable level of quadrant B thinking—and the judgment of choosing when to apply it most profitably—can enhance the effectiveness of the other thinking quadrants. What if you tend to avoid Quadrant B thinking but would like to develop this ability? A bookkeeping class would be good training. The planning exercises in Part 2 of this book for "producers" will develop quadrant B skills, as will the activities listed in Table 3.4.

Table 3.4 Activities for Practicing Quadrant B Thinking

- Cook a new dish by following the instructions in a complicated recipe.
- Use a "programmed learning" software package to learn something new.
- Plan a project by writing down each step in detail; then do it.
- Assemble a model kit or a piece of modular furniture by following the instructions.
- Develop a personal budget, then keep track of all expenditures for one month.
- Prepare a personal property list; then put it into a safe-deposit box.
- Set up a filing system for your paperwork and correspondence.
- Organize your desk drawer or clothes closet.
- Organize your records, disks, books, photographs, or other collections.
- Find a mistake in your bank or credit card statements or monthly bills.
- Be exactly on time all day. Read about time management and carry out one piece of advice, down to the last detail.
- Visit a manufacturing plant to observe how a product is made.

With the sense of sight, the idea communicates the emotion, whereas, with sound, the emotion communicates the idea, which is more direct and therefore more powerful.

Alfred North Whitehead

Characteristics of interpersonal quadrant C thinking

Quadrant C thinking is sensory, kinesthetic, emotional, people-oriented, and symbolic. It deals with awareness of feelings, body sensations, spiritual values, music, teamwork, nurturing, personal relationships, and communication. Nurses, social workers, teachers, trainers, and mothers of infants usually exhibit HBDI profiles with strong C-quadrant preferences. Quadrant C cultures are humanistic, cooperative, and spiritual; they are value-driven and feelings-oriented. Mahatma Gandhi, the Hindu social reformer, Lao Tsu, Chinese philosopher, Dr. Martin Luther King, Jr., and Princess Diana typify strong quadrant C people.

People who prefer quadrant C thinking have preferences for certain subjects in school: the social sciences, music, dance, and highly skilled sports. They would rather participate in group activities than work alone; thus they receive little encouragement in the typical engineering classroom for their most comfortable thinking mode. Teachers, nurses, counselors, social workers, and musicians generally have strong preferences for quadrant C thinking, although musicians and composers involve quadrant A thinking when they analyze musical scores or evaluate a performance. People with quadrant C dominances talk about "the family" or "the team" or "personal growth" and "values." Stereotypically, they are viewed as "bleeding hearts" or "soft touch" or "talk, talk, talk." Table 3.5 lists some learning activities preferred by quadrant C thinkers.

Table 3.5 "Quadrant C" Learning Activities and Behaviors
• Listening to others and sharing ideas and intuitions.
• Motivating yourself by asking "why" and by looking for personal meaning.
• Reading the preface of a book to get clues on the author's purpose.
• Learning through sensory input—moving, feeling, smelling, tasting, listening.
• Hands-on learning by touching and using a tool or object.
• Using group-study opportunities and group discussions.
• Keeping a journal to record feelings and spiritual values, not details.
• Doing dramatics—the physical acting out of emotions is important, not imagination.
• Taking people-oriented field trips.
• Traveling to other cultures to meet people and find out how they live.
• Studying with background music, or making up rap songs as a memory aid.
• Using people-oriented case studies.
• Respecting others' rights and views; people are important, not things.
• Learning by teaching others.
• Preferring video to audio to make use of body language clues.

⌛ **One-Minute Activity 3-3: Quadrant C Learning**

Circle the dots in Table 3.5 for those items that are easy for you *and* that you enjoy doing.

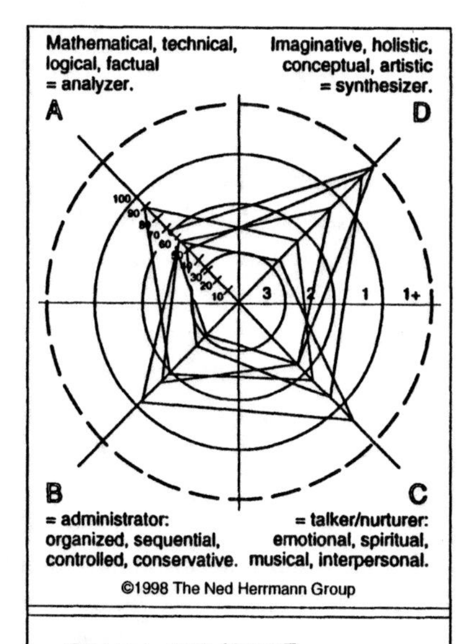

ADMINISTRATIVE
HARD WORKER
CREATIVE
PROBLEM SOLVER
OVERWORKED
UNDERPAID
UNDERAPPRECIATED
SYNTHESIZER

Figure 3.8 Team profile and quadrant C "definition of engineer."

Example: The results for the C group in the engineering student exercise are shown in Figure 3.8. This team used emotionally loaded terms like "overworked, underpaid, underappreciated." Their focus was on people; social interaction was important. This group did much talking and ran out of time for writing down ideas. Despite the strong D-quadrant preferences present, the behavioral characteristics of quadrant C thinking predominated, possibly due to strong interpersonal skills, as well as the mutual reinforcement the C thinkers found in working together. Students with a strong quadrant C dominance are rare in engineering, and students have few opportunities to learn in this mode. The influence of a strong quadrant B thinker in this group is seen in the neat presentation and grammatical symmetry.

Practice: With teamwork and communication skills being stressed by ABET, engineering faculty and engineering students must pay special attention to developing quadrant C thinking. What can you do if you tend to avoid quadrant C thinking but would like to develop these abilities and attitudes? A list of ideas is given in Table 3.6. We have found in our research that engineering students who became teaching assistants in creative problem solving classes without exception increased their preferences for quadrant C thinking. In this book, all team activities, communication exercises, and some of the tasks in the mindsets of the "explorer," the "artist," and the "producer" will provide practice in quadrant C thinking.

Table 3.6 Activities for Practicing Quadrant C Thinking

- When in a conversation, spend most of the time listening to the other person.
- Get involved in a new outdoor exercise activity or in a team sport.
- Play with a small child the way he or she wants to play.
- Think about what other people have done for you and find a way to thank them.
- Volunteer in your community: at the local animal shelter, working with senior citizens, Big Brother or Big Sister, in a reading or tutoring program, scouting, Little League, an environmental action group or a neighborhood improvement association, etc.
- Explore your spirituality. Become active in a religious fellowship.
- Join a church choir or a barbershop quartet. Compose a song; get someone to sing it.
- Savor a vegetable or fruit that you have never tasted before, or grow and use herbs.
- Grow flowers; make fragrant bouquets and bring them to someone who is lonely.
- Use art work, colors, and accessories to create a specific mood in a room.
- Take a seminar on how to communicate or express your feelings better.
- Make time for family meals—think up a heart-felt reason to have a special celebration.
- Play a musical instrument "playfully"; learn to enjoy a new style of music.
- Allow tears to come to your eyes without feeling shame or guilt.
- Get together with a friend (in person or by e-mail); share your feelings about an issue.

Characteristics of imaginative quadrant D thinking

Quadrant D thinking is visual, holistic, innovative, metaphorical, creative, imaginative, integrative, conceptual, spatial, flexible, and intuitive. It deals with futures, possibilities, synthesis, play, dreams, vision, strategic planning, the broader context, entrepreneurship, change, and innovation. A quadrant D culture is explorative, entrepreneurial, inventive, and future-oriented. It is playful, risk-driven, and independent. Pablo Picasso, the modern painter, Leonardo da Vinci, the Renaissance painter, sculptor, architect, and scientist, Albert Einstein, the physicist, and Amelia Earhart, aviation pioneer, are examples of strong quadrant D thinkers.

As a rule, indeed, grown-up people are fairly correct on matters of fact; it is in the higher gift of imagination that they are so sadly to seek.

Kenneth Grahame,
The Golden Age, *1895*

People who prefer quadrant D thinking enjoy school subjects such as the arts (painting and sculpture) as well as geometry, design, poetry, and architecture. Quadrant D students are attracted to the "art" and invention aspects of the engineering profession. Entrepreneurs, explorers, artists, and playwrights strongly prefer quadrant D thinking, as may scientists involved in research and development in medicine, physics, and engineering. Dominant quadrant D thinkers talk about "playing with an idea" or "the big picture" or "the cutting edge" and "innovation." They are talked about as "having their heads in the clouds" or as being "undisciplined" or "unrealistic dreamers." Table 3.7 lists learning activities preferred by quadrant D thinkers.

Table 3.7 "Quadrant D" Learning Activities and Behaviors
• Looking for the big picture and context, not the details, of a new topic.
• Taking the initiative in getting actively involved to make learning more interesting.
• Doing simulations and asking what-if questions.
• Making use of the visual aids in lectures. Preferring pictures to words when learning.
• Doing open-ended problems and finding several possible solutions.
• Appreciating the beauty in the problem and the elegance of the solution.
• Leading a brainstorming session—wild ideas, not the team, are important.
• Experimenting and playing with ideas and possibilities.
• Traveling to other cultures to have adventures and explore new places.
• Thinking about trends.
• Thinking about the future and making up long-range goals.
• Relying on intuition to find solutions, not on facts or logic.
• Synthesizing ideas and information to come up with something new.
• Using future-oriented case discussions.
• Trying a different way (not the procedure) of doing something just for the fun of it.

⌛ **One-Minute Activity 3-4: Quadrant D Learning**

Circle the dots in Table 3.7 for those items that are easy for you *and* that you enjoy doing.

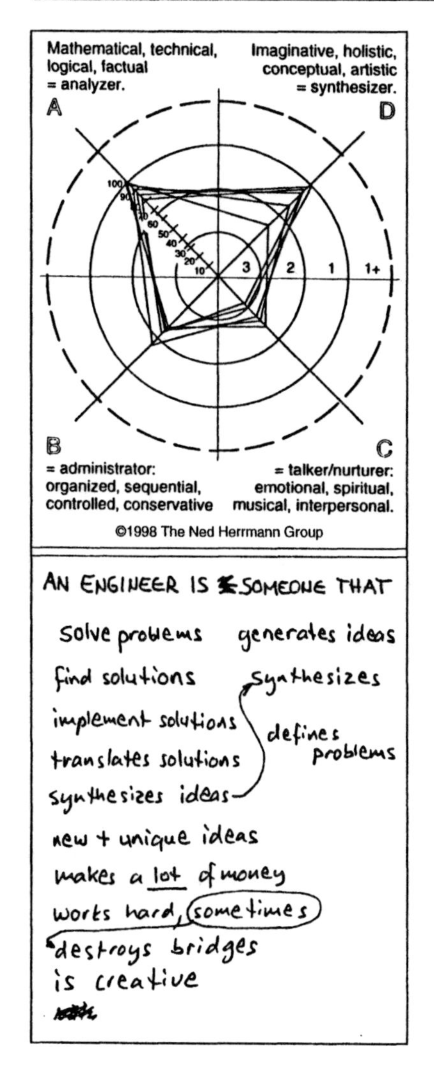

Figure 3.9 Team profile and quadrant D "definition of engineer."

Many people with strong quadrant D thinking preferences have persisted despite everything the educational system may have done to discourage them. They may feel like outsiders—different, odd, weird, or even crazy. In self-defense, they may have developed a chip-on-the-shoulder attitude, or they may have learned to enjoy independence: they are self-motivated and truly march to a different drummer which often makes it difficult to integrate them into a team. A strong quadrant D person may not be able to understand the language and "tribal" bonding between members of a left-brain dominant group but will thrive in an environment where creative ideas are appreciated and nurtured.

Example: Figure 3.9 shows the results for the more D tilted group in the exercise with the engineering students. As is typical for quadrant D thinkers, this team continued working "when time was up." When they finally returned to class after several reminders, they were visibly upset that they could not continue their brainstorming. Even a quick glance at their page shows that this group operated differently. They had many ideas all over the page; they changed their mind; they made connections with arrows; they underlined, and they had to give an explanation of what they had written, because some of it made no sense to their classmates—the "sometimes" did not belong to "working hard" but to "destroying bridges." In brief, this group brainstormed ideas without thinking about any structure, thus they too had problems with grammar. They had the most unusual idea: "An engineer sometimes destroys bridges." Their strength in quadrant A dominance is shown by "making big bucks." However, the informal exercise format—their flip chart was set up in the hall—enabled this group to freely express their D-quadrant thinking.

Practice: We have found that students who continue their involvement with creative problem solving maintain or increase their preferences for quadrant D thinking, whereas the average profile in quadrant D drops as students go through the engineering curriculum. What can you do if you want to develop your quadrant D abilities? Activities in Part 2 of this book for the "explorer," the "artist," and the "engineer" will exercise quadrant D thinking, as will many of the ideas discussed in Chapter 6. Table 3.8 on the following page lists ideas you can play with.

When several thinking modes are habitually accessed by an individual or purposefully available to a team, chances are increased that new ideas and concepts will be created and implemented in practical solutions and significant innovation.

Table 3.8 Activities for Practicing Quadrant D Thinking

- Look at the big picture, not the details, of a problem or issue.
- Make a study of a trend; then predict at least three different future developments.
- Ask what-if questions and come up with a lot of different answers.
- Allow yourself to daydream.
- Make sketches to help you memorize material that you are learning.
- Create a logo or a web page.
- Do a problem that requires brainstorming; find at least twenty possible answers, alone or in a small group.
- Appreciate the beauty of a design, building, appliance, or object.
- Play with Tinker Toys, Skill Sticks, or Legos. Use them to invent a useful gadget.
- Design and build a kite. Fly the kite the way it is meant to be flown.
- Attend an imaginative story-telling session; read a book of folk tales or myths; participate in role-playing games.
- Invent a gourmet dish for a "theme" dinner; decorate the table imaginatively.
- Take a drive (or walk) to nowhere in particular without feeling guilty.
- Take 2000 photographs without worrying about cost; try unusual shots.
- Imagine yourself in the year 2000, 2020, 2040.
- Play with modeling clay or finger-paints. Take an art class.

Multidominant (whole-brain) thinking and learning

Exercise: In the definition exercise, the fifth group of engineering students had very different HBDI profiles: A+, AB, DC+, CD+, and DA. Together they formed a whole-brain team with quadrant C being slightly lower. Their profiles and group result are shown in Figure 3.10. This definition is more complete: it has whole sentences; it is balanced on the page; it gives two options, and it even includes a sketch and a stab at humor. The team's lesser preference for quadrant C thinking shows in the terminology of "guy" and "individual." As you compare these five team results, keep in mind that this was a very brief exercise where the teams did not have time to revise their first draft. The results with these inexperienced groups do show that differences in thinking preference are expressed in different vocabularies and problem-solving approaches.

You see things and you say "why?" But I dream things that never were, and I say "why not?"

George Bernhard Shaw

We have just seen the characteristics of four distinct "ways of knowing." However, only about 7 percent of people have a single strong dominance; about 60 percent have a double dominance, 30 percent have a triple dominance, and only 3 percent have a "square" profile with dominances in all four quadrants. Each person represents a unique coalition of thinking preferences. Think of having a team of players inside your brain. You send out specialists for specific tasks; you send out one, two or maybe even three star players more often than the others, but to function well, the whole team is needed. Examples of famous people with double dominances (from a proforma analysis by Ned Herrmann) are:

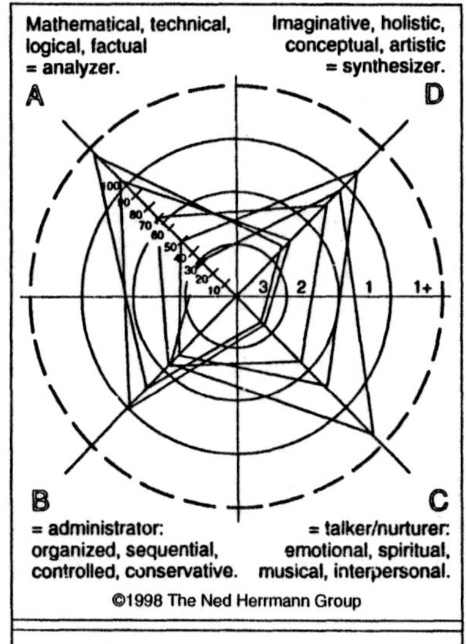

AN ENGINEER IS:

1. A GUY WHO DRIVES TRAINS...

–OR–

2. AN INDIVIDUAL WHO USES EXISTING KNOWLEDGE CREATIVELY TO SOLVE CURRENT PROBLEMS.

FOR THOSE WHO DON'T COMPREHEND THIS IS A STRIPED HAT.

Figure 3.10 Whole-brain team profile and "definition of an engineer."

A + B: Aristotle, Henry Ford, Margaret Thatcher.
B + C: Susan B. Anthony, Mother Theresa, Lech Walesa.
C + D: Shakespeare, Eleanor Roosevelt, Jim Henson, Mozart.
D + A: Galileo, Madame Curie, George Bernhard Shaw.
B + D: Sadam Hussein.

Examples of people with balanced or whole-brain thinking preferences are Ben Franklin, Sebastian Bach, Winston Churchill, Thomas Jefferson, Albert Schweitzer, and Chief Sitting Bull. Ned Herrmann found that no single profile is more prominent or more valuable than any of the others. People are happier and usually will do well when their activities and job requirements match their thinking preferences. If you have an avoidance in one of the quadrants, our advice is to make a career choice that will not require you to have to function in this mode for long periods on a daily basis—the frustration and energy level required would be too great.

Figure 3.11 shows that learning involves all four quadrants. We have what is called external learning taught from authority through lectures and textbooks ➛ quadrant A. We have internal learning through insight, visualization, synthesis, or a sudden understanding of a concept holistically and intuitively ➛ quadrant D. We have interactive learning through discussions and sensory experiences where we try, fail, try again with an opportunity for verbal feedback and encouragement ➛ quadrant C. Finally, we have procedural learning through a methodical, step-by-step testing of what is being taught, as well as practice and repetition to improve skills ➛ quadrant B. Effective teachers have discovered ways of incorporating all of these modes into their teaching strategies. This goal is difficult when instructors arecomfortable teaching in only one or two quadrants.

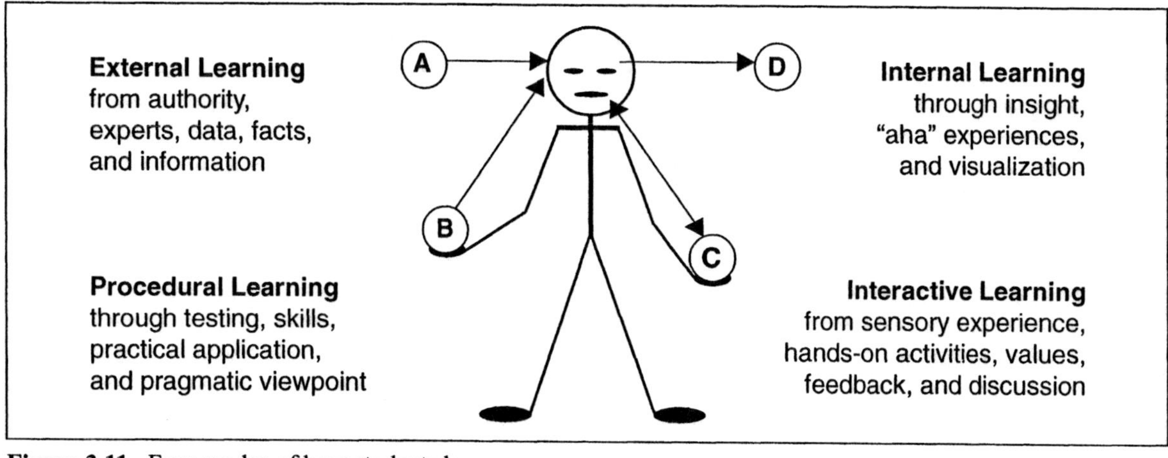

Figure 3.11 Four modes of how students learn.

Ten-Minute Team Activity 3-5: Whole-Brain Teaching

In groups of four people, select a concept in some subject (science, math, engineering, economics), then make up examples of how to teach the concept in all four thinking quadrants. Share your ideas with a larger group, or develop additional examples for several different fields.

What does all this mean for engineers?

He was living like an engineer in a mechanical world. No wonder he had become dry as a stone.

Simone de Beauvoir, The Mandarins

Engineering students with unusual profiles often ask if they should switch to another field. Although engineers on the average have a typical quadrant A dominant profile, companies—to succeed in the global marketplace—are increasingly seeking engineers who have strong quadrant D (innovative, strategic) and quadrant C (communication and teamwork) skills. This is why we are emphasizing these foundational skills in this book. Quadrant B engineers may gravitate toward administration, safety, and quality control. Quadrant D engineers may become entrepreneurs, inventors, and conceptual designers. Quadrant C engineers are needed for sales, teaching and training, writing technical manuals for non-technical audiences, customer relations, user-friendly and environmentally benign products, international relations and politics, just to give examples of the many opportunities open to "different" engineers.

As shown in Table 3.9 and Figure 3.12, engineering design requires a whole-brain approach. This has two implications: 1. Each engineer must develop fluency in all four quadrants—in essence be able to put on four different "hats" depending on the task at hand—analyzer, administrator, collaborator, and synthesizer. 2. We must purposely select different thinkers to make up a whole-brain multidisciplinary team. Such a team will be more productive if each member appreciates the contributions those with different thinking modes can bring to the team.

Sir William Halcrow, addressing the Institution of Civil Engineers, expressed the same principle this way: "The well-being of the world largely depends upon the work of the engineer. There is a great future and unlimited scope for the profession; new works of all kinds are and will be required in every country, and for the young [person] of imagination and keenness I cannot conceive a more attractive profession. Imagination is necessary as well as scientific knowledge."

Table 3.9 The Four Thinking Quadrants Involved in Planning and Designing a Bridge

Quadrant A:	Technical specifications. Financing. Practical project logistics.
Quadrant B:	Low-risk, efficient path for getting from Point x to Point y. How to build it.
Quadrant C:	Connecting people. Effect on communities and environment. Politics.
Quadrant D:	Future traffic projections. Different possibilities. Artistic design concepts.

A Analyzer

Applying mathematical models.
Calculating specifications.
Comparing alternatives.
Computing benefits and costs of solutions.
Drawing inferences from statistical information.
Drawing physical and mathematical analogies.
Evaluating and optimizing conceptual designs.
Formulating reasoned, analytical approaches.
Generating quantitative results.
Generating predictions based on math models.
Quantifying criteria for solution evaluation.
Performing preliminary engineering analyses.
Solving mathematical equations.
Separating factual data from opinion.
Taking principles and data to logical conclusions.
Verifying assumptions and arbitrary parameters.
Writing computer programs.
Writing project proposals and technical reports.

Synthesizer D

Brainstorming wild and crazy ideas.
Conceiving new approaches to design problems.
Creating an imaginative work environment.
Creating new models of system behavior.
Developing metaphors for projects and goals.
Developing several competing design alternatives.
Drawing solutions from fields outside engineering.
Framing problems in new formats.
Leading teams to innovative solutions.
Leading with vision; seeing the whole picture.
Looking for innovation and break-through ideas.
Presenting results in imaginative ways.
Redefining old problems with new insights.
Recognizing opportunities for improvement.
Visualizing new connections or arrangements.
Using crazy ideas as triggers to innovative concepts.
Sketching possible design solutions.
Synthesizing solutions from other engineering fields.

B Administrator

Checking drawings for errors.
Checking specifications against codes.
Collecting and safe-guarding project records.
Debugging computer programs.
Developing checklists.
Drafting bills of material.
Expediting design details.
Following design procedures.
Issuing change orders and tracking design changes.
Linking complex project plans and schedules.
Optimizing procedures.
Organizing and scheduling design projects.
Producing "as-built" drawings.
Synchronizing product and process design.
Supervising design drafters.
Taking action to implement design plans.
Tracking project expenditures.
Updating software; scheduling required training.

Collaborator 

Being sensitive to team members' feelings.
Brainstorming concepts with teams.
Building effective relationships with all customers.
Communicating effectively at all stages of design.
Continuously teaching yourself/others new techniques.
Cultivating enthusiasm.
Developing environmentally benign concepts.
Encouraging/training coworkers in new technology.
Enjoying teamwork.
Involving implementers of solutions in their creation.
Maintaining ethics and values.
"Seeking first to understand, then to be understood."
Seeking win-win solutions that benefit all parties.
Selling solutions and ideas.
Sensing customer needs.
Sharing goals and experiences.
Using senses and intuition to define the design problem.
Working toward synergy rather than compromise.

Figure 3.12 Engineering design requires a whole-brain approach (based on the Herrmann whole brain model, ©1998 The Ned Herrmann Group).

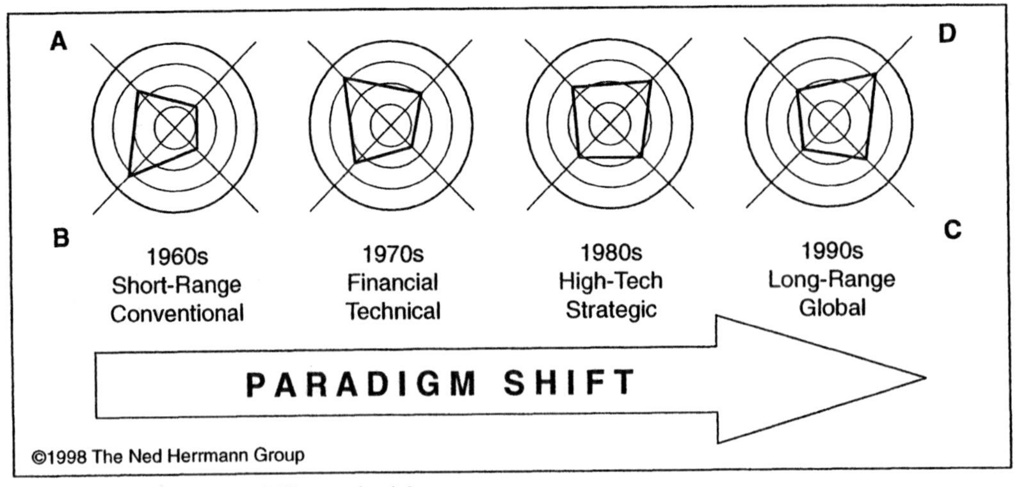

Figure 3.13 Thinking skills required for success.

Ned Herrmann investigated the changes in the average thinking preference profile of successful people in the past four decades. The progression of the paradigm shift is shown in Figure 3.13. Compare this figure with the average profile of the typical engineering faculty as shown on Figure 3.5 and with the fact that graduating seniors exhibit a profile very much like that of the faculty. The paradigm shift in ABET and recent curriculum restructuring efforts supported by NSF driven by the demands of industry should result in a shift to the right for many engineering curricula. The HBDI constitutes an ideal assessment tool to verify that this shift will happen in the desired directions.

The current typical engineering profile does not meet the requirements of industry for teamwork skills and an entrepreneurial, flexible, global outlook. Thus students who in the past were "encouraged" to leave engineering because they did not fit the mold are now sought by forward-looking industry. These students will still feel uncomfortable in many engineering classes, but if they understand their own thinking preferences and the mismatch with their peers, their typical professors, and the traditional, extremely analytical curriculum, they can develop strategies for coping and for optimizing their learning—they will have superior thinking modes for succeeding in the workplace of the future.

Our research over several years has found that the HBDI profile of many students changed between their freshman and senior years, and the magnitude of these changes for many students was startling. A majority of students became more left-brain (in essence"clones" of the faculty) because of the pressure of the intense curriculum. The HBDI profiles of students who received repeated encouragement and reinforcement in creative thinking (or those who "discovered" their creativity and found that right-brain modes were legitimate and useful) shifted to the right. Students who built quadrant C and quadrant D activities into their

A whole-brain team can obtain optimum problem-solving results if the team members learn to understand each other and appreciate the contribution each person can make to the team because of the differences and their strengths.

A common understanding of the HBDI model is empowering; it is valuable for team building, communication, training, management, knowledge creation, and innovation.

lives were also able to resist the left pull of the curriculum. Engineering freshmen with high quadrant D scores dropped out of engineering at a much higher rate than the average. The students cited the following reasons for leaving engineering: lack of challenge or support for right-brain thinking modes; lack of creativity; lack of teamwork; lack of synthesis and connections with real life; lack of holistic learning. A curriculum and classroom environment that is less hostile to these thinkers will do much in retaining these talented students in engineering.

It is a matter of grave concern that many engineering students are graduating with a sharp drop in the thinking modes required for a teamwork-based work environment. Engineering faculty members who are involved in curriculum restructuring are finding considerable resistance to change among their peers—getting faculty to buy in and make the necessary broad-based changes in engineering education is proving to be very difficult. The new ABET criteria are a start, but it will be many years before assessments will show if the restructuring efforts have lead to fundamental change or are just window dressing. Another source of resistance to change comes from the students. They, too, do not like change. Because they have figured out how to get good grades under the old paradigm, they see a new playing field as a threat. Students must be taught flexible thinking, creative problem solving, and teamwork skills so they can successfully cope with the changing world.

The knowledge creation model

Ikujiro Nonaka and Hirotaka Takeuchi, in their book *The Knowledge-Creating Company: How Japanese Companies Create the Dynamics of Innovation,* develop a detailed theory of the knowledge creation process as practiced in Japan. As we read this fascinating work, we were struck by the correspondence between the four-quadrant knowledge creation model and the Herrmann model. These connections—for *goals* and actions—are summarized in Figure 3.14. We believe this model can be applied to improve the learning process of study groups and work teams, with the following results:

Step 1 (quadrant C) ➛ a collective commitment to learning and to sharing experiences and information, as well as developing a "feel" for knowledge and cooperation.

Step 2 (quadrant D) ➛ an intuitive understanding of concepts, change, and innovation; an increased ability to think flexibly, use metaphors, and explore new ideas.

Step 3 (quadrant A) ➛ synthesis of new knowledge through combination and team synergy, and continuous improvement through feedback from within and outside the team, going far beyond current approaches in information processing.

The value of most products and services depends primarily on how "knowledge-based intangibles"—like technological know-how, product design, marketing presentation, understanding of the customer, personal creativity, and innovation— can be developed.

J.B. Quinn,
Intelligent Enterprise: A Knowledge and Service-Based Paradigm for Industry,
Free Press, New York, 1992

Step 4 (quadrant B) ➛ practical experiences from using the new learning. This new tacit knowledge can trigger a new learning cycle at a higher level, involving new teams and new areas of learning, ultimately spiraling up to involve upper levels in the organization (from the department level to top management) and then to the larger community, while yielding continuous innovation and competitive advantage at each level.

To better understand Figure 3.14, we need to consider some definitions. The term "learning" as used here is a whole-brain process—it is thus much more than just a passive acquisition of facts and information. Generally in Western culture, "knowledge" is used synonymously with data and information. Nonaka and Takeuchi call this *explicit* knowledge that can be expressed in words and numbers and systematically shared in the form of scientific formulas, principles, or procedures. They define another *tacit* type of knowledge as understood in Japan:

> Tacit knowledge is highly personal and hard to formalize, making it difficult to communicate or share with others. Subjective insights, intuitions and hunches fall into this category of knowledge. It is deeply rooted in an individual's action and experience, as well as in the ideals, values, or emotions he or she embraces. Tacit knowledge can be segmented into two dimensions: 1. It encompasses the hard-to-pin-down skills or crafts captured in the term "know-how." 2. It consists of mental models, beliefs and perceptions so ingrained that we take them for granted; they reflect our image of reality and our vision for the future.

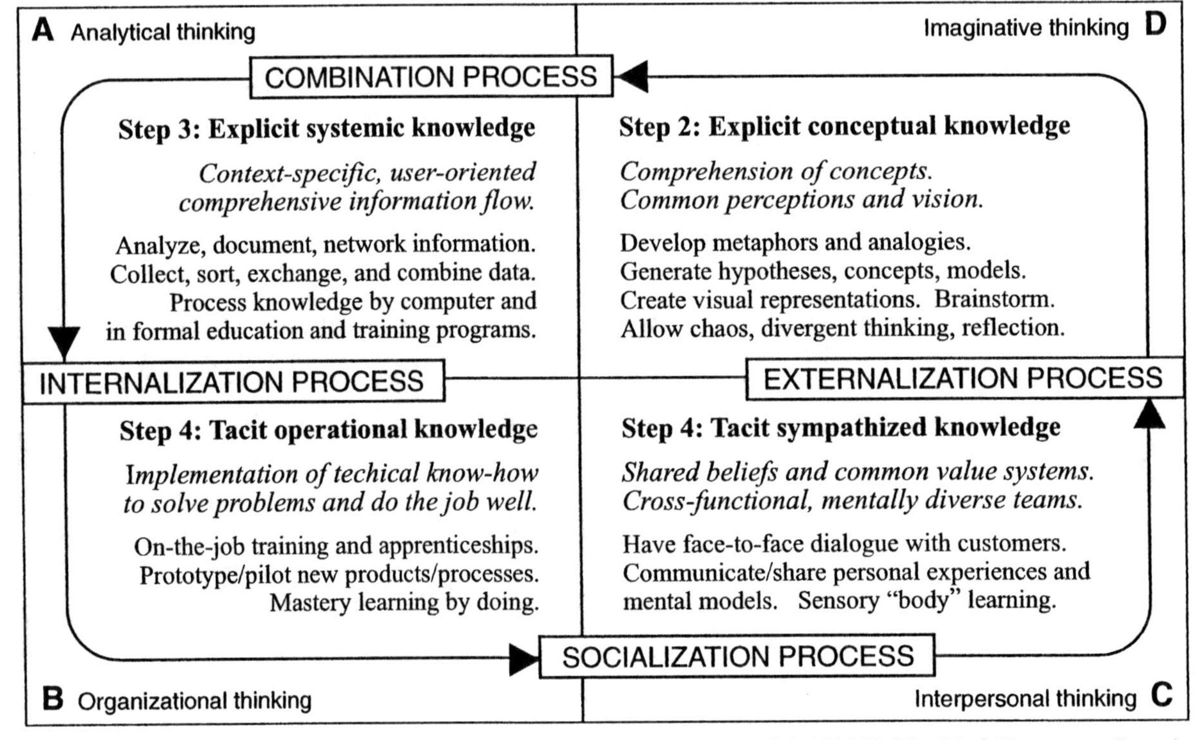

Figure 3.14 Knowledge creation superimposed on the Herrmann model (©1998 The Ned Herrmann Group).

Sympathize: [syn-, together + pathos, feeling]
1. to share or understand the feelings or ideas of another;
2. to be in harmony or accord.

From Webster's New World Dictionary

Ikujiro Nonaka and Hirotaka Takeuchi are presently professors at Hitotsubashi University in Japan. They both studied at the University of California at Berkeley, and Takeuchi later taught in the Harvard Business School before returning to Japan. So how did we make this "leap" from management to engineering education and learning (a case of personal knowledge creation)? It happened when the new understanding of the knowledge creation model combined with our tacit knowledge of the Herrmann model and with education/training in the university and in industry. Suddenly we "saw" the connections—the models are powerful tools for explaining problems and give insight into the changes we were trying to make through teaching the creative problem solving process.

Nonaka and Takeuchi say that "Western managers need to unlearn their old view of knowledge and grasp the importance of the Japanese view. They need to get out of the old mode of thinking that knowledge can be acquired, taught, and trained through manuals, books, or lectures. Instead, they need to pay more attention to the less formal and systematic side of knowledge and start focusing on highly subjective insights, intuitions, and hunches that are gained through metaphors, pictures, or experiences." Now look at this statement through the "lens" of the Herrmann model. The old mode is quadrant A type education; what is missing are the right-brain ways of thinking and learning (quadrants C and D), as well as the hands-on, experienced-based tacit "ways of learning" and sharing (quadrant B combined with quadrant C). Note that the new ABET criteria implicitly require engineering education and students to be able to function in all thinking quadrants and steps of the knowledge-creation cycle. The two models can diagnose problems in learning and teamwork; creative problem solving can be used to find the best solutions, as we will show with examples later in this section.

The essence of innovation is to re-create the world according to a particular ideal or vision. Creating new knowledge is also not simply a matter of learning from others or acquiring knowledge from the outside. Knowledge has to be built on its own, frequently requiring intensive and laborious interaction among members of the organization.

I. Nonaka and H. Takeuchi

Steps in knowledge creation and learning

The following paragraphs describe the characteristics of the four steps of knowledge creation and how they apply to learning. It is very important to understand the key idea in this dynamic model: *knowledge is created in the process of moving from one quadrant to the next*, especially where tacit knowledge is converted to explicit knowledge and vice versa in an interactive social process which happens most effectively at the team level.

Step 1 — Socialization

Through dialogue and discussion (within and outside formal classes and training programs), the organizational vision, mental models, and personal experiences are shared. The broad picture or context of the organizational vision must be part of the team's vision—students who have as their goal merely passing a course or getting an A are not reaching high enough; they need to see their project and their teamwork in the context

Sharing is not for the sake of sharing; it has to have a directed purpose and relatioship to the work in progress.

Ikujiro Nonaka (Ref. 18.16)

of their entire engineering education and their future work. Hands-on training and mentoring relationships are important for transferring tacit (subconscious) knowledge. We usually find it very difficult to express tacit knowledge in words—where we (our bodies and minds) just "know" how to do something—or we may not even be aware that we have this knowledge. But others, by observing and imitating our actions, can learn this knowledge from us. This is one of the benefits of teamwork. A key task for engineering designers at this stage is to build relationships with customers so these can share their tacit knowledge and experiences with the problem or current product that needs to be addressed with the new design. Customers do not usually verbalize their deepest needs; these can only be discerned intuitively through personal contact.

This brings up one disadvantage of distance learning—it cannot easily provide this one-on-one modeling for transmitting tacit knowledge. Teamwork and motivation are enhanced by attention to quadrant C socialization or "sympathized knowledge" creation. We have observed that teams need time for socialization and the building of mutual trust, so ideas can be shared without fear of ridicule in productive brainstorming sessions. A student team in a sailboat design project was dysfunctional because one member refused to share his previous sailing experience with his team members. His attitude was conditioned by the prevailing competitive grading system where hoarding knowledge is an advantage. Thus the model brings out one area that will require organizational change, if learning in the new paradigm is to be improved: team learning must be given equal, if not more, weight and reward than individual effort.

Step 2 — Externalization

Visualization and concept creation are the key elements: managers (or team leaders) must articulate their organizational vision in the shape of metaphors, followed by analogies and then models, to form an understandable foundation or externalization for explicit "conceptual knowledge" creation. The metaphorical concepts are developed through brainstorming by teams. Quadrant D integrative, flexible thinking (which is enhanced by training in brainstorming techniques) facilitates change and is a prerequisite for innovation. Metaphors are powerful tools for building an intuitive, common understanding of concepts, so the team will be a unit proceeding in the same direction and solving the same problem.

Experience serves not only to confirm theory, but differs from it without disturbing it, it leads to new truths which theory only has not been able to reach.

Dalembert, quoted in P.S. Girard's Traite Analytique de la Resistance des Solides

Visual components and analogies can be transmitted by distance learning. When done in the context of teams, a lively interaction between different minds will augment the synergy of creative ideas. For productive brainstorming, some team members should have explicit, others should have tacit knowledge of the problem. Some members should be drawn from outside the field—they will not have preconceived notions of what will not work and may thus generate particularly original ideas.

This is the rationale for using multidisciplinary teams in capstone design courses. We have found that engineering students need much encouragement for considering innovative ideas—they have already been conditioned to be practical and realistic, whereas a student from liberal arts, an elementary school kid, or other "outsiders" will have an unbiased, fresh approach able to bring forth original ideas for development.

Step 3 — Combination

Data, concepts, ideas, and solutions are analyzed, categorized, "practicalized," evaluated, and documented for structured information flow and preservation of new knowledge. This combination process can lead to synthesis and thus a new level of explicit, "systemic knowledge" creation. Information collected through two feedback loops leads to continuous improvement. The quadrant B loop leads to enhanced problem solutions and knowledge as people are applying new information and learning. The quadrant D loop achieves better solutions through additional creative thinking, brainstorming and synthesis that eliminates flaws as part of creative problem solving. One particular technique for this feedback loop is the Pugh method, a team evaluation procedure for design concept optimization, which will be explained in Chapter 11. In universities, we still have much to learn about using a team approach in information processing and learning. These skills are crucial in concurrent engineering. In product design, Step 3 results in a prototype.

When organizations innovate, they do not simply process information—from the outside in—to solve existing problems or adapt to a changing environment. They actually create new knowledge and information—from the inside out—to redefine both problems and solutions and, in the process, to re-create their environment.

I. Nonaka and H. Takeuchi

Explicit knowledge can be acquired through distance learning, from documentation of facts, data, and processes, and through analysis. In our Western educational and training programs we often overemphasize this quadrant A thinking in the traditional classroom or believe it is the only valid way of knowledge creation and learning. At this level, we acquire necessary "head knowledge" but we may still be quite unable to do the job. This is where we learn the "science" of engineering and design, but we have not yet learned the "art" (which is a synthesis of tacit "know-how" and intuitive, creative thinking).

Step 4 — Internalization

The new knowledge is applied to solve problems and do the job. Experience is gained through practice, on-the-job training, experimenting, and pilot programs. There is no shortcut to "getting your hands dirty," although having at-elbow coaches and support helps to accelerate the process, as will sharing of tips as people become proficient in using the new knowledge and training. The sharing of practical learning experiences starts a new cycle in dynamic knowledge creation and can spiral to the organizational level. This step of internalization and acquiring "operational knowledge" requires repetitive practice and the careful attention to detail, planning, and organization of quadrant B thinking; it is the direct opposite of externalization which uses quadrant D thinking. This step cannot be taught by distance learning but is an important benefit of

The activities in the text are designed to initiate tacit knowledge acquisition of the topics being discussed.

This tactic helps all learners but is crucial for quadrant B and quadrant C thinkers.

co-op education, internships, apprenticeships, and laboratory courses. Many companies who hire engineering graduates do not place them in engineering positions (a big disappointment to these graduates who think they are well prepared for their jobs): the new hires go through a period of being engineers-in-training so they can acquire tacit knowledge about engineering as well as about the company's culture. In a way, graduation is not the end but just the beginning of life-long learning (which will be a balance of tacit and explicit knowledge acquisition).

Few individuals or organizations will pay careful attention to all four thinking quadrants or steps in knowledge creation—excelling in one mode usually brings discomfort with opposite ways of thinking and processing information. When the HBDI model is understood and applied in knowledge creation and in creative problem solving, creative individuals and their ideas will be valued instead of ostracized or ridiculed by their more conservative, left-brain peers, instructors, and supervisors.

Whole-brain thinking can facilitate the interaction between explicit and tacit **knowledge practitioners**. The knowledge creation cycle does not happen automatically or in a vacuum. Whole-brain thinkers are ideally suited to function as integrators who can purposefully manage and lead the process by nurturing social interaction between tacit knowledge and explicit knowledge practitioners. They must form a strategic link between the idealized vision and the chaos of the real world faced by the front-line workers (or learners) as they deal with technologies, products, markets, and procedures. They must "engineer" a practical conceptual framework that workers (or students) can understand and integrate with their day-to-day experiences but which also connects to the broader organizational and external (even global) context. *Knowledge operators* function in the area of tacit knowledge generation—they can be salespeople interacting with customers; they can be experienced production line workers, skilled craftspersons, technicians, or hands-on experimenters with in-depth local knowledge (for example, test drivers living in a particular country). *Knowledge specialists* deal primarily with explicit, structured and technical knowledge not necessarily of immediate interest to the operators—they include R&D scientists, marketing researchers, design and software engineers, as well as finance, personnel, and legal staff. The knowledge creation process and teamwork are enhanced when mediators optimize the communication skills of the practitioners.

It is the dynamic process of interaction between individual and organizational spirals that fuels innovation and adds value, not information or knowledge per se.

For innovation, the knowledge creation process must interactively spiral through several individual and organizational levels, depending on the complexity of the project and the changes required. It might be interesting to apply the knowledge creation model to the process of change in engineering education. Did so many curriculum restructuring efforts in the past fail because steps in knowledge creation were omitted and prevented the process from spiraling to higher organizational levels?

⌛ **Activity 3-6: Applying the Knowledge Creation Model**

Apply the knowledge creation model to a current project that you are involved in, at work, in your engineering courses, or in a campus organization. Identify aspects of the project with each quadrant of the model. Which areas should receive more attention to strengthen knowledge creation, learning, and the innovation spiral? Share your findings with a group.

Illustrations

So far, we have presented mostly explicit knowledge about the knowledge creation model. Now we want to show four different applications that will illustrate the process.

1. San Francisco-Oakland Bay Bridge

Table 3.10 summarizes three knowledge creation cycles spiraling up during the planning and design of the bridge.

Table 3.10 San Francisco-Oakland Bay Bridge*

Round 1: Early History	
Socialization:	Public discussion and increasing traffic needs after World War 1.
Externalization:	Thirty-eight bold proposals and design concepts by 1928.
Combination:	Board of 3 distinguished engineers recommends analysis of preferred site for more detailed design and cost estimates. Benefits of bridge versus tunnel.
Internalization:	Focus on bridge failures with large cantilever designs.
Round 2: Bridge Authority	
Socialization:	Political efforts to circumvent the War Department and its requirements.
Externalization:	Creating a publicly-owned facility.
Combination:	Financing through revenue bonds; appointment of state highway engineer in charge of project; borings and analysis to find best location.
Internalization:	Detailed traffic studies; best route; California Toll Bridge Authority.
Round 3: Bridge Design	
Socialization:	Many experienced engineers and independent consultants work together on the project; Oakland's wishes to allow future port facility expansions are met.
Externalization:	1931: serious design work with many possible designs. Scenic beauty to be taken into account. Goat island tunnel bore to be larger than any in existence.
Combination:	Engineering experience and judgment play a key role in narrowing down the possibilities. Switch from cantilever to suspension concept for the San Francisco side of the bridge for economic, safety and aesthetic reasons. Budget: $75 million.
Internalization:	Special model tests are conducted since no previous experience with multiple-span suspension bridges existed.

Note: Construction was completed ahead of schedule and within budget in 1936.

* Model applied to information given in Henry Petroski, *Invention by Design,* Harvard University Press, 1996.

The San Francisco-Oakland Bay Bridge was designed and built in the context of past and contemporary engineering knowledge. When the Oakland side was damaged in the 1989 earthquake, new knowledge about the nature of earthquakes and the bridge's vulnerability became available. Although this crucial traffic link was repaired in a record 30 days, plans for the construction of a new span at a projected cost of $1.3 billion are now being developed in new cycles of knowledge creation.

Oakland Bay Bridge

The final design is a single-tower, self-anchored, asymmetrical suspension bridge, creating a "majestic portal," with a 15.5 ft wide bicycle/pedestrian path added along the South side.

See http://www.mtc.dst.ca.us/projects/baybridge/bbfin.htm for stunning pictures and updates on this exciting design project.

The Bay Bridge Engineering and Design Advisory Panel (EDAP) reviewed more than one dozen design options, and the public was invited to express its preference for these conceptual designs. In June 1997, EDAP recommended that design consultants be hired to further develop two options to the 30 percent design stage, so that more detailed information on seismic performance, cost, visual appearance and other issues could be obtained before the final decision was made. The two concepts were: (a) a self-anchored suspension bridge, and (b) a cable-stayed bridge—both with single or twin tower options. The location for the new bridge has already been chosen north of the existing span based on flexibility, fewer land-use conflicts, and enhanced vistas. The sidebar gives the decision on the final design made by the Metropolitan Transportation Commission in mid-summer 1998. The design process is projected to take four years, with a minimum of three years for construction. The last step in the project will be the demolition of the current bridge. It is fascinating to monitor the progress of the project on the Internet which has become a wonderful tool in the socialization process of getting the public involved.

2. Development of a Strategic Planning Document

The processes that were used to develop the strategic planning document of the William States Lee College of Engineering at the University of North Carolina at Charlotte can be described as cycles in knowledge creation. The demands for engineering education reform were merged with TQM and resulted in the Dean's vision of "a new way of doing business." This vision needed to involve all stakeholders, thus faculty dialogue was facilitated with several retreats during Fall 1992 (Step 1). A consensus concept (Step 2) emerged in that the first item in the strategic planning process should be the development of a "shared vision."

To collect ideas and information for this statement, several vision workshops were held during Spring and Fall 1993, involving faculty, administrators, students, and people from industry. The data were compiled and synthesized into a final vision document, discussed, revised, and then formally ratified (Step 3). This vision has now been implemented into all future planning processes of the College and its departments (Step 4). In a parallel cycle, the strategic planning committee appointed by the Dean in early 1993 began the formal planning process that would ultimately produce the planning document.

Steps 1 and 2: During Spring 1993, planning meetings were held in each department, in keeping with the "shared vision." An understanding of the concept of "a new way of doing business" led to the appointment of a faculty reward system task force to facilitate future changes.
Step 3: During 1994, the response from national corporate executives gave additional guidance to process and content of the developing strategic plan. Also, faculties were trained on how to make the knowledge shared in informal meetings explicit. Finally, during Spring 1995, the innovative strategic planning document was completed in a matrix format listing the vision, goals, constraints, strategies, and deliverables as the planning elements versus processes or systems elements such as "student learning and development, faculty development, and resource and community development." This was a living document which incorporated processes that achieve continuous planning, input, assessment, and improvement of all the activities focused on learning.
Step 4: Since that time, the plan has undergone several revisions, each time incorporating more of the parallel departmental plans, as this tacit knowledge is being gained and spiraled throughout the College. An important off-shoot was the creation of a new multidisciplinary course, Introduction to Engineering I and II, which focuses on team skills and involves many faculty members. These tacit experiences are teaching that it is possible to establish "a new way of doing business" despite the formidable inertia of an established system, and that it takes patient communication to keep the process going.

3. Education and Training Program in Industry

The Herrmann brain dominance and the knowledge creation models can be applied to anything that could be improved with creativity: programs, product design and development, methods and procedures, production, and services. The example of a training program demonstrates the use of these foundational thinking tools to analyze problems and find solutions. Had these models been integrated into the development of the program right from the start, the current direction would have been more innovative and less costly. Without a common knowledge of these models, change is very difficult to implement.

The half-life of education is 30 days — if you haven't used the new knowledge within a month, then you lose half of it.

Peter Merrill,
Do It Right the Second Time,
Productivity Press, 1997

A high-tech training program was developed as a joint effort between a software developer, a global manufacturing company, and a state government grant which funded the participation of college professors and students. Enormous challenges had to be faced and quickly resolved in an environment operating under severe time constraints, such as hiring and training the "right" faculty; integrating and certifying university and industry instructors; developing a curriculum, documentation, facilities, and procedures to handle frequent major software upgrades, to evaluate the progress of the trainees, and to bridge classroom theory with application on the job, all with the critical goal of getting employees to quickly ramp up to high productivity.

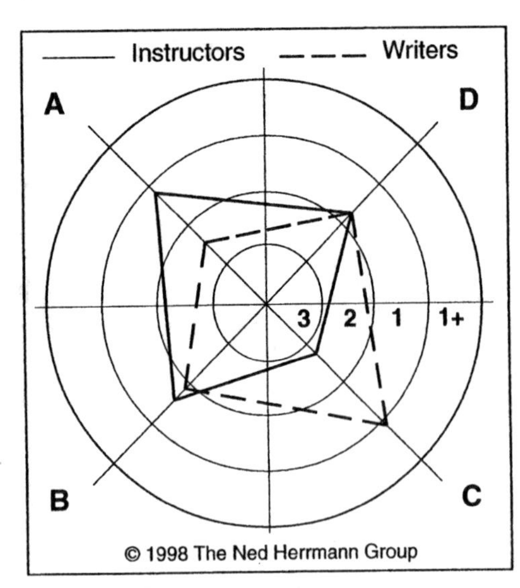

Figure 3.15 Average HBDI profile of instructors compared to manual writers.

Findings from the Herrmann model (HBDI). Although the organization emphasizes teamwork, a significant number of employees participating in the HBDI survey had a low preference for quadrant C thinking—not surprising for technical staff. On the other hand, the management team for the training program had strong quadrant D preferences and was able to cope well with frequent changes, chaos, and crises. The initial lack of structure observed in the overall "culture" of the training program was confirmed by the low average in quadrant B thinking of the original group. As shown in Figure 3.15, a mismatch was diagnosed between the quadrant A dominance of many instructors and the quadrant C dominance of the manual writers. This was difficult to resolve because of the communication barrier between the two groups. Also, key people involved did not attend an HBDI workshop.

Pilot classes of trainees were analyzed with the HBDI to show the instructors that they needed to teach to all thinking quadrants. Each class of six to ten students contained at least one individual with strong quadrant D thinking preferences, and in most classes, strong thinking preferences in all four quadrants were present. One strongly quadrant C dominant class required a different teaching approach (based on group learning) than the usual quadrant A delivery. Recommendations were:

- Continue to emphasize that instructors need to address the thinking preferences of the trainees in their classrooms.
- Review the teaching manuals to systematically sustain a four-quadrant approach. Include examples on how the new skills can be used to encourage innovation and achieve the organization's global vision.
- Offer a course in visualization and solid modeling to strengthen the quadrant D thinking of the trainees.
- Implement the HBDI widely to foster respect for quadrant C and quadrant D thinking to enhance future success of the organization.

Findings from the knowledge creation model. We have only recently begun to use this new model to better understand why certain aspects of the training program worked and why changes were needed.

STEP 1—Socialization: Informal meetings for sharing information and experiences are regularly scheduled by the manager, as are celebrations of successes. These provide an opportunity for knowledge operators and knowledge specialists to meet and arrange further dialogue for specific problem solving and information sharing. The manual writers were not spending enough time in the classroom to pick up on the discomfort the trainees experienced due to the lack of logical organization in the manual. Although the instructors were aware of this problem (since it bothered them, too), they were unable to share it in a quadrant C way that did not hurt the writers' feelings.

STEP 2—Externalization: One important management goal was to significantly shorten "ramp-up" time to full productivity—enabling the trainees to do their job quickly with the new software. Experiences in many large companies have shown that on the average, ramp up to full productivity takes at least eight months, and many employees can take much longer. Thus a time reduction would yield substantial cost savings. It was interesting to see how the ramp-up graph (see Figure 3.16) captured the imagination of management. It confirmed the importance of conceptualizing the vision in graphical or metaphorical form. The analogy of comparing the old wire-frame tool with an ax and the new solid modeler software with a chain saw also worked well.

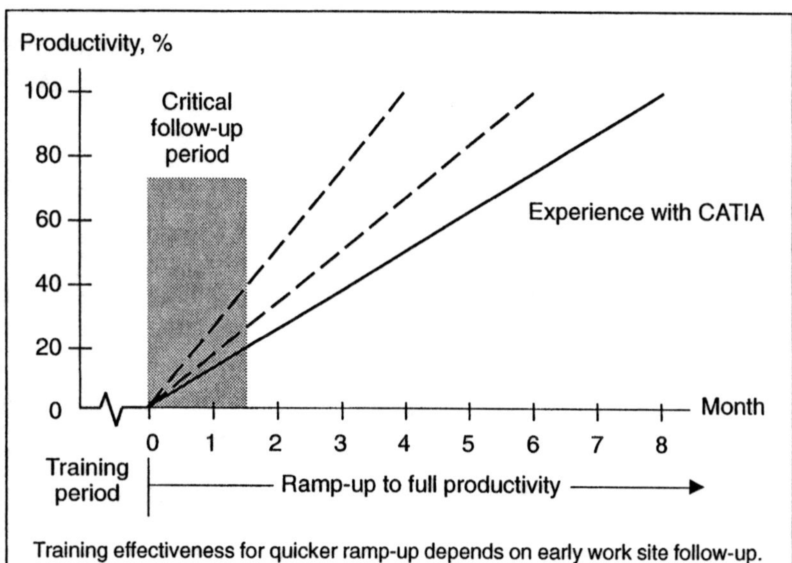

Figure 3.16 Visualizing the key training goal → shortening the time to full productivity.

ANALOGY

A chain saw used like an ax may not even result in minimally acceptable work —the chain saw has to be used in a new and different way. The worker is then capable of doing more things faster (or better) than with the old ax.

STEP 3—Combination: Within the old paradigm of classroom instruction, research and analysis provided information on optimum class size and configuration: 10 students, a senior instructor and a lab assistant, with each person having a workstation. Surveys gave information and feedback on training needs by different types of employees—this helped to streamline and custom design the curriculum as well as improve the manuals. A major organizational focus is on converting existing product information to "libraries" accessible with the new design tool; once these standard parts and components are available, designers will be able to focus on combining them in more innovative ways. Overall, the bulk of the training is concentrated in this step of teaching explicit knowledge.

STEP 4—Internalization: When it became obvious that explicit knowledge acquisition in the classroom (even when supplemented with hands-on lab exercises) was not sufficient to enable trainees to do their jobs, on-the-job training became imperative. Instructors—with their student assistants—were cycled to job sites to be available for at-elbow support

Western organizations are strongly oriented toward explicit knowledge, especially analysis, and the focus is on individuals.

Japanese organizations are oriented toward tacit knowledge, with emphasis on experience, and they focus on teams.

Today's globally competitive world requires an understanding of both approaches for integration and cooperation in multinational enterprises.

and just-in-time learning (and to themselves learn the tacit knowledge required for applying the new software in product design and development). To improve the process even further, instructors are now being paired with application engineer mentors. A continuing problem is the lack of tacit knowledge about the company's products by the software designers. To overcome this problem, the developers are being paired with experienced product designers.

EXPANSION SPIRALS: User groups are encouraged to share experiences to "spiral up" organizational knowledge, as well as extending it to supplier companies. The college instructors are planning to completely change their design courses when they return to their campuses—the explicit and tacit knowledge learned while in industry has changed the way they teach and has given them a conceptual understanding of the needs of industry. In essence, they have been prepared for introducing innovation in education.

Program analysis with the knowledge creation model resulted in the following insight and recommendations:

- The knowledge creation model reinforces the idea that the training program has to be a whole-brain process; it also values the chaotic conditions during program development; and it shows that documenting learning through both successes and "failures" is important.
- To kindle excitement and imagination at the front lines, the program needs a "catchy" metaphor to link the organizational long-term leadership vision with the responsibilities of the individual employees.
- The feedback loop from the users allowed rapid change in direction for continuous improvement in training.
- A flexible organizational structure is crucial. The interface with universities was very difficult when administrators insisted on following old procedures. Collaborative projects between industry, government, and universities can work when all three have the freedom to change.
- Examples, team exercises, applications, and user hints should be added to the training materials to strengthen tacit knowledge acquisition.
- Had the focus been on tacit knowledge, the effort of developing the traditional classrooms might have been on a much smaller scale.

For the new paradigm to be effective, it cannot be housed in an old setting, such as top-down or bottom-up management styles or a traditional hierarchical structure.

I. Nonaka and H. Takeuchi

4. Matsushita's Home Bakery

Knowledge creation is a new model that offers exciting possibilities for organizational management to achieve innovation. The case studies discussed in Nonaka and Takeuchi's book give a wealth of practical applications of the model which we cannot do in the limited scope of this chapter. However, we want to summarize one example: the design of a fully automatic bread-making machine and how this changed the entire company. The design is unique because it captured the tacit skills of one of Japan's foremost master bakers. Table 3.11 presents the four cycles; note how innovation eventually reached and benefited the employees.

Table 3.11 Matsushita Home Bakery Example*

Cycle 1—Team Level

1. Socialization:	Identify consumer dream of a home bread-making machine.
2. Externalization:	Vision of product as "easy and rich."
3. Combination:	Prototype product—but bread is not "rich."
4. Internalization:	Software developer becomes apprenticed to famous baker to learn the art of kneading dough (tacit knowledge) to make excellent bread, a process that takes many months.

Cycle 2—Team Level

1. Socialization:	When the developer interacts with the design team, she is eventually able to convey this new tacit knowledge …
2. Externalization:	… as a mental image of "twisting-stretch" motion.
3. Combination:	The improved design/prototype incorporates this motion—but it takes a year through trial and error to produce tasty bread.
4. Internalization:	The project is transferred to the commercialization division but still includes the original design team because of their valuable tacit knowledge of the product.

Cycle 3—Expanded Team Level

1. Socialization:	Share knowledge to reduce cost and identify opportunities for innovation.
2. Externalization:	Concept: Add yeast later in the mixing process (as was done traditionally).
3. Combination:	Design changes meet quality goals and reduce cost. Market delay is justified because the product meets "easy and rich."
4. Internalization:	Success of this product shifts the focus of the entire company toward creating products that meet customer needs and dreams.

Cycle 4—Organizational Level

1. Socialization:	Engineers' attitude toward new projects and customers changes.
2. Externalization:	Company now has vision of "human electronics."
3. Combination:	New products: integrated coffee mill/automatic coffee brewer, large-screen TV, and induction-heating rice cooker that all become bestsellers within the quality concept of "easy and rich."
4. Internalization:	The process spirals up to higher levels and eventually changes the entire company through collaborative concurrent engineering. "Human electronics" extends to the company's employees—a 150-hour work month to give them more time with their families. This policy was first implemented as a pilot in one division to yield tacit knowledge of what a 150-hour work month would require to maintain the same productivity. One of the chief resulting strategies was to eliminate time wasted in meetings.

*Facts from I. Nonaka and H. Takeuchi, *The Knowledge-Creating Company,* Oxford University Press, 1995.

⌛ 10-Minute Activity 3-7: Identify Metaphors for Concepts

Look at a magazine such as *Times* or *Newsweek* (or others containing advertising for consumer products). Look at the ads and find examples where a concept behind the product is taught with a metaphor. Example: Sprint's quality telephone connections by the "drop of a pin."

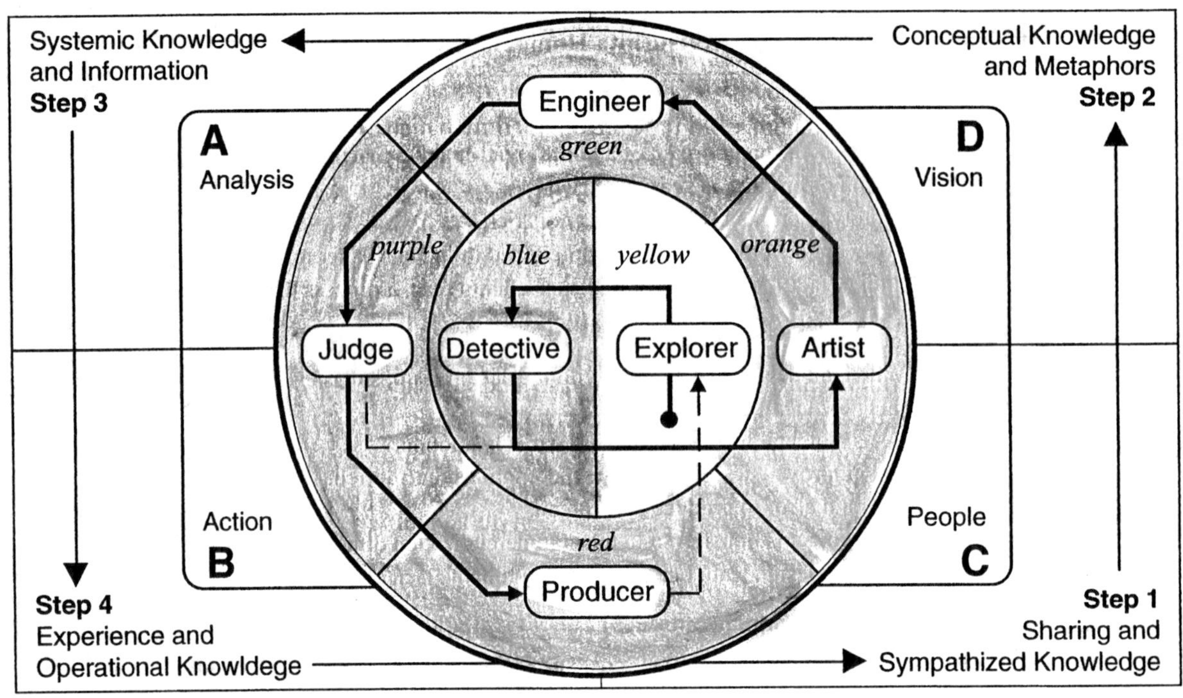

Figure 3.17 The creative problem solving model and associated metaphors discussed below are shown superimposed on the knowledge creation model and the Herrmann whole brain model (©1998 The Ned Herrmann Group).

Creative problem solving metaphors

When we researched the writing of a manual to teach engineers how to be more creative, we read many books on creativity, attended creativity seminars, and then developed a "whole-brain" problem solving model that integrates the needed right-brain thinking steps with the left-brain modes commonly preferred by engineers. Later, we discovered the Herrmann model and found that it explained *why* the structured creative problem-solving approach worked.

The thinking and activities required at each step in the process are visualized with a corresponding metaphorical mindset. The mindsets will always be indicated with quotation marks, to distinguish them from the professions. The "explorer" and "detective" discover and investigate the "real" problem and its context and then define it as a positive goal. The "artist" brainstorms many ideas; the "engineer" synthesizes better ideas; the "judge" determines the best solution, and the "producer" puts it into action. The process iteratively cycles through all four thinking quadrants, as shown in Figure 3.17. Each color is associated with one of the mindsets—the range of the colored band identifies the primary thinking quadrants used in the double-dominant mindsets. Please note that our creative problem solving color scheme differs from the colors that Ned Herrmann uses for the four quadrants of his HBDI model.

ENHANCEMENT

Use colored pencils in Figure 3.17 to shade the respective areas as indicated. (Do not use ink pens or markers as they may bleed through the page.)

Figure 3.17 shows the relationship of the creative problem solving model to the knowledge-creation steps. When we applied the model of knowledge creation to our creative problem solving model, we were able to make an improvement that will be noted by those who have used our earlier books. The process now starts in quadrant C and cycles counter-clockwise (reversing the sequence of "detective" and "explorer" and resulting in the more logical sequence of divergent, then convergent thinking for problem definition). Starting in quadrant C also makes the creative problem solving model fall in line with the Kolb learning cycle and Bernice McCarthy's 4MAT model (Ref. 3.7). Each step of the creative problem solving model will be discussed and illustrated in detail in Part 2 of this book, and its specific application to design and innovation will be demonstrated in Part 3.

People have always had distinct preferences in their approaches to problem solving.

Why then is it only now becoming so necessary for managers to understand those differences?

Today's complex products demand integrating the expertise of individuals who do not innately understand one another.

Today's pace of change demands that these individuals quickly develop the ability to work together.

Rightly harnessed, the energy released by the intersection of different thought processes will propel innovation.

Dorothy Leonard, Harvard Business School professor, and Susaan Straus, organizational consultant

We first learned about different problem-solving mindsets from Roger Von Oech, but then we added the "detective" for data analysis to the "explorer" to indicate that both left-brain and right-brain thinking are required for complete problem definition. We also invented the "engineer" to bring out an important step in creative problem solving—idea synthesis and optimization—that is different from and intermediary to the "artist" and the "judge." And we changed the "warrior" into the "producer" as requested by students and teachers. These metaphors make it easier to remember the type of thinking that we need to use at each step of the creative problem-solving process. Let's visualize these metaphors:

1. "Explorer" When we are looking at the big picture or context of a problem or want to discover its opportunity aspects, we need to think like an explorer. Imagine being armed with field glasses, a compass, and a large notebook, keeping a sharp eye out for ideas. Imagine a character like Indiana Jones. Among famous explorers we have Margaret Mead, Christopher Columbus, David Livingstone, Jacques-Yves Cousteau, Thor Heyrdahl, Roald Amundsen, and the astronauts and cosmonauts.

2. "Detective" If we are struggling with a problem that involves some difficulty, we need to think like a detective. Imagine seeing Agatha Christie's Hercule Poirot (or Miss Jane Marple) or Sherlock Holmes. From television, we have Sgt. Columbo, J.B. Fletcher (in "Murder She Wrote") or Virgil Tibbs ("In the Heat of the Night"). Can you see these detectives walking around in the dark, with a flashlight or matches? Can you see them using their minds to evaluate clues? Can you hear them asking questions? Problem definition ends when the "explorer" and "detective" come up with a positive problem definition statement based on the analysis of the collected information.

3. "Artist" Now picture yourself as an artist, a Picasso or Michelangelo, Diego Rivera, Georgia O'Keeffe, Grandma Moses, or Jim Henson. This is the stage where we brainstorm to come up with a multitude of creative ideas—here we need to think like an artist—the more imaginative, the better. See yourself standing in front of a large sketch pad, furiously sketching ideas and fragments of ideas. Perhaps you can imagine a team of artists collaborating on an idea collage.

4. "Engineer" Next, conjure up a new image in your mind, that of an engineer, an inventor, a designer, a tinkerer—a whole team of engineers working together. They sit in front of a drawing board or work station; they examine and play around with all kinds of wild ideas, with a view toward combining them to get more practical, comprehensive, and optimized ideas. The team borrows ideas from nature and force-fits them to synthesize innovative solutions to their design problem.

5. "Judge" Now, in your mind's eye, enter into a courtroom. In front of you sits the judge, gavel in hand and ready to render a verdict as to which ideas and solutions are best and should be implemented. As "judges" we set up criteria to evaluate ideas and solutions. "Judges" look for flaws and then try to overcome them with additional creative thinking in the "artist's" and "engineer's" mindset.

6. "Producer" Finally, it is time for the "producer." A producer is a jack-(or jane)-of all trades: maker, doer, mover, parent, organizer, builder, executor, director, practitioner, planter-grower-harvester, seller, general, leader, manager, implementer. The "producer" is responsible for putting out quality products. "Producers" need courage; they take risks with creative ideas and innovation. They need good communication skills for managing teamwork successfully. In essence, the producer is in charge of a new round of creative problem solving, where the problem is the implementation process. Another image that is important here is the picture of a person who is ready to fight or stand up for an idea. The "producer" must be persistent to carry the project to a successful conclusion while working to overcome opposition and other adversity.

Let's play around with these six roles or mindsets for a moment and look at different scenarios or skits.

- What would happen if we left out the "explorer" or the "detective"? We can still come up with many good ideas and find a solution, except that the solution may not fit the real problem and its context. So we may still have a problem.
- What would happen if we left out the "artist" *and* the "engineer"? This happens each time you take the first idea that comes to mind and rush to implement it without looking for alternatives. It happens when the "judge" appears too soon in the process—when you get an idea and instantly tell yourself: "This won't work; this is a dumb idea." It happens when we ridicule the ideas of others. Most analytical problem solving approaches leave out the "artist" and the "engineer."

We see the world not as it is, but as we are.

Stephen R. Covey

Without this playing with fantasy no creative work has ever yet come to birth. The debt we owe to play of imagination is incalculable.

Carl Gustav Jung

- What happens when we have the "artist" but not the "engineer"? Without the "engineer," the "judge" may be getting only half-baked ideas to evaluate. The "engineer" is needed to make good ideas better, to make wild ideas more practical, and to develop optimum solutions.
- What happens when we leave out the "judge"? Without the "judge," we will not be able to select the best idea or find the flaws of ideas.
- What would happen if we only had the "producer"? Could this work? It might in rare cases, just because the enthusiasm and energy of "producers" could carry it off—if they are lucky enough to pick a solution out of the blue that actually would solve the problem. You can probably think of times when you and your friends took this approach—young people have a marvelous knack for getting excited about ideas. But most of the time, ideas must be evaluated by the "engineer" and the "judge" to prevent "producers" from taking reckless risks.

To create the best conditions for coming up with a good solution to the problem, it is best to follow the process in the sequence that lets the brain do the best thinking, whether we are alone and have to solve a problem quickly or whether we have a team to help and several weeks or even months to work out an optimum solution.

The remaining three chapters in Part 1 show applications of the three mental models to teamwork (Chapter 4), communications (Chapter 5), and creative thinking (Chapter 6). These chapters constitute a valuable resource for years to come in your education and future workplace.

Resources for further learning

The following books and articles are just a sampling—see your library or bookstore for additional titles and periodical. Also, check scientific and popular journals for reports on the latest brain research results.

3.1 Rick Crandall, editor, *Break-out Creativity: Bringing Creativity to the Workplace,* Select Press, Corte Madera, California 1998. This small paperback has been published for the Association for Innovation in Management and gives practical advice on how to be more creative.

3.2 Richard Felder, "Matters of Style," ASEE PRISM, December 1996, pp. 18-23. Professor Felder, a chemical engineering professor at North Carolina State University, is applying collaborative and cooperative learning in an integrated curriculum. Watch for future publications on his results.

3.3 ✓ Ned Herrmann, *The Creative Brain,* Brain Books, Lake Lure, North Carolina, 1990. This "whole-brain" book was the main resource for this chapter. It explains the theory and development of the four-quadrant model of brain dominance and contains applications to many different areas of life. This is a must read for anyone who is involved in teaching or management.

3.4 ✓ Ned Herrmann, *The Whole Brain Business Book: Unlocking the Power of Whole Brain Thinking in Organizations and Individuals,* McGraw-Hill, New York, 1996. This book contains much insight and practical advice on how to use whole-brain thinking to enhance leadership, teamwork, and creativity in your organization to optimize productivity.

3.5 ✓ Dorothy Leonard and Susaan Straus, "Putting Your Company's Whole Brain to Work," *Harvard Business Review,* Reprint 97407, July-August 1997, pp.110-121. This article contains a discussion of the Herrmann Model and the Myers-Briggs Type Indicator.

3.6 Monika Lumsdaine and Edward Lumsdaine, "Thinking Preferences of Engineering Students: Implications for Curriculum Restructuring," *Journal of Engineering Education,* April 1995, Vol. 84, No. 2, pp. 193-204. This article describes a longitudinal study at the University of Toledo.

3.7 Bernice McCarthy, *The 4-MAT in Action, Creative Lesson Plans for Teaching to Learning Styles with Right/Left Mode Techniques,* Excel, Barrington, IL, 1983. This model is based on the Kolb learning cycle and covers all four thinking quadrants—it is a whole-brain approach to teaching and learning.

3.8 ✓ Ikujiro Nonaka and Hirotaka Takeuchi, *The Knowledge-Creating Company: How Japanese Companies Create the Dynamics of Innovation,* Oxford University Press, New York, 1995. The authors show, through a theoretical model and many case studies (which include organizations in the U.S.) how Japanese companies create new knowledge and use it to manufacture successful products and develop innovative technologies.

3.9 J. William Shelnutt and Kim Buch, "Using Total Quality Principles for Strategic Planning and Curriculum Revision," *Journal of Engineering Education,* Vol. 85, No. 3, July 1996, pp. 201-207. This article summarizes the process used to develop the planning matrix in the College of Engineering at the University of North Carolina at Charlotte.

3.10 Roger Von Oech, *A Kick in the Seat of the Pants,* Harper and Row, 1986. Four roles of the creative process are presented: explorer, artist, judge, and warrior, together with interesting stories and exercises.

Whatever your profile, there are other normal people like you somewhere in the world. Celebrate your uniqueness!

With effort and practice, you can change and develop your thinking preferences.

Ned Herrmann

"Claim your space!"

Assessing your thinking preferences

The Herrmann Brain Dominance Instrument (HBDI): Knowing one's thinking preferences is very useful—the HBDI is a powerful tool to gain insight into why we do things the way we do and why we have problems in communicating with people who think differently from us. We can set goals, practice specific skills, and expand the range of our thinking repertoire to become more situationally whole-brain thinkers and effective problem solvers. We strongly recommend that you complete the HBDI. It can be obtained from this book's authors or from Herrmann International, 794 Buffalo Creek Road, Lake Lure, NC 28746; phone 828/625-9153, fax 828/625-1402. It can be downloaded from the web at

www.hbdi.com. The cost (in 1999) is $59 plus postage for an individual HBDI evaluation and an informative interactive packet of materials. Discounts for students and other groups are available. Optional services include team and organizational analyses and in-house workshops. Student versions are now being field tested for secondary school students.

Differences in learning styles for students: Classroom group activities are more beneficial when students with different learning styles work together. You may obtain a preliminary idea about your learning style preferences from Activity 3-8.

Activity 3-8: Learning Preference Distribution

Tabulate the total number of circles for each learning preference from One-Minute Activities 3-1, 3-2, 3-3, and 3-4 in this chapter. Then add up the total for all the responses and calculate the percentage contribution for each quadrant:

Quadrant A	=	______	=	______ %
Quadrant B	=	______	=	______ %
Quadrant C	=	______	=	______ %
Quadrant D	=	______	=	______ %
Total number of responses	=	______	=	100 %
Average score per quarter (divide the total by 4)			=	______

Evaluation: The area with the highest score is likely the quadrant of your strongest thinking preference, especially if the score is much higher than the average (or greater than 35%). Check to be sure your calculated percentages total up to 100. Check over all four lists of learning activities and underline the one activity you dislike the most. The location of your least favored activity is often in your least preferred quadrant. If your highest score does not vary much from the next two or three, you are likely to be multidominant or whole-brained.

Implications: The ranking according to the calculated percentages indicates where you will have to make a special effort and where you may have unique abilities and interests to contribute to your team. Since students tend to prefer visual (quadrant D) learning activities, even if quadrant D is not a strong thinking preference, this brief assessment will often be skewed and is not a substitute for the HBDI, the only instrument available for obtaining an accurate brain dominance profile.

Example 1: Number of Circles: A = 12; B = 3; C = 7; D = 8. Total = 30.
Percentages: A = 40%, B = 10%, C = 23%, D = 27%. Average/quarter = 7.5 circles.
From this result, it can be assumed that this student is an analytical learner. This student should be grouped with others having high preferences in quadrants B, C, and D.

Example 2: Number of Circles: A = 6; B = 7; C = 2; D = 5. Total = 20.
Percentages: A = 30%, B = 35%, C = 10%, D = 25%. Average per quarter = 5 circles.
From this result, it can be assumed that this is a multi-dominant learner, with a low preference in quadrant C. The scores are not sufficiently spread out to determine the ranking of the dominant quadrants.

Example 3: Number of Circles: A = 3; B = 1; C = 3; D = 2. Total = 9. Average = 2.
No valid conclusions on learning preferences can be drawn from only a few data points.

Exercises

3.1 Interview

Interview an engineer, teacher, and person in business about their experiences with creativity at work. Do they have a supportive climate? What would they change? Write up a summary and discuss the factors that you think are important to encourage creative thinking and innovation on the job. What kind of questions would you ask during a job interview to gauge the creative climate of a prospective employer? What "clues" would you look for to identify the mental preferences of the interviewer?

3.2 ✓ Metaphors

A metaphor is a figure of speech in which one thing is talked about as if it were something else. Example: "Break the cocoon—fly with creative thinking" (contributed by Dawn Rinehart, an engineering student). Kim Steger (another engineering student) developed an essay using the growth of a rose to illustrate creativity. Make up five different metaphors for creative thinking. If you are in a group, compare your examples with those of your group members.

3.3 ✓ Analogies

An analogy is a comparison to something that is similar. You could say, "My college dorm is like a maze with no exit, with occupants always hungry and scrounging for food." Or, "My university is like a supertanker—huge and powerful, with a set course that is tough to change." Or, "My organization is like a beacon of light, helping people avoid rocks and shipwrecks." Think of five additional analogies along the lines of the examples, but using only positive statements if possible.

3.4 ★ Optional Study: Exodus ★

Chapter 3 in the Book of Exodus (Bible, Old Testament) makes an interesting study in four-quadrant thinking. Moses exhibits strong quadrant B thinking—very appropriate when one's business is safeguarding sheep. The chapter shows two phases of God trying to move Moses out of quadrant B thinking toward creative thinking. First, diagram or sketch the action of verses 1 through 6 as they relate to the four thinking quadrants. Next, diagram or sketch the action of verses 7 through 22 as they relate to the HBDI model.

The brain is designed to be whole, but at the same time we can and must learn to appreciate our brain's uniqueness and that of others.

A balanced view between wholeness and specialization is the key.

Ned Herrmann

3.5 ★ Applying the 5-Step Creative Problem Solving Model ★

Imagine that you have been given the assignment of leading a team to design a new, large children's play apparatus, similar to those at some McDonald's restaurants. The sponsors have mentioned goals of more inviting, more fun, safer, and less costly, but have given you the freedom to design anything that is competitive with existing designs. Use the 5-step creative problem solving model to outline your approach. List the questions that you would ask at each step, activities the team would undertake, and the expected generic outcomes of each step. You are not expected to design the apparatus, but only to plan the outline of the method you would use to solve the design problem.

Don't pick people who are like you. Pick people you can trust. Pick people who see things differently and will challenge you and the group.

Cheryl Cook, Pennsylvania Rural Development, USDA

3.6 ★ Identifying the Mental Models in an Existing Design ★
Research a case study describing the design of something and identify the three mental models:

a. the thinking quadrants used at various stages in the design process;
b. the steps and cycles of knowledge creation;
c. the stages or mindsets of the creative problem-solving model.

Examples you might want to use are the original Ford Mustang (1965), the SR-71 spy aircraft, the Brooklyn Bridge, the English Chunnel, the Boeing 777, the original Xerox machine, or Disney World.

Chapter 3 — review of key concepts and action checklist

Important HBDI concepts:

- HBDI profile results are value-neutral. This is not a test, and there are no right or wrong answers. All profiles are unique and valuable.
- The brain is specialized and situational—different modes are used for different tasks.
- Thinking in a less preferred mode takes more energy.
- When we understand the value of the different thinking modes, we learn to appreciate the power of diversity in teamwork.
- Each mode has its own way of problem solving and vocabulary. Thus understanding these differences helps improve communication.
- A whole-brain team made up of people with different strengths can obtain optimum problem-solving results.
- With effort and practice, thinking preferences can be changed and developed. We can build new structures in our brain and use our brain more effectively.
- Each person is a unique coalition of all four thinking modes, but different occupational groups exhibit characteristic generic profiles.

Review the descriptions of the four quadrants in the Herrmann model and write down four major characteristics for each mode:

Quadrant A: ______________________________

Quadrant B: ______________________________

Quadrant C: ______________________________

Quadrant D: ______________________________

The steps in the knowledge creation model:

Step 1 (quadrant C): Team members share beliefs, mental models, experiences, goals. They "know" their customers; there is sensory, intuitive, and kinesthetic learning.

Step 2 (quadrant D): Team members develop metaphors for goals and concepts; they brainstorm to make the tacit knowledge explicit so it can be understood by all.

Step 3 (quadrant A): Information is collected, shared, combined, synthesized, analyzed, documented, and networked within the context of learning or problem solving. This is explicit knowledge and comprehensive information flow with feedback.

Step 4 (quadrant B): Experience is gained with application of the newly created knowledge; people learn "know-how" in pilot programs, apprenticeships, and prototyping. The process is then spiraled up to a higher level in a new knowledge creating cycle. This is tacit "learning by doing" or on-the-job training.

For superior learning, knowledge creation, and innovation, do not skip any of these steps.

The words "theory" and "practice" are of Greek origin—they carry our thoughts back to the ancient philosophers by whom they were contrived, and by whom they were also contrasted and placed in opposition, as denoting two mutually conflicting and mutually inconsistent ideas.

[This fallacy] based on a double system of natural laws retarded for centuries the development of physical science, notably mechanics.

William Rankine

Steps and mindsets of the creative problem solving model:

1. Problem definition:	"Explorer" + "Detective"
2. Idea generation:	"Artist"
3. Idea optimization:	"Engineer"
4. Idea judgment:	"Judge"
5. Solution implementation:	"Producer"

Action checklist

☐ Do you and the people you work or study with understand the HBDI model? Do they understand and appreciate their own thinking preferences and that of others? If you have not yet done so, arrange to obtain your own HBDI profile (and, if possible, those of your team).

☐ Are you habitually paying attention to the knowledge creation process in your study or work groups to yield better learning and a high-quality project outcome?

☐ Be on the lookout for problems that your team can solve as you study the creative problem solving process in Part 2 of the book. The hands-on application will give you tacit knowledge for superior learning.

☐ What time of day are you most creative? Use this time to think up and jot down ideas, then take one afternoon a month to further explore one of these creative ideas. Block off this time in your calendar right now for the next six months.

Teamwork

What you can learn from this chapter:

- What do you know about teamwork? Teams in concurrent engineering.
- Advantages and disadvantages of teamwork. The difference between homogeneous and heterogeneous teams.
- Team development: individual traits required, stages of development, roles of members, and dealing with conflict.
- Examples of functional and dysfunctional teams.
- Management guidelines for developing self-directed teams.
- "Know-how": forming whole-brain project teams based on the HBDI; managing your team for productivity; evaluating team members.
- Further learning: references, exercises, review, and action checklist.

What is teamwork?

In the introduction of this book as well as in the preceding chapter, we have talked about the importance of having teamwork skills for the workplace of the twenty-first century. In this chapter, we will show you how to develop effective project and engineering design teams. Let's start with a diagnostic quiz: What do you know about teamwork?

Five-Minute Activity 4-1: Diagnostic Quiz on Teamwork

You can do this as an individual or with a small group (from two to four people). Quickly go through the quiz on the next page. If there is a disagreement in the group (or if you can't make up your mind), mark both "true" and "false." The purpose of the quiz is to identify common myths about teamwork that you may have.

Evaluation: Check your answers against the key at the bottom of the quiz. A perfect score means that you have a good understanding of teamwork. If you or your group missed more than three, you will benefit from studying this chapter.

Why is teamwork important in engineering? Figure 4.1 shows the conventional product development and manufacturing process. Compare this sequential process with concurrent (simultaneous) engineering which is a team approach as depicted in Figure 4.2. In the traditional method, marketing people specify what they "think" customers want,

Teamwork: Diagnostic Quiz

	True	False
1. In effective teams, members have a chance to demonstrate their unique talents and skills.	❑	❑
2. Most people perceive teams as a means for self improvement and personal development. Thus, a team's purpose is to enhance the individual and to help all team members achieve their potential.	❑	❑
3. Individual brainstorming is more effective than team brainstorming.	❑	❑
4. Individual brainstorming is more efficient than team brainstorming.	❑	❑
5. Most inventors work in isolation.	❑	❑
6. An effective team does not have members who are outsiders; each person is an expert in the project or problem area.	❑	❑
7. When people choose their own team, the team members usually will exhibit similar thinking styles.	❑	❑
8. Once a person has learned how to function in a mentally diverse team, these skills can be transferred to working well in any other team.	❑	❑
9. Teams high in quadrant C thinking are automatically good teams.	❑	❑
10. Teams low in quadrant C thinking cannot become good teams.	❑	❑
11. Teamwork happens automatically when people work together on a project.	❑	❑
12. "Chitchat" or social interaction are not important to team development.	❑	❑
13. It takes time and effort before a dozen people working together will become a productive, well-functioning team.	❑	❑
14. Conflict does not occur in an effective team.	❑	❑
15. An individual's personal achievements are more important than the collective accomplishments of the team.	❑	❑
16. A team's effectiveness is strongly influenced by its ability to set goals and its relationship to management.	❑	❑
17. The team leader is more important than the team members.	❑	❑
18. A team can never have too many meetings.	❑	❑

Answer Key: 1-T, 2-T, 3-F, 4-T, 5-F, 6-F, 7-T, 8-T, 9-F, 10-F, 11-F, 12-F, 13-T, 14-F, 15-F, 16-T, 17-F, 18-F.

Figure 4.1 In traditional engineering, people work in isolation, with little or no interaction between departments and their functions (from Lucas Engineering and Systems, UK, in Ref. 4.5).

Figure 4.2 In concurrent engineering, people involving all functions work in teams; customers and suppliers are involved right from the start of the product development process (from Lucas Engineering and Systems, UK, in Ref. 4.5). See Appendix A for information on the QFD "House of Quality" shown on the chart.

without direct input (either tacit or explicit) from the customers. These specifications are then "thrown over the wall" to the designers who pass on their concepts to production engineering where the plans by this time most likely contain many arbitrary decisions which in turn are further changed by manufacturing—with the result that the manufactured product does not meet customer expectations in type, cost, or quality.

Concurrent engineering with its teamwork and up-front problem solving during the conceptual design stage leads to significant time and cost savings and a product that can successfully compete in the marketplace. Thus engineers must be able to work and communicate with people with diverse priorities, thinking styles, and knowledge. The Accreditation Board for Engineering and Technology (ABET) recognizes the need and value of developing effective multidisciplinary teamwork skills in its new Engineering Criteria 2000 (see Table 1.2). Engineering schools will now have to demonstrate that their students have learned these skills.

Cross-functional teams are at the heart of the knowledge creation process, both to generate knowledge and then to disperse it throughout the organization. According to Nonaka and Takeuchi, (Ref. 3.7), "project teams with cross-functional diversity are often used by Japanese firms at every phase of innovation. [Also] in most companies there are four or five core members, each of whom has had a multiple functional career. For example, the core members who developed Fuji Xerox's FX-3500 have had at least three functional shifts, even though they were only in their 30's at that time." These functions included R&D, planning, production, quality control, sales and marketing, and customer service.

Teamwork has strong cultural roots in Japan but is much more difficult in our more individualistic Western culture, because we have different expectations about teams. The American Society for Quality Control worked with Fortune 500 companies to find out how Americans actually view teams. Their study showed that

- Teams are seen from a personal, internalized, practical perspective.
- Teams are perceived as "a means for self-improvement and personal development—a vehicle for individual fulfillment and success."
- Teams are thought "to enhance the individual, to support further achievement, and to help team members achieve their personal potential."

Thus, for teams to be effective, members must be given a chance to use their unique talents and skills. The goals of the team must include aspects that will benefit the organization as a whole as well as the individual team members. In the following section, we will look at the benefits and liabilities of using teams in creative problem solving.

Advantages and disadvantages of teamwork in creative problem solving

Let's begin with a demonstration. This activity requires a group of five or more people.

⌛ ✓ Ten-Minute Group Activity 4-2: Idea Generation

First, each person in the group works alone. Take a sheet of paper and write down as many uses for a telephone book as you can think of. Stop after three minutes. Next, the group prepares a common list on a blackboard or flip chart of all the ideas thought up by the group members. Each person gets a turn to quickly share ideas. The others can cross duplicates off their lists. If you get a new thought as you hear the ideas presented by the others, jot it down on the back of your sheet. After all ideas from the first round have been recorded, repeat the process with the new ideas, but now go through the group in reverse order. Finally, give the group an opportunity to express additional ideas that come to mind as each person contemplates the group list.

Teams must make sense for the organization and the individual and become a truly value-adding experience for everyone. Until more organizations realize this, we will continue to have spotty success in navigating "the uncharted waters of teamwork."

Michael D. Jones, President of ASQC, Milwaukee, Wisconsin

Now let's look at what usually happens during this exercise. At the end of the first three minutes, the average person will have generated ten ideas. For the group as a whole, many of these ideas will be duplicates, but a good number will be unique. Note the many surprising ideas from the second list and the general brainstorming—seeing or hearing other people's ideas helps your mind think of additional ideas, some original and some that are combinations or elaborations of previously mentioned ideas. Thus this illustration of the **idea trigger method** demonstrates one of the benefits of teamwork—the interaction between minds increases the output of ideas. Although, when seen in terms of staff time, individual brainstorming is more efficient, the output of a team is usually more effective because of this idea interaction and synergy.

Another benefit of teams is the amount of explicit information and tacit experience available for problem solving. Leonardo da Vinci is an example of a person who knew almost everything that was known in his time. Today, it is rare of an individual to thoroughly know everything in an entire field, such as chemistry or literature or electrical engineering. We know that Eli Whitney invented the cotton gin, Alexander Graham Bell the telephone, and Jonas Salk a vaccine for polio. Why don't we talk about the inventor of the Boeing-747 airplane? One reason is that it was developed by many teams of engineers. Also, about 90 percent of patents for inventions are not for completely new products but for improvements of existing patents. Most inventors do not work in isolation—they build on the ideas of others. The invention of the cotton gin is an example. The original idea came from Catherine Littlefield Greene who supported Eli Whitney financially on the project; she also improved the prototype and shared in the royalties from the patent.

With today's knowledge explosion, it is no longer possible for a single person to know all the data connected to a problem. This is why teams are often used for problem solving. Consider three individuals with brain dominances A, B, and D+ C, as shown in Figure 4.3. Each person differs in background and experiences and thus has a different explicit and tacit knowledge base as well as different ideas and biases on particular subjects and problems. When the three people work together, group (or team) W now contains a large background of which only a small amount is in common. The possibility is increased that new, creative combinations of ideas will occur—the ideas and suggestions of one person can stimulate the imagination of the other team members. Also, a diversity of people working together compensates for bias and thus can achieve better judgment.

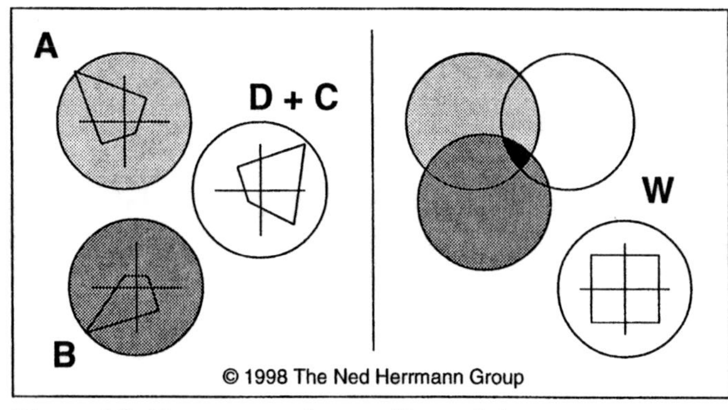

Figure 4.3 Team approach to problem solving.

Heterogeneous teams consisting of differences in mental preferences are capable of higher and more effective creative output than homogeneous teams consisting of similar mental preferences and same gender.

Ned Herrmann

Team W in Figure 4.3 represents a team that has been purposefully put together to form a "whole brain." People who are not aware of the thinking preference model will usually form teams that are homogeneous—everyone on the team will have a similar HBDI profile (for example, see Figure 3.6 for a quadrant A dominant group). Homogeneous teams are able to reach consensus quickly, since they communicate in the same thinking quadrant, share a similar "world view," and have a common problem solving approach. This can be an advantage when time is of the essence. However, homogeneous groups rapidly develop a group culture or "tribe" that can blind them to important issues and worthwhile ideas from outside the group. In hundreds of workshops and consulting experiences, Ned Herrmann has found that homogeneous teams reach an "obvious" or "adequate" solution quickly, whereas heterogeneous teams will come up with several superior, innovative solutions (and take a much longer time doing it). When we talk about teamwork in the remainder of this book, we will always mean a heterogeneous team, unless we specifically identify the team as being homogeneous. Figure 3.10 shows a composite HBDI profile for a heterogeneous group.

Heterogeneous teams, when first put together, will find it difficult to communicate, especially when members have strong opposing thinking preferences without a whole-brain "mediator" or "translator" present. Team communication will be discussed in detail in Chapter 5. Although teamwork has been assigned to quadrant C in the Herrmann model (because of the importance of personal interaction and communication), this does not mean that teams high in quadrant C will automatically make good teams—emotional factors that are not effectively dealt with may be a hindrance. Also, the team may do much talking but be unable to stick to its assigned task and bring problem solving to a successful conclusion. Conversely, engineering teams that are low in quadrant C can be successful if they pay attention to communication with customers, mutual interaction, and sharing (Step 1 in the knowledge creation cycle). In general, this will be easier if at least one or two people on the team have a preference in Circle 1 (see p. 53) for quadrant C thinking, even if this is not their strongest preference. Table 4.1 summarizes the advantages and disadvantages of using teams for creative problem solving.

Once individuals experience the stimulation, excitement, and creative outcomes of heterogeneous team membership they can participate in a heterogeneous team made up of different members without going through an elaborate learning curve—the skills are transferable to other group situations.

Ned Herrmann

Team development

Our school systems train students to work alone by rewarding individual achievement. Public education since the early part of the twentieth century has mainly been geared to producing docile assembly line workers who will "check in their brains at the factory gate upon entering." However, U.S. companies are finding that higher-level thinking skills and teamwork are needed to increase productivity and competitive quality for the global marketplace. Because collaborative learning and working

Table 4.1 Advantages and Disadvantages of Using Teams for Creative Problem Solving

Advantages

1. More knowledge is available to help solve the problem; people with different expertise and thinking skills can be brought together from different departments within the organization to form cross-functional teams.
2. People interact with one another; ideas are used as stepping stones to more creative, better solutions. The team members are encouraged to build on one another's ideas. When this process really clicks, productivity and effectiveness increase due to synergy.
3. If there is one "best" solution for a particular problem, the team has a good chance of finding it. Teams have an advantage in identifying opportunities and in taking greater risks (and thus increasing the chance that innovation will occur).
4. People who take part in the problem-solving and decision-making process are usually more willing to accept the solution than if the solution were developed by an individual and imposed by the "voice of authority."
5. The team members learn from each other—both explicit and tacit knowledge are transferred when knowledge operators interact with knowledge specialists.
6. The team provides an encouraging environment for developing leadership skills.

Disadvantages

1. A greater investment in effort and total personnel time is needed, not just for solving the problem but also for team development.
2. In general, the team process has a low efficiency—a large number of ideas may be generated, but only a few of these will be truly good solutions.
3. The people making up the group or team may not get along with one another. Unresolved conflicts, negative emotions, and hostility will lessen the creative idea output of the team, unless the team learns to appreciate the value of diversity and can creatively develop win-win outcomes.
4. Teams can suffer from the "group think" phenomenon. The group exhibits extreme conformity and peer pressure. Independent ideas are not allowed, and group members may be intimidated by a leader or a vocal minority. Or they may be anxious or over-eager to please a manager and thus become "yes" persons instead of autonomously pursuing the best problem-solving outcome.

in teams are not widely taught and used in schools, businesses and industry are spending vast amounts of money and effort to train their employees on how to work in teams. Teamwork requires cooperation at all levels in an organization—in essence it demands a cultural paradigm shift between the traditionally more adversarial roles of management, employees, and labor unions. This change is not easy, because it must happen on many different levels, from cultural values to organizational structure to the attitude of each individual. One successful example of this process is the Saturn automobile plant in Tennessee which generally operates outside the rules General Motors has with the United Auto Workers union. The issue of organizational change is addressed in Chapter 18 and can also be investigated in a study of total quality management (TQM). Appendix F presents a summary of TQM principles. What we want to explore here is how to develop good teams.

Individual characteristics

What traits in individuals should we look for when we make up a creative problem-solving team? Table 4.2 lists some desirable traits. Not everyone in the team will have all of these, but they should be present in the team as a whole. Note how many of these characteristics involve quadrant D thinking (the right-hand column). Macho males—influenced by the social culture around them—may be embarrassed about expressing creative ideas when these involve feelings. People with aesthetic interests tend to be concerned with form and beauty in their surroundings: because they want to go beyond just a practical solution to an elegant solution, they often achieve a higher-quality product.

Table 4.2 What Makes a Creative Team?

• Intelligence: high intellectual standards. • Expertise in problem area or related fields. • Variety of experiences outside the problem area; broad interests; multidisciplinary. • Willingness to test assumptions. • Self-discipline; strong work ethics; commitment. • Perseverance and concentration. • Skill for dialogue and candid debate with customers and coworkers. • Self-confidence and self-esteem; self-motivation. • Enthusiasm and energy.	• Openness to new ideas; eagerness to learn. • Ability to toy with ideas; originality; tinkering. • Tolerance for ambiguity; flexibility. • Willingness to take risks; no fear of making mistakes. • Ability to defer judgment. • Curiosity, inquisitiveness; imagination; creativity, resourcefulness; vision. • Humor and impulsiveness. • Aesthetic interests; knack for elaboration. • Willingness to consider multiple approaches and look for the "unobvious."

Effective team members, like creative people, are not simply the result of good genes. It takes attention, self-awareness, and hard work to develop and enhance the skills required to become an effective team member. Like the dancer Fred Astaire, who would practice for hours to make a short routine seem effortless and graceful, a team member who communicates with ease may well have consciously practiced active listening skills for months or years. The first step is acknowledging that there are skills to be learned, since many people feel that teamwork is simply working together, and they may never have considered which skills might enhance teamwork. Fortunately, the prize that awaits the diligent and persevering is likely to be rewarding beyond anything they might have imagined. Team members on highly productive teams rate the experience as among the most satisfying in their lives.

Although developing such personal skills can be a lifelong endeavor, beginning to cultivate a few such skills can make a significant difference. Table 4.3 outlines eight basic personal team skills needed for maximum effectiveness. As you read through these skills, rate yourself on how well and how often you practice each one. If the list highlights a skill that you rarely practice or do poorly, you may want to consciously improve it through practice in your next series of team meetings.

Table 4.3 Eight Basic Personal Team Skills Seen from Different Perspectives

SKILL	HOW IT SHOWS UP			
	Personally	**To the Team**	**In Productivity**	**To Management**
Proactivity	I take personal responsibility for the team's success.	A feeling that "we're all in this together, whatever it takes."	The team is inspired to take on and meet challenges.	The team is seen as cohesive.
Reliability	I follow through on commitments.	"If she says she'll do it, you can bank on it."	The job gets done well and on time.	Team and personal reliability are apparent.
Participation	I contribute to discussions and share the load of leadership and organization.	"She's always here, always on time, and always has something of value to contribute."	Participation leads to synergy that leads to increased productivity.	The team is seen as dynamic, responsive, and productive.
Active Listening	I "seek first to understand, then to be understood."* I give careful attention and rarely interrupt.	Team members feel understood; they are willing to listen to other points of view.	Greater understanding of diverse points of view leads to better, more creative solutions.	The team is seen as creative, innovative, and synergistic.
Coaching	I appreciate the unique skills and strengths of others, and I encourage using them for the team's benefit.	Team members feel valued; they do their best for the team.	Members work together and apply a diversity of strengths to achieve much more than each person alone.	The team is seen as responsive to both internal and external suggestions for improvement.
Communicating	I communicate clearly and effectively, both interpersonally and in formal presentations.	Interpersonal communication becomes efficient, and misunderstandings are avoided.	Little time is wasted on clarifying misunderstandings; the team functions smoothly in step.	Team presentations are seen as models of clarity and effectiveness.
Giving Useful Feedback	I respectfully give team members useful feedback on their contributions or tactfully point out any actions disruptive to the team.	Team members provide useful feedback on behaviors which might otherwise not have been acknowledged.	The team collectively learns and grows.	Both the team and individuals exhibit abilities to adapt, change, and grow.
Accepting Feedback and Responsibility	I acknowledge and respect the feelings and observations of others (whether or not I agree) and carefully consider making changes.	Team members can hear and understand the feedback and can act on the information in an environment in which they can learn and grow.	The team becomes capable of handling its own internal problems quickly and with minimal loss of productivity.	Both the team and its members are known for quickly resolving problems that arise and moving on to complete the team's charge.

*From Stephen R. Covey, *The 7 Habits of Highly Effective People,* Fireside, New York, 1989.

Stages in team development

It may take as long as two years before a dozen multidisciplinary people working together on a product development project will become a well-functioning, effective team. However, the time required for team development depends on the frequency and intensity of team interaction. Some teams develop quickly through the predictable and necessary stages of growth, while others take longer. These stages are sometimes referred to as *forming, storming, norming,* and *performing* (Ref. 4.6).

1. **Forming:** When the group is first formed, the group members still act as individuals; they do not contribute to the group as a whole but look out for themselves, as humorously illustrated in Figure 4.4. They are merely an assembly under a manager. "Chit-chat" helps initiate the group's socialization process. Members cautiously get to know each other, explore limits of acceptable group behavior, and tentatively try to define the group's purpose and goals, but irrelevant discussions and complaints distract the team from accomplishing much of value.

2. **Storming:** Realization that the task is difficult and that not much progress is being made prompts disagreements, blaming, and impatience with the process. Some members try to do it all on their own and avoid collaborating with team members. Others question the value and purpose of the team and of any work being done, especially if they feel they know more about the problem area and how to solve it. Tension, disunity, and jealousy result.

Figure 4.4 Systems thinking (Boeing graph, used by permission).

3. **Norming:** When the team's objectives are worked out collectively, the common problems or goals begin to draw the individuals together into a group, although the sense of individual responsibility and autonomy is still very strong. Conversations among group members now extend beyond neutral subjects to organizational and budget matters. After achieving some successes in problem solving, the group begins to realize that team development is important, and the individuals as well as management participate in sharpening communication and other team skills. Team responsibility begins to develop, and the team presents a united front to outsiders. Value questions and motivation are discussed. Thus, having begun to understand each other in the storming process, members now start to accept each other and the team rules or "norms." They become more cooperative, try to avoid conflict, and work out their differences, leaving time for constructive work.

Groups reach consensus on the best solution no one can disagree with. Teams reach consensus on the best solution everyone can agree with.

Charles Henning, president of Innovation-Productivity-Quality (IPQ)

4. Performing: The members have accepted each other's strengths and weaknesses, and they have defined workable team roles. With the ability to diagnose and work through team problems, the team now becomes an effective, productive, and cohesive unit. Members start to feel attached to the team and confident of its abilities. The team becomes involved in problem solving in a wider area within the organization. Team members feel united and strongly bonded; they are now open to outsiders and seek contact with the wider community to extend their influence. The purpose of the team is seen in the context of the organization's goals and connected to its broader tasks and responsibilities. Dreams and visions are shared; new ideas and personal differences are evaluated and worked out based on the common vision. The team may become self-directed. Team members fully share accountability for the team's actions, and they operate from a basis of trust and mutual respect.

Most problem-solving teams are assembled for short-term projects. This does not permit the team to build rapport over the span of two years or more as might be the case for a product development team. But the principles of good teamwork and the four stages of team development outlined here still apply. Special attention will need to be paid by the team leader or facilitator at the outset to establish mutual trust and understanding through some team-building activities such as taking out time for leisurely introductions of all team members, discussing the team rules, and providing the "big picture" and motivation—in other words, not skimping on the Step 1 socialization in the knowledge creation model. Additional guidelines for team success (including conflict resolution) are discussed in the following two sections.

Roles and responsibilities of team members

William Golomski, a senior lecturer in business policy and quality management at the University of Chicago, lists the following roles and responsibilities of team members:

- The opinion seeker asks for clarification of values behind issues and tests for agreement.
- The encourager accepts, praises, and agrees with the contributions from others or builds group warmth and solidarity.
- The opinion giver states opinions or beliefs.
- The coordinator clarifies relationships among tasks, ideas, and suggestions.
- The harmonizer attempts to reconcile differences, relieves tension, and helps team members explore and value differences.
- The gatekeeper regulates the flow of discussion and encourages participation by directing conversational traffic, minimizing simultaneous conversations, quieting dominant members, and eliciting participation from quiet members.

Management is an intellectual process that provides leadership and an environment in which people are willing to work together toward an end purpose.

Paraphrased from David I. Cleland and Harold Kerzner

- The initiator/contributor proposes new ideas and methods.
- The information seeker tests the factual accuracy of suggestions.
- The information giver offers information, facts, and data.
- The elaborator diagnoses problems and adds relevant details.
- The orienter summarizes, raises questions about the team's direction, and defines the position of the team in the organization.
- The evaluator/critic examines team accomplishments in light of standards and goals.
- The energizer prods the team toward action and decisiveness.
- The procedural technician distributes materials and obtains equipment.
- The **recorder** (or note taker) records suggestions, ideas, decisions, and outlines discussions.
- The team **process observer** provides feedback on group dynamics.

Different members usually function in more than one role, or a particular role may be assumed by various members at different times. The roles of recorder and of process observer are especially important and should be specifically assigned to designated team members.

Dealing with conflict

The storming stage can be resolved more quickly if the team, right from the beginning, agrees to ground rules that will help it deal with the inevitable conflicts that will arise when different people work together. We tend to think that this is not necessary (and we may even feel foolish about setting up such rules). Two common behaviors—if not acknowledged and overcome—can be harmful to a team's success: (1) we automatically tend to discount emotional arguments, especially if we have the typical analytical engineer's mindset, and (2) we tend to ignore people who have unorthodox ideas, and this frequently may cause them to withdraw from making contributions during meetings. In either case, the team will lose the benefits of diverse thinking.

In a *Harvard Business Review* article, Dorothy Leonard and Susaan Straus give this advice about managing "creative abrasion" (Ref. 3.5):

Teams can learn to use conflict creatively to develop superior win-win solutions.

1. Make sure that everyone on the team understands the relevance of honoring one another's differences in thinking style.
2. Make the team's operating guidelines explicit. For example: "Anyone can disagree about anything with anyone, but no one can disagree without stating the reason" and "When someone states an objection, everyone else should listen to it, try to understand it, treat it as legitimate, and counter with their reasons if they don't agree."
3. In meetings, allow time for both divergent discussion to uncover imaginative alternatives and convergent discussion to select an option and plan its implementation. Innovation requires both types of discussion, but people who excel at different types can, as one manager observed, "drive each other nuts."

To become flexible, quality-conscious, and thence competitive, the modest-sized, task-oriented, semi-autonomous, mainly self-managing team should be the basic organization building block.

Tom Peters,
Thriving on Chaos

4. Use the language of the mental models to depersonalize the tension and conflict that diverse thinking preferences can cause in a group. This understanding is a powerful tool for defusing anger since each style brings a uniquely valuable perspective to the process of innovation, just as each style has some negatives associated with it.

A practical technique for dealing with negative comments is the baseball metaphor, "Three strikes and you're out!" This policy is explained to the team early on and becomes part of the ground rules. If a group member (usually an inexperienced, new person on the team) makes another negative remark after two reminders, he or she will be asked to leave the team's meeting, because negative thinking inhibits creative thinking, mutual trust, and respect among the team members. The team leader must be prepared to eject anyone who continues to be critical. The metaphor and peer pressure make a powerful combination; only very rarely will a "strike three" need to be called. Other effective approaches for overcoming negative thinking are discussed in Chapter 6. Also, the section on negotiation presented in Chapter 5 shows how creative problem solving can be used to work out win-win solutions to conflict.

It is normal for a team to experience a regression to the storming stage when a new team member is added to an existing team. Until the team members get to know the new member and accept his or her style and unique contributions, the new member is likely to be met with caution or even hostility. This phenomenon can be expected particularly if this new member is assigned to the team by management to overcome some perceived shortcomings, or if the existing team has already developed a high level of trust and support among its members. The team must then redevelop norms of behavior that include the participation by the new member. Only then can the team resume the performing stage.

Examples of dysfunctional and functional teams

This section presents summaries of the outcomes of nine different teams (ranging from students and faculty to industry and medicine). As you read through these examples, try to identify the factors in the team process that are being illustrated. What made the teams successful? Why were the teams dysfunctional, and how could the problems have been prevented or overcome? What can you learn from each example?

1. Creative Problem-Solving Team: A five-member heterogeneous freshman team formed based on the HBDI had two members who were foreign students and had difficulties in communicating in English. The instructor was apprehensive on how this would affect the performance of this team. However, their final project outcome and presentation were excellent. One of the two students commented: "Before I took this course, I was 'speechless' and did not know how to participate in a discussion. Now it is different; I have self-confidence in my communication with people, and I know how to participate in a group. Everybody in my team helped me speak without feeling scared in front of people."

Conversation is the laboratory and workshop of the student.

Ralph Waldo Emerson

2. Multidisciplinary Student Team: A team of four students in a total quality systems class worked on a semester-long team project. Two students majored in industrial/organizational psychology, one in business, and one in engineering technology. After two weeks, two students reported to the instructor that they simply could not work with the technology student because he dominated discussions and would not listen to the ideas of others. They wanted to remove him from the team. The instructor asked the two complaining students to give it another two weeks before making any changes. Then he asked the technology student to pose a new ground rule for possible adoption by the team: Each person was to make his or her point only after stating the other person's point to his or her satisfaction. After two weeks, the team reported that things were improving; they wanted to stay together and did not need any intervention. The team went on to complete a fine project.

3. Project Team in a Fluid Mechanics Class: A week before the end of the quarter, one of the members of a sailboat design project team of juniors told the instructor that he would turn in a minority final report, since he did not agree with the design of the other two members. This student had refused to share his considerable knowledge of sailboating with the others in his group; also, because of this knowledge, he was not willing to consider the creative ideas of his team mates. The result was that neither of the two project reports was of good quality—one design was not practical; the other not innovative. These students did not have a thinking profile assessment or training in team development. This is a typical outcome of what can happen when students are not given the tools needed to learn teamwork skills.

4. Student Team in Industry: A team of 12 engineering students (juniors) in a ten-week summer quality improvement internship at a textile plant experienced difficulties when two older male students began to dominate the team in various ways. These two students, who were strong in A and B quadrant thinking, sometimes summarily dismissed ideas from C/D thinkers in the team, primarily the females. A company-administered peer evaluation and feedback session brought the feelings of the female students to light, and the situation began to improve. The team ended the internship quite successfully, and the team members, particularly the two males, reported that they appreciated the team and the growth opportunities they experienced. Overall, the team members reported that the experience was the highlight of their academic careers.

5. Curriculum Development Faculty Team: A team of faculty developed a new course sequence for all freshman engineering students involving multidisciplinary teams and conceptual design. The team had to overcome a number of challenges of widely different types, from devising a viable concept that held the courses together to solving the logistics nightmare of faculty observation and scoring of final team presentations for 450 students. The strong A/B thinking preferences of two of the team members balanced the C and very strong D preferences of two other members. The members learned to appreciate and value each other's strengths and to depend on each other to get the job done. The course sequence was highly successful and now enjoys the support of the faculty at large—many had previously thought "it couldn't be pulled off."

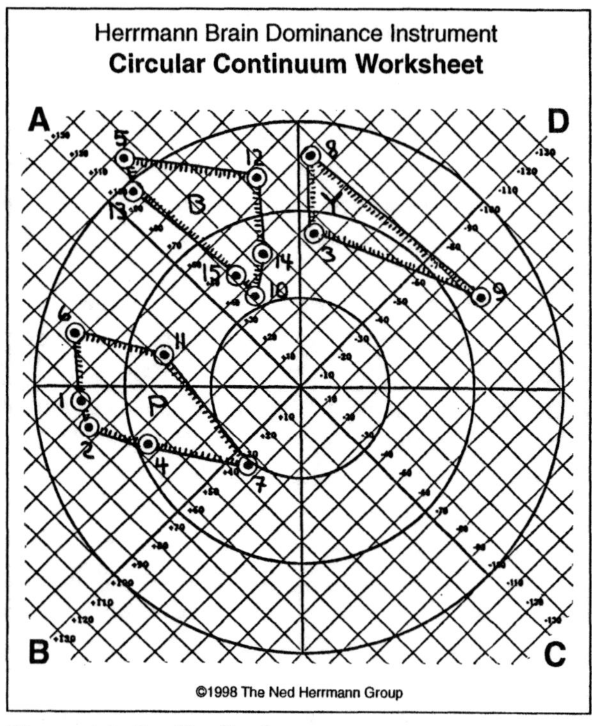

Figure 4.5 Profile tilt of a team that excluded three members who were "different."

6. Leadership Team in Industry: A team of 15 people involved in developing advanced solid modeling software and training programs for a major auto company had a one-hour HBDI workshop held in a small conference room with a long table that just barely seated 12 people. Thus the consultant prepared purple name tags for the A+B dominant group of six and seated them on one side of the table, blue tags for the strongly A dominant group seated on the other side of the table, and three yellow tags for the three chairs placed at the end of the table along the wall, for the quadrant D thinkers. The "purples" and "blues" arrived on time and took their seats. Then they noticed the three empty chairs at the back and said rather derisively: "Oh yeah, the oddballs." It was quite obvious that these three were treated as "outsiders" by their left-brain colleagues. After the presentation, the team leader talked to the consultant for 90 minutes about the problems the team was having, particularly with the strongly quadrant D dominant individual (#9 in Figure 4.5). The success of the entire team within the organization depended on having innovative ideas but was hampered by personal animosity and a competitive spirit. A few months later, the leader reported that the entire climate in the group had changed. The contributions of the quadrant D people are sought and valued; they are treated with respect and have truly become a part of the team, thus increasing the team's productivity.

7. Shaped-Hose Kaizen Team at Aeroquip Corporation, Forest City, NC. This nine-person team was formed for an intensive 3-day Kaizen Blitz problem-solving project to reorganize a production line that manufactures hoses shaped to fit particular automotive applications. *Kaizen* means continuous improvement in Japanese and *Blitz* is lightning in

German). The team of line operators, shift leaders, maintenance, and management spent the first day planning, defining goals, studying the present system, and assigning responsibility for action plan items. The second day, the team identified specific action items and began to implement them. These actions included reconfiguring the production area to a U-shaped cell, defining a different loading and unloading procedure, reducing the number of types of shipping boxes, reorganizing the storage area layout and tool templates, relocating the SPC (statistical process control) data collection equipment, reducing inventory, and relocating mandrel storage. On the third day, the team completed Kaizen implementation reports, revised standard work practice instructions, and made a presentation to staff and management. Major results were:

- Elimination of unnecessary inspection of outside hose diameters.
- Reduction of inside/outside diameter checks from 100 to 18 percent.
- Reduction of inventory from 27,000 parts to 5,539.
- Increase in throughput from 108 to 197 pieces per person-hour.
- Reduction in setup time from 11.5 minutes to 4 minutes.
- Reduction in cycle time from 81 minutes to 32 minutes.

Buy-in of production personnel was achieved by having them involved on the team and in all stages of implementation.

8. Surface Mount Technology Process Improvement Team, Solectron Corporation, Charlotte, NC. This 14-person team from a Malcolm Baldrige award-winning company examined and improved a solder paste screening process using the tools of continuous quality improvement in W. Edwards Deming's Plan-Do-Check-Act cycle. First, they examined the corporate objectives and set deployment goals which included 25 percent cuts in defect rate and cycle time, a cost saving of $2 million, and a 40 percent gross profit margin on investments. Then they conducted a SIPOC analysis (examining suppliers, inputs, products, outputs, and customers). They also identified stakeholders of the process (those who would "notice" if the process failed), analyzed root causes of machine utilization and efficiency, and conducted a Pareto analysis of downtime. A benchmarking study revealed that the process operated at typical levels for downtime but was well below the level of best practices in the industry. Using the tools of process mapping, statistical process control (SPC), stakeholder input, and cause-and-effect diagrams, the team made changes in the process which yielded significant results:

- Defect-free kits (products) increased from 50 percent to 78 percent.
- Changeover time decreased from 82 minutes to 29 minutes.
- Machine utilization increased by 50 percent, overtime by 35 percent.
- Projected savings are $3.5 million/year with an outlay of $200,000 (for the team members' time and training for line personnel).

These last two case studies were presented at the Charlotte Area Team Excellence Showcase sponsored by the North Carolina Quality Leadership Foundation and the Charlotte Section, American Society for Quality, University of North Carolina at Charlotte, 4/21/98.

In a well-organized system, all the components work together to support each other.

W. Edwards Deming, quality expert for manufacturing industries (Japan and U.S.)

9. Hospital: A regional hospital in the Midwest was having problems with a dysfunctional team of surgeons, and a consultant was hired to help solve the problem. The team fell apart when the "whole-brain" physician retired—he had previously facilitated communication in the group as well as the group's interaction with other professionals in the hospital. The three remaining doctors on the team had very different brain dominances, and the very strong double-dominant A+B individual with his rigid, schedule-oriented approach annoyed his more flexible colleagues. As an outcome of the meeting (and as a result of these doctors now understanding and appreciating their differences), they spontaneously agreed to be more cooperative and assist each other in the operating room. They realized that together they formed a whole brain and thus would be better able to meet emergencies and solve problems to the benefit of their patients. Their working climate improved to such a degree that the consultant was called back to work with another group of doctors who were facing important hiring and leadership issues.

Ten-Minute Activity 4-3: Diagnosing a Dysfunctional Team

Think back to an experience you have had with a dysfunctional team. Briefly describe the problem and how it could have been solved (or prevented) with the knowledge you now have about teamwork, the knowledge creation model, and the HBDI. If you are in a class or workshop, share your brief case study with the class or group.

Team management guidelines

Successful teams do no happen by accident. We will briefly look at the overall context and factors that need to be considered for proper team management. Teams and managers in their organization have a joint responsibility for establishing goals and commitments, in selecting the team structure and composition, and in providing a supportive climate.

Goals: The team must be given a charge or assignment with a clear, achievable, significant goal, purpose, or mission aligned with the organization's vision and values. Often, objectives or tasks are specified; sometimes, the team members are empowered to work out the objectives themselves. All team members must understand and agree with the customer-driven objectives that will direct their efforts.

Commitment: Both management and the team are committed to the teamwork concept, to the problem-solving process and results, to standards of excellence, and to dedicating their efforts to the good of the organization. The team members are committed to work hard and to nurture a positive team spirit. The team must have assurance that its results and recommendations will be taken seriously.

To survive in a difficult environment (such as war), teamwork is essential. It takes old-fashioned hard work to grow into a good team. True leadership is looking out for the welfare of the team's members; it is not self-serving.

Jonathan C. Henkel, lieutenant colonel (retired), U.S. Army, Vietnam conflict veteran

Structure: The objectives determine the structure and scope of the team, preferably across departmental lines and with the major stakeholders affected by the problem represented directly or indirectly. Experienced teams can be self-directed. New teams may be guided by a facilitator who is a coach and listener, not an authoritarian manager. Shared leadership will develop naturally, as the stages in the creative problem solving process demand, and as different members take over appropriate roles for short or longer time periods. The process is accelerated if supported with mutually developed criteria. All team members are treated as equals; facilitators serve the team's interests, not their own agenda. Facilitators must have a positive attitude toward the team and its competence.

Selection of team members: The members of the team are selected for their diverse thinking skills, personal characteristics, expertise, cross-functional experiences, and other abilities they can contribute to the team and the problem-solving task. They should have good communication skills, including "constructive differing." Habitually negative thinkers and people with a hostile attitude should not be selected to serve on a team, even if they qualify otherwise. Team members who are familiar with the creative problem solving process and are able to respect each other's contributions and thinking skills are especially valuable.

Climate: A collaborative, not competitive, climate must be maintained within the team, with the support from management. The team results—not individual glory and self-advancement—are important; credit for the problem-solving results will only be given to the team as a whole. Team members know that they are accountable for making good decsions to achieve team success. Team members support each other—they give and accept positive feedback as well as constructive criticism for continuous improvement. Management supplies the needed resources (time, finances, facilities, networking) and recognition for the team's contributions in the organization. As we shall see in Chapter 18, innovative teams must be given special protection and time to develop viable solutions from their creative ideas without interference or criticism.

In the following section, you will find the outline of a procedure we have successfully used to form whole-brain student teams. The remainder of this chapter discusses practical "how-to" tools to make teamwork more efficient and interaction among team members more effective.

How to form whole-brain project teams

We have found that teams who are purposefully formed to be whole-brained have better project outcomes and less chance of being dysfunctional than if they are randomly assigned or if they are put together based solely on achieving multidisciplinary teams. A facilitator responsible for forming teams, the class instructor, or the HBDI evaluator can conduct the team building activity, depending on the circumstances. The process is described here so that you can understand the rationale and then use the procedure as leader or manager of a large project.

Preparation (for a class or training workshop)

1. Homogeneous grouping: As soon as the HBDI results are available, the students are organized into homogeneous "color" groups, with each group having the same number, except for the "multi" category that is reserved for the most whole-brain "extra" students. Assign students to their homogeneous groups based on similarity in thinking profiles. Select the same number of groups and "colors" as the number of students that will make up the whole-brain teams. We have had good experiences with teams of five. For large classes, teams may have to be bigger to cut down on the number of projects to be evaluated. As an example, Figure 4.6 shows profile "tilt" for a class of 31 electrical engineering freshmen grouped into five *homogeneous* teams with N = 6, where N is the number of planned *heterogeneous* teams with five students each (and one team with six students). Depending on the mix of HBDI profiles present, the colors are assigned roughly relating to the strongest thinking preferences, using this scheme:

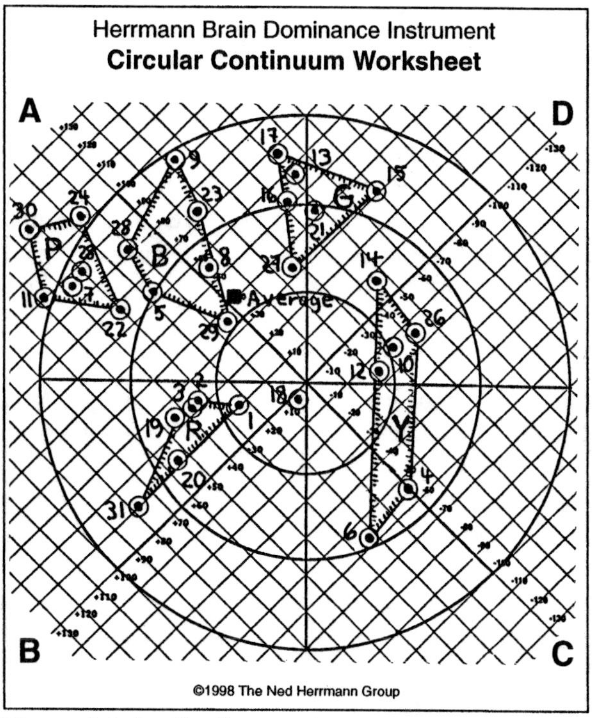

Red:	Quadrant B dominant.
Purple:	Quadrant A + quadrant B double-dominant.
Blue:	Strongly quadrant A dominant .
Green:	Quadrant A + quadrant D double-dominant.
Yellow:	Quadrant D dominant, or double dominant in quadrants D + C (see Fig. 4.6).
Orange:	Quadrant C dominant (rare in engineering).
Multi:	Whole-brain (#18 in Figure 4.6).

We mark slips of papers with each student's name and the respective color dot. "Multi" is indicated by a circle drawn in pencil. Do not use name tags to avoid stereotyping people with a particular mental preference (the color dot is strictly for grouping purposes and relates only approximately to particular thinking quadrants or profile tilt).

Herrmann Brain Dominance Instrument
Circular Continuum Worksheet

©1998 The Ned Herrmann Group

Figure 4.6 Profile tilt example. The tilt coordinates are calculated as A-C, B-D from the HBDI scores.

2. Team Rosters and Team Leaders: Prepare the team rosters, as shown in Table 4.4. Assign the team leaders of the day. These are the students with lowest C thinking (usually with HBDI scores less than 40). This assures that these students who may be uncomfortable working and talking with others will not be concentrated in any one team. Also, this assignment gives them an opportunity to "stretch." Enter one leader per team, on the respective color line. Then duplicate the roster.

Table 4.4 Preparation of Team Rosters

Color	Team 1	Team 2	Team 3	.....	Team N
Red	R ________	R ________	R ________	.....	R ________
Purple	P ________	P ________	P ________	.....	P ________
Blue	B ________	B ________	B ________	.....	B ________
Green	G ________	G ________	G ________	.....	G ________
Yellow	Y ________	Y ________	Y ________	.....	Y ________
Orange	O ________	O ________	O ________	.....	O ________
Multi	M ________	M ________	M ________	.....	M ________

3. High Quadrant C People. Now select N students with the highest scores in quadrant C from your HBDI database. These team members are given an orange "dot" along with their regular color assignment. We very rarely have enough engineering students with high scores in quadrant C to make up a full "orange" homogeneous team. In Figure 4.6, only #4 and #6 were primarily quadrant C dominant.

4. Doubling Up. If you are short one or two people to make up a full color team, students who have strong double dominances can be given two dots at this time and will thus fill a double role. In the example of Figure 4.6, #15 would have been a candidate for yellow and green, #28 and #5 for purple and blue, had such a doubling been necessary.

Team building activity (in class)

1. After students or workshop participants have had a presentation on the HBDI model and have received their profile packets, they are given their color assignment and asked to stand along the sides and back of the room. Team leaders are called out, given their copy of the roster (with duplicate), and assigned a table. They are now in charge of the team "draft" and need to raise their hand if they are looking for a member of a particular color. Once a member joins their team, they need to enter the name on the appropriate line(s).

2. The orange "dots" are invited to join a team, one per team. Any other "double dots" present in the class would joint a team at this time as well. All need to enter their name on two lines, according to their double dots, and there must be no duplication of colors in the team.

3. The remaining students can now join the teams, one color at a time, and only one student of each color per team. Also, close friends should not join the same team, as such a subgroup will be a detriment to the development of the whole team. But within this constraint, this method

leaves students some choice on which team to join. They feel welcomed into their team since the team is looking for a member of that particular color. The method solves the problem of how to deal fairly with minority and female students who are often assigned to teams arbitrarily for diversity's sake; here they have some choice (within the color constraint) about the team they want to join. Say: "Let the remaining reds join the teams. Leaders, if you need a red, raise your hand." Then continue in the same way with purple, blue, green, and yellow.

4. The "multis" (if any) are now invited to join any team they want. They can also "pinch-hit" for any slot that has become vacant because a student has dropped the class. Each team should now have all colors filled in—except multi, depending on how many students were available in that category. "Trades" (for the same colors) should not really be necessary, except in unusual circumstances. The team leaders make sure all names have been entered on the roster; it is turned in to the instructor or facilitator, and the teams keep the duplicate for their own records.

5. Students or participants can now overlay their profile transparencies to get a team composite. If time remains in the class period, the new teams can get better acquainted. They may want to discuss how they will be able to meet together for their future team assignments; they need to exchange e-mail addresses and phone numbers.

Forming a whole-brain work team

Once a team has been formed based on its project assignment, the team needs to verify its composite HBDI profile. If the team is very homogeneous, it must make a special effort to seek out an additional member or two who would be able to bring the missing thinking preferences to the team. These people would not necessarily have to be experts in the project area. On technical teams, the missing thinking skills will most likely be in quadrant C, followed by quadrant D. Development teams may be short of quadrant B thinkers. People who prefer quadrant B and C thinking are most likely found among the administrative assistants, secretaries, and clerks, the quadrant D preferences among the "outsiders" and loners in the organization—not the persons who would normally be sought out to join a team. Ned Herrmann also recommends that teams try to have people of both genders to add a broader array of thinking styles and valuable "differences" to the team.

It is not possible to create the optimum whole-brain team with only one gender involved. The differences are both subtle and profound. You cannot go to where you want to be mentally [in your organization or team] without both males and females involved in the process.

Ned Herrmann

If it is not possible to add people for balance to the team on a permanent basis, efforts should be made to include them during the crucial brainstorming and other sessions when particular thinking skills are needed in the creative problem solving process. At the least, the team must make sure it pays attention to the more unfamiliar thinking modes—typically quadrant C for engineers, and very likely also quadrant D.

Tools for organizing and managing your team for productivity

I studied the lives of great men and famous women; and I found that the men and women who got to the top were those who did the jobs they had in hand, with everything they had of energy, enthusiasm and hard work.

Harry S. Truman

A highly productive team doesn't happen by chance, no matter how skilled the members. In fact, the work of a team involves extra effort over that of individuals to take advantage of the team's creativity and synergy. Certain team tools have been shown repeatedly to reduce the team overhead and keep the team productive. If your team operates without these tools (which mostly involve quadrant B organization), you risk blunting the team's effectiveness and dissipating its energy. These tools are:

- ❑ A well-defined team charge or mission statement
- ❑ A timetable or project plan
- ❑ Team ground rules
- ❑ Team member roles
- ❑ Meeting agendas
- ❑ Meeting minutes
- ❑ Running task list

Team Charge or Mission Statement. A sponsor or manager may charge a team to accomplish a certain task or set of tasks. Frequently, however, the sponsor does not have enough information to define the task well. It becomes the team's job to re-articulate the mission in detail and in terms the members understand. Then the team must present the refined mission statement to the sponsor for approval. This process is typically very helpful in defining the problem to be addressed, acknowledging expectations of the team, limiting the scope of the task(s), and specifying constraints under which the team is to work. Regardless of whether the team defines its own mission or works under the charge of a sponsor, a well-defined mission statement can keep the team focused on the right objectives. A team mission statement should include

- ❑ A statement of the overall purpose of the team and any particular problems to be solved;
- ❑ A reference to the customers or stakeholders the team is serving, specifically mentioning their contacts or representatives;
- ❑ A statement of scope and a list of constraints within which the team is to work (limited geographic area, specified manufacturing plant, other qualifying assumptions);
- ❑ The time frame in which the team is to complete its work;
- ❑ The budget within which the team is to work;
- ❑ A list of deliverables expected of the team, with anticipated completion dates (i.e., plan of action, progress report, conceptual designs, "best" solution, and final report).

These elements of the mission statement should be concise and clearly understood by all team members. A one-page mission statement is usually sufficient except for the most complex team charges.

A Timetable or Project Plan. Once the team has refined and agreed to its charge, a plan of action is in order. This usually takes the form of a Gantt chart showing the beginning and ending dates of the team's work and all the deliverables expected with their projected dates of completion. Each detailed task necessary to complete the project should be listed, with beginning and ending dates. As the project progresses, the person responsible for each task should be shown on the plan. Project planning software tools can be useful, but teams should avoid overly complicated programs. See Chapter 15 for easy-to-use templates.

Look for planning tools that are flexible and allow changes easily, since a plan is only useful if it is kept up to date. A cardinal rule is never to show a Gantt chart in a presentation and make the comment that it is out of date, or that the team is behind schedule. The Gantt chart, or any other planning tool, should always answer the question, "What do we have left to do, and how will we accomplish it in the time remaining?"

Team Member Roles. Whether assigned roles for team members rotate or are fixed by the team charge, time spent in defining these roles so the team clearly understands the responsibilities involved will prove very useful. Although some special roles may evolve from the particular team charge and activities (as outlined in the earlier section on team development), the following basic roles are recommended for any team:

Note Taker—the person who summarizes task reports from team members, records team decisions, lists tasks assignments (dates and persons), and sends this summary along with the agreed-upon agenda for the next meeting to all team members.

Process Observer—the team member who observes how well the team meeting process and the tasks assignment procedures are working and who may make suggestions for improvements (like an internal auditor). This role is often neglected but can be crucial for inexperienced teams. Although it is tempting to permanently assign this role to a whole-brain member who has good communication skills, team learning and development may be served better if all members can gain experience with this role.

Teams are groups of people who deal with problems with confidence, professionalism, and a can-do attitude.

Leader—the spokesperson for the team and the person who typically calls meetings, reports to management, and conducts the meetings.

Meeting Leader—the team member who assumes the leader role for a particular meeting. Sometimes this person is chosen for a series of meetings because of his or her expertise or thinking preferences.

Team Gound Rules. Spending time up front to define and discuss the rules under which your team intends to operate can save much time later on. These rules, which must be accepted by each person, offer the team members more efficient and satisfying ways to interact with each other. Table 4.5 is an example of a set of team rules adopted by a student team. Your team should feel free to add or subtract rules from this example.

Table 4.5 Ground Rules to Encourage Team Synergy — Example

To promote team productivity and harmony, we agree individually and collectively to abide by the following rules until the team amends or rescinds them:

1. We agree to treat each other with respect and courtesy the way we want to be treated.
2. We agree to make team decisions by consensus.
3. We agree that any team discussion may be shared outside the team unless a team member asks that it be treated confidentially.
4. We agree to be on time to each team meeting, and we will notify the meeting leader in advance when we will be late or absent.
5. We agree individually to complete work assigned to us on time, to notify the team leader in advance if the work cannot be completed as scheduled, and to send the work by other means if we cannot attend the meeting on the work's due date.
6. We agree to attend team meetings every __________ (weekday) at ________ (time) at the following location: __________________ .
7. Each one of us agrees to check e-mail daily and to notify other team members promptly by e-mail of any significant developments in the work of the team.
8. We agree that we will share and rotate the roles of meeting leader, note-taker, and process observer. The roles will be assigned at the end of each meeting for the following meeting.

Sometimes the rules are useful even in their infraction. Let's say, some team members are frequently late for team meetings. This could lead to a re-examination of the "on time to meetings" ground rule and a discussion of why some members are late and whether the meeting time or the location is working against promptness. Rarely does a team revoke a ground rule such as this—typically the discussion leads the team to find a win-win solution for all.

⌛ Ten-Minute Activity 4-4: Team Meetings

As a team, brainstorm a win-win solution for a team that has discovered one or more of its members cannot reasonably make the scheduled team meetings (for example due to different work schedules).

Meeting Agendas. Teams that make the last item of every meeting agenda the tentative agenda for the next meeting have a running start at making each meeting shorter and more productive. Special items can be added by the designated team leader or meeting leader when the meeting reminder notice, the notes of the last meeting, and the revised agenda are sent out (usually by e-mail). All tasks assigned for reporting in the next meeting should be listed as separate agenda items, with the person responsible for the task. The agenda should also schedule the process observer for reporting on the meeting process and possible improvements.

A tactic called the **parking lot** makes staying on track with the agenda much easier. The meeting leader simply lists any discussion topic brought up but not on the agenda on a flip chart page titled "Parking Lot." At the end of the meeting, these items are added to the agenda if there is time, or they are added to the agenda of the next meeting. In this way, the team members bringing up these items are assured that they are heard and discussed without disrupting or side-tracking the agreed-upon agenda. This is a very useful strategy to keep the peace between the B-quadrant and the C- and D-quadrant dominant team members.

Meeting Notes. The team member designated as the meeting note taker assumes responsibility for recording the business actions of the team. These notes should not be verbatim "minutes" but should include:

- ❏ A record of the meeting date, time, and attendees.
- ❏ A summary of the progress reports of team members with assignments made at previous meetings.
- ❏ Decisions of the team.
- ❏ New assignments—by team member and date to be completed.
- ❏ Comments and suggestions from the process observer, along with actions the team wants to take based on these observations.
- ❏ The date, time, and agenda for the next meeting, including designations for the meeting leader, note taker, and process observer if these are to change for the next meeting.

Running Task List. A running task list kept on a flip chart in view of the team during the meeting allows members to get a feel for the size and number of tasks facing the team. As an agenda item near the end of the meeting, members volunteer (or are assigned) to complete these tasks by the agreed-upon time. Assigning responsibility at the end of the meeting in this fashion assures team members that the tasks are apportioned fairly and that each will be done in a timely manner.

Team Activity 4-5: Team Process, Part 1

1. With your team, develop and refine a mission statement for an upcoming project that everyone can support. Have it reviewed by the team sponsor, and make any necessary changes. Copy and distribute the final team mission statement to each team member.
2. Create a list of possible ground rules. Discuss and refine them until each team member agrees to abide by them. Have them typed up and distributed to each team member.
3. In your next three meetings, experiment with having a team member serve as process observer. After three meetings, have an agenda item to discuss the effectiveness of a process observer and whether to continue this monitoring process.
4. Practice using written agendas in your team as described in this section. Prepare a brief questionnaire for each team member asking about what is working with the agendas and what is not. Solicit ideas for improvement. Discuss the results as an agenda item at the next team meeting.

Team Activity 4-6: Team Process, Part 2

1. Use the running task list for three team meetings, then discuss the results as a specific agenda item at the following meeting. Decide how the process might be improved. Note whether tasks were assigned equitably among team members and if they were completed on time.
2. Use the "parking lot" for three meetings, then discuss the results as a specific agenda item at the next meeting. Decide how the process might be improved. Note specifically whether the team kept on track during meetings using this process.

Team member evaluation

Teams in industry are evaluated based on results and the bottom line—if they are not accomplishing their goals or progressing according to plan, they will be disbanded (and the members may even lose their jobs).

Evaluating the performance of student teams presents a different problem. Both the instructor(s) and the students are caught in a system that is driven by grades, not necessarily by the amount of learning that is happening in the process. Complicating the matter is the allocation between the team's performance and individual effort—on the project and in the course. How much of the course consists of teamwork? Are you in a freshman course whose objective is teaching teamwork skills? Are you in a traditional course, with a team project added (which may constitute 25 percent of the course work or be equivalent to one exam)? Or are you in a capstone design course, where the project *is* the course (and where it is assumed that students already have teaming skills)? Most instructors will give final grades that have an individual and a team component, and students are usually informed at the beginning of the course on how their grades will be determined.

Students have one big problem with the team grade. If the team members do not contribute equally to the team's effort, students feel that it is not fair for everyone to profit equally from the team's grade or reward. To address this and give students an incentive to participate fully on the team, the peer contribution rating form shown on the right can be used. Students are given a copy of the form at the beginning of the team project. If the course is set up to focus especially on developing teamwork skills, the team may be evaluated once or twice during the early part of the project to give the members feedback on their performance. Only the final evaluation should be counted toward the grade. We have found that for student teams who have developed synergy, the ratings will often say that everyone contributed equally. The rating form is useful primarily for giving student teams some power to deal with members who are not performing or pulling their weight on the team. But if this form is used too often, this could be counterproductive to team building: improvement and cooperation must be given sufficient time to develop.

Peer Contribution Rating Form

Purpose: This form is used to allow team members to rate the contributions of fellow team members. The results during the term are used to identify problems and give the team an opportunity to improve. The results at the end of the course may be used in determining individual performance grades. Your input will remain anonymous and will not be revealed to anyone else on the team.

Instructions:

1. Fill out this form, sign it, place in a business envelope, and return it to your instructor by the due date.
2. Evaluate each member according to his or her contribution to the team effort. Circle the appropriate response on the following scale: **P = poor, A = adequate or average, T = tops.**
   A. **Quality**—value and quality of contributions, suggestions, opinions, ideas, creativity.
   B. **Quantity**—participation, sharing of responsibility, attendance at team meetings, willingness to do his or her share of the work, preparation for meetings.
   C. **Attitude**—if poor, indicate the nature of the perceived problem (confrontational, negative, indifferent, lazy, bossy, non-cooperative, etc.) in the space at the bottom of the form.
   D. **Contribution** (in percent) to the entire team's work of each team member. The total of all contributions must equal 100%.
   E. **Yes or No**: "Would you choose this individual to be on your next team?" If no, offer one or two constructive ideas on how the team member could improve, using the back of the form.
3. If desired, you can also highlight one or two outstanding contributions made to the team by a particular member. Use the space at the bottom of the form (or the back, if you need more space).

Your Name ______________________ Team Name/No. ______________________

Full name of team members	A **Quality**	B **Quantity**	C **Attitude**	D **%**	E **Yes or No**
1. ________________	P A T	P A T	P A T	__	yes no
2. ________________	P A T	P A T	P A T	__	yes no
3. ________________	P A T	P A T	P A T	__	yes no
4. ________________	P A T	P A T	P A T	__	yes no
5. ________________	P A T	P A T	P A T	__	yes no
6. ________________	P A T	P A T	P A T	__	yes no
7. ________________	P A T	P A T	P A T	__	yes no
8. ________________	P A T	P A T	P A T	__	yes no
				100	

Problems:

Praise:

Resources for further learning

4.1 David I. Cleland and Harold Kerzner, *Engineering Team Management,* Krieger Publishing, Malabar, Florida, 1990. This book includes discussions on the ambiance of team management, communications, leadership, motivation, planning and organizing, and decision making, especially as applied to developing high-performing technical teams.

4.2 ✓ Stephen R. Covey, *The 7 Habits of Highly Effective People: Powerful Lessons in Personal Change,* Simon and Schuster, 1989. Values and character development are central to this book. It outlines a pathway for living with integrity. The principles provide the security which encourages change and gives wisdom for using opportunities brought about by change. It also develops personal characteristics that are important to teamwork.

4.3 ✓ GOAL/QPC and Joiner Associates, *The Team Memory Jogger: a pocket guide for team members.* A quick, inexpensive, pocket-sized reference for becoming an effective team member, starting teams off on the right track, getting work done, documenting the team's work, and handling conflict and uneven participation. For address, see Ref. 4.6.

4.4 Tom Peters, *Thriving on Chaos: Handbook for a Management Revolution,* Knopf, New York, 1987. Managers today confront accelerating change with constant innovations in computer and telecommunications technology. This book gives practical guidelines for survival, flexibility, and empowering teams.

4.5 Paul G. Ranky, *Concurrent/Simultaneous Engineering (Methods, Tools & Case Studies),* CIMware Limited, Guildford, Surrey, England, 1994. This book explains CE/SE by giving both system analysis and design models as well as practical methods, tools and solutions.

4.6 ✓ Peter R. Scholtes, *The Team Handbook,* Joiner Associates, 3800 Regent Street, Madison, WI 53705-0446, 1988. This is a practical guide aimed toward quality improvement teams, with many useful tips and tools.

4.7 J. William Shelnutt et al., "Forming Student Project Teams Based on Herrmann Brain Dominance (HBDI) Results," *ASEE Annual Conference Proceedings,* June 1996, Washington, DC.

Exercises

A recent survey showed that 66% of seniors in engineering think they are ready to work in a team, but only 12% of employers find that they are ready.

4.1 "Loosening Up" a Team

Get a group of people to stand in a circle facing each other. One person pretends to throw a ball to someone else; that person "catches" the imaginary ball and passes it on to another person. Call out the name of the person before throwing the ball. This is a good exercise to do when people don't know each other well yet. Then use a sound instead of a ball to pass around; The person who receives a "rooster's crow" repeats it and then makes up a new sound to pass on, for example the wail of a siren. Alternately, this activity can be done with a prop, such as a bandanna. It is handed to a neighbor in the circle by saying: "Sue, this is a bandanna." Sue replies: "No, no, no—this is a tourniquet (and she does a little demo). Then Sue in turn hands it on by saying: "Anthony, this is a tourniquet." He motions: "No, no, no—this is an oil rag," etc.

Training in teamwork is needed because our culture is individualistic and confrontational.

4.2 ✓ Sharing Your Vision of the Future

Work in a small group. Stretch your imagination with the following three situations. Jot down your thoughts and then compare them with each other. If possible, develop a group composite vision.

a. Picture a perfect day for yourself, six months from now. Where will you be? In front of a group of your peers, giving a great speech? Running a race and winning? Playing the lead part in a performance? What will you look like? How will you feel? Will you feel great about yourself because you've broken a bad habit or made other improvements?

b. Now picture yourself five years from now. What have you accomplished in the past five years? Did you obtain a college degree? What new things might you have learned? Are you in a continuing education program? What have you accomplished in your personal life? Have you changed as a person? Have you grown and matured, not just physically, but also spiritually? What kind of family and friends will you have?

c. From where you are today, write as many endings as you can for this sentence: One of the things I'd really like to do during the next ten years is... Now look over your list of ideas and mark three or four that are most important to you. Ask yourself: What can I begin doing now to make one or more of these dreams a reality? Include one activity in your weekly schedule that will move you toward achieving your dream.

4.3 ✓ What Have You Learned So Far about Teamwork?

Look over your initial answers to the True-and-False Quiz. Have all your questions and myths been cleared up as you studied the chapter? Write a brief paragraph describing the three most interesting, useful, or surprising things you have learned in this chapter, giving the reasons for your choice. What is an important question you still have?

4.4 ★ Team Analysis ★

1. Read through your HBDI packet (both Part I and Part II, as well as the individual profile interpretation). Make sure you have a good understanding of the four thinking quadrants and your strengths and weaknesses in each quadrant, as they relate to problem solving and teamwork. Note that each quadrant contains clusters of different modes and thinking abilities, and each person has a unique set of preferences and competencies even if the overall profile is similar.

2. Make a list of your strengths. How will these be able to contribute to the success of your current team project? Similarly, make a list of those

modes that you would like your team members to have in order to make up a whole-brain team.

3. Get together with your team members and compare your answers to Item 2. What are the combined strengths of the team? What are the team's weaker areas? The team will have to "stretch" and pay special attention so that these areas will not be neglected as you do your project:

a. Calculate the average HBDI profile of your team by adding up the scores for each quadrant and dividing by the number of members.
b. Plot the profile "tilts" of the members of your team and connect the extreme "dots" with a line to form a polygon.
c. Overlay the profiles of all team members.

From these three items, you should gain some insight into the strengths and weaknesses of your team.

4. Write a brief team report: Summarize and discuss the team's average profile and its characteristics (both strengths and weaknesses) based on the insight gained from the discussion in Item 3. Attach the team profile "tilt" sketch. List all team members and indicate who did the planning, writing, calculations, plotting, etc., and how you solved the problem of dividing up the work.

4.5 ★ Individual Report on Your Project Experience ★

At the end of your project, have a debriefing with the team to review how the team performed based on its team HBDI profile. How did the profile (and the members' individual strengths) help the team achieve its project goals, overcome difficulties, and improve communication?

Write an individual one-page report on the following topics:

a. Project outcome: Discuss the insight gained from the HBDI and its application to the team's project process and outcome, based on the debriefing.
b. Personal application: How has awareness of different thinking modes helped you develop as a person? How has the model helped you understand and communicate with people who are different from you? How has it helped in teamwork?

A high level of conversation during an unstructured task alone was not a good predictor of team performance. Having someone in the group who contributes a lot of ideas was needed.

Successful teams require individuals who can verbally contribute and support their ideas and position.

Marla R. Hacker, engineering professor, Oregon State University, in ASQ Quality Progress, *January 1999.*

4.6 ★ Using Knowledge Creation in Study Groups ★

The knowledge-creation model can be used to optimize the learning process of group study. The steps shown in Table 4.6 do not require an equal amount of time, but they should not be skipped. Under each step, study groups need to pay attention to the items that apply to the task at hand and to how long the group has worked together. When these steps are being followed, the members will find that over time the group is developing into an effective team. Explicitly apply the steps in Table 4.6 in your next group study project. Then write a brief summary on the process and the results. Observing processes and thinking about the results is training in metacognition.

Table 4.6 ✓ Using the Knowledge Creation Cycle to Improve Group Study

Step 1: Socialization Process

__ Meet in a congenial environment.
__ Motivate each other to support and enhance learning. If necessary, develop some ground rules such as "no criticism or put-downs" or requiring positive ways to resolve (or accept) conflict.
__ Decide on the leadership role—one solution is to rotate so each member gets a chance to develop skills with the support of the others.
__ Understand and appreciate mental diversity. Accept that some group members initially may not be comfortable with teaming.
__ Share goals: what does the group need to accomplish?
__ Share previous experiences that might be relevant to the subject being studied.
__ Make a commitment to the group to work hard, be honest, responsible, and willing to contribute and listen to each other.

Step 2: Externalization Process

__ Develop metaphors and analogies to make concepts taught in class or in the textbook easier to understand by everyone.
__ Generate hypotheses and models of the problem(s). Sketch visual representations.
__ Brainstorm different possibilities and approaches to solving the problem. Allow divergent thinking as well as reflection.
__ Look at the big picture and context of the problem—are these relevant to the assignment?

Step 3: Combination Process

__ Collect and network all needed data and information to solve the problem. Redundant information at this point is fine; it accelerates learning by the group.
__ Analyze the problem. Do the known facts support the theory?
__ Carry out the calculations. Evaluate the solution—do the quantities make sense?
__ Seek new combinations with the information learned previously or obtained outside of class.

Step 4: Internalization Process

__ Check the problem-solving process and the results.
__ Check if the concepts and procedures have been learned: everyone does a similar problem and then compares answers. If necessary, repeat with additional drill to gain "operational" skills.
__ As a next step, apply the principles to a more difficult problem or situation for mastery learning.
__ Neatly write up the problem sets that have to be turned in. Also think about the evaluation of the *team process* and discuss what has been learned. If required, summarize in a paragraph and submit.

Chapter 4 — review of key concepts and action checklist

Advantages of teamwork: A team has a mix of knowledge, experiences, and thinking skills available which can interact synergistically for solving problems. Homogeneous teams (where members have similar HBDI profiles) can communicate easily and reach consensus quickly—an advantage when time is of the essence—but the outcome is usually an adequate, but not a "best" solution. Heterogeneous teams need to learn how to appreciate their differences; they are then capable of achieving superior performance. Multidisciplinary teams composed of the key stakeholders make it easier to have the solution accepted and implemented. Industry (especially for concurrent engineering) requires that graduating engineers are able to function in multidisciplinary teams.

Good Advice for Teamwork:

The secret to success is to learn to accept the impossible, to do without the indispensable, and bear the intolerable.

Nelson Mandela, President of South Africa

Team development: Personal team skills can be learned and practiced. A team will typically go through four stages of development: forming, storming, norming, and performing. Members of a team take responsibility for various roles to keep the team functioning efficiently. Techniques are available that can help a team through the storming phase quickly and resolve conflicts. We can learn from the experiences of successful as well as dysfunctional teams.

Team management guidelines: Teams must have an achievable goal or mission and be committed to the teamwork concept and to the problem-solving process and results. Its specific objectives or tasks are customer-driven and determine the structure and scope of the team. Members are selected for the abilities they can bring to the problem-solving task. Management gives assurance that the team's output will be seriously considered for implementation and supplies the necessary resources and recognition. The team maintains a collaborative climate.

Tools for organizing and managing a team for productivity: Organizing tools increase a team's productivity. They include (1) a well-defined team charge or mission; (2) a timetable or project plan; (3) team ground rules; (4) assignment of team roles of leader, meeting leader, note taker, and team process observer; (5) using meeting agendas, notes or "minutes," and a running task list. An anonymous evaluation form can be used to rate the contribution of team members if required to identify problems at midterm or to help assign a performance grade.

Action checklist

☐ Ask your team members to tell you the one positive thing that you could do that would —in their view—make you a better team member. Then do it (realizing that it may take several weeks of practice before this becomes a habit).

☐ If you are faced with having to solve a significant problem, make sure you are using a cross-functional, mentally diverse team.

☐ If you are currently in a team, check to see if you are using the four organizational tools that can make the team more productive: team charge or mission, timetable or project plan, team ground rules, and assignment of team roles.

☐ If you must regularly attend meetings, try to increase productivity by using: (a) agendas where the last item becomes the agenda for the next meeting, and (b) the "parking lot" technique to deal with interruptions.

5

Communications

> What you can learn from this chapter:
> - Teamwork depends on good communication: do you "ruffle feathers" when you communicate?
> - What makes a good communicator? Being a good listener. Examples of communications difficulties for "different" people.
> - Tools: Negotiating a win-win outcome. The 30-second message.
> - Communication in engineering design: criteria.
> - Overview of formats for design documents and oral technical presentations, including proposals, reports, summaries, tables, and graphics.
> - Further learning: resources, exercises, review, and and action checklist.

In this chapter, we will first look at general principles of good verbal communication, since it is a key for building an effective team. The second half of the chapter will focus on design communication and summarize the complete set of formats of design documents (given in Chapter 17) that model the 12-step design process presented in Chapter 14.

Verbal communication and teamwork

Do you ruffle people's feathers when you communicate? The single most frequent reason by far why people are fired from their job is because they do not get along with their colleagues or their bosses. Having good communication skills—knowing how to listen and how to interact with people positively—is very important to productive teamwork. With creative thinking and applying the mental models we can improve communication (both verbal and written).

> ⧖ ✓ **Fifteen-Minute Activity 5-1: Don't Frustrate... Communicate!**
> The Ned Herrmann Group has prepared a "Foursights" poster on whole brain communication. It begins with an assessment of the "hot buttons" that annoy or frustrate thinkers in the respective four quadrants. Check all the items in Table 5.1 that you think are making your communication less effective. Many of these may be a result of very strong preferences in other quadrants or an avoidance of thinking in a particular quadrant.

Table 5.1 Barriers to Effective Communication (or What Can Drive People Crazy)

"How our communication can be perceived by others who have thinking preferences that differ from our own."

Barriers to communicating with Quadrant A:

- ❑ Inarticulate, "off the track" communication.
- ❑ Excessive "chatter."
- ❑ Vague, ambiguous approaches or instructions.
- ❑ Illogical comments.
- ❑ Inefficient use of time.
- ❑ Lack of facts or data.
- ❑ Inappropriate informality.
- ❑ Overt sharing of personal feelings.
- ❑ Impression of not knowing the "right" answer.
- ❑ Fear of challenge or debate.
- ❑ Lack of quantitative "proof" or facts for ideas.
- ❑ Lack of clarity.
- ❑ Excessive use of hands or gestures.
- ❑ Unrealistic or "touchy feely" approaches.

Barriers to communicating with Quadrant D:

- ❑ Repetition.
- ❑ Too slow paced.
- ❑ "Playing it safe" or "by the book."
- ❑ Overly structured, predictable.
- ❑ Absence of humor and fun.
- ❑ Lack of flexibility, too rigid.
- ❑ Inability to "get" concepts or metaphors.
- ❑ Drowning in detail.
- ❑ Too many numbers.
- ❑ "Can't see the forest for the trees."
- ❑ Inability to talk about intangibles.
- ❑ Narrow focus.
- ❑ Resistance to new approaches.
- ❑ Dry, boring topic or style.

Barriers to communicating with Quadrant B:

- ❑ Unknown or absence of a clear agenda.
- ❑ Disorganized.
- ❑ Hopping around from subject to subject.
- ❑ On and on and on and on.
- ❑ Unpredictable.
- ❑ Too fast paced.
- ❑ Unclear instructions or language.
- ❑ Too much beating around the bush.
- ❑ Incomplete sentences.
- ❑ Lack of closure.
- ❑ Not letting a person finish their thoughts.
- ❑ Lack of practicality.
- ❑ Too many ideas at once.
- ❑ Unexpected "off the wall" language.

Barriers to communicating with Quadrant C:

- ❑ Lack of interaction.
- ❑ No eye contact.
- ❑ Impersonal approach or examples.
- ❑ Dry or "cold" unenthusiastic interaction.
- ❑ Insensitive comments.
- ❑ No time for personal sharing.
- ❑ Low recognition or praise.
- ❑ Lack of respect for feelings.
- ❑ Overly direct or brusque dialogue.
- ❑ Critical, judgmental attitude and voice.
- ❑ Being cut off or ignored.
- ❑ Lack of empathy for others.
- ❑ Avoidance of face-to-face communication.
- ❑ All data, no nonsense.

⌛ **✓ Fifteen-Minute Activity 5-2: Improve Your Communication**

How do others see you? It is often helpful to ask others with whom you have communication challenges to go through the list and indicate which traits you have that form a barrier to effective communication, since you may be unable to perceive these habits yourself. Next, find a supportive friend. Look over the results of Activities 5-1 and 5-2. What do the results say about the way you deliver your messages? Pick three items that you want to change. Make a plan with your friend on how you can address and overcome the chosen communication "flaws" and how your progress will be monitored and encouraged over the next three weeks.

What makes a good communicator?

The first answer that comes to mind would be to avoid doing all the things listed in Table 5.1 (or do their opposites). We will talk about four-quadrant communication and share examples later in this section. First, we want to look at the broader picture and basic concepts.

Good communication is more than fluency with words—it depends on our attitude. When we communicate, we do not merely pass on facts and know-how—the package includes feelings, values, hopes, and dreams, and our attitudes are expressed in our body language. Communication is easier when people have a common language, thinking preference, culture, and memories. During times of change, we cannot assume that we have many common bonds or that they operate as reliably as during "business as usual." Yet good communication is especially critical during times of change, when our success depends on our ability to "sell" our ideas and solutions. Our communication must not only be transmitted and received, it must be understood and acted upon.

Perhaps the biggest barrier to communication is the assumption that it has taken place.

Ron Meiss, communications consultant

Communication, reduced to the basics, involves a sender, a receiver, and a message, as diagrammed in Figure 5.1. As the sender, we must encode our message to attract maximum attention and generate the desired motivation. The receiver must have enough time and information to decode the message properly. Then the receiver in turn becomes the sender and transmits feedback so that both parties can verify that the message has been properly understood and will result in the desired action or change. Both the sender and the receiver must be aware that the message is affected by two sets of screens or filters as well as by direct interference. Filters are internal signals that can distort the message. Many different filters can be involved such as language, culture, values, bias, memory, previous experience, emotions, expectations, paradigms, time pressures, lack of speaking and listening skills, motives and agendas, attitudes, physical well-being, and brain dominance.

The interference affecting a message directly is often defined by a technical term—the signal-to-noise ratio. The S/N ratio is an analogy borrowed from the field of radio signals and has to do with interference to the message through influences in the physical environment surrounding the speaker and listener. It indicates how clearly a signal is coming

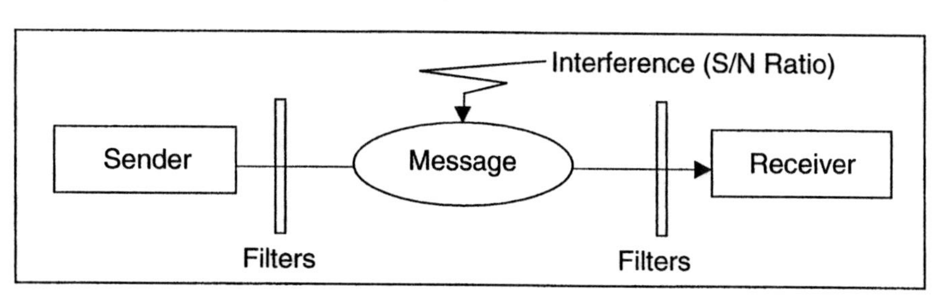

Figure 5.1
Factors affecting the transmission of a message.

through in an environment filled with competing signals, which are called "noise." Thus the clarity of a verbal message can be affected by the background noise in a room or other distractions (for example, a ringing telephone, a secretary entering the room, people talking, a passing siren, a thunderclap, music, or a blaring television set). Sometimes filters are used to screen out the noise, thereby improving the transmission of the message. Some people have a mindset that can ignore annoyances in their environment much easier than others. Strongly quadrant C persons are very sensitive to sensory (and especially auditory) stimuli.

What you are, stands over you the while, and thunders so that I cannot hear what you say.

Ralph Waldo Emerson

One-way communication from a speaker to a listener appears to be simple and easy. At first glance, taking the time for feedback seems to complicate matters. Why can't a boss simply tell employees what to do, either verbally or by memo? For routine tasks, this approach may be adequate, but in new situations, increased variability occurs in the way the message is understood as well as in the values and priorities of the different people involved. For this reason, two-way communication—even though it is slower and often messy—becomes especially important in times of major change. Teamwork is built on good communication, and for long-term benefits, these skills must be carefully nurtured.

In most circumstances—barring a very unpleasant or boring situation—people do not like change. If priorities and values must be changed in order to innovate, intense communication is necessary. But in any organization, communication across disciplines is unusually difficult because people's minds are not particularly eager to learn new jargon and techniques—something to remember when working as part of a cross-functional team. People may not know or want to admit that learning and growth are needed. They do not like to be perceived as ignorant and will be reluctant to ask questions. But as we have seen, sharing is an essential first step in the knowledge-creation cycle, and effective communication is the key to keep the knowledge-creation process spiraling.

What is good communication? When a group of 14-year-old students from Detroit inner-city schools brainstormed this question, they came up with the list of characteristics shown in Table 5.2.

Table 5.2 To Be a Good Communicator

- Get to know people before judging.
- Respect others; appreciate them as they are.
- Spend time together. Talk one-on-one.
- Take time to listen to each other's point of view.
- Don't try to be the leader (or the person in control) all the time.
- Learn how to have a "fair" fight.
- Be yourself; be comfortable; be open, kind, and caring.
- Be understanding and supportive.
- Have a positive attitude.
- Watch body language.
- Keep a sense of humor.
- Have communications umpires.

> ***Communication Ideas from Dale Carnegie:***
>
> - ***Remember people's names.***
> - ***Talk in terms of the other person's interests.***
> - ***Respect other people's opinions.***
> - ***Ask questions; don't give direct orders.***
> - ***Admit when you are wrong.***
> - ***Let others save face.***
> - ***Give heartfelt praise and honest appreciation.***
> - ***Make the other person feel important.***
> - ***Be courteous and encouraging. Smile.***

The bottom two items on the list are intriguing, don't you think? These students felt that their school environment and their lives would be much improved if they—and their teachers—were taught communication skills. The students came up with many of the same ideas that Dale Carnegie promoted in *How to Win Friends and Influence People* (see sidebar). Can you think of other ideas for improving communication? Since thinking about interpersonal relationships is a quadrant C ability, it is not surprising that this subject is neglected in our left-brain-biased education systems and by people in technical fields who have strong analytical thinking preferences. Research has shown that good communication lowers stress. Our minds work better and can think more creatively when we are not under stress. We can control how we react to the environment around us by maintaining a positive, caring attitude. We can learn and practice good communication skills, such as listening and giving thoughtful feedback.

Being a good listener

If we want the listener to hear us, we not only must speak loud enough to be heard, we must use the correct language. We must attract the listener's attention and clearly speak in terms of the listener's interests. If we want the message to be understood, we first must know something about the listener's thinking preferences, as well as the level of previous knowledge and cultural experiences. To get a response, we must be specific and invite a response.

A Harvard study in the 1970s found that 9 percent of communication time is used for writing, 16 percent for reading, 30 percent for speaking, and 45 percent for listening. We can speak at a speed of about 120 to 140 words a minute, yet we can hear as much as 600 words a minute (if we concentrate). A study at the University of California at Los Angeles found that 7 percent of a verbal message comes from words, 38 percent comes from tone, pitch, inflection, rate, and emphasis, and 48 percent of the message comes from body language. Thus listening involves more than just paying attention to the words of a message. Since we learn best when we ask questions and discover the answers for ourselves, how can we get our listeners to ask questions?

> ***When I listen,***
> ***I have the power.***
> ***When I speak,***
> ***I give it away.***
>
> *François Voltaire,*
> *French philosopher*

What makes a good listener? Table 5.3 summarizes some characteristics. These hints concentrate mostly on the things the listener can do to improve communication, both to ensure getting the message and to give an appropriate response. In Chapter 10, we will learn more about judgment and how to be constructive, not negative.

Table 5.3 Characteristics of Good Listeners

- They want to hear what others have to say.
- They want to help with the problem.
- They accept the feelings of others as genuine (but understand that feelings are transitory).
- They trust the other person to think and solve their own problems.
- They listen to understand and do not judge negatively.
- They do not "correct" the message or change the subject.
- They focus on the goal, not minor issues.
- They know that first impressions or appearance can be deceiving.
- They do not finish the speaker's sentences!
- They do not jump to conclusions.
- They do not prejudge: "I've heard this before; it's boring; it's too hard."
- They pay attention; they maintain eye contact (in Western culture only—elsewhere it may be considered rude); they smile if appropriate.
- They give feedback with appropriate body language to show they are listening.
- They ask questions.
- They can summarize the facts and meaning of what was said.
- They can pick up on the nonverbal message.
- They are not distracted by unconventional behavior or anger.
- They are in control of their own behavior and focus on solutions.
- They can give supportive feedback (as well as constructive criticism if asked).

Team Activity 5-3: Body Language

Investigate what body language is used in different cultures to indicate that the person is listening. How is nonverbal feedback given for agreement or disagreement? Try to interview people from at least three different cultural backgrounds: Native American, Asian (Chinese, Japanese, Malayan, Indian, Pakistani, Korean), Latin American, North African, Central African, South African, Middle Eastern, or Eastern European.

Communication and thinking profiles—examples

Bain dominance has a large impact on communication between people. In general, people with similar HBDI profiles and similar occupations find it easy to communicate, whereas people with diagonally opposite profiles (with different occupations and gender) find it the most difficult to communicate, unless they are strongly motivated to make it work.

1. Married couple

A young couple with an HBDI profile pattern very similar to that shown in Figure 3.15 (on page 76) had a difficult first year of making adjustments in marriage. Both were well-educated, with advanced college degrees and had lived independently as young adults. He was an electrical/computer engineer; she was an elementary school teacher. Things did not improve after the husband got his Ph.D. and took his first job. Having started their family, they were strongly committed to finding a solution to their continuing communications problems. After two years of counseling, it dawned on the husband that he really had to *talk* to his wife! At this point, their communication (and their marriage) improved noticeably. Both are making an effort to communicate in the mode of the other's strongest thinking preference, especially when discussing crucial issues. Ned Herrmann has found that about 85 percent of couples marry spouses with opposite thinking profiles. When they learn to communicate because they understand and value the differences, the marriage has a good chance for success; if the communications difficulties remain, the marriage does not last or is not happy. A similar dichotomy in the average HBDI profiles has been obeserved for two groups of professionals in academe: engineering faculty and career counselors.

2. Father-daughter conflicts

A quadrant A father (engineer) often bemoaned the lack of logical thinking in his quadrant C daughter; she in turn was often hurt by his "cold fish" approach, such as not sending flowers when she was in the hospital, or giving a birthday gift late and making sure she knew how much he spent on it. A quadrant B father who was estranged from his quadrant D daughter was relieved to find after he understood the HBDI model that she was not crazy nor did she things purposely to annoy him (in occupation, clothing, life style)—she was just different. Now it is OK for her to see a gorgeous view and for him to note the fly on the window.

Employees are expected to have greater ability to communicate and "sell" their own ideas—orally, electronically, and on paper—not only among fellow employees, but to suppliers and customers.

James Braham, senior editor, Machine Design

3. Team in industry

A work team at Ford Motor Company in advanced vehicle technology asked for an HBDI workshop. This team had worked together for some time so that the members knew each other well, thus providing ideal conditions to do a small experiment. The members were formed into homogeneous groups with the following average HBDI scores:

Group	A	B	C	D
The Reds	52	112	72	51
The Purples	101	97	30	56
The Blues	117	83	32	51
The Yellows	61	58	68	111
The Greens	90	83	43	79
The Oranges	65	82	74	78

The team members were seated according to their groups. They were asked to rate the degree of difficulty of communicating with each of the other groups as a whole, ranking them from 1 (easy to communicate) to 6 (very difficult to communicate). The results were very interesting. All groups with strong thinking preferences found it difficult to communicate with each other (rankings of 5 or 6), including the Purples and Blues who were quite similar but had a low score in quadrant C. It was surprising to find that the Yellows and Reds had difficulties, even though both had a secondary preference in quadrant C. All teams found it easy to communicate (rankings of 1 or 2) with the whole-brained Oranges, and the Yellows and Reds also found it easy to communicate with the Greens (all three groups had a preference in quadrant D). Two of the Oranges left the team soon after the workshop. It is recommended that this team try and find new members who are whole-brain thinkers.

4. Engineering consulting company

The operations manager of a small company in environmental engineering asked that her entire company be assessed with the HBDI and given a one-hour workshop. Almost everyone participated on a voluntary basis. The results showed a widely scattered but even distribution across the A and D quadrants in their profile "tilts" and three individuals clustered at a distance from the others in the B-quadrant portion of the graph. When one of the owners saw this result, he said: "Uh-oh, I guess we have to get rid of these"—pointing to the three lonely dots at the bottom. This again shows how people who are "different" are perceived and treated as outsiders. But the owner came to realize that these three were the people who actually kept the company running on a daily basis, who set schedules, billed the clients, paid the employees, and kept everything in order. He not only needed to keep them, he needed to give them more support and encouragement! Actually, the company overall did well and had recently gone through a successful merger—no surprise with the many engineers who had flexible quadrant D thinking preferences.

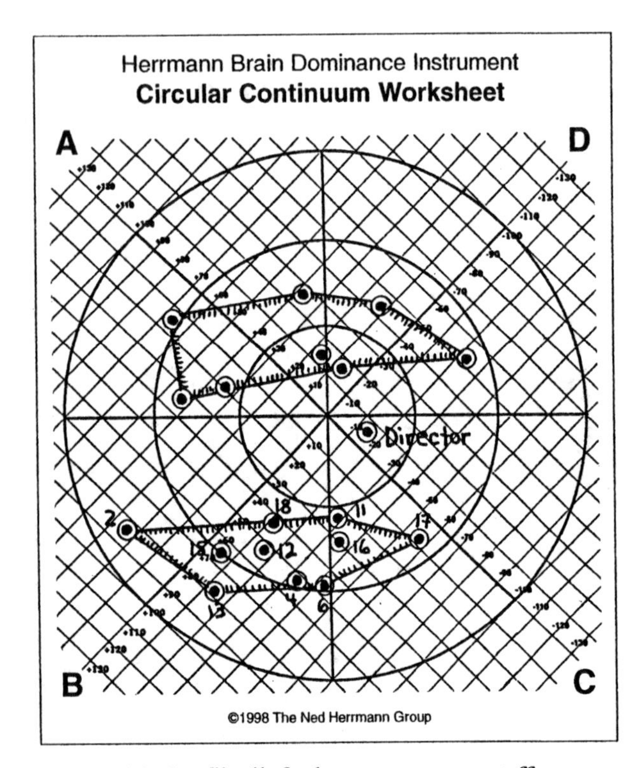

Figure 5.2 Profile tilt for human resources staff

5. Human resources staff

The staff of the benefits office in an educational institution scheduled a retreat to improve the office climate. The retreat included an HBDI workshop. Figure 5.2 shows that the staff was almost evenly divided into two "tribes"—a strongly cerebral and a strongly limbic group, with the director right in the middle. When asked if she spent a lot of time mediating between her staff, she laughingly said, "Yes, eight hours a day, five days a week." Now, with new people on the staff, she wants to schedule another HBDI workshop.

6. Quadrant B manager, quadrant D staff

A manager of a group of advertising people in Toronto faced the problem of how to get his D-quadrant staff to file travel reports and expense forms on time (quadrant D thinkers "hate" paperwork). The group was very competitive, so he began offering a monthly prize to those who performed best in timeliness and completeness. This worked for a while, until the group became bored. The manager is now looking for new ideas.

Two practical communication tools

Before we present detailed information and sample formats for all types of technical design communication, we want to summarize two communications tools: a model for negotiating win-win outcomes to a conflict, and practical hints on "how to get your point across in 30 seconds." Both tools can be used to improve communication not just within a team but also in the team's interaction with others.

Let us never negotiate out of fear, but let us never fear to negotiate.

John F. Kennedy, 1961 Inaugural Address

Negotiating a win-win outcome

Often, our interaction with people is more complicated than sending and receiving messages, especially when a conflict exists that needs to be resolved. In such a situation, we have to negotiate to come up with a solution. Negotiation is done all the time. Negotiating does not have to be an adversarial battle but can be a productive problem-solving process, as demonstrated by Roger Fisher and William Ury of the Harvard Negotiation Project in their book, *Getting to Yes—Negotiating Agreement Without Giving In.* According to the Ury-Fisher model, there are basically three approaches to negotiation: soft negotiation, hard negotiation, and principled negotiation. Here is a brief summary:

Soft negotiation. One person wants to avoid personal conflict and makes concessions quickly to reach an agreement, but as a result, this person may eventually feel exploited and become bitter. The participants are friends or family members, or they may have an employer-employee relationship. The balance of power is unequal; one of the parties has a much larger investment or deeper commitment to maintaining the relationship than the other. This person is trusting and flexible; he or she will make offers and change positions to resolve the conflict, and he or she will yield to pressure. This person will reveal the bottom line and will accept losses as the price of peace and agreement. The individual's behavior may be guided by cultural pressures or significant personal values. This person is most likely a strong quadrant C thinker who wants to "win" the negotiation through accommodation.

Hard negotiation. The situation is perceived as a contest of wills. Both parties want to win, at almost any cost. This process is exhausting and can cause serious harm to personal relationships. The participants are adversaries on an equal footing; they demand concessions as the price of maintaining their relationship. They are inflexible; they distrust each other; they make threats, apply pressure, and mislead as to the bottom line. They demand one-sided gains as the price of agreement—compromise is out of the question. They dig into their positions and thus find it very difficult to yield and thereby "lose face" or status. This mindset, too, may be strongly shaped by cultural influences; these people most likely are quadrant B thinkers who try to "win" the negotiation through intimidation and control. Union negotiations with management traditionally have followed this pattern.

Principled negotiation. Issues are decided on merit. Both parties work toward a superior outcome that will benefit everyone concerned. People listen to each other and try to understand the other side's position. If conflicting interests persist, the solution is worked out based on fair standards and goals. The participants are problem solvers; they seek an optimum option that will be mutually agreeable, efficient, and amicable. They are able to separate the issues from personal feelings. Through creative thinking, they invent options for mutual gain, and the final decisions are based on agreed-upon objective criteria. The participants are open to reason and will yield to principles; they use whole-brain thinking and consider values, relationships, the context, the long view, the facts, as well as the mutual benefits and risks in arriving at the best solution through cooperation as equals. Developing win-win outcomes requires flexibility and a positive attitude. Benefits are maximized for both parties; self-respect is maintained, and relationships are strengthened. Also, in today's rapidly changing world, negotiation is a continuing process of working for improvement.

In collaborative decision making (or principled negotiation), the negotiators are able to put themselves into the other side's shoes.

William Ury

A successful outcome in negotiation depends on good communication skills. However, we can often observe that serious errors in communicating are committed by negotiators, be it in labor and management relations or even in the international arena. Can you think of examples for the following situations?

1. Negotiators are not talking to each other or are not understood. Instead, they are playing to the gallery or constituents.
2. Negotiators are not listening; they are not paying attention to what is being said because they are thinking of what to say next.
3. Negotiators are speaking different languages—a situation that lends itself to misunderstanding and misinterpretation.

When a translator is involved in negotiations, special care must be paid to the different cultural meanings that can be attached to words after they are translated. It always amazes us how even common words can have a considerable difference in meaning in another language. Also people can speak different "languages" even if they use the some tongue—if their cultural background and experiences are different, or as we have seen, if they have different strong thinking preferences. Table 5.4 gives some guidelines for successful communication and negotiation. Ultimate success is not defined in terms of getting your way but in terms of building partnerships and effective teamwork. At the close of each negotiation, reflect on what you have learned. This will empower you to become an even more successful negotiator.

Table 5.4 Guidelines for Negotiation and Communication

1. Listen actively and acknowledge what is being said. Provide feedback from the point of view of the other person or group by stating their position in positive terms.
2. Speak to be understood. Look at the others as partners for solving a joint problem. The more important the decision, the fewer people should be involved. Two is best for a "summit" meeting.
3. Don't condemn. Describe the problem in terms of personal impact. "We feel discriminated against" is better than "you are a racist or oppressor." Try not to provoke a defensive reaction or anger; instead, stick to the objectives.
4. Take the long-term view and build relationships. It is possible to "win the skirmish and lose the war!"
5. Follow creative problem solving: do not judge too soon, look for options and alternatives, do not assume a fixed pie (either/or) concept or act in pure self-interest. Brainstorm—alone, with the other party, or with other interested people—then do a creative evaluation to find the best options. Develop a list of objective criteria.
6. If you are negotiating from a weak position, have a Plan B. This way, you will not be tempted or forced into accepting a plan that will put you too much at a disadvantage.
7. What if the other party won't play and follow the rules of principled negotiation? In this case, do not attack the opposing position—look behind it. Do not defend your ideas or take the attack personally. Instead, invite criticism and advice. Listen and agree as much as possible. Restate an attack on you as an attack on the problem. Reframe the opposing position by using what-if questions. Build on the proposed idea; make it easy for the other party to gain honor or a good way out. Discuss the cost of drawn-out disagreement. Most of all, treat everyone with respect.

10-Minute Team Activity 5-4: Negotiation

In a group of three, analyze a current negotiation, for example, in labor and management, or on the international scene. Identify the type of negotiation being used. Cite supporting evidence. Discuss how creative thinking could be introduced or strengthened in the situation. How could this affect the outcome? Or describe a case in which you were able to mediate a dispute. What strategies helped you to be successful?

How to get your point across in 30 seconds

This technique is relevant in our busy times. We are living in the information age and are being bombarded with messages from everywhere. Think back—by how much would you say your junk mail has increased over the last two or three years? If you are on a computer network and can't read your e-mail for a week, how many messages will have piled up? Just to cope and preserve our sanity, we are learning to "tune out." In this kind of environment, where our messages have to compete with a lot of information "noise," how can we make sure that we are being heard? How can we become efficient communicators without wasting our efforts?

Milo O. Frank, a business communications consultant, has written a poignant book on *How to Get Your Point Across in 30 Seconds—or Less*. We have found this approach very useful and would like to share some of the important concepts of this technique which will enable you to

- ❑ Focus your thinking, writing, and speaking.
- ❑ Be logical and concise; have better meetings and interviews.
- ❑ Improve listening and keep conversations on track.
- ❑ Make better presentations; be more successful in "selling" ideas.
- ❑ Use questions and answers to make a point more effectively.
- ❑ Have increased self-confidence and achieve your objectives.

This approach is especially helpful when you want a specific response from people—when you are asking them to do something for you, or when you want them to react and get involved on some issues. Why 30 seconds or less? Why not one minute or two, or even five? We would like to submit the reasons listed in Table 5.5 for your consideration.

Table 5.5 Why Messages Should Be 30 Seconds or Less

- Memos and letters of request are too long—just check over your junk mail or phone solicitations.
- The attention span of the average person is 30 seconds.
- Doctors listen to their patients for an average of only 19 seconds before they start making a diagnosis and proceed with the physical exam (according to research done at Michigan State University).
- You are allowed to add an explanation of 100 words to your credit report.
- E-mail messages are more effective when they are sized to fit on a computer screen without scrolling.
- TV commercials do a good job of getting their message across in 30 seconds (or even 15 seconds for example in a Super Bowl half spot).
- TV news "sound bites" are 30 seconds long or they do not get air time. Reporters spend about 30 seconds introducing the subject. Then the topic or sound bite is shown, followed by a summary not exceeding 30 seconds.
- Most importantly, if you can't say it in 30 seconds, you probably are not thinking about your message clearly. You may need more time to present supplementary information (if asked), but the main thrust of your message should be very concise. President Abraham Lincoln's Gettysburg address and President George Washington's inaugural speech are brief but extremely effective messages.

The following discussion will give you the steps for preparing a 30-second message. Preparing such a message takes much thinking and creative problem solving and can easily take an hour or more, especially for beginners. Thirty-second messages can be verbal or written—they can be telephone requests and messages left with answering machines or secretaries; memos, letters, fax messages, and thank-you notes; abstract for scholarly papers and work proposals; formal presentations at meetings; interviews; a request or sale solicitation; social situations with superiors, chance meetings, and giving toasts. The 30 seconds in an elevator may be all the time you have to present a creative idea to your company's president. And you may only have the 30 seconds of a commercial break on TV to present an urgent request to a family member.

Preparation

As you prepare your message, you must determine your objective, your audience, and your strategy.

Objective: What do you want to achieve? Why? You need to have a single, clear-cut, specific objective.

Audience: Who can get you what you want? Know what your audience is going to want from you.

Approach: How will you get what you want? Brainstorm different ideas, then select the one that meets the objective and audience best—in form as well as content. Ask yourself: What's the basis of my game plan? What is the heart of my message? What is the single best statement that will lead to what I want? How will this statement relate to the needs and interests of the audience? Then select the most appropriate form: phone call, memo, newspaper ad, formal presentation, etc. The cartoons in Figures 5.3 and 5.4 illustrate the process.

Figure 5.3 First approach.

Blondie (9/14/91), reprinted with special permission of King Features Syndicate.

Figure 5.4 Second approach.

Blondie (9/15/91), reprinted with special permission of King Features Syndicate.

What is the difference between the first and second approach? Why is the second approach successful? Do you suppose the list of benefits directed at the "audience" has something to do with it?

Message

After you have settled on the approach you want to use, you need to work on the three parts of the message: hook, subject, and close (or metaphorically spoken of as "hook, line, and sinker").

Hook: To get attention, state the hook in the form of a question. You may use humor (at your own expense only) or a visual aid. The hook should be a bridge connecting the audience to what you want. If you have a very brief message, the entire message can be the hook.

Subject: Answer who, what, where, when, why, and how as they relate directly to your explicit or hidden objective. Does the message correspond with your approach? Is it relevant to your audience?

Close: This is the bottom line. Be forceful or subtle in asking for what you want, depending on your audience and how well you know them. Demand a specific action within a stated time frame, or ask for a reaction through the power of suggestion. The first three paragraphs in this book's preface are a 30-second message with an indirect close—we are asking students and design engineers to buy and study this book to learn to be effective problem solvers. In a widely competitive environment, quality (even in communication) has to meet ever-expanding standards and expectations. Use the checklist in Table 5.6 to create an effective message that will be remembered.

Table 5.6 Effective Communications Checklist

Based on the Herrmann Whole Brain Model, ©1998 The Ned Herrmann Group

Quadrant A—Clarity __ Do you have concise facts? __ Are you providing quantitative data? __ Will the audience have the same understanding of your words as you do? __ Are the arguments or analysis supporting your position logical?	**Quadrant D—Imagery** __ Are you painting a creative word picture or metaphor to be remembered easily? __ Are you using a colorful, imaginative visual aid? __ Are you providing the context, a look at the future, or the "big picture"? __ Are the concepts sound or clear? __ Are you addressing the problem of change?
Quadrant B—Action Plan __ Does your request ask for well-planned, orderly implementation? __ Are you providing the necessary details? __ Is your message well-organized, neat, and in an appropriate format? __ Do you know when to stop?	**Quadrant C—Emotional Appeal** __ Are you reaching the heart of the audience by sharing emotions? __ Are you relating personal failures, experiences, and examples? __ Are you user-friendly and building relationships?

Presentation

In oral presentations, style and appearance, "acting," and mode of speaking are important since they help transmit the meaning of the message.

Style and appearance: Give some thought to your personal style and image. Monitor your body language. Practice delivering your message in front of friends who can critique you in a supportive way. Better yet, have someone videotape your presentation, then use critical thinking to evaluate your performance. Examine your facial expressions, eye contact, posture, gestures, and tone of voice. Check your appearance—do you know what kind of clothes make you look your best? Please yourself, but realize that in some situations it does matter what others think. Being considerate of others has preference over your own tastes. Wear clothing that will draw the audience's attention to what you are saying, not to itself. Good taste in clothes shows that you care about other people and about yourself. Be clean and well-groomed. Observe the rules of etiquette—social interactions are more comfortable when everyone knows what is acceptable behavior.

Your appearance and style speak louder than words.

Acting: Are you conveying a positive attitude? If you "act" friendly, this will make you feel friendly. Smile. Focus on different people in the audience while you speak. Do not read off a script or memorize your speech. You may feel that you are being asked to pretend, to do play-acting. To some degree, that's what good communicators do. A prime example is former U.S. President Ronald Reagan.

Mode of speaking: It helps to show surprise, puzzlement, or concern in your facial expression and voice as you speak. Do not use distracting body language (like pulling on your fingers or jingling coins in your pocket). If you speak in a monotone, learn to modulate your voice. Use strategic pauses. Practice breathing and relaxation techniques prior to the start to reduce your stress level and thus have your voice sound more natural. Start on time. Respond directly to questions from the audience, but don't get carried away. Finishing on time is much appreciated.

Example of a 30-second message:

We would like to share with you one final, important thought about communication. What do we do when we are operating a piece of equipment and suddenly find that we are in trouble because something is not working right? The first thing we should do is go and read the manufacturer's instructions. The best teaching on communication and relationships is given by Jesus Christ in the Sermon on the Mount (as recorded in the Gospel of Matthew, Chapters 5 through 7). It all comes together in the Golden Rule:

Treat others as you would have them treat you.

Communication in engineering design

Engineering design is the *communication* of a set of rational decisions obtained with creative problem solving for accomplishing certain stated objectives within prescribed constraints.

Rather than being an afterthought in design, communication is integral to the design process, as will be shown in detail in Part 3 of this book. The effectiveness of a design can be impaired by faulty communication. Engineers have long used drawings or schematic diagrams to convey precise information since relationships among mating parts or components can be portrayed more efficiently in visual rather than verbal terms. Rarely is a drawing alone sufficient, however. Even if it is possible to communicate the selected design alternatives accurately in a drawing, the rationale, or justification, for selecting each alternative almost always requires verbal amplification.

Is it really necessary to include the rationales for decisions in design communication? Certainly they are frequently omitted in the design package, whether intentionally or by oversight. Sometimes a "bare-bones" drawing is all that is provided. The design may lose its persuasive power and ability to sell itself, when the design rationales are omitted, particularly in the early stages of design, when major decisions are not yet "locked-in." After all, if the designer has done a good job in making design decisions and has selected the best alternative after a thorough analysis of performance of all options against the objectives, the rationale is simply a summary of the results of this process. Omitting this information might invite needless questioning about the decision and perhaps repetition of the selection evaluation process. Cases of "re-inventing the wheel" can sometimes be traced to lack of information on why the previous designer made particular choices. Faced with such uncertainty, engineers may feel compelled to re-examine the alternatives and go through the design process all over again.

When the competing alternatives are closely matched, there may be a tendency to omit rationales for decisions to avoid undermining confidence in the decision. This would be a mistake, since the recipient of the design would lose valuable information that may allow or dissuade a design change later on. More often than not, the designer may *gain* confidence and favor by bending over backward to describe the competing alternatives at least as well as the one selected.

All designs must be sold; most must be sold many times. At virtually every stage of design, designers must present their work for review and approval to go forward. It is simply not enough to present an unelaborated drawing and expect it to sell itself. A concise and easy to follow summary of the evolution of the design which includes the decision rationale is essential. Therefore we will take effective design to mean *effective communication of the design decisions along with their rationale(s).*

Criteria for effective technical communication

Clear thinking becomes clear writing.

William Zinsser, On Writing Well, *1994*

Engineers are known to value communications which get the point across quickly and unambiguously. But, in addition to clarity and concision, several other criteria are important, such as accuracy, precision, thoroughness, organization, audience focus, credibility, and timeliness. Each of these is summarized in the following paragraphs.

Clarity. All engineering information must be unambiguous and leave no doubt about the intent of the communication. Well-prepared engineering drawings are excellent tools to present clear descriptions of design decisions. In addition to the details of any design or analysis, the context of the information is also important so it can be used effectively with the proper perspective. This means that background information on a design project, the sequence of design steps, or information on competitive designs may be as important as the results of design decisions.

Concision. Engineering communications should avoid extraneous details that do not contribute to thoroughness or clarity. This means that interesting sidelights, or anecdotes, or personal observations generally have no place in engineering communications (unless the purpose is to elaborate on the history or background of a project). The quality of an engineering communication is as much measured by what is left out as by what is included. Too much verbiage and too many optional details are distracting to the purpose of engineering information. Concision also refers to condensing the material so that it is presented in compact form such as a table or graph. Preparation of clear and unambiguous graphical information is an art which can be learned. Edward Tufte, a Yale University professor, has written several excellent books on "information design" or using graphical information effectively (see Ref. 5.12).

Accuracy and precision. Accuracy refers to providing correct information, with precision to the level of uncertainty in the information. For example, a certain light bulb may be designated as a "100 watt" bulb, which is an accurate nominal descriptor for that class of bulbs, but the level of precision may really be 100 watts plus or minus 2 watts. Here the precision would be 2 parts in 100. Another way this is expressed in engineering nomenclature is by the number of significant figures in a result, such as 100.0 watts. Such a designation typically implies that the figure is precise to within one part in the last digit, here one part in 1000 (0.1/100). It is considered poor form in engineering to list figures with more significant digits than are merited by the supporting data. A device may have a readout of five digits, but be accurate only to the first three, depending on the range of the instrument under the particular conditions of use. Thus a calculator result of 26.75642 horsepower should be written as 26.8 horsepower if the dynamometer accuracy is plus or minus 0.1 horsepower in the range being used.

The ability to communicate orally and in writing, mathematically, and graphically is the key to success for practicing engineers.

Thoroughness and logical organization. An engineering report should present all information needed in a fashion that is easy to follow. The reader should not have to guess how the information is arranged or in what order. A table of contents, list of illustrations, and list of appendices help provide the organizational logic of reports. Report formats may differ by engineering discipline or by corporate policy. Making the logical organization clear to the reader is more important than any particular report format. Later in this chapter we discuss workable formats for design project proposals, progress reports, and final reports.

Audience focus. Focusing on the intended user of the information provides the yardstick by which we determine what is appropriate for inclusion in a drawing, a graph, a table, or a report. It also determines to some extent the medium and the vocabulary we use. For example, the drawing of a product for the marketing department would look quite different than the drawing for production shop floor personnel. Verbal presentations, too, call for different approaches with different audiences—a concept proposal presented to investment bankers would call for a different approach, style, vocabulary, and dress than a presentation on the same product to a group of computer programmers.

Credibility. Before investment bankers, managers, or customers "buy into" a design, they want assurance that the design is well done and that the product will do what it is supposed to do. They want to know that the decisions made were the best possible from a wide range of alternatives, and that the concepts have been tested. Achieving this credibility is rarely simple. First, designers have to do all the homework—the research, analysis, synthesis, and testing of ideas. Then the design communication has to convey that this work has been done and done well. Many design decisions are not obvious when looking at a product or drawing; thus the most credible designs invariably involve creative ways of showing their features. Presenting the best points of a design may be a marketing ploy; but engineering communication goes one step further. We must point out any areas of limitation or weakness. This canon of ethics applies to engineers the way prescription drugs carry warnings on side effects. For example, engineering designers make sure users know the limitations on the safe loads carried on elevators, engine speeds in automobiles, power limits on loudspeakers, and safe g-loads on aircraft wings. Paradoxically, including such design product limitations (or even downright weaknesses) in engineering design communications usually serves to *increase* the credibility of the designer.

Responsible engineers are expected to present a balanced, objective appraisal of their work, because they hold a public trust. When they do so, the credibility of their designs is enhanced.

Timeliness. Engineers rarely work alone. Most of the work is done in teams within organizations. Everyone must work together to meet the project schedule. Sometimes the time available is compressed due to commitments to customers or competitive pressures. These pressures are felt in design communications as well. A late testing report may delay release of a product and cause loss in market share even though the product has better performance than the competition. A late set of shop drawings may delay the production schedule and cause the company to miss the peak selling period. The timing of communications should be planned as carefully as the other deliverables in the design process.

What is written without effort is in general read without pleasure.

Samuel Johnson

Formats for written design communication

As mentioned earlier, no standardized formats exist that fit every company's set of design communications. We are presenting generic formats that are able to illustrate well accepted principles for reports, summaries, tables, graphics, and verbal presentations.

Reports

The following example formats for a set of design communications could serve as the reporting basis for an entire engineering *design project.* The designations DP-1, DP-2, etc., refer to documents which are part of the twelve-step design process discussed in Chapter 14. Some of these documents are only one or two pages, and some incorporate within them other documents on the list. This set of documents can serve as assignments arranged such that students will be able to consolidate results from each step into successive steps without rewriting everything. A sample of each format is shown in detail in Chapter 17.

■ **DP-1 Project Concept Statement** — This paragraph presents the project title, purpose, goals, sponsor, other stakeholders, and the intended users of the design product, done in the earliest stages of design.

■ **DP-3 Survey of User Needs** — This is a reporting form for a potential user eliciting information on preferences, opinions on problems with present designs, possibility of use for a new product, or desired features in a new product. Producing and administering effective survey instruments is a complex subject fraught with possibilities of error, both in their design and interpretation of results. Piloting new surveys on small samples of the population can help avoid some of these errors. Even deciding on the appropriate population to be surveyed merits considerable thought, especially if stratification of the population, whether intended or not, may bias conclusions derived from the sample. The choice of an appropriate sample size depends on the level of uncertainty that will satisfy you. For sampling errors of 5%, appropriate samples would be 80, 278, 370, and 384 for total populations of 100, 1000, 10,000, and essentially infinite, respectively. (See Reference 5.11 for a discussion of designing and administering surveys, including sample sizes.)

In engineering design, the designer uses three types of knowledge: knowledge to generate ideas, knowledge to evaluate ideas, and knowledge to structure the design process.

David G. Ullman,
The Mechanical Design Process, *1992*

■ **DP-5 Design Problem Analysis** — This short report presents the entire context of a design problem, together with supporting tables of design objectives and constraints, written so that unnecessary constraints or assumed solutions are avoided. The analysis culminates in a short directive sentence guiding the design effort. In Chapter 7, these are referred to as the briefing document and the problem definition statement.

■ **DP–6 Design Project Plan** — This graphic or tabular presentation of tasks required to complete a design project is arranged to show sequential relationships, the personnel responsible for each task, and the work time required to complete each task (see Chapter 15 for details).

■ **DP-6A Design Project Proposal** — This report summarizes the case for commencing a design project, including project concept statement, design problem analysis, and design project plan.

■ **DP-8A Design Decisions** — This is an annotated list of decisions constituting the design, including the alternatives considered for each decision and the rationale for each (such as research, calculations, and analyses). This material forms the essence and bulk of the design. Also see Chapter 16 for economic analysis guidelines and templates.

■ **DP-8 Design Project Progress Report** — This report summarizes the progress of the design project to date, with special emphasis on progress toward making the decisions constituting the design. This report always includes an up-to-date project plan, showing how the project will be completed in the time remaining.

■ **DP-10 Test Plan** — This describes the purpose, objectives, and steps planned to test aspects of an engineering design. The test could be an evaluation by another, independent, analytical method, a computer simulation, or a test of a prototype. In some cases it could also be an evaluation by experts or potential users, in which cases the plan would include specific questions asked of the expert reviewers.

■ **DP-11 Evaluation Results Report (Report of Design Review)** — This report describes the test and evaluation of a design, including the test plan, the results, conclusions, and recommendations for further design iterations. As a design review, this report would include an assessment of the success of the design in meeting the objectives (derived from knowledge of the needs of the potential users). Such a report could be written at any stage of the design project, using best estimates of the degree to which the design meets performance objectives.

When you follow a standard format, you have more time to concentrate on the content and quality of the message.

■ **DP-12 Final Design Project Report** — This is a comprehensive report describing a design project from initiation through evaluation. It includes the project concept statement DP-1, design problem statement DP-2, design project plan DP-6, design decisions DP-9, and final design review DP-11.

Although this documentation process appears to be mostly one-way, to inform, use it as an opportunity to get feedback.

Then learn from it. Continuously improve the current design project as well as the process for doing future projects.

Summaries

Two important summaries are the executive summary and the design concept descriptions (see Chapter 17 for sample formats).

■ **DP-6B Executive Summary** — An executive summary is similar to an abstract, except that it is never simply a summary of report topics, as are some abstracts. The executive summary presents a condensed version of the essentials of an entire report, usually within one page. Numbered here to be a part of the design proposal report, it is also included in progress reports (DP-8) and final reports (DP-12).

■ **DP-7B Design Concept Descriptions** — These are brief summaries of the main features of the design alternatives used in the Pugh matrix evaluation (both in Phase I and Phase II).

Tables

Tables can be very effective means of delivering information accurately and concisely. If the intent is to show a trend or a relationship between two or more variables, however, a plot or bar chart would be a better tool. Titles for tables are typically placed at the top of the table—titles for figures are usually placed below the figure. The title should refer to the values within the body of the table, not to any column or row headings. Formats for the following example tables appear in Chapter 17.

■ **DP-2 Table of Design Constraints** — This table lists the constraints imposed on the design by the design sponsor (not necessarily the user), or by engineering codes, or by applications environments, or by competitive pressures. The table includes a way to measure each constraint (an operational definition) and a quantitative limit or range for each.

■ **DP-4 Table of Design Objectives** — This table lists the performance objectives of the design that typically come from surveys of potential users, marketing experience, or a benchmarking analysis (see Chapter 7 for a description of techniques used in industry). Similar to design constraints, performance objectives must be measured by an operational definition. They are typically expressed as some desirable attributes which are to be maximized (or minimized in the case of a negative attribute). Target values for each objective help assess the gap between present design solutions and the new product. This in turn helps to identify the technologies which might be able to do the job.

■ **DP-8C Bill of Material** — Bills of material are tabular summaries of component parts of assemblies, showing the quantity and complete specification for each component. Frequently they are included as part of an assembly drawing, usually as a table along the right lower side above the drawing title block.

Graphics

Most engineers prefer to use graphics to convey information whenever possible to take advantage of their powerful abilities to portray relationships between mating parts or variables and to detect and illustrate trends. Indeed, engineering drawings are seen as almost synonymous with design. While we cannot provide a complete review of engineering graphics, we present examples of five types of drawings typically included in an engineering design report. Students need to be aware that with high-tech software (such as solid modeling) and the rapid advances being made, the client/customer may require a design to be submitted and transmitted electronically for review, checking, and manufacturing.

The distinction between engineering as we understand it today (sequential product and manufacturing engineerig) and simultaneous engineering is that communication has to be simultaneous, not sequential nor sporadic. Communication is the key to improved relationships and performance at all levels throughout the organizations.

K. Clark and T. Fujimoto
Product Development Performance, *1991.*

■ **DP-7A Concept Drawings** — These information-packed drawings are creative presentations of the design features of various approaches (concepts) for meeting the needs of the users of the product. Several such drawings would typically be done early in the design process for comparison of competing concepts (Pugh matrix Phase I). Liberal annotation of concept drawings helps bring out the features of the design.

■ **DP -7C More Detailed Concept Drawings** — These are similar to the earlier concept drawings (DP-7A) but show more details of the optimized and synthesized design options from Phase II of the Pugh matrix.

■ **DP-8B Assembly Drawings** — A complete assembly drawing shows all of the components of the assembly in their proper positions with respect to each other. Usually such drawings identify each component with call-out numbers (numbered "lollipops") which are summarized in a bill of material (DP-8C) above the drawing title block.

■ **DP-9 Detail Drawings** — At the production (or tolerance design) stage of design, each component of an assembly is drawn on a separate sheet to facilitate parts manufacture. These detail drawings are named and numbered consistently with the assembly drawings.

■ **DP-12B Sales Drawings** — Once the design is essentially complete, drawings showing the design features which appeal to users are crafted for use in marketing. These are renderings in which design details are not as important as the functional characteristics seen by the user, such as appearance and operation, although sometimes sales drawings show details of innovative features as well.

Plots. Most spreadsheets (Excel or Lotus 1-2-3, for example) offer a wide variety of plot options which enable the designer to portray relationships between variables, typically in the system or production design stages. While it is easy to get a plot from such programs, making a truly effective plot requires extra work and attention to detail. Check to make sure that your plots have the attributes listed in Table 5.7.

Table 5.7 Checklist for Producing a Quality Plot

- ❑ A plot title beneath the plot, usually referring to the y-axis variable as a function of the x–axis variable.
- ❑ Titles on each axis.
- ❑ Numerical annotations on each axis.
- ❑ Unambiguous magnification factors (say, in thousands of tons rather than tons x 10^3).
- ❑ Units on each axis (feet, grams, psia, joules, etc.).
- ❑ Multiple plots identified with legend or annotation on each.
- ❑ Axis scales which are easy for the reader to follow (not necessarily easiest for the writer to plot or what comes out of the spreadsheet automatically).
- ❑ Axis scales which cause the plots to fill most of the plot space.
- ❑ For a series of plots which are to be compared with each other, use the same axis scales for each.

Bar graphs. Bar graphs can be used effectively to present discrete data, such as sales by year or defects by shift. Again, spreadsheet programs provide a wide variety of useful graphing tools, but these rarely produce the graph wanted on the first attempt. In general, *avoid use of pie charts*, since it is very difficult for readers to compare relative sizes of slices. Instead, use vertical or horizontal bar charts. Check to make sure that your graphs have the attributes listed in Table 5.8.

Table 5.8 Checklist for Producing a Quality Bar Graph

- ❑ A graph title beneath the graph, usually referring to the bar length variable as a function of the discrete variable, such as "sales by year."
- ❑ Titles on each axis.
- ❑ Numerical annotations on the bar-length axis.
- ❑ Separate identifiers for each bar on the discrete variable axis.
- ❑ Horizontal bars when the bar identifiers are lengthy for easier reading.
- ❑ Units on the bar-length axis ($, hours, feet, etc.).
- ❑ Avoid three-dimensional bars (unless the thickness of the bar has a physical meaning) since they make comparisons of the bar lengths difficult.
- ❑ Bar length scales which are easy for the reader to follow (not necessarily easiest for the writer to plot or what comes out of the spreadsheet automatically).
- ❑ Unambiguous bar length scale magnification factors (say, in megapascals rather than pascals x 10^6).
- ❑ Bar length scales which cause the longest bar to extend nearly the length or width of the plot space.
- ❑ For a series of graphs to be compared with each other, use the same bar length scale for each.
- ❑ If the bars have no natural progression, such as successive years, arrange bars in order of bar length to form a Pareto chart, useful in setting priorities (see Appendix C).

Oral technical presentations

Effective oral presentations are driven by two primary considerations:

- Subject matter or content chosen with a focus on audience needs.
- Presentation structure designed to optimize understanding and retention of the information.

A brief discussion of these two consideration is supplemented by formats for three oral design project presentations.

Focus on your audience. Just as in real estate where the three most important factors are location, location, and location, in oral presentations the top three are audience, audience, and audience. The speaker should clearly understand the purpose of the communication with the audience. That audience may range from one or two people (perhaps a decision maker such as a chief engineer and product manager) to several hundred people for a general technical presentation for educational purposes. The needs and expectations of the audience should direct both the content and structure of the presentation. In planning your presentation, ask yourself questions such as those listed in Table 5.9.

Table 5.9 Questions for Gauging the Needs of an Audience

- What is the overall purpose of the presentation? Is it to inform, convince, educate, seek dialogue, or prepare the audience to make decisions?
- What use is the audience likely to make of the information in the presentation?
- What decisions hinge on the information?
- What particular information does the audience need?
- What type of relationship do you have with the audience in general or with particular members?
- How much time has the audience made available to hear the presentation?
- What are audience expectations as to media, format, and structure?
- What are the backgrounds of the audience members and their level of understanding of any necessary technical information?
- What actions do you want the audience to take?

The more diverse the audience, the more difficult the planning process, due to different levels of understanding and differing needs for using the information. Sometimes it becomes necessary to target particular members of the audience (typically decision makers) at the risk of leaving some of the needs of other audience members unmet.

Structure your presentation for understanding and retention. Oral technical presentations are much more than condensations of written documents. An effective oral presentation relies on the attention and memory mechanisms operating when people yield the structure of information gathering to the speaker, as outlined in Table 5.10. When reviewing a written document, readers may refer back and forth among the

Table 5.10 How to Optimize Audience Understanding and Retention of a Message

Your first job is to **command attention,** to provide a hook to draw people away from their preoccupations. This hook can establish a connection with the audience by verbalizing your knowledge of the audience:

- State the purpose of the presentation in audience terms.
- Acknowledge the needs and expectations of the audience.

As soon as you begin, your listeners start to form an impression of your competence and credibility, based more on **nonverbal cues** than the words used. Thus, you must:

- Stand erect, with a bearing commanding respect.
- Seek appropriate eye contact with people in the audience.
- Speak clearly and with calm authority from confidence in the subject matter.
- Project energy, enthusiasm, and competence.

The **memory capacity** of typical listeners is limited to 3 to 5 main points, and then only if the points are repeated and reinforced. Thus, you must:

- Preview the 3 to 5 main points to have the listeners anticipating them as they are presented.
- Continuously tie the points to the structure of the presentation and to each other to provide as many links as possible in the listeners' mind to reinforce their retention.
- Provide written summaries (handouts) if you have many sub-points. Do not overwhelm your audience with too much detail; this weakens your presentation's strength to provide emphasis.
- Summarize the main points at the end of the presentation to give an additional link and closure.
- Request specific actions desired of members of the audience (such as approval to proceed, funding, review comments, or suggestions).

People prefer to use their **dominant thinking** and learning styles to assimilate information: the personal touch of verbal description for quadrant C; orderly, sequential tables or lists for quadrant B; the concision of a mathematical model for quadrant A, and the summarizing power of a graph or a chart for quadrant D. Thus you must:

- Take advantage of as many of these mechanisms as practical for each point, to appeal to a variety of learning styles and to reinforce the learning for each listener through multiple mechanisms and links.
- Point out the intended inference or conclusion to be derived from each table, plot, or chart pre•sented. Coach the audience in getting the message!
- Use bullet slides or transparencies as a visual aid to help the audience digest the information. Be careful not to write too much at each bullet. If you are tempted to read the slide to the audience (a deadly bore), this is a sure sign that you have put too many words at each bullet.

In this information age, everyone is **too busy for overtime presentations.** If the presentation takes too long, it is likely to be cut short. This will omit the valuable concluding sections and clarifying questions, leaving an impression of disorganization and incompetence. An overtime presentation is generally a failure on many counts. You risk having the target audience leave before they get the information they need to make decisions, defeating the entire purpose of your presentation. Therefore, you must:

- Plan (and rehearse) your presentation to finish in the allotted time.
- Plan time for questions at the end.
- Manage unsolicited questions to avoid running over your allocated time.
- Never omit the summary and closure, even if you must cut some details of the presentation. The summary and closure will leave the audience with the feeling that the process was completed in a competent manner, even if some things were not covered.

contents as needed to clarify and compare. Typically in oral presentations, however, the sequence, pace, and emphasis of information are chosen by the speaker, for better or worse. The challenge then is to structure information flow for optimum audience understanding, retention, and learning. Also, you will need to draw and focus the listeners' attention, as summarized in Table 5.10, and you must time your presentation carefully. These items expand on hints given for the 30-second message.

When using visual aids, check out the room and equipment ahead of time:

- **Do you know how to operate the available projector and any other equipment you are planning to use?**
- **Will you need a microphone?**
- **Will there be enough seating for the expected audience?**
- **Will most people have the screen located to their left?**
- **If you show a video clip, is it set to go at the touch of a button?**
- **In case of equipment problems, do you have a backup plan for making your presentation?**

Formats for Oral Presentations

The following generic example formats for design project proposals, design project progress reports, and design project final reports are given in Chapter 17.

■ **DP-6C Design Project Proposal Presentation** — A proposal presentation briefs design project sponsors (and representatives from marketing, sales, and manufacturing) on the approach, constraints, weighted objectives based on user needs, and tentative project plans. This information helps them decide whether to authorize the continuation of the project. This presentation supplements the written Design Project Proposal document (DP-5) and emphasizes the aspects important to the sponsors, in particular the design problem statement. It affords an opportunity for sponsors to redirect aspects of the approach of the design team.

■ **DP-8D Design Project Progress Presentation** — This presentation summarizes the status of the project and its accomplishments to date. The audience will probably include design sponsors along with representatives from marketing and manufacturing. Typically, this report emphasizes the set of overall design concepts considered, with a discussion of the features of each and the criteria for evaluation (the design objectives) with their relative weights. It also discusses a list of other design decisions in progress. If a final concept has been selected, a major function of this presentation is to relate the rationale for that decision for buy-in by the sponsors. A revised project plan (such as a Gantt chart) shows how the project will be completed in the time remaining. This progress presentation again gives project sponsors the opportunity to redirect some of the design decisions in progress.

■ **DP-12A Final Design Project Presentation** — The final design project presentation summarizes the results of the project for its sponsors and others from marketing, sales, manufacturing, and finance. Its purpose is to brief decision makers on the relative success of the final design at meeting the objectives derived from user needs and to seek approval to go into production. The major focus here will be to establish credibility for the decision-making process of the design. Designers must establish that credibility by presenting a balanced picture of the alternatives considered, the trade-offs involved, and the efficacy of the final decisions in relation to user needs. Here is where the potential of the oral presentation far exceeds that of the written report. Part of a stakeholders' concern in making a decision to proceed on a project is whether the designers have used a creative approach to obtain superior solutions, examined all reasonable alternatives, exercised good judgment, and followed prudent precautions with public safety and resources. These intangibles can be more important to the stakeholders than technical expertise, but they are hard to examine on the basis of a written report. Decision makers may depend on gut feelings about the credibility, integrity, and creativity of the designer(s) gained from the oral presentation.

Summary of example design communication formats

Table 5.11 presents a summary of the examples of design communication formats that are compiled in Chapter 17. These formats are more than suggested ways to organize design communications. Taken as a whole, they represent a model of the 12-step design process itself. These formats can form a set of standard documents in an engineering design office; they can be used as a set of assignments for a capstone course in engineering design, and they prepare first-year students for what's ahead.

Table 5.11 Summary of Design Communication Formats in Chapter 17
Format numbers are associated with the 12-step design process of Figure 14.1

DP-1	Project Concept Statement	Short written report
DP-2	Table of Design Constraints	Table
DP-3	Survey of User (Customer) Needs	Short written report
DP-4	Table of Design Objectives	Table
DP-5	Design Problem Analysis (Briefing Document)	Short written report
DP-6	Design Project Plan	Chart
DP-6A	Design Project Proposal	Formal written report
DP-6B	Executive Summary	One-page written report
DP-6C	Design Project Proposal Presentation	Verbal presentation
DP-7	Modified Pugh Matrix Format	Pugh matrix
DP-7A	Concept Drawings (Pugh Matrix Phase I)	"Formal" sketch
DP-7B	Design Concept Descriptions	Short summary statements
DP-7C	Concept Drawings (Pugh Matrix Phase II)	"Formal" sketch
DP-8	Design Project Progress Report	Formal written report
DP-8A	Design Decisions	Written summary report
DP-8B	Assembly Drawings	Formal drawing
DP-8C	Bill of Material	Table
DP-8D	Design Project Progress Presentation	Verbal presentation
DP-9	Detail Drawings	Formal drawing
DP-10	Test Plan	Short written report
DP-11	Evaluation Results Report (Report on Design Review)	Short written report
DP-12	Final Design Project Report	Formal written report
DP-12A	Final Design Project Presentation	Verbal presentation
DP-12B	Sales Drawing	Artistic rendering
DP-12C	Final Design Project Evaluation by the Design Team	Evaluation form

Resources for further learning

There is one language that is known to all technical disciplines and to nontechnical people alike, and that is plain English. Blessed is the engineer who uses plain English as often as possible. Keep it simple.

Sidney Love,
Managing and Creating Successful Engineered Designs, *1986*

5.1 Kenneth Blanchard and Spencer Johnson, *The One-Minute Manager,* Morrow, New York, 1982. This book teaches goal-setting, praising, and reprimanding as one-minute communication; it makes an interesting companion piece to Frank Milo's book.

5.2 ✓ Dale Carnegie, *How to Win Friends and Influence People,* Simon & Schuster, New York, 1937. This book (available in paperback) provides perhaps the most widely used advice on how to get along with people and have them accept your ideas. The text is dated now, but the advice is still valid. Dale Carnegie courses have been very successful in teaching people public speaking skills. Two executives of Dale Carnegie & Associates, Stuart R. Levine and Michael A. Crom, have published a current version: *The Leader in You: How to Win Friends, Influence People, and Succeed in a Changing World.*

5.3 Suzette Haden Elgin, *Success with the Gentle Art of Verbal Self-Defense,* Prentice-Hall, Englewood Cliffs, New Jersey, 1989. This book by a noted communications consultant, focuses on replacing patterns of verbal abuse with courteous and effective communication. The book includes many interesting exercises and an extended bibliography. This is just one of many books available on "how to deal with difficult people."

5.4 ✓ Roger Fisher and William Ury, *Getting to Yes—Negotiating Agreement Without Giving In,* Houghton Mifflin, Boston, 1981. This book presents a concise, proven, commonsense method of negotiation what will help you get along with people while pursuing your goals.

5.5 ✓ Milo O. Frank, *How to Get Your Point Across in 30 Seconds—Or Less,* Simon & Schuster, New York, 1986. The author presents his discovery of the 30-second message that is at the heart of effective communication.

5.6 William J. Kolarik, *Creating Quality: Concepts, Systems, Strategies, and Tools,* McGraw-Hill, New York, 1995. This text was one of the resource books for the technical communications section.

5.7 ✓ J. Campbell Martin: *The Successful Engineer: Personal and Professional Skills—a Sourcebook,* McGraw-Hill, New York, 1993. Intended for upper-level engineering students, it discusses many topics relevant to personal and professional growth, including communications.

5.8 ✓ Judith Martin, *Miss Manners: Guide for the Turn-of-the-Millennium,* Simon & Schuster, New York, 1989. This large soft-cover "Definite Reference for Civilized Behavior" gives explicit, practical, and entertaining advice on social, business, and personal etiquette.

5.9 ✓ *New Testament* (any easy-to-read version). The gospel in your strongest thinking quadrant is a good place to start if you have never read the Bible before. The Gospel of Matthew written by the factual tax collector is for quadrant A thinkers; the Gospel of Mark is an "action" account written for quadrant B thinkers; the Gospel of John is symbolic for quadrant C thinkers, and the Gospel of Luke, the physician, scientist and artist of his day, is written with a whole-world outlook for quadrant D thinkers. The teachings of Jesus Christ not only have a lot to say about the relationship between God and human beings but also between people themselves.

5.10 William Strunk, Jr., and E.B. White, *The Elements of Style,* third edition, Macmillan, New York, 1979. Eighty-five pages of examples are given for improving written expression for clarity and brevity in this classic "little" book.

5.11 Linda A. Suskie, *Questionnaire Survey Research—What Works,* Association for Institutional Research, Florida State University, Tallahassee, Florida, 1996. This book provides practical advice on how to ask the right questions and sample representative groups to obtain useful results.

5.12 ✓ Edward R. Tufte, *Visual Explanations, Images and Quantities, Evidence and Narrative,* Graphics Press, Cheshire, Connecticut, 1997. This book is an excellent resource for making effective graphs.

Toastmasters International— This organization is an outstanding resource for learning and honing speaking and communication skills. It is dedicated to helping its members improve their ability to speak clearly and concisely, to develop and strengthen their leadership and executive potential, and to achieve whatever self-development goals they may have set for themselves.

Exercises

5.1 ✓ Definitions

a. The concept of democracy is frequently in the news lately. Do some research—how is this concept understood in three or four different parts of the world, such as Central or South America, Eastern Europe, Western Europe, China, Haiti, India, Sri Lanka?

b. Make up your own definition of power (in a relationship) and compare it with those of your friends and family members.

c. Make up your own definition of negotiation. Then ask a male and a female friend each to define the word also. Note the similarities and differences among the three definitions.

5.2 Disagreement

Next time you have a serious argument or disagreement with a person close to you, approach the situation differently. Take a time-out to identify at least ten factors and goals involved in the situation on which you are in agreement. Return to the problem at hand—do you find it easier now to focus on a cooperative solution?

A person who gets up to speak in public and a person who is too fearful to get up to speak have one thing in common: they both have fear. But a good speaker reacts to that emotion in a special way— by accepting it and making it work for him or her.

Jan D'Arcy, speech and video consultant

5.3 ✓ Thinking Preferences

Write a 30-second message about something that is important to you. Write it in four different ways—to reach quadrant A, B, C, and D thinkers in turn. Then combine these approaches to reach all four quadrants at once. This exercise could be used as a team assignment.

5.4 Oral Thirty-Second Message

This activity requires three or more people. Each person prepares a 30-second message on a common topic (or alternately, on a topic of choice). Then each person presents the 30-second message to the others. The audience has to give positive feedback on what they think worked especially well in the message. Incorporate tips from Table 5.10.

5.5 ★ Proverbs and Communication ★

In a group of three, research proverbs that have to do with communication. Discuss under what circumstances they may be true, and when they are a gross simplification. Example: "Sticks and stones may break my bones but words will never hurt me." What is required to make this a true statement? Under what circumstances is this false? Can some of the proverbs be identified with particular brain quadrants? Make up four different definitions for communication, one for each thinking quadrant.

5.6 ★ Questions of Etiquette ★

This is a group assignment. Discuss the following scenarios and the proper way to respond. You may need to consult a book on etiquette.

a. You are in the hall talking to a colleague, when a visitor—a good acquaintance of yours—walks by and greets you. This person does not know your colleague.
b. You are conducting a business meeting. Some people from the outside have been invited to attend, and they are about to enter the room.
c. You have an appointment with someone. You realize that you will be delayed.
d. You have dialed a wrong number.
e. You have someone in your office who made an appointment to see you. Suddenly, the secretary interrupts you to tell you that you have an important telephone call.
f. You are speaking or writing to a person who has a professional degree or an affiliation after the surname. Make up some specific examples. How would you address these people?
g. Julia Montez Smith is married to Sidney W. Smith. How should she be addressed in her personal life and in her workplace? What if she were divorced or widowed?
h. Imagine that you are the chief executive officer in a company. Come up with the ten most important "rules" or guidelines for projecting a well-polished image to your customers and the community.
i. Develop a list of five important etiquette rules for people on e-mail.

If a message is important, make it redundant. Transmit in two different formats and paths at optimum times.

5.7 ✓ Revising a Piece of Technical Writing

Using the steps outlined in Table 5.12, improve the quality of a one-page technical briefing, proposal, or summary. If you are working in a team, divide up the tasks according to the strongest thinking preferences of the members, then have the entire team do the final review. If working alone, you will need an "outside" review for Step 8.

Table 5.12 How to Create a Quality Written Technical Communication

Writing the first draft: Write fast: Get your ideas down quickly—don't pause to correct spelling or grammar, or the best thoughts will slip away. Use mindmapping if you are familiar with the technique.

Steps for checking and revising your document: Practicing engineers today cannot expect to have a secretary to help with producing their documents. The quality of the finished product will be your responsibility. The task becomes easier when each revision has a specific target:

1. Check for technical accuracy. Are all numbers correct? Are graphs and drawings unambiguous?
2. Write for clarity. Avoid jargon and sentences that are longer than 20 words if possible. Write to a level a bit lower than your general audience. Again, avoid ambiguity.
3. Check for good organization and logical development. Organize with headers and subheads.
4. Check each paragraph. The first sentence should introduce the subject of the paragraph.
5. Check for concision. What can you delete (words, sentences, digressions from main topics)?
6. Add transitions (words or brief sentences) to connect different topics and thoughts.
7. Check for errors in grammar, punctuation, and spelling.
8. Review: Have a competent "outsider" give critical comments. Revise your work if necessary.

Producing your document: Use an appropriate type style, page layout, and format. For example, **bold** or a contrasting color are more pleasing than underline. Do not use more than two or three text fonts—too much creativity (or sloppiness) will detract from the technical content. Print in final form. A day or so later, do a final critical appraisal with the "eyes of the targeted reader." If you are sending out multiple or bound copies, check for quality and completeness (pages can get lost, out of sequence, dog-eared, or misaligned).

Chapter 5 — review of key concepts and action checklist

Verbal communications: Because of our thinking preferences, we may have habits and filters that keep us from communicating effectively. Do you ruffle people's feathers when you communicate? To make our message understood, we have to communicate from the receiver's viewpoint. We must develop good listening skills. With creative problem solving, we can use principled negotiation to work out win-win solutions. The Golden Rule summarizes the essence of good communication and relationships: Treat others as you would have them treat you.

The 30-second message: To create an effective message, first consider the objective, the audience, and an appropriate approach (format). Then use a hook to get attention, prepare a clear message with facts, plans, emotional appeal (if appropriate) and visual content. Close by asking for an action or reaction.

Written technical communication: Engineering design is communication, and the effectiveness of a design can be impaired by faulty communication. Drawings and mathematical analyses are effective ways engineers use to communicate; however, the rationale for design decisions must be supplied in writing to create confidence in the design.

Criteria for effective technical communication: Both verbal and written presentations should be prepared with these criteria: clarity, concision, accuracy and precision, thoroughness and logical organization, audience focus, credibility, and timeliness.

Formats: Accepted formats should be followed in written technical communication, including project proposals, customer surveys, problem statements and briefings, design project plans, progress reports, final reports, test plans, etc., as well as for summaries, tables, and graphics. Verbal presentations can be a key in "selling" the design to the company's decision makers, thus a focus on audience needs and an approach that enhances understanding and retention are very important: 1. Draw attention. 2. Use the right body language. 3. Reinforce three to five main points. 4. Address all thinking styles. 5. Finish on time.

Action checklist

☐ Does your communication address all four thinking quadrants? Analyze a recent presentation you made for a general audience to gauge your strong (and weak) points. Then make a conscious effort to address people in modes that are in your area of least preference.

☐ Identify people with whom you have frequent interaction—the people who are most important to you. What are their strongest thinking preferences (based on your understanding of the HBDI model and the "clues" in their behavior)? To which quadrants might you be "deaf" or "blind" unless you pay special attention?

How's Your Vocabulary?

Anyone can, and everyone should, learn how to speak his/her language well. Nothing on earth gives a person away to others more quickly than the way he/she speaks.

Earl Nightingale

☐ Enter your ideas for new products or inventions in a bound notebook, with each entry dated and signed. Together with the standard design documentation, this creates a complete record of the origin of an idea for legal purposes involving patents (see Chapter 18).

☐ Next time when you have to "sell" an idea, make sure the person you think is the decision maker actually is. This is not always obvious in an organization—thus try to get this "insider information" ahead of time.

☐ E-mail messages are usually rather informal. However, they are still communication. Read through your outgoing e-mail messages at least once to check for clarity and eliminate spelling errors, before sending them off. Remember, e-mail is not a private, but a very public form of communication. What image are you conveying about yourself?

☐ Don't let poor grammar and spelling skills keep you from a successful career in engineering. It is never too late to improve your writing skills. Have a friend who is competent in grammar and spelling analyze your writing. Then do a Pareto analysis and find your most frequent mistakes. Every two weeks, concentrate on correcting one item. Within a few short months, your skills will have improved noticeably.

6 Mental Blocks

What you can learn from this chapter:

- Removing false assumption barriers: "I am not creative"; an intelligent mind is a good thinker; play is frivolous.
- Removing habit barriers: there is only one right answer; looking at a problem in isolation; following the rules.
- Removing attitude barriers: negative thinking, fear of failure or risk avoidance; ambiguity.
- Encouraging creative thinking. The benefits of constraints.
- Further learning: references, exercises, negative thinking project, review, and action checklist.

Creativity is looking at the same thing as everyone else and thinking something different.

Albert Szent-Györgyi, Nobel Prize-winning physicist

In the first five chapters of Part 1 of this book, we presented explicit knowledge about the foundational skills needed for success and innovation in the rapidly changing world of the twenty-first century. We want to wrap up Part 1 by bringing the focus back to creative thinking—a common thread in all the topics discussed so far: paradigm change, visualization, mental models, teamwork, and communication.

A "rockhound" going home after a day of walking the rugged, rocky beaches along Lake Superior will have pockets full of pebbles. Some of these are obviously beautiful banded or "mooned" agates—they have been polished by the natural action of storms and waves for many years. Others look chipped, pock-marked, rough, and ugly. Yet, once the top layers are removed by being churned with grit and polish for two or three weeks in a rock tumbler, the beauty hidden in these rough stones appears. The objective of this chapter is to provide some of this "grit" so you can "polish" your creative thinking skills. Remember, to acquire tacit knowledge, you must apply the techniques you will find in this chapter. If you are already creative, you will discover them to be useful tools to "polish" and encourage others to become more creative.

What we believe about creativity has a major impact on how much creative thinking we do and how we encourage others to express their creativity.

Removing "false assumption" barriers

False Assumptions

We believe every one is creative and can learn to be more creative—we can learn to use the D-quadrant thinking abilities of our brain more frequently and more effectively. Believing otherwise is a major barrier to creativity, with serious consequences. Let's illustrate. If a state agency is seeking proposals on testing and assessments that can better identify talented and gifted students, the underlying assumption is that only some students have (or are born with) these exceptional talents. But if the agency were to seek proposals on how to improve classrooms and teaching to encourage creative thinking and creative problem solving, the underlying assumption is quite different: creativity can be nurtured and developed in all students. Which approach would yield greater benefits to the state's children? We will now look at some of the false assumptions people have about thinking and creativity.

⌛ ✓ **One-Minute Activity 6-1: Group Problem**

Circle the group you think is the most creative.

NASA Engineers High School Teachers Homemakers

College Students First Graders Journalists

Movie Producers Abstract Painters Auto Mechanics

Here are some statistics that will help you evaluate your answer to Activity 6-1. When individuals at various ages were tested for creativity, the results were as follows: At age 40, two percent were creative. At age 25, two percent were creative. At age 17, 10 percent were creative, but at age 5, over 90 percent were creative. All were people who had never been taught how to nurture their creativity. Thus as a group, first graders are the most creative, because they have not yet learned the mental blocks to creative thinking; they can still let their imaginations run free. When shown a sketch of two circles, one inside the other, they come up with imaginative answers—adults usually see the geometric figures only. Homemakers were also found to be very creative because the job requires much flexibility and improvisation in handling many different tasks and small children—often simultaneously. But, with proper use, creative ability is independent of age! False assumptions can be likened to prejudice. Have you ever thought to yourself: "I am not creative"? This is a false assumption, because we have an astounding potential to be creative and can learn to polish our creative talents and use our whole mind better, as we have already seen in the previous chapters.

Here is another false assumption: "An intelligent mind is a good thinker." According to Edward de Bono, highly intelligent persons who are not properly trained may be poor thinkers for a number of reasons:

- They can construct a rational, well-argued case for any point of view and thus do not see the need to explore alternatives.
- Because verbal fluency is often mistaken for good thinking, they learn to substitute one for the other.
- Their mental quickness leads them to jump to conclusions from only a few data points.
- They mistake understanding with quick thinking and slowness with being dull-witted. If "exploratory" is substituted for "slow," the benefits of slower, deliberate thinking become apparent.
- The critical use of intelligence is usually more satisfying than the constructive use. To prove someone else wrong gives instant superiority but does not lead to creative thinking; it destroys it in the critical individual as well as in all within "hearing" distance.

✓ Ten-Minute Activity 6-2: Poor Versus Good Thinking

Look over the "bullets" listed above. Write a short paragraph with specific examples on one of the items listed and how it relates to your experience. Then share your insight with two other people.

Another false assumption prevalent in the business environment, in our schools, and sometimes even among parents, is that "play is frivolous." Play is very important to our mental well-being. Play with your family members, especially with young children. Play pretend games, play pretend ball. Play word games. Play around with words by yourself or in groups; playing around with words will lead you to play around with ideas. Also play with blocks and other materials; construct models of concrete items as well as models that represent abstract concepts and ideas. The Moebius strip is an example of a very practical idea that was considered to be only a plaything—an abstract mathematical concept—for many years, but now is used to reduce wear on continuously moving tapes and conveyor belts since the configuration has only one side and thus only one edge for even wear.

Make yourself a Moebius strip. Play around with it. What would happen if you cut it into two strips lengthwise? What would happen if you cut it into thirds (three strips) lengthwise?

Humor is related to play and is very beneficial to creative thinking because it turns the mind from the usual, expected track. Thus funny ideas may lead to unusual combinations—they can be stepping stones to creative solutions. Have you seen the orange "smiley" faces used in Michigan highway renovation projects in the last decade? Someone with a sense of humor as well as a good portion of quadrant C thinking must have been behind the idea, asking how funny signs could be used to cheer people through construction areas. Humor relieves stress, tension, and monotony because it switches the mind out of a sequential mode "laterally" to new tracks.

Creativity is a learned response to a situation, drawing from within the necessary energy, information, and other resources necessary to solve a problem.

William J. Riffe, director of manufacturing systems engineering, Kettering University.

Another useful technique for playing with ideas is asking what-if questions. Roger Von Oech, in his book, *A Whack on the Side of the Head,* tells the following story:

> A few years ago, a Dutch city had a trash problem. A once-clean section of town had become an eyesore because people stopped using the trash cans. Cigarette butts, candy wrappers, newspapers, bottles, and other garbage littered the streets. The sanitation department became concerned. One idea was to double the littering fine from 25 to 50 guilders for each offense. This didn't work. Increased patrolling didn't work. Then someone had an idea: What if trash cans paid people money for putting in trash? This idea, to say the least, whacked everyone's thinking. The what-if question changed the situation from a "punish the litterer" problem to a "reward the law-abider" problem. The idea, however, had one major fault—if implemented, the city would quickly go bankrupt. However, the people did not reject the idea but used it as a stepping stone instead. They came up with the following "reward": The sanitation department developed electronic trash cans that had a sensor on the top for detecting when a piece of trash was deposited. This activated a tape recorder that would play a recording of a joke. Different trash cans told different jokes. Some developed quite a reputation for their shaggy dog stories; others told puns or elephant jokes. The jokes were changed periodically. As a result, people went out of their way to put their trash in the cans. Soon, the town was clean again. Other cities copied the idea. They found that cans just saying "dank u zeer" when something was deposited had the same beneficial effect.

Why don't we ask what-if questions more often? First, according to Roger Von Oech, we're not taught to do it; we are not in the habit of doing it. Then it is a low-probability technique—you have to ask many what-if questions and follow many different stepping stones before you come up with a truly practical idea. You can practice asking what-if questions as a daily fun exercise that may lead to some unexpected, useful ideas. Here is a question to get you started: What if people were only two feet tall? Play around with this idea for a while. What would be the effect on energy consumption? On overcrowded cities? On the world's food supply? What would our homes look like? This exercise is not as preposterous as it may appear at first. We may get an appreciation for the idea that "bigger is not always better." Also, we may gain insight into the world of a toddler or a person in a wheelchair.

We can use computers to ask what-if questions. We can simulate many situations in virtual reality; we can investigate different conditions instead of running actual experiments in engineering, physics, biology, mathematics, and the social sciences. Computers really let us be inquisitive; here we can take risks and explore situations that would be too dangerous (or too expensive) to do in real life. The purpose of this—besides gaining a thorough understanding of underlying principles—is to find the best way, the optimum solution, to a given problem.

Four-quadrant "cure" for false assumptions:

C. Spend time with creative people—share your assumptions about creativity.
D. Play with ideas, analogies, metaphors. Use humor. Ask what-if questions.
A. Get the facts about creativity; decide which would be reasonable and practical for you to adopt.
B. Practice new creative thinking and problem solving modes.

Removing "habit" barriers

⌛ ✓ **Two-Minute Activity 6-3: Symbols Problem**

Circle the figure that is different from all the others. Explain the reason for your choice.

a. □ b. ⌘ c. ◆ d. ✚ e. ▲

Reason:

Were you able to find reasons to justify each of the figures as "different from all the others"? This problem illustrates that different answers can be "correct" or appropriate, depending on the questions being asked or the criteria being used. Most people stop after finding one answer, because we have not been trained to look for alternatives. Thinking of alternatives is not easy for most people because of a mental block we have learned in school at an early age:

There is only one right answer

> ***Nothing is more dangerous than an idea when it is the only one you have.***
>
> *Émile Chartier, French philosopher*

When we solve problems, we must not assume that a problem will have only one right answer. This is a serious mental block when we are dealing with other than purely mathematical problems. Therefore, don't stop after the first answer—investigate to see if other answers would be better depending on the circumstances. This is especially important when dealing with ideas! How do we know that the idea that we have is best if we have nothing else to compare it with? We have an expression for this type of thinker—a person with a one-track mind. This brings us to a related mental block:

Looking at a problem in isolation

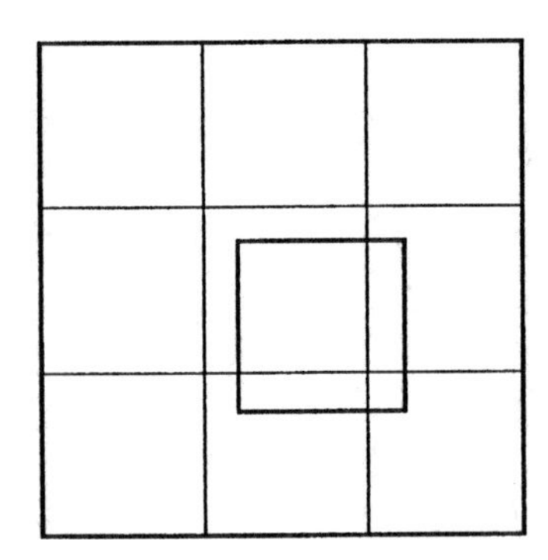

> ⌛ ✓ **One-Minute Activity 6-4: Grid Problem**
> How many squares do you see?

How many squares do you "see" in the figure on the left? You can use a mathematical formula to find the total number of squares in a grid with n squares along a side:

$$\Sigma = 1^2 + 2^2 + 3^2 \ldots + n^2$$

Here we have an $n = 3$ square, with a smaller square superimposed. This adds three squares to the figure (the new square as well as two smaller squares cut from the existing squares). But $1 + 4 + 9 + 3 = 17$ is still not a complete answer, because you have at least one other square in the problem, and that is the spelled-out word in "squares." Did you use another way to look at the problem? Did you determine that you are really looking at a town planning map, where the smaller square indicates the location of *one* city square in a grid of city blocks? Perhaps you saw an infinite number of squares in your mind, if you imagined that you were looking at the top view of a three-dimensional cube or column.

> **"The context is never irrelevant, unless you're dead."**
>
> **How do you think this statement is true (or not true) for the work of engineers?**
>
> **Cite examples to support your view.**

How many squares do you see, not just in the sketch at the top of the page, but in your surroundings: on your desk or clothing, on the ceiling or on the floor, and perhaps when looking outside the window? Before we can find answers to a problem, we must first determine the context and the boundaries. We must find out if the problem is part of a larger problem. We must look at the whole situation. Having a very narrow view of a problem is a mental block.

Especially when people have become experts in their work, they are naturally more narrowly focused on what they know so well. They tend to forget to look beyond the familiar to new horizons. Thus people who can take a multidisciplinary approach (or a wider, "softer" focus) are usually very valuable to their organizations, especially in an increasingly more global environment. We need to get into a habit of looking not just at a leaf or a branch of a tree—we need to look at the whole tree and the whole forest, and perhaps even beyond. We need to take the time for the long-term view—the wide-angle lens. Some decisions we make may have consequences that last beyond our lifetime. When dealing with problems, we need divergent as well as convergent thinking.

This brings up an activity to illustrate another mental block.

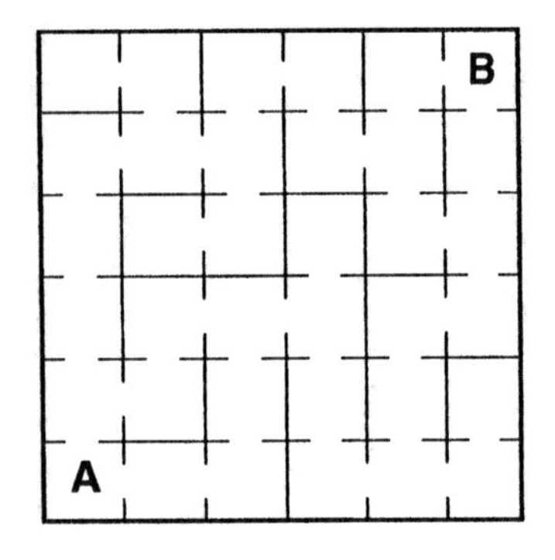

> ⌛ ✓ **One-Minute Activity 6-5: Maze Problem**
> Sketch a path from A to B.

In trying to find a path through the maze on the left of Activity 6-5, did you get to a dead end and had to backtrack? That's one problem-solving strategy—sometimes we need to know what doesn't work. Or, did you start at B and then went backward? This also is good strategy: we look at our goal and then try to find the best way to get there. Did you trace a path through the maze? Most people, when given this type of problem, think this is what they are *supposed* to do. But what did the problem ask you to do? Why not draw a line from A to B, straight through the maze, or around it? What about folding the page to put A right next to B (for a minimal path)? Other possibilities are drawing a line from A to B in the instructions, or from the "a" in maze to the "b" in problem in the activity's title. This problem illustrates overcoming this mental block:

Following the "rules"

We have to make sure we do not make up our own rules and barriers where none exist. Before we can come up with novel ideas, we must question existing constraints. Especially managers and administrators need to develop a habit of looking at the purpose of paperwork and procedures. Are the "rules" we put on others and ourselves really necessary or helpful? How often are we encouraging our coworkers, our friends or family members to look for unusual solutions to problems, even in a combative situation? "Following the rules" can develop into a judgmental, critical attitude that is a mental block discussed in the next section. When we are afraid of questioning arbitrary criteria, we may miss opportunities for creative thinking, improvement, and innovation. Are we following hidden paradigms and rules when we insist on conformity? Exercises 6.7 and 10.8 will explore how to deal with the "following the rules" barrier existing with authoritarian governments.

"Ownership" creates a natural block to a person's willingness to play… it influences the ability of people to look beyond the current reality.

Jim Pierce, engineering consultant

Sometime we follow rules when the original reason for the rules no longer exists. An example is the computer keyboard. Have you wondered why the letters are arranged the way they are? When the typewriter was first invented, it soon developed a problem: the keys were jamming frequently because the typists worked too fast. The obvious solution was to slow down the typists. This is the reason some of the most frequently used letters (such as A, S, L) are typed with the weaker fingers, and E, N, T have been placed above or below the main level. But why are we still using this inefficient keyboard when we now have equipment that works faster than any operator can type? Actually, better keyboards have been invented, but only a few people make the initial effort to install and learn the new system. Today, we need to examine many of our work routines and our ways of teaching and learning in light of the power now available in calculators and computers. Do our traditional ways still make sense? Change is possible: just think a few years back when Great Britain switched to the metric system of measurements.

At this point we need to pause and clear up a false assumption: "Creative thinking and breaking the rules let you act in an undisciplined or even unlawful manner." Yes, creativity is an unstructured activity, and yes, we need to break "rules" when playing with ideas. However, creativity blossoms better when it has some boundaries and direction. This is why we have the steps in creative problem solving. Thus we also follow rules of etiquette and acceptable behavior as we interact with people. Rude behavior creates stress, and stress creates chemicals in the brain that keep it from thinking creatively. Courtesy and consideration for others are important parts of a creative environment. Thus we need judgment to discern which "rules" have to be suspended to let quadrant D thinking flourish and which rules provide the security that enhances creativity or the security that make a civilized society possible. The following case study illustrates what happened when a designer stepped outside a conventional paradigm.

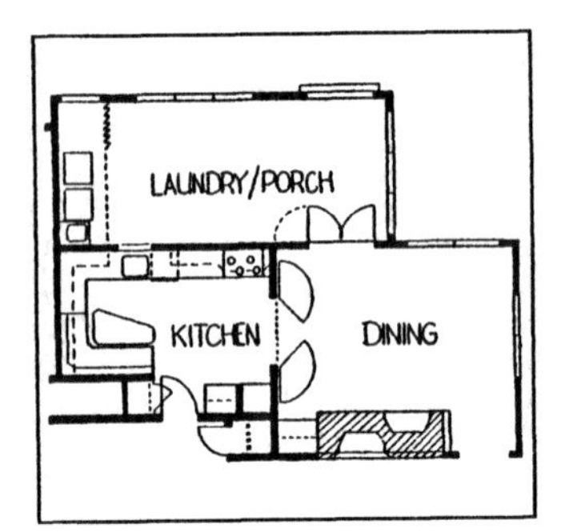

Figure 6.1 Original kitchen/ laundry layout.

Case study: kitchen design

Problem: A young professional couple in California had a comfortable home they liked very much, except for the kitchen, which was old-fashioned and had an impractical and potentially dangerous floor plan (with the cooking area and wall oven impinging into the main traffic lane) as shown in Figure 6.1. When the house burned down in a firestorm, rebuilding it presented an opportunity for improvement.

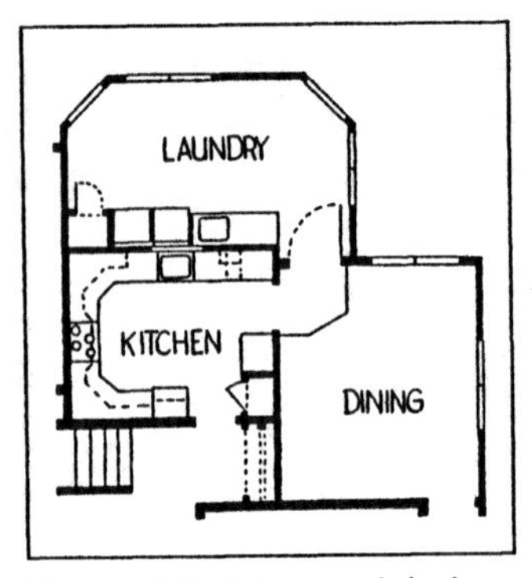

Figure 6.2 Improved design.

Conventional solution: The house was redesigned along the lines of the original plan, since it suited the couple's life style. The improved layout for the kitchen and laundry/garden room is shown in Figure 6.2. At his point, all the people involved (the designer, the couple, the interior decorator, and the architect) were still bound by the old paradigm—the original configuration of the house. The new plan was acceptable since it had an improved layout and safer traffic pattern.

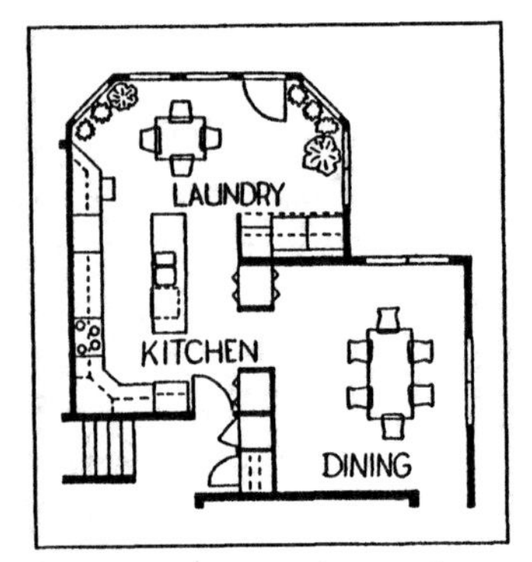

Figure 6.3 Creative garden kitchen design.

Creative solution: The couple showed off the plans to friends who immediately questioned the kitchen design and suggested some "wild" ideas—they had no investment in the old plan. Thus they were able to see different solutions. When these changes were described to the designer over the phone, she was unable to visualize them. However, just the idea that other, better solutions were possible sent her back to the drawing board. She tried different layouts, but nothing seemed to click until she removed the dividing wall between the kitchen and garden room to create a large, open space. Immediately, the most logical place for the laundry equipment became clear. When the kitchen island was turned perpendicular to the garden room (not a "logical" but an "intuitive" solution), the entire plan suddenly fell into place. It was easy to accommodate the food and dish pantries, the desk, the wall ovens, the large sink and dishwasher, the cook top, and the refrigerator for an efficient work triangle easily accessible from the breakfast area as well as from the formal dining room. In less than two hours, the new plan was designed, drawn, and faxed to the couple, who loved it (see Figure 6.3). When the architect saw the new design, he was surprised and commented, "Why didn't we see this solution sooner?" He incorporated it into the house plan with some additional improvements. The couple thinks that this new kitchen has increased the resale value of the house by $20,000 without adding to the construction costs.

Four-quadrant "cure" for habit barriers:

C. Be courteous to others and consciously broaden your outlook by seeking and sharing ideas.

D. Develop an adventurous mind; look for many "different" alternatives.

A. Analyze the reasons for "rules" and decide which should be suspended.

B. Develop fluency in all four modes of thinking; then make their use a habit in communication and decision making.

✓ One-Minute Activity 6-6: Dot Problem

What do you see in the figure on the left?

This little exercise has two purposes. Have you learned to think of alternate answers? In your imagination, were you able to "see" other things besides just a small black dot? The second purpose is to show that it is our human nature to notice small negatives more easily than large positives. Do you realize that the figure is 99.9 percent white? Why do we focus on the black dot that covers only one one-thousandth of the area shown? An unexpected proof of this happened with this illustration when the printer of an earlier edition removed the dot, thinking that it was a blemish. The discriminating ability of the human mind is very important to survival but often misused because of our attitude.

Removing "attitude" barriers

The group of mental blocks we want to consider next is more difficult to deal with because these barriers involve our attitudes and emotions. They require improvements in our quadrant C thinking. Here is the first of these attitudes:

Negative, pessimistic thinking

Learn to appreciate your own creative ideas and nurture the creativity in others.

Creative ideas are fragile— handle with care!

Negative thinking, criticism, sarcasm, and put-downs are mental blocks that not only inhibit creative thinking in the person using these blocks—they have the same effect on all those coming in contact with the negative thinker and are thus doubly destructive. It is so easy for us to focus on small shortcomings of an idea (or of people) rather than appreciate or recognize or compliment the good features or qualities. When we are presented with new ideas, we should make a real effort to react positively. A judgmental attitude, including your own inner "critical voice" can be powerful barriers against expressing creative thoughts.

We must be especially careful when interacting with children. Do you know a typical child receives about 150 negative reactions for each positive reinforcement, within the family as well as outside the home? If you must reprimand a child, try to give encouragement and praise, before and after the negative statement. Even then, keep the negative statement neutral; do not attack the person, only the undesirable action. Also, focus on positive goals, not prohibitions. When we see an idea as "different" or "interesting," we kindle our curiosity and are thus directed toward further investigation. Quick judgment tends to be a negative judgment; thus, take time to make a thoughtful, creative evaluation when judging ideas (those of others, as well as your own). Spend as much or more time looking for the good points and the interesting aspects, as you do on the flaws.

To overcome a spirit of criticism and negative thinking, look at things as being different or interesting— not good or bad!

Edward de Bono

One of the biggest causes of dysfunctional, unproductive teams in business, industry, and in schools is the influence of pessimistic, overly critical people on the team. How can you deal with this problem, short of firing or side-lining these otherwise highly qualified persons? Edward de Bono has invented a tool, the "six colored thinking hats," that make it possible to encourage certain thinking modes at different times and for specific tasks, while limiting others. Thus, the WHITE hat is used for facts, figures, and objective information (quadrant A thinking). The GREEN hat stands for creativity and new ideas (quadrant D thinking). The BLUE hat is in control of the other hats and the thinking steps (or problem solving process)—it is thus a quadrant B mode. The RED hat is used to bring out emotions and feelings (including anger and fear)—it involves quadrant C thinking. Two hats take care of two types of outlook: the YELLOW hat focuses on positive constructive thoughts (primarily quadrants A and D), and the BLACK hat is used to express logical negative thoughts and cautions (mainly quadrants A and B). Edward de Bono recommends that the yellow hat always be used ahead of the black hat when evaluating new and creative ideas. This technique will allow time for positive evaluation (when negative comments are not allowed). But the pessimists know that they will be given a "black hat" period when they can bring forth their critical objections for a legitimate hearing and evaluation. Thus the six hats are a useful method for limiting negative thinking in meetings and creative problem solving.

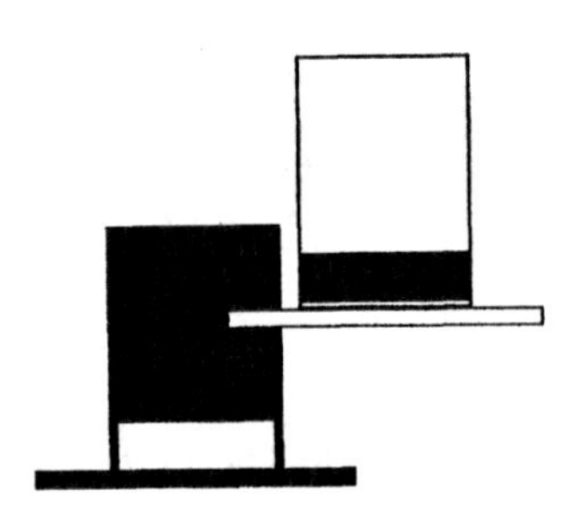

Negative thinking is not the same thing as constructive discontent. Discontent can be used as a stepping stone and motivator to find a better way to solve the problems that are bothering us. To do this, we need an attitude that looks at problems as merely being temporary inconveniences: "Let's get to work to change and improve the situation!" The focus must be on the future and imagining the ideal situation, then playing with different scenarios on how we might get there. We are probably not conscious of how much negative thinking we do routinely and the influence it has on our success. The "investigation of negative thinking" exercise at the end of this chapter is a tool to help diagnose the frequency and types of negative thinking we do.

Here is another important barrier that involves our emotions. It prevents creative thinking as well as action:

The turtle only makes progress when it sticks its neck out!

Risk-avoidance or fear of failure

Lack of risk taking is also expressed when we are overly pedantic, nit-picking, fussy, or anxious. The risk involved in ideas is not the same thing as physical risk taking. We all know from personal experience that teenagers especially do like to take physical risks, for example, when driving, experimenting with things (including drugs), and other activities. You have to use good judgment to decide when to take a risk. You would not jump from a ten-story building—you know what the physical law of gravity would do to you. The risks we are thinking about here are things like speaking out in a group when you have an idea, even though it may be "hooted" down. It is learning something new where you may fail at first until you become good at it, or standing up against peer pressure to get involved in serious studying and excellence, because it is your future that is at stake. Yes, you have to stick your neck out when you are championing a creative idea; you also need a thick shell, and you have to be persistent in getting to your goal—you can expect your critics to make "turtle soup" out of you and your idea.

No pessimist has ever won a battle.

General Dwight D. Eisenhower

What do you think—is a person who misses in two out of three tries very successful? Well, a 0.333 batting average is among the best. Do you realize that it took hundreds of failed experiments for Thomas Edison and his team to invent an improved incandescent light bulb? His vision was a bulb that would work in a city-wide electric system—the many bulbs patented years earlier by other inventors just wouldn't do for his purpose. Mistakes can be triggers or stepping stones for creative ideas and superior solutions. We can learn from mistakes—thus we should not be afraid to make them. Japanese companies that are very quality-conscious "applaud" the appearance of a flaw in their assembly lines instead of hiding the defect. They recognize these occurrences as good learning experiences and real opportunities to make improvements.

We can learn from and use mistakes. Here is a classic example from industry. The 3M Company encourages creativity in its employees, and its researchers are allowed to spend about fifteen percent of their time exploring creative ideas and projects. Some years ago, a scientist by the name of Art Fry decided to make use of this time to deal with a small irritation in his life. He sang in the church choir and used small bits of paper to mark his pages in the hymnal. Invariably, these pieces of paper would fall out and end up on the floor. He remembered that a colleague, Spence Silver, had developed an adhesive everyone thought was a failure because it did not stick very well. Art Fry played around with this adhesive and found that it made not only a good bookmark but was great for writing notes because it would stay in place as long as needed, yet could be removed without damage. The resulting product's trade name is Post-it. It has become one of 3M's most successful office products, although it failed at first to generate sales in its test markets. Also, Art Fry had to invent and build a machine in his basement to produce the blocks of sheets since the traditional 3M products came in rolls only. The story of how these problems were overcome creatively is told in Chapter 12. To get a creative idea implemented takes persistence; we may have to turn early failures into success at several points when moving an idea from the original dream to the marketplace.

Failure is a necessary and productive part of the innovation process.

Jack V. Matson, Director, Leonard Center, Pennsylvania State University

Jack Matson was an engineering professor at the University of Houston when he did research in the area of encouraging student creativity. He found that those students who made more mistakes initially in a project ended up being the most successful. As a student, you may have to make a decision—will you play it safe and follow conventional paths in your projects, or will you risk failure for the chance to really come up with some especially creative solutions? Our task as instructors (or parents) is to find ways of evaluating the progress of our students (or children) so that they are not being penalized for early failure but only assessed on the total of their learning at the end. This raises the question of how the quality of teaching affects learning. How is teaching to be evaluated—does "early failure" apply to the instructional process and curriculum as well? How (or should) we give teachers and school systems the opportunity to experiment to become better educators?

When we try many different approaches, three things happen: we increase our chance of failure, we increase our chance of winning, and either way we learn!

Do you know that a very popular model of diving board was a failed airplane wing? You may have heard of it by its trade name, Duraflex. The fear of failure leads to other associated mental blocks: not being willing to take the responsibility for independent thinking or being passive and incurious—letting life simply pass by—in essence leading to mental "laziness" that refuses to ask questions and does not want to get involved in finding solutions to problems. Such a person may learn to become a chronic complainer and negative thinker without the motivation to become part of the solution or to develop mental flexibility— we could use the image of a "couch potato" mindset to describe such a thinker.

Here is another mental block that acts on our emotions:

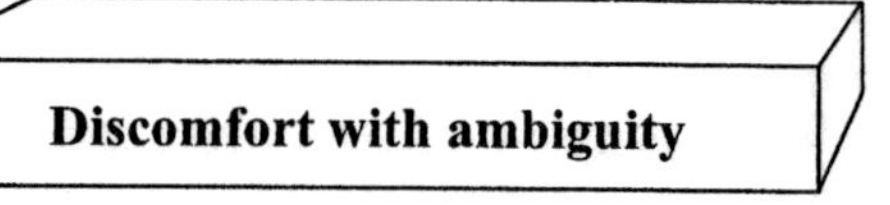

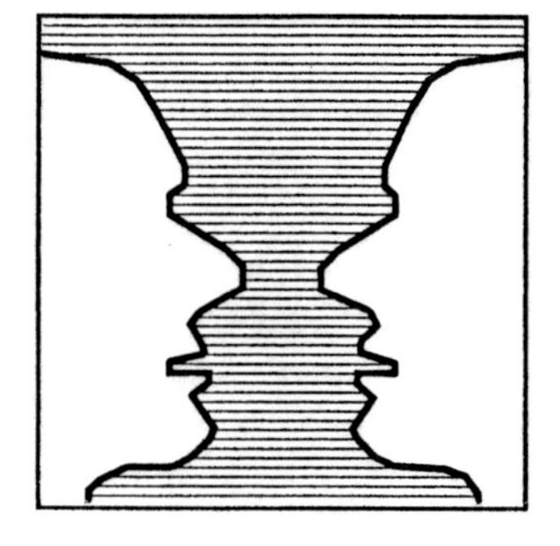

We know from experience that ambiguous situations may lead to serious misunderstandings and conflict, and thus we try to avoid them. Visual images as well as verbal information can contain ambiguity. For example, what do you see in the simple sketch shown on the left—a vase or goblet, or mirror images of a face in profile? What if you turned the drawing around? Can you imagine a bell hanging from a rafter, or do you see a candlestick? Quadrant A and especially quadrant B thinkers are uncomfortable with ambiguity; these minds prefer things to be black or white, not various shades of gray. You may be familiar with the fascinating drawings of M.C. Escher; they illustrate the conflict that can arise between visual cues and the brain's interpretation of the situation.

Here are three ambiguous statements from job references (from Robert J. Thornton, Lehigh University, "Lexicon of Internationally Ambiguous Recommendations (LIAR)," *Detroit News,* February 8, 1988, pp. D1-2). Can you discern two different meanings for each statement?

- I am pleased to say that this person is a former friend of mine.
- In my opinion, you will be fortunate to get this person to work for you.
- I can assure you that no person would be better for this job.

High school and college students are often uncomfortable with ambiguity. They do not mind having a lengthy homework assignment, as long as it comes with detailed, specific instructions and just the right amount of information. They dislike having to make assumptions or dig up information on their own to solve problems. They do not realize that they are getting a great opportunity to cope with real-life ambiguous situations. Even though most of us may be uncomfortable with ambiguity, we sometimes need to explore such situations or ideas because they can be a source of especially creative trains of thought. Therefore, don't be in a hurry to resolve an ambiguity; take the time to look at the situation from many angles. Ask more questions, and especially give your subconscious mind a chance to process the ambiguity.

If you are in an ambiguous situation, use it to get more information.

Paradoxes serve a similar purpose; they can be whacks that can get us to think in a new direction by putting things into a different context. Jesus Christ used many paradoxical statements in His teachings to get people to think. For example, He told His disciples that "Whoever wants to save his life will lose it, but whoever loses his life for me will find it" (Matthew 16:25, NIV Bible). Paradoxical thinking can lead to creative thinking and inventions. For example, Corelle™ dishes are "unbreakable" china. Could the idea of a "water-repellent sponge" lead someone to invent a new way of separating chemical solutions?

Creativity is being able to solve problems in the mind that don't even exist yet.

Brian Webb, engineering student

The three mental blocks that we discussed in this section—negative thinking, the fear of failure, and ambiguity—involve our attitudes and emotions. They, too, can be removed with practice.

Four-quadrant "cure" for emotional barriers:
C. Encourage the people around you with positive comments.
D. Take risks with learning; creatively use failure as a stepping stone to success.
A. Use ambiguous situations to get more information.
B. Practice out-of-the-safe-keeping-box thinking daily.

Encouraging creative thinking

Besides overcoming the mental barriers in quadrant A, quadrant B, and quadrant C thinking, we can encourage creativity with improvements in quadrant D thinking in several ways:

- We can be an example; we need to practice and exhibit creative thinking in our life and in our work, so others can see it in action. Then we also must balance imaginative thinking with mental toughness to get results from our creativity.
- We can recognize and encourage creative thinking in others—Ned Herrmann found that praise increases the use and preference for specific thinking modes.
- We can build a favorable environment for creative expression.

Be an example

To check on your creativity and mental blocks, Table 6.1 lists seven important points made by Harold McMaster, a noted inventor in the field of glass-making from Toledo, when he spoke to students and engineers. We need to discuss one other factor needed to achieve our goals in creativity and thinking, and that is *mental toughness.* This, too, is learned, not inherited. Top athletes have learned this well. Personality style is unrelated to mental toughness: you can be introverted or extroverted, energetic or low-key—this has no bearing on your success. The traits of

Table 6.1 Requirements for Creative Thinking
✓ Be curious—look at the frontiers of knowledge.
✓ Obtain a solid foundation in the field you're working in.
✓ Invent to satisfy a need.
✓ Look for new ways of doing things; take the familiar and look at it in another way.
✓ Question conventional wisdom.
✓ Observe trends, look for opportunities, then work hard.
✓ Realize that most progress is made in small steps through continuous improvement.

mental toughness in competitive sports are described in *Mental Toughness Training for Sports: Achieving Athletic Excellence.* We believe that they apply equally to success in learning, in thinking, in problem solving. Mental toughness provides a context of discipline for the expression and implementation of creativity. The traits are listed in Table 6.2. The lack of discipline and motivation and the resulting chaos in one's life can be a serious barrier to applied creativity.

Table 6.2 Traits of Mental Toughness

1. You are self-motivated and self-directed; you are doing *your* thing.
2. You are positive but realistic—you build up, you praise, you are optimistic; your eye is on the goal, not on possible failure.
3. You are in control of negative emotions—so what if the environment or your coworkers are not perfect or make mistakes. You may need their forgiveness at times, too. Reacting with anger does not solve problems; it makes solutions more difficult.
4. You are calm and relaxed under pressure— you deliberately see the opportunity, not the crisis.
5. You are energetic and ready for action; you are determined to give your best performance and do the best job that can be done.
6. You are persistent; you have a vision. You know what you want to achieve; temporary setbacks do not daunt you.
7. You are mentally alert and focused—you are in control of your concentration. You can use divergent or convergent thinking in response to what the situation requires.
8. You are self-confident—you believe in yourself and know that you can perform well. You are well prepared; you have past successes; you can do it again, even in a new, unfamiliar context.
9. You are responsible for your own actions and behavior—you take responsibility for your own thinking skills and ideas (or, when working in a team, for the results of the group effort). You will see the project through, and you are accountable for the outcome and consequences.

When you adopt a disciplined mental attitude about creativity, you will be calm and relaxed, you will have fun and enjoy your activities, you will have energy, and you will be in control. Developing mental toughness and discipline is hard work, because you have to build different structures in your mind and form new habits, but it will require less effort as you practice and as these attitudes become more automatic. We encourage you to build these structures to shelter your inner creative environment, because with discipline, the mind is sharpened for its tasks, and you will be encouraged by the results you will be able to achieve!

Encourage others

How do you recognize creative thinking in someone else? It is possible to look for creative expressions in people who do not appear to be "gifted" or "talented" in special areas. Table 6.3 lists questions that you can ask to identify traits of creative thinking. We encourage the creativity of others by giving positive feedback and "playing" with ideas together—in essence providing a nurturing, supportive social environment. We can also encourage creativity when we create a stimulating physical environment for people of all ages.

Table 6.3 Recognizing Creative Thinking

1. Look for the unexpected—is the idea something different and original? Did it overcome some of the mental blocks discussed in this chapter? Does it show mental flexibility?
2. Is the idea, product, or solution an unusual or new combination or synthesis of ideas? Does it improve an existing idea or product?
3. Does it have a potential for further creative development, even though the idea appears to be "useless" in its present form?
4. Does the idea or product "feel" right? Is it a logical answer to a problem?
5. Does the idea or product respond to the situation or context; does it solve the need well?
6. Does the solution have a sense of wholeness, beauty, or elegance?
7. Does the solution work on several levels? Does it stimulate thinking and invite other applications?
8. Did the person listen to "something inside the head"?
9. Was the solution a result of brainstorming (alone or with others)?

Build a creative environment

Protection from harm while having the freedom to explore plays an important part in the mental development and creativity of children. Babies and toddlers need opportunities to touch, to feel (with hands and mouth), to taste, to smell, and to manipulate many different objects, so that the young brain can make many neural connections and thus develop to its full potential. Parents must provide a stimulating environment for their young children that is free from physical hazards and unnecessary restraints. Such an environment does not have to be expensive—it only has to invite imagination! Children will come up with amazing ways to invent and play with cardboard boxes of all sizes, wood blocks, paper, crayons, rags, all kinds of odds and ends. Shut off the television! Insted, read to children, frequently, and from a wide variety of subjects. Let children help with cooking and other household projects. Take them to the public library; attend the story hour. Take them to the zoo, to parks, to nature programs, to sandboxes, to the beach or a mountain creek. Let children get their hands and feet wet in mud puddles (and join in the fun). Then talk about all these fascinating experiences.

Resources for further learning

It is difficult to pare down the list of books published on creativity in recent years. Most of the books listed in Part 1 give additional references that discuss the nature, discovery development, and application of creativity. Many organizations offer workshops, conferences, and institutes for creativity and creative problem solving—search the Web to see a current listing of upcoming activities. We recommend that you attend a creativity "camp" or similar event—if you are a quadrant A or B person, you will experience a new world; if you are a quadrant C or D person, you will be encouraged by the networking. The following books relate to topics discussed in this chapter.

The person who teaches your child to talk teaches your child to think.

Jane M. Healy,
Endangered Minds, *1991*

6.1 Edward de Bono, *Lateral Thinking,* Harper and Row, 1970. This book is highly recommended for teaching flexible thinking skills. A more recent book by the author, *Serious Creativity,* Harper Business, New York, 1992, is now available in paperback; it builds on twenty-five years of practical experience with lateral thinking and the deliberate use of creativity.

6.2 ✓ Edward de Bono, *Six Thinking Hats,* Little, Brown & Co., Boston, 1985. Six distinct modes of thinking are identified with six colored "thinking hats." Using these hats helps focus discussions, improves communication and decision-making, and increases the productivity of teams.

6.3 Robert Fritz, *Creating,* Fawcett, New York, 1991. This author clearly differentiates between creativity (or thinking creatively) and creating what one really wants. Creating is a skill that can be mastered. When this skill is used in music or painting, the results are artistic; in technology, the results are inventions; in business, the results are production, and with people, the results are improved relationships. This book contains practical ideas and questions.

6.4 Robert Fulghum, *All I Really Need to Know I Learned in Kindergarten: Uncommon Thoughts on Common Things,* Villard Books, New York, 1989. This warmhearted bestseller contains many low-key stories of positive thinking. Similarly, we recommend the *Chicken Soup for the Soul* books.

Thinking back to our stimulating experiences in the Herrmann Learning Center, a creative environment for adults is surprisingly like the one described for young children—except that some of the toys may be a bit more sophisticated.

6.5 Martin Gardner, *aha! Gotcha—Paradoxes to Puzzle and Delight,* W.H. Freeman, San Francisco, 1982. This book presents a fun collection of puzzlers from logic, probability, numbers, geometry, time, and statistics.

6.6 Peter Jacoby, *Unlocking Your Creative Power,* Ramsey Press, San Diego, 1993. This small volume is a light-hearted guide for leading you to discover your own creativity.

6.7 James E. Loehr, *Mental Toughness Training for Sports: Achieving Athletic Excellence,* Greene Press, New York, 1982. This book shows how winning athletes develop the mind to do their best. These principles and exercises can be applied to learning self-discipline for thinking tasks.

6.8 Ruth Stafford Peale, editor, *Guideposts* magazine. This monthly publication presents tested methods for developing courage, strength, and positive attitudes. For example, see "The Choir Singer's Bookmark," by Arthur L. Fry, January 1989, pages 7-9. We also recommend *The Power of Positive Thinking* authored by her late husband, Dr. Norman Vincent Peale.

Creative thinking exercises

6.1 Paradoxes

Take a few minutes to make up some paradoxes. A paradox is a contradictory statement. See if you can use your paradoxes as "jazzy" book titles. Examples are: warm ice, a soothing rock concert, unbreakable glass, soft stones, dry rain, a calm tornado, a timid hero, dreadful happiness. If the last three were book titles—what would these books be about? Can you see where writing paradoxes is useful? Wouldn't it be great to have "warm" ice for treating a sprain, so it would have the benefit of coldness without making the skin numb?

6.2 Definitions and Slogans

Make up some slogans to encourage creative thinking. Pick one that you especially like; write it on a notecard and place it where you will see it frequently, such as on the dashboard of your car, your bathroom mirror, your refrigerator door, or over your TV set.

If there are two courses of action, always take the third.

Old Jewish saying

6.3 ✓ Learning from Failure

Describe an incident in which you made a mistake "with class" or when an initial failure was used as a stepping stone to success. Include humor or a cartoon if possible.

6.4 Good News Bulletin

You are in charge of writing an "all good news" bulletin. List at least ten items that you would feature in the bulletin and make up appropriate headlines. Do you find that this is more difficult to do than thinking of "bad" news? Why?

6.5 Positive Thinking

Make up five different, imaginative scenarios of turning a bad situation into a positive outcome.
Example 1: Last summer, I was bitten by a snake while on a hike. The good Samaritan who came to my rescue was so kind and caring—we are now married. *Example 2:* My car's not working. So what—I can walk to the store and get in my exercise at the same time.

6.6 ✓ What-if Questions

Make up some interesting what-if questions.
Examples: What if everybody decided to be perfectly considerate for just one day? What if transportation became so advanced that cars and highways became obsolete—what kind of transportation could this be? What if computer viruses got out of hand and made computers unreliable to use? (When we first posed this last question in 1993, the "millennium bug" was not yet in the news. It certainly has added some relevance to the what-if question.)
By yourself or with a friend, play around with the example problems or use some of your own what-if questions. Can you come up with ten (or even twenty) ideas of what might happen for each question?

6.7 ★ ✓Authoritarian Environment ★

If you live in an authoritarian environment (strict parents, teachers, boss, or government system), think about what steps you can take to overcome the "follow the rules" barrier, yet live at peace with the authorities. This is a preliminary assignment to get your thinking started. Exercise Problem 10.8 will continue this assignment.

Project: Investigation of Negative Thinking

Week 1—General data collection and preliminary analysis

Instructions: Duplicate the following two pages; then use the Negativism Score Sheet to increase awareness of your own negative thinking. This investigation can also be useful if you have to work with people who have a habitually negative, critical outlook. Keep the tally sheet with you at all times, together with a pencil. Each time you catch yourself being negative or experiencing one of the items listen in the tabulation below, make a hatch mark on the sheet. Add up the total for each day for seven days.

Example results: We did this assignment for two weeks with a class of 14-year old students from the inner city of Detroit. The results showed these students doing better with less incidence of negative thinking during the second week since their outlook improved as they became aware of their own negative thinking. We also found—quite surprisingly—a trend of lowest scores on Sunday and Monday, with highest scores (more frequent negative thinking) in midweek.

Preliminary analysis: After Week 1, write down your conclusions about your data on the Negativism Score Sheet. What do your results show about your thinking patterns? What do you think are the major causes of your negativism? Write down the six most important categories in 1(b) of the Analysis Worksheet (see the table below for ideas). Use the seventh category for "miscellaneous."

Week 2—Detailed data collection, analysis, and application

Instructions: To investigate your negative thinking habits in detail, keep detailed scores for one week on your negative thinking according to the categories. Add up the category totals and daily totals.

Analysis: Compare your Week 1 and Week 2 totals. What do you conclude from these results? Then construct a Pareto diagram using the data from Week 2 (see Chapter 7 for an exampls of a Pareto diagram).

Applications: Describe two things that you want to change to become a more positive thinker. Make a plan on how you would accomplish this change, then start implementing the plan (see Chapter 12). Also, during this week, purposely turn a negative situation into a positive outcome.

Report: Write a one-page summary about the results of this project, including applications and insight.

Examples of Types of Negative Thinking	
1.	Judging others: being critical, nitpicking, looking for "wrongs" and "flaws."
2.	Avoiding people or situations; procrastinating.
3	General complaining, whining, moaning (from habit).
4.	Lack of self-discipline.
5.	Angry and spiteful; looking for trouble and revenge.
6.	Expressing sarcasm, scolding, intolerance, impatience.
7.	Using abusive language and profanity; lack of respect for others.
8.	Down on self: dejected, pessimistic, fearful.
9.	Receiving put-downs, sarcasm, negative feedback from others.
10.	Imagining slights; having a "self-pity party"—seeing life as unfair.
11.	Lack of vision, positive goals, meaning, hope; sad, seeing "no way out."

Negativism Score Sheet (Data Collection)

Lumsdaine, Lumsdaine and Shelnutt, *Creative Problem Solving and Engineering Design,* page 172, ©1999 McGraw-Hill.

For each day, make a hatchmark every time you catch yourself doing negative thinking, using sarcasm, or putting yourself or someone else down.

Starting date: ____________________

Day 1	Total =
Day 2	Total =
Day 3	Total =
Day 4	Total =
Day 5	Total =
Day 6	Total =
Day 7	Total =
Total	**Week 1:**

Day 8								Total =
Day 9								Total =
Day 10								Total =
Day 11								Total =
Day 12								Total =
Day 13								Total =
Day 14								Total =
Total								**Week 2:**
	A	**B**	**C**	**D**	**E**	**F**	**Misc.**	

Analysis Worksheet

Lumsdaine, Lumsdaine and Shelnutt, *Creative Problem Solving and Engineering Design,* page 173, ©1999 McGraw-Hill.

1. Analysis of Week 1 Results:

a. My conclusions about the Week 1 scores are:

b. The six most important types of negative thinking that I seem to be doing are:

A =

B =

C =

D =

E =

F =

2. Analysis of Week 2 Results:

a. My conclusions from comparing the Week 2 totals with the Week 1 results are:

b. Construction of a Pareto diagram to identify the most frequent causes of my negative thinking:

Misc.

3. Application:

a. I will become a more positive thinker by:

b. I will become a more supportive person by:

Chapter 6 — review of key concepts and action checklist

Obstacles to creative thinking: false assumptions. "I am not creative" is a false assumption. Just about everybody born without a severe mental impairment is creative and with some training can learn to be more creative. Other examples of false assumptions are, "Intelligent minds are good thinkers" and "Play is frivolous."

Mental blocks that we have been taught: habits. "There is only one right answer" can be overcome when you look for several alternatives. "Looking a problem in isolation" can be overcome when you look at the context and the big picture. "Following the rules" can be overcome if you examine the reasons for the rules and paradigms.

Mental blocks that involve emotions and attitudes. To overcome "negative thinking," look at things as being different or interesting, not good or bad. To overcome the "fear of failure," look at mistakes as wonderful learning opportunities and stepping stones to more creative thinking and success! Use "ambiguity" as a prompt to explore the different angles. Be persistent in looking for better ideas! Be courteous!

Encourage creativity in yourself and others. Be a role model; let others see how you express your creativity within the boundaries of mental toughness and self-discipline. Encourage creativity in children and adults.

The basic aim of education is not to accumulate knowledge, but rather to learn to think creatively, teach oneself, and "seek answers to questions as yet unexplored."

Jim Killian,
former president of MIT

Action checklist

☐ This chapter is full of action items and prompts. Skim back over the chapter and highlight those "things to do" that would benefit you most. Then select one that has a high potential for developing your thinking and behavior, and do it over the next three weeks.

☐ Place your watch on the opposite arm from where you usually wear it. Each time you look at it, let it serve as a reminder to see if you can incorporate creative thinking into your current activity.

☐ Schedule a one-hour "playing with ideas" time into your weekly calendar. Then keep your appointment! Keep a notebook with your most creative ideas. Be flexible, not perfect!

☐ Make a conscious effort to praise the creative thinking of another person, even if the occurrence annoys you. Example: Your seven-year old sister uses the tea strainer for cleaning the cat's litter box (when she can't find the regular tool to do the assigned chore).

☐ Clip cartoons and funny stories for a personal humor file—get your family to help. Review it when you are bogged down with a problem.

Review of Part 1

The questions below will help you check up on your learning—on your own or as an early midterm review or exam to gauge your progress if you are using this book in a formal class.

Quadrant A and Quadrant B Type Questions

1. List five benefits of having creative problem-solving skills.
2. You are designing a toy for a small child. Four different mental languages are needed to do this task well. Explain how and why each language is used.
3. Describe the thinking characteristics of each brain quadrant of the Herrmann model. Give an example of behavior for each quadrant.
4. Define the difference between explicit and tacit knowledge and give two examples of each.
5. List the steps and associated mindsets of the creative problem-solving process.
6. (a) Why is a positive attitude important when you are solving a problem? (b) Why is it important in teamwork?
7. Describe the communications process, naming filters and noise that can affect the message between sender and receiver.
8. Describe the characteristics of a learning environment that encourages creativity. Then select an organization you are familiar with and use your list of characteristics to analyze its creative climate and performance.
9. Discuss two reasons why intelligence does not necessarily result in good thinking.
10. How would you overcome the mental block, "There is only one right answer?"

Quadrant C and Quadrant D Type Questions

11. Write a sound bite or television commercial to remind people to think creatively. Start by brainstorming several ideas; then try to combine them into one "best" idea.
12. Write a creative explanation of "paradigm shift" and include your own illustration.
13. Answer the paradigm shift question about one of the following: (a) elementary school; (b) junior high or high school; (c) college. What is impossible to do today but if it could be done would fundamentally change the educational systems in the United States? Example: Replace the traditional eighth grade with a year of community service and just-in-time teaching. Students will learn hands-on skills, project management, cooperation, communication skills, and success in implementing creative problem solving as they renovate neighborhoods and are involved in environmental projects and other constructive, real-life activities.

Review of Part 1 continued

14. Select an intangible word, foreign word, name, number, or date; develop a story or an image that can be visualized or sketched to help remember the item.

15. Write a humorous scenario of a communications difficulty in a team, and how you would resolve it. Alternately, write a dialogue that illustrates the application of a technique that will stop a pessimistic person from continuing with critical comments.

16. Design a collage (or quilt) which illustrates and summarizes the characteristics of the Herrmann four-quadrant model of thinking preferences, using symbols and imagery rather than words.

17. Design a "clip art" symbol for each of the six creative problem solving mindsets.

18. Write a creative skit that illustrates the use of the "parking lot" technique.

19. Make a contour drawing of your left hand (holding the pencil in your right hand). Then switch hands and repeat. In both cases, do not look at your drawing but keep your eyes firmly on the hand being drawn. Then discuss the feelings you experienced while doing this activity, as well as the artistic results.

20. Brainstorm the problem: In what ways can people have fun at a party without using alcohol? List at least twenty different, positive ideas not involving risky behavior.

Whole Brain Questions

21. Brainstorm and then describe a home environment that would foster creativity in people of all ages. Do two versions, one where money is not a limiting factor, and one where you have to be creative without spending any money. Then do an analysis of the results: which version had the more creative ideas? Did having the constraints restrict or encourage your creative thinking?

22. Develop a whole brain question (and give possible answers) on any topic covered in Part I. Identify how it addresses each thinking quadrant.

23. Develop an action item for any topic covered in Part 1 of this book—something that will involve explicit as well as tacit knowledge creation.

24. Review of learning:
    a. In your view, what is the most important concept that you have learned from each chapter? Briefly describe why these are the most important to you.
    b. Then try to connect your chosen concepts with a visual representation (sketch, flow chart, whatever). Can you discern an overall trend or common themes?
    c. Write down two important questions you still have related to Part 1 of this book.

25. "Whether you are looking at your job, your organization, or your personal life, creativity does not just happen—plan for it!" What does this statement mean to you, personally? Write up a concrete example that would illustrate and implement this principle.

Part 2

The Creative Problem Solving Process

THE CREATIVE PROBLEM SOLVING PROCESS

PROBLEM DEFINITION

BRAINSTORMING

SYNTHESIZE IDEAS

IMPLEMENTATION

7

Problem Definition

What you can learn from this chapter:

- Problem definition: finding the "real" problem requires both right-brain and left-brain thinking; each mode performs different tasks.
- Traits of the "explorer's" mindset. Future view and trend watching. Techniques for problem exploration. Contextual problem solving.
- Traits of the "detective's" mindset. Tools for identifying root causes. Resource assessment, briefing, problem statement, incubation.
- Applications: Case study and design project guidelines.
- Further learning: references, exercises, review, and action checklist.

In Part 1 of this book, we presented the fundamental thinking skills of creativity and visualization, teamwork, and communication, and we introduced three mental models—thinking preference, knowledge creation, and creative problem solving. In Part 2, we will examine the creative problem solving process in detail—one step per chapter—and connect it to engineering design. The key word here is *process;* this means we will go beyond theoretical discussion of the model to application or "doing" and then evaluating what we have learned. Chapter 11 will introduce the integration of two steps—the Pugh method for evaluating and optimizing concepts (designs or ideas).

Now, to open the discussion of problem definition, here is a story.

> Somewhere in the Middle East, a man owned 17 camels—his entire wealth. He had three sons who helped him in his transportation business. While on one of their trips, the father fell ill at an oasis. He called the sons to his side and told them his last will: the oldest son was to have half the camels, the middle son one third of the camels, and the youngest one ninth of the camels (which represented a fair share of the time each had helped the father in the business). Then the man died. After the burial, the sons were faced with the problem of how to divide the camels according to their father's wishes. The discussion soon centered, rather heatedly, on how to kill and cut up some of the camels to come up with the specified shares. At this moment, an old man arrived at camp, hungry and thirsty, and with a camel in the same condition. The old man listened to the argument for a while and then offered to help solve the dilemma by giving them his camel, if they would provide shelter and food for him for the night. The sons agreed.

You are either part of the solution or part of the problem.

Attributed to Eldrige Cleaver (circa 1968)

During the night, the oldest son decided he better leave with his share of nine camels before the old man—or his brothers—had a change of heart. Later, the middle son woke up. When he noticed nine camels gone, he hastened to take his share and departed with six. In the morning, the youngest brother, noting that his siblings had already helped themselves to their inheritance, took the two camels of his share and bid farewell to the old man, with thanks for his wisdom. The old man then resumed his journey with his well-fed and rested camel.

What was the "real" problem in this story? We have several "apparent" problems: the hungry traveler, the father's death, the dividing of the inheritance without bloodshed. And why were the sons in a hurry to depart with their camels? Was it because they knew they had a better deal than what the father had specified—for example, with nine camels, the oldest had more than half of seventeen. The "real" problem in the story was that the father's math did not add up—a fact that was recognized by the old man who was then able to profit from the situation.

Wei ji.
The Chinese symbol for crisis is made up of two words: danger + opportunity.

Overview and objectives

A problem has two aspects, although one may be more apparent: difficulty (or danger), and opportunity (or challenge). It is easy to overlook the opportunity aspect when dealing with an emergency. Yet once the crisis has been dealt with, we can seize the chance to introduce a policy of continuous improvement or creatively make a fundamental change leading to true innovation. Dealing with the two aspects of problems requires two different mindsets: the "detective" to address the crisis and the "explorer" to exploit the opportunity. Table 7.1 compares these two approaches. Depending on the type of problem we are dealing with, we may need to concentrate on one or the other track; however, if we iterate between both modes we will have a better probability that we will identify the "real" problem and its context. Ideally, we recommend following the knowledge creation cycle by beginning with the "explorer" for Steps 1 and 2, followed by the "detective" for Steps 3 and 4.

Problem:* *[pro-, forward + ballein, to throw]
1. a question or matter to be thought about or worked out.
2. a matter, person, etc. that is perplexing or difficult.

Definition:* *[de-, from + finis, boundary]
1. a defining or being defined; a determination of the boundaries, extent, or nature.
2. a statement of the meaning of a word, phrase, etc.
3. (a) a putting or being in clear, sharp outline (b) a making or being definite or explicit.
4. the power of a lens to show (an object) in clear, sharp outline.
5. radio and television: the clearness with which sounds or images are reproduced.

Webster's New World Dictionary

Table 7.1 Whole-Brain Problem Definition

Left-brain "detective"	FOCUS	Right-brain "explorer"
Assigned a problem or crisis. Something is not working right.	**Type of problem**	Find or identify a "mess." Uncover a problem or opportunity.
Autocratic chain-of-command ➛ Who is responsible? Who is the expert?	**People**	Cooperative teamwork ➛ include people not too close to the problem; people from related and other fields, not experts.
What is so terrible about the situation?	**Feelings**	What would be nice, if it could be done?
Narrow scope ➛ focus on task.	**Scope**	Wide scop ➛ explore change.
List facts already known. Determine what data are needed.	**Facts**	Look into the context; set goals. Imagine the ideal situation.
Determine constraints and limits: time, budget, staff, resources.	**Boundaries**	Keep limits in the back of the mind; seek to overcome the boundaries.
Use existing tools and methods. Traditional approach: analytical, sequential, convergent thinking.	**Problem solving paradigm**	Look for new paradigms, trends, and alternatives; use divergent, intuitive, sensory, flexible thinking.
Search for root causes and clues. Collect and analyze data.	**Tasks**	Seek out the context and trends; make connections; look to the future.
Problem: bikes are stolen. Experiments: test chains/cables with hacksaw and bolt cutters. Conduct a customer survey. Conclusion: better locks are needed.	**Example**	See the stolen bikes in context— a systems problem in bike security. Consider changes in bike design, bike parking, and bike registration. Brainstorm other uses for security systems.

The main assignment or strategy for "explorers" and "detectives" is to find the *real* problem. It is surprising to realize how much effort is wasted in families, groups, and entire organizations because no time is taken to carefully define a problem. The very first step here is to agree or accept that a problem exists. If you or other people concerned in the situation deny that there is a problem, nothing will happen to improve the situation or find a solution. Problem definition makes sure that everyone involved understands the situation and works on solving the same problem, because what may appear to be the problem may not be the real problem. Let's look at three examples. It is Monday morning, and a child you know does not want to get out of bed. The child moans and complains about a stomach ache. Is the problem an upset stomach? If you take the time to investigate the situation, you may find that the real problem is the bully that is tormenting the younger kids at school or

How are you going
to see the sun
if you lie on your stomach?

Ashanti Proverb

some peer pressure issue. Here is an example from science. Let's say you live in a house near a dairy farm. You and your neighbors are having a severe problem with too many houseflies. How are you going to solve this problem? What is the real problem? Do you want to design a more effective fly trap, or do you want to look at the broader picture of improving the health of the community by preventing the flies from breeding? The solution that you will be seeking will depend on what you see as being the real problem. You need a clearly defined objective! Look at a mouse or mole trap. What is the purpose of such a trap—to trap these rodents or to kill them? If you only want to trap these "critters," you will design a different trap than if you want to kill the "vermin."

Figure 7.1 The "explorer's" mindset.

The "explorer" for divergent thinking

The main objective of "explorers" is to discover the context of a problem. They look for opportunities. They must have a sense of adventure and an eye for the "far" view (as illustrated in Figure 7.1). How can we be "explorers" in ordinary, conventional surroundings? The "explorer's" mindset is a matter of developing and practicing quadrant D thinking skills and attitudes and being a curious, continuous learner. The information we gather as "explorers" will prepare us to recognize and solve problems. "Explorers" use quadrant D thinking to speculate about futures, possibilities, long-term effects, and other far-ranging aspects that may be connected to the problem. "Explorers" also use quadrant C thinking to investigate how the problem affects people. Is a communications barrier associated with the problem? Do people need special training to solve the problem? Why do you suppose we like to link the "explorer's" mindset with the color yellow?

During problem definition, especially when a problem is very complicated, people may get discouraged—the mountain of data collected makes the problem look overwhelming. This is why it is important to use the "explorer's" mindset—it is needed to get a divergent view and perspective on the context to balance the narrow, convergent, and often negative thinking of the "detective." If you are involved with a difficult problem, you must take steps to prevent a negative attitude. First, you are in charge of your life, and you can make decisions to make your life better. You can ask yourself: "How does the problem relate to the goals I have for my life? Is it my responsibility to do something about the problem? Do I have talents and abilities that will help me find a solution?" Sometimes, the answer here is "no." You may need to turn a problem over to people who are trained to treat it. For example, you cannot solve the problem of a friend who is suicidal—you need to get the help of others. But many times you will find that you are able to do more than you give yourself credit for. Go for it! A positive attitude helps your mind be creative. These exploratory habits have a tremendous benefit for our minds by keeping us mentally fit, because the new neural connections we make as learners will prevent a decline in our brain's functioning as we get older.

Breakthrough ideas are most likely to occur when you are actively, confidently searching for new opportunities. They occur to those who are prepared.

Denis E. Waitley and Robert B. Tucker

Trend watching, or how to anticipate the future

Studying trends can help us see the development of problems in a wider context and time frame. With this information, we may be better able to devise appropriate solutions. Studying trends lets us identify areas for action, markets, and future products or services. The purpose of trend watching is to discover opportunities and problems to solve.

Studying trends is also important in career development. Where are the opportunities? What new technology—if you get into it quickly—will give you a competitive advantage? As a student, you need to watch trends to make wise choices in the courses you take. Watch for areas of rapid development. Learn as much about the newest technology as you can. Here we can share a personal experience. Midstream in Monika's undergraduate mathematics program, the university decided that two computer courses would now be required for graduation. Students could elect to graduate under the "old" or the "new" rules. Because of her growing family (and because no one counseled her otherwise or taught her to look at trends), she decided to take the easy way out—she graduated without a single computer course. When she interviewed for jobs, she was in for a jolt. Invariably the first question asked by the interviewers was: "What computer courses have you taken?" Look into courses being offered by newly hired assistant professors or faculty returning from working in industry (and these courses may be outside your department). Attend seminars, colloquia, or other types of lectures by guest speakers; they may give you a glimpse of coming new paradigms. You will have to search out these important learning opportunities on your own—nobody will "make" you attend or reward you for the effort.

In an article on "How to Think Like an Innovator" in the May-June 1988 issue of *The Futurist* magazine, Denis E. Waitley and Robert B. Tucker, two California consultants on personal and executive development, offer some ideas on how to become a good trend watcher. They are summarized in Table 7.2. They also have this to say:

Where the telescope ends, the microscope begins. Which of the two has the grander view?

Victor Hugo,
Les Misérables

> In studying America's leading inventors, we were constantly struck by how well informed they were on a broad range of current events, issues, and trends, both within and outside their particular fields. A knack for trend watching is one of the inventor's secret skills. It is one of the things innovators do to make their own luck. Innovators ride the wave of change because they constantly study the wave. Successful information gathering is not something we are born with; it is a skill that can be developed.

Table 7.2 How to Become a Good Trend Spotter

1. **Audit your information intake.**
   Cut down on mental "junk" food—make informed choices about what you currently read. What sources should you add? Innovators may spend as much as a third of their day reading.

2. **Make your reading time count.**
   Read articles that contain ideas—and take notes as you read. Look for what is different, incongruous, new, worrisome, exciting, unexpected. Seek to broaden your world view.

3. **Develop your front-line observational skills.**
   Become a people watcher. Listen in on conversations to find out how people think and feel. How do the main topics of conversation change over time? Listen in on some talk shows (radio, television, or "chat rooms"); what perspectives do you pick up?

4. **Ask questions.**
   Take the initiative to ask questions in all kinds of situations. As an engineer, can you really know what the customer wants, if you don't ask the right questions?

5. **Adopt the methods of professional trend watchers.**
   One of these is John Naisbitt, author of *Megatrends.* His organization does content analyses of 300 daily U.S. newspapers. Adopt this method for your junk mail—how is it different from last year's? Look for developing trends. This goes for the popular culture as well (movies, MTV, videos).

6. **Find opportunities.**
   Look at current activities and interests for ideas that may appeal to others. Search for solutions to negative trends and offer a means of prevention. Watch for patterns that can tip you off to new opportunities. Even when a present trend is against you, it can be used to come up with a breakthrough idea to counteract it. Also, watch what the competition is doing and do it better, with added value.

EXAMPLE:
A manager from Lansdale Semiconductor, Inc., a company which was not doing well in the overcrowded chip industry, decided to go against the trend of developing increasingly sophisticated technology. Instead, they concentrated on making outdated lines of integrated circuits for the military. It did not take long for this company to take the lead in after-market sales of obsolete chips.

Tools for "explorers"

In today's world, we can no longer work in an isolated corner. We have to learn to develop a global perspective. Holistic, imaginative thinking can lead us to more thoroughly understand a problem than is possible with the strictly analytical approach we have been taught in school. "Explorers" lead the knowledge creation process through the first two steps: 1. Through discussion and sharing, they discover the mental models people have about the problem. 2. They develop a metaphor or concept that will make the broader aspects of the problem understandable and will provide direction for the "detective" activities. This is especially important when dealing with an unstructured, ill-defined problem. Several techniques are available to assist "explorers" in this process.

1. Networking. Recently, we heard a student say: "I don't want to know about the instructor's experiences in China or in industry—I'm in this class to learn fluid mechanics." How sad—this student missed several networking opportunities for the future: (a) getting a good recommendation or job referral from the instructor; (b) getting valuable job information on trends, opportunities and preparation; (c) getting important cultural tips for future success—the student may someday work for a manager who is from a different culture, or his company may send him overseas. People and what (or whom) they know outside the immediate problem area can be a very important long-range investment to our future success; thus take time to make and nurture personal connections.

Looking to the future is not easy, especially for young people—but it is crucial that you develop and practice this ability because your survival is at stake.

2. Searching the Web. Spend an afternoon "surfing" to explore the problem area as well as related topics. You may gain new insight into the context of the problem and discover possible solutions. Jot these ideas down. Also, you may come across people who have dealt with similar problems and might be willing to give advice.

3. Keeping an idea file. A good habit for "explorers" is to collect interesting information and ideas. Have a small notebook or stack of note cards handy. Good ideas about many things may cross your path or pop into your mind at any time, and if you record and file them, you will have a gold mine available when you need some "thought starters." Also, scan a news or business magazine periodically for trends and jot down any ideas that come to your mind as you do this.

4. Modeling a problem. Ned Herrmann developed an unusual technique for problem exploration. Problem modeling is a group activity: the members construct a model—a physical representation—of the problem. A plethora of arts and crafts materials, construction toys, machine parts, tools, and objects from nature are invitingly displayed in a workroom. The participants select the materials they want—then they begin constructing a model for an unstructured problem that needs a creative solution. The assignment is to visualize all aspects of the problem. This right-brain activity gives the participants a surprising amount of insight into the problem as well as ideas about potential solutions.

5. Doing a patent search. The U.S. patent file of around 5 million patents is the world's largest storehouse of technological information, and 4 million of these patents are not described anywhere else in the literature. You will get a feel for the state of the art in the problem area. Examining patents is a thought-starter tool that also provides historical background. How did earlier people solve a particular problem? You will discover "how things work" and you might get leads on innovative companies (either to work for or to help with a particular project assignment you have as a student). Table 7.3 gvies some hints on where to start an exploratory patent search.

Table 7.3 Hints for Doing an Exploratory Patent Search

The following steps are for problem "explorers" only. If you are thinking of patenting an invention, your patent search needs to be more thorough (see Reference 7.8).

1. Use the *Index to the U.S. Patent Classifications.* Its general terms guide you to the correct classification numbers for over 400 classes and 110,000 subclasses. Also explore synonyms for your subject area. For more information, use the *Manual of Classification.* It provides detailed definition on classes and subclasses. To obtain a proper field of search, check the classification definitions that clarify the brief phrases given in the manual.
2. With class and subclass numbers, you can use the on-line computer at a patent depository library to get a printout of all patent numbers relating to the subclass. This list enables you to go to bound volumes and begin exploring the individual patents. Keep an organized worksheet to record your progress. Many university libraries have a government documents section with partial patent files. Check out the resources available on campus for on-line patent searching. At Michigan Tech, this computer is located in the school of business, not in engineering or the library. Check out the on-line tutorial of the Engineering Library at the University of Texas-Austin at **www.lib.utexas.edu/Libs/ENG/PTUT/ptut.html**.
3. To get a feel for new inventions (and just for fun), browse through the *Official Gazette* issued every Tuesday by the U.S. Patent and Trademark Office.

6. Morphological creativity. This structured method is very useful for dealing with a complex problem since all elements of the problem are identified and presented in all possible relationships, together with the values sought. The problem is then synthesized to at most seven parameters, with seven components each in a two-dimensional matrix format with movable columns. The selection of the primary objective of the problem solving helps to identify the specific elements of the problem and will then focus the problem down to a manageable level. It is a complicated method that involves brainstorming and categorizing all the different factors—but the concept of this method can be used by "explorers" to get new views about a problem. Occasionally, problems are set up as a three-dimensional matrix. These matrices can be worked out first through discussion and writing, and then the most important relationships can be brainstormed with a group. An advantage of this technique is that the process usually yields not only valuable insight into the problem but also a number of different solutions.

Technology is neither good nor bad, nor is it neutral. It has short-range and long-range impacts. Impacts may differ according to the scale at which a technology is applied. Technology always entails trade-offs. In short, technology has different results in different contexts.

Melvin Kranzberg, founding editor of Technology and Culture

7. Synectics. This approach starts by considering how each group member understands the problem submitted by the client or expert. A brainstorming panel begins to generate a large number of possible goal-wishes or objectives, and the client or expert selects the alternative that is closest to a plausible solution to the real problem. Then the panel concentrates its problem-solving activities in that particular direction. Synectics is a very complicated method that requires a skillful leader and special training. It employs analogy, paradoxes, and "excursions" to stimulate creative thinking. This technique is not only used to explore problems, its main application is for developing imaginative solutions to problems.

Contextual problem solving

The analytical mind can spot "right" answers, but it takes a very creative mind to ask the right questions.

Before we move on to data analysis and the "detective's" mindset, we need to take a quick look at another responsibility of "explorers"—contextual problem solving. How do most people learn to solve problems? One way is through experience, another is in school. We have seen in Part 1 that our schools teach mostly analytical problem solving, which works well when we are dealing with narrow, well-defined problems.

However, in the real world, many problems are ill-structured. Many technological solutions to problems are causing new and bigger problems because the larger context was not defined properly. An example is the construction of the Aswan High Dam in Egypt—an overpowering and inappropriate solution to the country's lack of energy sources and need for annual flood control. Systems thinking (which involves the context) is difficult for most of us, because we are trained to look at problems in isolation. To illustrate what we mean by contextual problem solving (perhaps best contrasted with plug-and-chug problem solving), we will discuss three examples.

Example 1— How many buses?

An army has to move some soldiers to a different location. If a maximum of 39 soldiers and their gear fit safely into one bus, how many buses are needed to move 1261 soldiers?

(a) 31 (b) 32 (c) 32.33 (d) 33 (e) 34

When this word problem is presented without multiple-choice answers, most students will do the long division of dividing 1261 by 39 to get a result of 32.33 or 32-1/3. When the students think about what the mathematical solution means, they will give 33 buses as the answer (since it is impossible to have one-third of a bus). In multiple-choice tests, some students will either guess at (c) or "deduce" that (d) is the right answer, just because of the way the answers are structured; thus this test will not show if they understand the word problem, if they can do long division correctly, or if they can make sense of the answer.

This problem was used in an article by *Newsweek* magazine to illustrate that U.S. students do not do as well in contextual problem solving as students from other countries (who spend more time on concepts and applications than on drill). To us, the problem also illustrates a basic flaw of multiple-choice testing. We have as many as one-third of our students say that the answer is one bus (if you have enough time to make many trips) or, "Let the soldiers walk and use the bus for the gear" (if the distance is not too far). The best answer very much depends on the context or situation, and students should be encouraged to ask questions about the problem or put down the reasoning behind their answers.

Example 2 — Engineering analysis

What does contextual problem solving mean in engineering? Here is a typical example at the first-year level:

> A 16-ft beam that weighs 8 lb_f per foot is resting horizontally. The left end of the beam is pinned to a vertical wall; the right end is supported by a cable that is attached to the vertical wall 8 ft above the left end of the beam. There is a 250-lb_f concentrated load acting vertically downward 4 ft from the right end of the beam. Determine the tension in the cable and the amount and direction of the reaction at the left end of the beam.

Students need to be taught to solve these types of problems—that is not the question. The problem is that they are *only* taught these well-defined problems. What is the context of this particular problem—why is the beam 16 feet long? Why does it have to be supported with a cable? Is there a better engineering solution for accomplishing the same purpose? The problems increase in complexity as the students advance, but the basic lack of attention to the overall context, to problem definition, and to looking for alternate solutions, persists into graduate school.

I always view problems as opportunities in work clothes.

Henry J. Kaiser

Example 3 — Truck economics

Here is a problem from a math review course for engineers in industry.

> The cost C of gas, oil, maintenance, and depreciation for running a certain truck is [50 + S/8] ¢/mile when it travels at a speed of S mph. A truck driver earns $10/hour. What is the most economical speed at which to operate the truck?

To solve this problem, the "cookbook" approach is to find the derivative of cost with respect to speed, set it equal to zero, and thus determine the speed in miles per hour (mph) that yields the minimum cost, or

$$C = 50 + S/8 + 1000/S$$
$$d(C) = 1/8 - 1000/S^2$$
$$S = 89.4427 \text{ mph}$$

Most students will give this answer; rarely will they round off to 89 or 90 miles per hour. But how can a driver keep the speedometer at 89.4427 mph? Does the answer agree with the assumptions made in the model? What factors come into play at high speed? Even though these calculations are mathematically "correct," the answer makes no sense. Also, the engineering principle of using the precision appropriate to the situation (discussed in Chapter 5) is being ignored.

When we use math software to give a graphical output to better understand the problem, we obtain the curve in Figure 7.2. We can see that an increase from 30 to 60 mph makes a noticeable difference in cost and should thus be made whenever traffic conditions allow, but the 2 ¢/mile

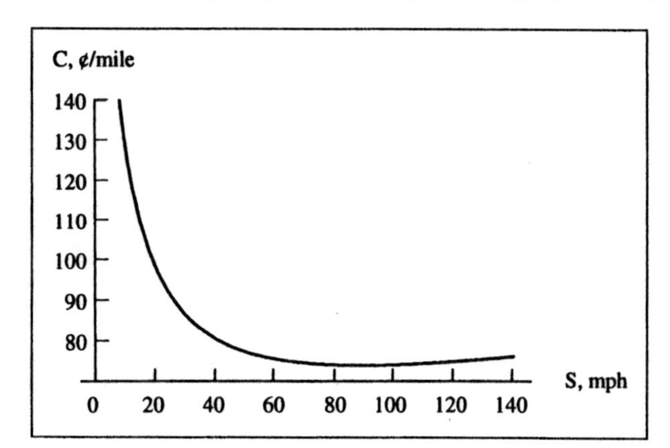

Figure 7.2 The optimum speed for driving a truck (plotted with Mathematica).

increase that would result in going from 60 to 90 mph is not justifiable. This analysis shows that the most economical speed would be at the legal speed limit for trucks. Yet a student who will give this as an answer may very well be marked "wrong" in a typical class, even though this student would have had a better understanding of the "real" problem and its context.

Many efforts are now under way for changing the way engineers are educated. Table 7.4 shows how Professor Edward Lumsdaine has changed the teaching of one course, Heat Transfer for Electrical Engineers, from analytical problem solving to contextual problem solving. As a result, students learned more; they performed better on tests (with a class average shift from C to B, even though the tests were harder); they gained self-confidence; they did not drop out or fail; they participated in class; and they developed an improved understanding of the subject and its connections to other fields. Many students complained more because the class did not fit into their usual paradigm. On the other hand, it kept at least two creative students from dropping out of engineering—they loved the challenge. The instructional team used the Pugh method on the course syllabus to identify the topics that are of greatest benefit to the students (see Chapter 11).

Table 7.4 Two Ways of Teaching Heat Transfer

Analytical Approach	Contextual Approach
• Students must know the fundamentals.	• Students must know the fundamentals.
• Minimal computer use.	• Extensive computer use.
• Only one "correct" solution expected.	• Multiple solutions/alternatives expected.
• Right-or-wrong answers.	• Contextual problem solving.
• Narrow focus on course or discipline.	• Multidisciplinary focus.
• Pure analysis—no design content.	• Application to design is central.
• Students work alone.	• Students work alone and in teams.
• Problems are fully defined.	• Problems are open-ended (less defined).
• Students spend much time substituting in equations (plug-and-chug).	• Students spend much time thinking critically and asking what-if questions.
• Learning is teacher-centered.	• Learning is student-centered.
• Students fear risk; failure is punished. Learning from failure does not occur.	• Students are encouraged to examine causes of failure for continuous improvement.
• Quick idea judgment.	• Deferred idea judgment.
• Artificial, neat problems.	• Real-life, "messy" problems.
• Isolated, disconnected learning; students learn no communication skills.	• Students are required to make a verbal presentation and a written project report.
• Left-brain thinking only; the creative problem-solving process is not used.	• The creative problem-solving process with its different mindsets is emphasized.

Figure 7.3 The "detective's" mindset.

The "detective" for convergent thinking

"Detectives" deal with crisis and danger—their job is to look for root causes. An important objective that guides the activities during the problem definition phase is to collect as much information as possible that is related to the problem. As a "detective," collect this data even if you think it is not very important or if you think you already know what the problem is. You are not to be a judge at this early stage, so do not decide too soon whether something is or is not important. A whole toolbox of methods is available to collect data, depending on the type of problem, the organization's problem-solving culture, and your expertise.

What attitudes must "detectives" have? They are looking for information that is hidden—to find it, they must be persistent; they must think logically about where and how to find the desired information and clues. Figure 7.3 is a humorous illustration of this mindset. A methodical, quadrant B approach combined with quadrant A analytical thinking is required. "Detectives" must be on the lookout for explicit as well as tacit knowledge about the problem. This requires careful questioning of two different groups of people—the knowledge specialists and the knowledge operators. Why do you suppose we like to associate the "detective's" mindset with the color blue?

Tools and techniques for "detectives"

"Detectives" have a veritable toolbox of techniques available that can help in identifying the root causes of problems. This section contains a brief summary of some of these tools; additional information can be found in the Appendixes and in the reference books at the end of this chapter. Engineering students need to be aware that they may have to go "back to school" and obtain training in these tools almost as soon as they start work in a manufacturing company since these techniques are very rarely included in a traditional engineering curriculum.

A problem is an imbalance between what should be and what actually is.

Paraphrased from the Kepner-Tregoe definition

1. Asking questions (Kepner-Tregoe approach). Detectives ask questions about who, what, where, when, why, and how much. Long lists of questions have been published to help in this process of data collection. Answer these questions in as much detail as possible so that you will have reliable data to define the real problem. In the Kepner-Tregoe method, the problem is defined as the extent of change from a former satisfactory state to the present unsatisfactory state, and finding the causes of the deviation should help solve the problem. It also helps to describe the problem in terms of what it is *not*. Thus the Kepner-Tregoe method is very good for finding the boundaries of a problem. As you collect data

Table 7.5 A List of Questions

- What makes this a problem? How big is the problem? How long has it existed?
- What makes this problem different from other problems?
- What events caused this problem? How did it get started? How was it discovered?
- Who has been involved, why, and in what way? Where is the problem located?
- What changes (in surroundings, equipment, procedures, personnel) occurred that could possibly be related to the problem?
- What are the specific causes of the problem—what is your evidence? How are these causes related?
- Does the problem pose a threat to people, your entire organization, or your community? In what way?
- Does the problem have long-term or only short-term effects on individual people, on the community, or on the environment? How?
- How complex is the problem? How are the different parts related?
- Is the problem connected to other problems? In what way?

and information about your problem, be sure to also include what it is *not* or things that were already tried and did not work. Table 7.5 gives a sample of questions useful for problem definition.

2. Surveys. Manufacturing and service companies depend on surveys to collect data on "the voice of the customer" (see pages 139 and 332). This data can then be analyzed and visualized with a Pareto diagram (see Appendix C). This approach makes an interesting exercise for students since the causes of a problem identified through the survey are usually surprisingly different from what the students originally think is the *real* problem. For example, a survey on bicycle locks found that the biggest problem was not the theft of bikes but that the bike owners could not open their own locks. This particular insight changed the direction of problem solving. A survey on toasters found several instances of kitchen fires caused by toasters. Although the frequency was low here, the costs and potential dangers were high, and thus fire prevention became one of the important design criteria in the design of an improved toaster.

3. Statistical process control (SPC). Manufacturing companies use a number of analytical methods for collecting specific data about problems. One of these approaches is called statistical process control. SPC uses seven different tools: check sheets, histograms, cause-and-effect (fishbone) diagrams, Pareto diagrams, scatter diagrams, process control charts, and additional documentation. Japanese companies give all their employees (from top management to shop floor workers) much training in statistical process control. These tools are methods for finding the causes of problems by making graphs of the data and then analyzing the results. Appendix C briefly summarizes and illustrates these SPC tools.

4. FMEA. Ford Motor Company, for example, uses two specific methods to analyze causes of failures. Failure mode and effects analysis (FMEA) explores all possible failure modes for a product or a process, whereas fault tree analysis (FTA) is restricted to the identification of the system elements and events that could lead to or have led to a single, particular failure. The FMEA allows engineers to assess the probability and effect of a failure. By identifying potential problem areas, an FMEA conducted early in the design process can aid in preventing defects and in planning appropriate test programs. Identified causes of failures are ranked according to frequency of occurrence, severity, and ease of detection. An FMEA can be used for services as well as for products—see Appendix D for more details. On a simple level, you use this type of analysis when you are thinking about how to prevent your suitcase from being lost by an airline. You know you need to have identification inside, as well as a sturdy tag outside, and you must verify the destination tag the airline affixes to the bag when you check in. But if you absolutely require the bag on arrival, even the small probability of a "failure" will lead you to pack lightly, so you can carry your bag on board. FMEA analysis results in action to eliminate major causes of failures.

The form of made things is always subject to change in response to their real or perceived shortcomings, their failure to function properly. This principle governs invention, innovation, and ingenuity; it is what drives all inventors, innovators, and engineers. And there follows a corollary: Since nothing is perfect, and, indeed, since even our ideas of perfection are not static, everything is subject to change over time.

Henry Petroski,
The Evolution of Useful Things

5. FTA. H.A. Watson of Bell Laboratories invented fault tree analysis to evaluate the safety of the Minuteman launch control system; this deductive analysis requires considerable information about the system. It graphically represents Boolean logic associated with the development of a particular system failure (see Appendix E). The FTA considers a single undesirable event and directs activities toward eliminating the event by controlling all the factors that could contribute to the failure. On a simple level, you would use this kind of thinking, for example, when the lights frequently go out in your house. First you might check to see if your house is the only one that is dark, or if the entire neighborhood is affected. If you are the only one, you would seek for causes, such as a faulty circuit breaker or frayed power line to your house. If you find that the problem is in your fuse, you will need to investigate what caused it to blow—do you have an appliance that is leaking current? The FTA results in recommendations and corrective actions. Both the FMEA and FTA make valuable contributions in problem definition because they help to differentiate and identify causes and effects.

6. Experiments and Weibull analysis. Sometimes, experiments are conducted to get the data needed to answer the list of questions and define the problem accurately. These experiments are not "trial and error" but carefully designed using special techniques and statistical methods. Weibull analysis is a technique used by manufacturing companies where the results of testing products to failure are plotted on a log-log paper. Cumulative failures (in percent) are graphed versus a product life parameter such as hours of operation or miles driven.

7. Benchmarking. When warranty claims and complaints about a product need to be analyzed, or products and services are evaluated against the competition, benchmarking techniques are employed. The "House of Quality"—the first step in a very structured procedure called quality function deployment (QFD)—is useful for collecting warranty data and comparing critical product quality characteristics against the competition (see Appendix A). The purpose is to improve the quality of a product's components above the level of the best competing product for those areas identified for continuous improvement. Another technique that uses benchmarking is the Pugh method for design optimization and concept evaluation—see Chapter 11. A typical benchmarking process is outlined in Appendix B. We use benchmarking when we compare ourselves to a role model, whether in sports, in music, in a scholastic effort, in growing flowers, or in being a good parent. When we set goals or benchmarks of what we want to achieve, we have taken the first step that will help us reach the goal. Collecting the necessary data for benchmarking is a challenging and often time-consuming task for "detectives."

8. Introspection. When time is too short to do in-depth data collection and analysis, we can engage in a few minutes of quiet introspection. We dig into our memories to bring up any information that we already have about the problem and jot it down for sharing with the team, before the team collectively works out the problem definition statement and moves on to brainstorming.

Five-Minute Activity 7-1: Trends

In groups of three, discuss in what ways can studying trends and acting upon this information help young people today prevent a future economic, medical, social, or environmental calamity (select one of the topic areas)?

Ten-Minute Group Activity 7-2: Questioning a Current Problem

In a team of three or four, select a small but annoying problem that is common to the group (or a problem currently under discussion in your organization). Then go through the list of questions in Table 7.4 to find which ones could yield valuable data and insight into the problem. Don't get sidetracked by answering the questions—just judge the usefulness of each question for obtaining relevant data about the problem.

Steps to complete problem definition

Four items remain to be done in preparation for brainstorming: the resource assessment, the briefing, the positive problem definition statement, and the incubation period.

Resource assessment. The following factors are also relevant and need to be considered during the problem definition process:
Time — Is the problem an emergency, or do you have much time to find causes and good solutions?
People — Should you try to solve the problem yourself, or can you and your team find other people to help you solve the problem?

Resources — What about finances? Does it appear that the problem will take much money to fix? Do you have this money? If you don't have money in the budget for this, you may have to take a different approach and either concentrate on solutions that do not take money or include fund raising as part of the problem.

The resource assessment can be done in the form of a **force field analysis**. Here, the situation or problem is analyzed in terms of supporting and hindering forces (and their strength) on the way toward achieving a satisfactory state or solution. During idea generation—the next step in creative problem solving—ways are sought to strengthen support and eliminate or minimize the obstacles.

The briefing document. The information collected about the problem is now assembled in a briefing document (see Table 7.6) for distribution to the problem-solving team. If the team has a facilitator, he or she may do this task. Although the data collection file may be substantial, the briefing document should be brief—at most a page or two for all but very complex problems—see also Document DP-5 in Chapter 17.

Table 7.6 Items to Be Included in the Briefing Document

a. Background and context of the problem, with a view on trends (if applicable).
b. Specific data collected about the problem, and results from data analysis.
c. Things that were tried but did not work.
d. Thoughts on possible solutions that have come to mind (as an attachment).
e. Conclusions: What is the real problem?
f. The problem definition statement expressed as a positive goal!

The problem definition statement. This statement is important since it will direct the thoughts of the brainstorming team toward solutions. This goal can be quite specific and even "impossible"—a big dream or wishful thinking. "How can we serve our customers better?" most likely will result in mundane ideas, but "How can we provide instant service?" will force the mind to seek unusual or innovative ways to reach the goal. In your team (or alone), play around with several versions of the statement before selecting the best one. Use a dictionary and thesaurus for concise or alternate meanings of words and to find synonyms.

It is important to brief the team ahead of the scheduled brainstorming—otherwise this activity will surface during the session, thus interrupting the creative thinking process. During the briefing, the team members can ask questions to make sure they understand the problem. They may want to share additional insight into the problem. Also, the problem definition statement can be paraphrased until all team members are able to clearly visualize and understand the goal of the problem solving.

The team needs to verify that the problem definition is closely related to the team's charge and objectives (as discussed in Chapters 4 and 5). Otherwise the team will not be solving the *right* problem.

Incubation. Now a time-out is called. The mind needs an incubation period, so it will be prepared to generate innovative ideas. An overnight period makes a good time-out. Otherwise, organize a refreshment period with some relaxing or creative activities. The subconscious mind cannot work on a problem if you are consciously thinking about it! Albert Einstein and Thomas Edison both played a musical instrument when they were stuck. Can you think of an instance when your subconscious mind came up with a solution to a problem while you were busy with something else? When you are assigned a project, do not wait until the last moment to start. Leave enough time to incubate the problem.

Sometimes, our subconscious mind will suddenly pop up an idea on a problem when we least expect it. Such unexpected illuminations about the problem must be written down immediately for sharing with the team ahead of the brainstorming session. These creative "aha" ideas are easily forgotten if you do not jot them down. If you are thinking of some well-known ideas, write these down, too. This process is called *purging.* It should not be skipped during incubation because the mind has to be cleared of mundane solutions before it will be able to come up with truly novel ideas. The **collective notebook method** is designed to collect the ideas of the individual team members during an incubation period that may stretch over several weeks. At the briefing, the members are instructed to daily jot down all ideas and thoughts that come to mind on the problem. The notebooks are then collected by the team leader who prepares a summary of the results, with the most interesting ideas selected for further exploration and brainstorming. If you are facing a nagging problem that you just don't know how to handle, try the notebook approach, alone or with a concerned family member or friend.

⌛ **✓ Ten-Minute Activity 7-3: Problem Definition**

Make up a positive problem definition statement and then paraphrase it several times. Do the different versions help you improve on the original definition? Look up the precise meaning of each noun and verb used in the definition in a dictionary.

Hands-on activity for problem definition

To learn the tacit aspects of the material in this chapter, you must conduct an exercise that will let you practice this initial step in the creative problem-solving process. We will first present a case study summarized on the following two pages. Then we will give step-by-step guidelines on how you can organize and conduct problem definition for your project.

Case Study — Curling Iron

Problem finding: A group of four students in a heat transfer class had to pick a topic for their team project. They talked about different possibilities. When one person mentioned that she was dissatisfied with the performance of her hair curling iron, the group decided to investigate the problem further to see if it would make a suitable design project.

Problem context and data collection: The team did a patent search, looked at a popular curling iron for benchmarking data, and investigated merchandising journal articles for trends in curling irons and other developments in personal care products. Doing a customer survey was a requirement in the project (see also example format for user needs survey in Figure 17.3). The students developed the questionnaire shown in Figure 7.5 and collected the data.

Data analysis and Pareto diagram: The results of the customer survey showed that price was the major determinant in choosing a curling iron, followed by features and necessity. Brand name ranked a distant fourth. Almost all respondents used the iron at a high setting, about 15 percent at a medium setting, and only a very few at a low setting. Close to 80 percent are willing to spend between \$6 to \$15 on the iron. About 50 percent would use the iron at home, followed by gym, school, and work. The identified problems and frequency were plotted in the Pareto diagram shown in Figure 7.4.

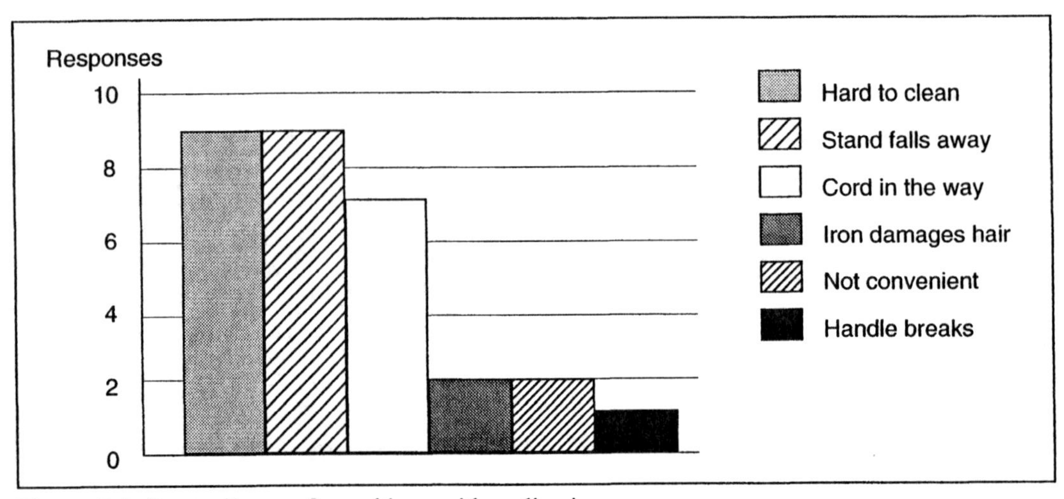

Figure 7.4 Pareto diagram for problems with curling irons.

Briefing document and "go" decision: The customer survey brought out several design flaws in today's curling irons, as summarized in the Pareto diagram. Quick warm-up time was mentioned as a desirable feature, as was portability. The survey showed that there was much room for improvement; thus the team decided to go ahead with the project.

Problem definition statement (as a positive goal): Design an improved curling iron that meets the customer's needs, using heat transfer analysis for optimum performance.

CURLING IRON SURVEY

1. How old are you? 5-13 14-17 18-25 over 25

2. Which of the following apply to your hair?
   _____ fine _____ permed _____ short
   _____ coarse _____ color-treated _____ shoulder length
   _____ thick _____ long

3. How many curling irons have you bought or been given?
   1 2 3 4 5 6 more than 6

4. Do you buy a curling iron for its
   features brand name price necessity?

5. How long does a curling iron usually last?
   < 6 months < 1 year 1-2 years 2-5 years > 5 years

6. What length of hair have you used an iron on?
   short shoulder length long

7. How much would you pay for a good curling iron? $ ________

8. Do you use anything else to curl your hair?
   hot rollers hot sticks permanent other ___________

9. Where do you use or would you use a curling iron if you could?
   home school car bus gym work

10. How often do you use your curling iron?
    1/month 1/week 2–5/week 1/day more

11. How much time do you spend curling your hair? _____ minutes/day

12. About how much time is spent on each curl that curls the way you want? _____ seconds

13. How much time total do you spend getting ready in the morning? _______________

14. How important is it for your hair to looks good ? (5 = very important)
    1 2 3 4 5

15. How many temperature settings do you use? low medium high more

16. What do you like least about your curling iron? ___________________________________

17. What problems (if any) do you have with your curling iron? ________________________

    ____

Figure 7.5 Example of a customer survey form.

This simplified procedure can be used in introductory creative problem solving classes, or where the emphasis is on learning the thinking *process*.

Engineering design courses will likely require the more rigorous process and *documentation* given in Part 3 of this book.

Guidelines for a simple design project

1. Select your team: Form a heterogeneous, multidisciplinary team if possible. Not all members need have expertise in the problem area. Engineering teams benefit if they can include students from other fields. Cultivate a positive attitude and expect the team to do well. Assign the roles of team leader, note taker, and process observer. Facilitating this team activity will give practice in leadership skills—the team at this point will most likely be in the storming stage.

2. Choose a problem topic: The objective is to select a problem that you can take through the entire creative problem solving process as you study each chapter in Part 2 of this book. You may have several options about choosing a problem topic:

a Brainstorm problems with your team within the context of your class. Have them vote a secret ballot to make the final selection.

b. For one week pay attention to items at work, at school, or around the house that are not working well and could benefit from redesign. When you are thinking, "I wish someone would invent a gadget to do this task" or, "Why hasn't anybody thought of doing this in a better way," these are tips that can lead to a good design topic.

c. Work on an assigned topic (recommended for an inexperienced team or a very diverse group of people as this will save time and arguments). People learn more about the process from a problem in which they are not too closely involved—otherwise they get carried away by the results and lose sight of the learning process.

d. Work on a sponsored design project or a design competition.

3. Focus the problem topic: Do not select a problem that is too large for your first project. On the other hand, do not narrow down the topic too much in the early stages or you will limit the creative possibilities in the solutions that will be generated. If you have a good topic but need to expand the problem, you can ask a series of "what is this about" questions. This technique invites divergent, contextual thinking.

When we find the real problem, we can eliminate the root causes instead of merely treating the symptoms.

Example of a diverging chain of questions:
"What is this problem about?"
Answer: "Housing."
"What is housing about?"
Answer: "Being warm and cozy."
"What is being warm and cozy about?"
Answer: "Feeling loved, cared for, and safe."

Note how this chain has brought out aspects of the problem that involve not only a physical need but also emotional needs. It helped us get the bigger picture. We must encourage our customers and other people involved with the problem to express the needs or "dreams" that are important but often remain unspoken.

REMINDER

Hands-on practice is important for learning and experiencing the creative problem-solving process.

Take the time to apply your theoretical knowledge in a team project.

At other times, to obtain a solution, we have to break problems down into smaller parts through convergent thinking. If we want to "squeeze" a problem, we can use a chain question process by asking "why?" Such questions can bring out the real reasons why people have a problem or what is important about the problem.

> ***Example of a converging chain of questions:***
> "Why do you want to improve your budgeting procedure?"
> Answer: "Because I'm always late in paying my bills."
> "Why are you always late?"
> Answer: "Because I have a habit of procrastination."
> "Why do you procrastinate?"
> Answer: "Because I hate paperwork."
> "Why do you hate paperwork?"
> Answer: "It requires quadrant B thinking which I avoid."

Chain questions let us eliminate rationalization; we can zero in on the real motivation underlying a problem. Here, the real problem is a mismatch between the task and the person's thinking preference, not the budgeting procedure or the habit of procrastination.

Your problem can be expanded or contracted as the team plays around with the problem definition. Do not be concerned about perfection. If the problem is too narrow, the team will likely add divergent ideas during brainstorming. If the topic is too broad, narrower subtopics can be selected during the idea evaluation stage. But just to cut down on the amount of work and the time required for the team project, select a reasonably narrow topic—one that may generate around 40 (not 300) brainstorming ideas that will result in four or five major design concepts.

4. Collect information and customer data: Here, the notebook method is useful. Begin with a library, patent, and Web search. The team can discuss this preliminary information to gain a feel for the important aspects of the problem that can then be addressed in the survey. The survey can collect different types of information, either purely quantitative data, or "weighted" data—in which people can indicate not only if they have a problem but how severe the problem is by ranking it as 0 for no problem, 1 for a small problem, 2 for a moderate problem, and 3 for a severe problem. In this case, the replies can be tabulated as total points, or they can be stratified into the number of answers in each category. Stratified data collection can give better insight. The team can have quite an interesting discussion of what questions should be included in the survey, how the collected data can be compiled efficiently, and how it should be analyzed and interpreted. (Also see pages 139 and 332.)

5. Analyze the data; make a Pareto diagram; prepare the briefing: The data collected with the survey can then be summarized. The ranked frequency of identified problems and causes can be plotted and displayed in the format of a Pareto diagram (see Appendix C). The team can sketch the diagram on a large chart and post it in a prominent place in the room as a reminder of customer needs that are to guide all phases of creative problem solving. The team can now prepare a summary paragraph or briefing on what they think is the real problem; it should be typed up and handed to each member to keep in the project notebook.

Develop the mental habit and flexibility to be able to change your focus on a problem all the way from a close-up to a bird's-eye view.

6. Develop the problem definition statement: Each person can suggest a problem definition statement as a positive goal or objective, based on the briefing. As a group, spend about 10 minutes playing around with various ideas for the problem definition statement, then converge this activity to a single, best, synthesized statement—one that has the general agreement of the team. Each person in the team should now have a clear understanding of what the real problem is. Close the meeting by posting the final problem definition statement on a flip chart, and make sure that each team member writes down this statement in the notebook. Now the team is prepared for incubation. Remind each person to keep the notebook handy for jotting down any ideas that come to mind about the problem or solutions before the team meets again for brainstorming.

Resources for further learning

Large corporations may have their own manuals for problem definition and data analysis, usually as part of company-wide quality control efforts. For example, Ford Motor Company publishes its own manuals on FMEA and FTA. Since national professional engineering societies are beginning to offer workshops for training in FMEA and FTA, reference material should become more widely available on these two analytical procedures. If you work for a company that is introducing new methods, take every opportunity to learn about these techniques by attending the training classes. Appendix F on TQM has an additional list of interesting books on the topics of quality, innovation, and manufacturing that include in-depth discussions on problems and solutions.

7.1 Myron S. Allen, *Morphological Creativity: The Miracle of Your Hidden Brain Power,* Prentice-Hall, Englewood Cliffs, New Jersey, 1962. This book presents the principles of morphological creativity; the technique is demonstrated by the organization of the material in the book.

7.2 Don P. Clausing, *Total Quality Development: A Step-by-Step Guide to World-Class Concurrent Engineering,* ASME Press, Fairfield, New Jersey, 1994. This book addresses the problem of quality in engineering and is written for technical readers.

7.3 W. Edwards Deming, *Out of the Crisis,* MIT Center for Advanced Engineering Study, Cambridge, Massachusetts, 1982. Quality in manufacturing is discussed by one of the early leaders of the quality movement in Japan.

7.4 Eliyahu M. Goldratt and Jeff Cox, *The Goal: Excellence in Manufacturing,* Creative Output, Milford, Connecticut, 1984. This novel presents the steps and concepts of problem solving in the context of manufacturing.

7.5 Herman Kahn and Anthony J. Wiener, *The Year 2000,* Macmillan, New York, 1967. This book provides a detailed example of the "science" of forecasting and the interpretation of trends. It is fun to observe how recent history and developments agree or disagree with their scenarios.

7.6 Charles H. Kepner and Benjamin B. Tregoe, *The Rational Manager,* McGraw-Hill, New York, 1965. This book thoroughly explains the Kepner-Tregoe method of problem solving. Even if you want to use more creative methods, the Kepner-Tregoe approach is excellent for initial problem definition, learning to ask the right question, and analyzing data, as well as for identifying potential problems during the solution implementation phase.

7.7 John Naisbitt, *Megatrends: Ten New Directions Transforming Our Lives,* Warner Books, New York, 1982. This is required reading for learning more about the "science" of trend watching. A follow-up with ten new forces shaping our future is: John Naisbitt and Patricia Aburdene, *Megatrends 2000: Ten New Directions for the 1990's,* Morrow, New York, 1990.

7.8 David Pressman, *Patent It Yourself,* third edition, Nolo Press, Berkeley, California, 1991. This book contains useful hints and forms for those who want to apply for their own patents.

7.9 George M. Prince, *Practice of Creativity,* Macmillan, New York, 1970. Although the main topic is Synectics, this book has useful comments for anyone who has to attend committee meetings. It also discusses the importance of the briefing document.

7.10 Denis E. Waitley and Robert B. Tucker, *Winning the Innovation Game,* Fleming N. Revell, Old Tappan, New Jersey, 1986. This book emphasizes creative thinking, innovation, and managing change; it shows how to obtain possible breakthrough ideas from observing trends.

Humanity lies in our urge to explore the world. It lies in our unique drive to understand the nature of the universe within which we live. It lies in our capacity to question the known and imagine the unknown.

Margaret Mead,
A Way of Seeing

Exercises for "explorers"

7.1 ✓ Exploring a Toaster

Think about designing a better toaster. What problem does a toaster solve? Imagine being a toaster. Make statements such as: "I have to take bread slices into myself." "I have to heat bread uniformly, without burning." "I have to kick the toast out at the right time and then shut off." "I have to keep a cool skin." Write five more statements like these—the purpose is to really identify with the problem.

7.2 ✓ The Greenhouse Effect

Many scientists are predicting that our climates are getting warmer. Brainstorm some positive outcomes or opportunities. For example, more air conditioners will be in demand (but will not be allowed to use freon).

New cosmetics providing better protection from the sun will be needed. Can you think of other markets, products, or new paradigms that may result from the greenhouse problem.

7.3 ★ Landfills ★
If you are not yet recycling your garbage, you may need to make some major changes within the next ten years, as most of the industrialized world will run out of space for landfills. What trends are you predicting—in government regulation, in business opportunities, in many people taking personal responsibility?

7.4 ★ Technical Knowledge ★
It has been predicted that all the technological knowledge we have today will represent only about 1 percent of the knowledge that will be available by the Year 2050. What are the implications of this (a) for education and schools, (b) for the workplace, (c) for libraries, (d) for book publishers, (e) for authors, (f) for business, or (g) for the Internet? Brainstorm one of these topics and see if you can come up with an opportunity that you or your group might want to develop.

When a gasoline truck overturned in front of The Tire Shop in Hancock, Michigan, Tom Riede, one of the employees, stopped a gushing gas leak by inserting and then inflating an inner tube. His quick thinking and creative use of a common material of his trade averted a major environmental disaster in this small community.

7.5 ★ Read about Exploration ★
Read a biography of an explorer or a book written by an explorer. What made this person be an explorer? What are some of the most striking personal characteristics? What were his or her goals and rewards? Or read about a team exploration effort, such as the Voyager space program or the Mir space station. How did the project grow and change from the original idea to final execution?

Exercises for "detectives"

7.6 ✓ Time Use Analysis
Over a period of three days, complete a detailed log on how you are using your time (in 15-minute chunks). Then do an analysis to determine which activities waste the most time. Make a Pareto diagram to find "the 20 percent that cause 80 percent of the trouble." Make a plan to eliminate the top three time wasters (one at a time).

7.7 ★ Cause-and-Effects Analysis ★
Select an item that you are using in your daily life that is not functioning properly. Examples: the front door "howls" when the wind blows above 10 mph; your bicycle's kickstand sticks in one position; your alarm clock fails to ring at least once a week; your computer has developed a strange quirk, or your car is pulling to the right when you are driving down a straight road. Make up a cause-and-effects chart (fishbone diagram) that identifies all the factors that could possibly be involved in causing the problem.

7.8 ★ Learning More About Data Collection and Analysis ★
Select one of the following tools described in the Appendix: QFD, SPC, FMEA, FTA, or Benchmarking and prepare a brief report to your class or group. Use additional resources, and if at all possible, obtain an actual case or example from industry.

7.9 ✓ Briefing Document Samples
As a team project, obtain samples of briefing documents from three different organizations. How was the data collected and presented? In what way could you improve the problem definition statement?

Chapter 7 — review of key concepts and action checklist

What is the real problem? Problem definition is a whole-brain process, involving exploration of the broad view as well as data collection and analysis. Table 7.7 is a summary to help determine when to use creative problem solving (depending on the type of problem).

Your role as "explorer": Be adventuresome. Have a habit of exploring new ideas, hobbies, fields. Watch for trends and opportunities. Look for improvements in products, procedures and services. Learn to be a contextual problem solver.

Table 7.7 Analytical or Creative Problem Solving?
Emergency: Use an authoritarian approach and predetermined procedures. Example: Fighting a fire and evacuating a building. A sudden crisis can produce creative solutions (see sidebar on the opposite page), but trained crews are needed to handle all aspects, such as cleanup, traffic control, environmental monitoring, evacuating a neighborhood, setting up shelters, coordination, etc.
Routine, well structured problem: Use standard methods and procedures. Examples: Building an ordinary warehouse. Getting to work under normal traffic conditions. Specifying standard components, manufacturing processes, and assembly.
Operational, tactical problems: Solve analytically to deal with crisis aspect *and* creatively to deal with the opportunity aspect. Creative problem solving will prevent superficial solutions to deep-seated problems and long-term solutions to short-term problems (such as hiring a permanent employee for a temporary work overload). It will enable us to be proactive and innovative and can prevent future crises. Example: Inventing a new manufacturing process for long-term rust prevention in your product. Pacifying dissatisfied customers by offering free repairs. Using analytical tests to determine the root causes of the rust problem.
Strategic problem: Use creative problem solving for important, long-term problems. Example: a new paradigm being adopted by your competitor could seriously affect your future business.
Unstructured, elusive, ambiguous problems: These poorly understood problems that may involve changing conditions require creative problem solving. Example: Dealing with disgruntled employees, seeking new customers, or developing new products.

Your role as "detective": First, accept that a problem exists; then ask questions to find the root causes of the "real" problem. Use appropriate methods to collect and analyze data and record the information in a notebook. Assess the resources. Prepare a briefing document; converge the problem down to a positive problem definition statement. Brief the team, then observe an incubation period before moving to the next creative problem solving phase ➛ idea generation.

Guidelines for problem definition in a hands-on project: Doing a project is important for acquiring tacit knowledge in being "explorers" and "detectives" and in using the problem definition tools:

1. Make up a heterogeneous problem solving team.
2. Choose the problem topic.
3. Focus the topic: If needed, use divergent thinking to look at the bigger picture or convergent thinking to break the problem into smaller parts.
4. Collect data through library or Web searches; make up a customer survey form and conduct the survey. Record all information in a notebook.
5. Analyze the data; rank the root causes of the problem and plot a Pareto diagram.
6. Prepare a briefing document: summarize the data and the conclusions about the "real" problem; write a positive problem definition statement.
7. Have a time-out for incubating the problem in the subconscious mind.

Action checklist

☐ Apply what you are learning about creative problem solving. If you are not part of a conceptual team design project, choose a problem at the periphery of your life as your exercise topic. If you are not too closely involved in the problem, you will be better able to evaluate the process as a learning experience.

☐ To practice the mindset of an "explorer," take one afternoon a month to look around in a subject you don't know anything about, by reading, speaking to people, visiting exhibits, or attending a lecture.

☐ Read regularly outside your own field. Try exploring different subjects—you will be surprised at how some will turn you on and lead you to discoveries, new interest, and increased creativity.

☐ As you read or listen to the daily news, look for trends that are developing in many areas, not just in your own community or in your own area of study or expertise.

☐ Observe these tips for your next incubation period: Relax! Stop working on the problem. Give your subconscious mind a chance to work. Go have fun. Listen to your favorite music; strum the guitar. Play ball with your friends; pull some weeds, rake your leaves. Do aerobic exercises, swim, or go for a walk in the woods—physical activity is very good for your creative mind.

This I know.
This I believe
with all my heart.
If we want a free
and peaceful world,
if we want
to make deserts bloom and
man grow
to greater dignity
as a human being—
we can do it!

Eleanor Roosevelt

8 Idea Generation

What you can learn from this chapter:
- The goal and history of brainstorming.
- Traits of the "artist's" mindset; the four rules; the role of constraints.
- How to lead a classic brainstorming session.
- Other brainstorming methods to accommodate different conditions.
- What to do when the team is "stuck" and can't think of creative ideas.
- Resources for further learning: books and tools, exercises, warm-up example; review, and action checklist.

Problem definition, the topic of the preceding chapter, constituted a complete knowledge-creation cycle. With brainstorming, or generating many ideas for solving the problem, we begin another cycle. The incubation period in between is important for preparing the mind to switch from the left-brain "detective" to the right-brain "artist." We have two different knowledge-creation "frames" acting here:
1. Learning about brainstorming in this chapter is one kind of knowledge creation—where we are attempting to cover the entire cycle from tacit to explicit back to tacit knowledge.
2. *Doing* idea generation in a creative problem-solving project involves mainly the transition from Step 1 to Step 2 (or the conversion of tacit knowledge to explicit knowledge in the problem topic or subject area).

Visualize being in a storm. You are being pelted by rain or sleet, and you feel the awesome power of the wind. Wouldn't it be wonderful if this energy could be harnessed and put to good use? In a way, when we brainstorm, we want to provoke a storm of ideas. A gentle breeze just won't have the same result. Brainstorming procedures are like a harness that attempts to direct and optimize the energy in idea generation.

The first of our senses which we should take care never to let rust through disuse is that sixth sense—the imagination.

Christopher Fry,
English actor and playwright

Now, what exactly is brainstorming? It is a group approach to creative thinking. The verbal method now known as classic brainstorming was developed in 1938 by Alex Osborn in his advertising business and came into widespread use in the 1950's as a group method of creative idea generation. The best number of people for a verbal brainstorming group is from three to about ten. Brainstorming does not work for all types of problems all the time, but its successes have made it a valuable problem-solving tool. It is easy to learn, and it gets more productive with practice. People frequently mistake routine, undirected, critical discussions in meetings with brainstorming. As you will see, brainstorming requires careful mental preparation. Although it is a creative, freewheeling activity, definite rules and procedures are followed. We will teach you the classic brainstorming method and then discuss a number of variations that have been developed for special conditions. Some of these techniques can be used by people working alone.

Figure 8.1 The "artist's" mindset.

The role of the "artist"

Generating novel and innovative ideas is at the heart of the creative problem-solving process. The metaphor of an artist illustrates the mindset required (see Figure 8.1). What do artists do? They create something new, something that first existed only in their minds. With the "artist's" mindset, your task in creative problem solving is to transform information into new ideas. This is the time when you can break out of your usual mold. Go to town with your quadrant D imagination and your quadrant C feelings! Welcome eccentric, wild, weird, crazy, off-the-wall, out-of-the-box ideas. In brainstorming, this process of using the imagination—this mental activity of coming up with anything but mundane ideas—is called "freewheeling." This means we impose few restrictions on ourselves or our team members on the types of ideas that can be expressed. Why do you suppose we like the color orange to represent the "artist's" mindset?

The four rules of brainstorming

Brainstorming is easy to learn because it only has four rules. These four rules are important principles, so fix them firmly in your mind!

1. Generate as many solutions as possible—quantity counts.
2. Wild ideas are welcome—be as creative as you can be.
3. "Hitchhiking" is encouraged—build on the ideas of others.
4. No criticism is allowed—defer judgment until later.

Rule 1: Generate as many solutions as possible. Quantity counts! The more ideas you generate individually and collectively, the better the chance that you and your group will come up with an innovative solution. Don't give long explanations along with your ideas, just toss them out quickly using key words only. Be brief!

Rule 2: Wild ideas are welcome. This point cannot be overemphasized. The more odd, weird, impossible, or crazy ideas are generated, the better the chances of coming up with a truly original solution in the end. The only boundary here is to avoid words and ideas that are hurtful or offensive to your team members because the stress that is caused will inhibit creative thinking along with undermining the team spirit.

Rule 3: Hitchhiking is encouraged. Ideas do not have to be completely new; it is perfectly fine to expand, build, or "hitchhike" on other people's ideas. Idea pinching is allowed! You can also apply this process when you use aspects of an unvoiced offensive or risky thought as a stepping stone to a better, more creative idea to share with the group.

Rule 4: Do not judge ideas. Do not put down ideas or the people who express them (including yourself)! However, humor, favorable exclamations, laughter, and applause are approved responses. In brainstorming there are no dumb ideas or right and wrong answers. Brainstorming is a deferred-judgment activity—idea evaluation and critical judgment come later in the creative problem-solving process.

Freedom versus control

Strongly left-brain thinkers may be particularly uncomfortable with two aspects of brainstorming: sharing ideas (quadrant C) and wild ideas (quadrant D). Give yourself permission to play and express all kinds of ideas. Brainstorming is fun! Be surprised by the freedom of the "storm."

Problems cannot be solved by thinking within the framework in which the problems were created.

Albert Einstein

Idea sharing: When brainstorming is a team activity, you as an individual cannot "hog" your own ideas or take credit for them. The interaction that occurs between the minds of the team members is important. Share all your ideas—someone else may use your idea as a stepping stone to another idea, which in turn is used by a third person to come up with something new—and you may just use that idea to think of something even better. But don't wait for the perfect idea. Look for successive steps forward! Idea sharing is not easy for some of us. We do not have much training in this type of thinking because it is strongly discouraged in our schools. Brainstorming is different—it is teamwork, and you are supposed to make use of the ideas of others. Remember that the information you have for solving a problem is not complete and not identical to that of your team members. When you collaborate with the others, you will be surprised to find what the "team mind" can achieve.

Wild ideas: Having or sharing wild ideas may make you feel ridiculous, or you may feel that others will laugh at your ideas. Please do not be self-conscious; everyone in your team is in the same boat. As you learn and practice creative thinking through a conscious effort, it will become easier to express wild ideas and overcome a "business as usual" mindset. Wild ideas are valuable at this stage because the normal forces later in the creative problem solving process will tend to make them more practical, especially during the "engineering" and judgment phases.

Conversely, strongly right-brained people may find it hard to accept constraints and follow procedures. Procedures can ensure that the process will be as efficient and productive as possible. For example, as we have seen in Chapter 4, the rules of etiquette and common courtesy make for a more congenial atmosphere in the team and thus enhance creativity.

How broad or narrow are our constraints? This may determine the number of options we can envision for a specific design.

Constraints: Constraints can both help and hinder brainstorming. They attempt to contain the "storm" within a specific goal or problem area. If they are too rigid right from the beginning, a vigorous "storm" can't develop, or many creative ideas may be rejected out of hand as not "fitting" the problem. Yet outrageous ideas can be the impetus to especially innovative solutions. Thus we recommend that any constraints included in the briefing be evaluated. Must they be present at this point, or could they be introduced more profitably at a later stage in creative problem solving? In general, a limited number of carefully thought-out constraints may not significantly affect creativity. The problem definition statement is a constraint: it provides direction and a target for idea generation, as well as boundaries. But team members should also have permission to push the boundaries—this is when breakthrough ideas may appear.

Design constraints need to be examined carefully because creative thinking can be used to eliminate the need for some of the constraints, particularly if these are arbitrary. Such constraints represent an opportunity for quality improvement and innovation. An example happened in the design of aircraft gas turbines, where for many years the distance from the disk to the root of the blade was chosen based on steam turbine practice. The resulting high disk temperature severely restricted the choice of material, resulting in high cost and limited strength. About 10 years later, Rolls Royce increased the distance by making the disk smaller. Simultaneously, the extended root of the blade was hollowed out. This decreased the temperature at the rim of the disk and allowed the use of better materials; this also resulted in a large reduction in the weight of the rotor. Thus, be on the lookout for assumptions and unspoken constraints such as, "We've always done it this way."

Brainstorming in engineering design: Brainstorming is used in many stages of engineering design. A newly formed design team may want to brainstorm team rules and how the performance of each team member should be evaluated. Brainstorming can be used to develop the customer survey and the design goals, separately or in complex procedures such as QFD (see Appendix A). Brainstorming is a main technique for generating conceptual design options (which will then be evaluated with the Pugh method that uses additional brainstorming to optimize designs). As we shall see, brainstorming is used in the judgment phase to help a team develop a good list of design evaluation criteria. Therefore, it is essential that students learn to be comfortable with both the thinking skills and the procedures needed for productive brainstorming.

Planning and leading a verbal brainstorming session

We are going to present the procedure for brainstorming from the leader or facilitator's point of view. Since you are studying this subject, it is quite likely that you will be the best-trained person in a group and thus will be "elected" to lead the brainstorming session in your organization, in your circle of friends, or in your family. We will summarize the preparations needed for a brainstorming session. Then we will go through the step-by-step procedure of conducting a brainstorming session listed in Table 8.1. We strongly recommend that you immediately practice these principles with a team.

Preparation

This involves selecting a team, choosing a location, scheduling, and preparing the materials needed.

Team members: Several people with quadrant D preferences should be part of the brainstorming team. If you are brainstorming an engineering design concept, the team should represent (directly or indirectly) such stakeholders as customers, sales, process engineering, design, and manufacturing. Make sure that each team member receives the briefing document ahead of the brainstorming session if possible.

Location: People are able to think more creatively if they are in an unfamiliar location. If it can be done, find a place off-site, with beautiful, relaxing surroundings. At the minimum, select a room that is "different"—not the room regularly used for meetings. People should be seated in a circle or U-shaped arrangement, not facing each other across a long conference table. If you must use a conference room and cannot change the arrangement of table and chairs, enhance its atmosphere by having classical background music, colored posters on the wall, and perhaps flowers and a snack with enticing odors. In Japan, brainstorming "camps" are used regularly and have been found to be very effective. In the U.S., we have seen "creativity camps" or "adventure excursions" advertised.

Many ideas grow better when transplanted into another mind than in the one where they sprang up.

Oliver Wendell Holmes

Scheduling: Brainstorming is exhausting; thus do not schedule more than two topics (or a three-hour period). Morning sessions are usually more productive, before people have become involved with their daily problems and routines. Schedule a sufficient time so people will not be pressured or hurried by later appointments. Again, the "camp" idea removes the time pressure—thus try to incorporate this idea into your scheduling. The theme can be carried through to your session announcements, agenda, and schedule reminders to the team members.

Materials: Obtain and set up the necessary equipment: easels, flip charts, markers, note cards, and visual aids or props to stimulate creative thinking. For long sessions also provide some refreshments. A tape recorder is useful to capture ideas and comments that may not get written down during the session. For a large team (a dozen people or more), you may want to have an assistant who can help write down the brainstormed ideas. Student teams, to keep down the expense, can use newsprint, butcher paper, or 4x6 notecards (and a heavy pen) to write down ideas.

Procedure

Table 8.1 outlines the procedure used for verbal brainstorming. Each item will be discussed below.

Table 8.1 Procedure for Leading a Brainstorming Session
1. Brief the team on the background of the problem; then post the problem definition statement.
2. Review the four brainstorming rules.
3. Explain the brainstorming procedure that will be used.
4. Do a creative thinking warm-up exercise.
5. Conduct the brainstorming.
6. End the session; collect all ideas.
7. Thank and dismiss the participants.

1. Briefing: Give the team a few minutes for social interaction and for each person to comfortably stake out a personal "space" in the seating arrangement. Turn on the tape recorder and open the session with a review of the briefing, inviting the team members to share any insight or ideas that came to their minds during incubation. Jot down solution ideas on the flip chart, sequentially numbering each idea. Then post the problem definition statement developed in a previous meeting. The team can clarify and modify it as needed in a brief discussion. Make sure that anything distracting, either on people's minds or in the room's environment, is taken care of before the actual brainstorming starts.

2. Review the rules: Review the four brainstorming rules and the "three strikes and you're out" policy for preventing negative thinking (p. 101).

3. Explain the procedure: In a small team of three to five members, ideas can just be called out as fast as they can be written down on the flip chart or on large sheets of paper posted on a wall. Alternately, ideas can be written down on 4x6 notecards (one idea per card). These cards need to be spread out on the table so the ideas remain visible to the team as brainstorming continues. In larger teams with up to a dozen members, people can take turns speaking. The other participants *must jot down all ideas* that flash into their minds on a note pad or note cards, so they won't forget these while they await their turn to share. Arrange a signal—such as a raised hand or snapped finger—to be used when someone has a modification or addition to an idea that has just been presented. Such hitchhiking is given priority. The combination of two posted ideas will be counted as a new idea. Ask for brief statements only; the "engineering" phase—the next step in creative problem solving—will provide opportunity for elaboration. Explain that all ideas will be numbered and recorded. Set an initial time limit of 20 to 30 minutes (depending on the complexity of the problem). Optionally, adding a quota can increase the number of ideas that will be generated. Example: "Let's see if we can come up with 40 ideas in 15 minutes."

4. Warm-up exercise: Conduct a 5-minute warm-up in creative thinking using a simple, familiar object (brick, pencil, popped corn, ruler, coffee cup). Some experts recommend that classical background music be turned on at this time and played until the end of the brainstorming to encourage the use of the right hemisphere of the brain. This exercise is a mini brainstorming session: jot down the ideas on a flip chart—they do not need to be numbered or be sequentially arranged. At first, mundane ideas will be expressed. Once the wild and humorous ideas come forth and the team members relax with laughter, their minds are "primed" and you can immediately move to brainstorming the defined problem.

5. Brainstorming: Ask the team members to start sharing ideas. They can begin by bringing out obvious, well-known ideas—these have to be purged first before the mind will be able to bring out some really new, creative ideas. This process is also known as "load dumping." Make sure all ideas are written down by yourself or an assistant. If the flow of ideas is very slow at the beginning—or when it slows down later—you as the facilitator can encourage the process by throwing out an outrageous or humorous wild idea that can serve as a stepping stone. Or the team can start on a spree of wishful thinking by asking what-if questions. If things still are not rolling, the session can be interrupted for a brief "excursion" for relaxation, then started again by using a force-fitting technique. This should start the flow of creative ideas. Don't rush into this; two or three quiet periods to allow reflection can be beneficial.

Experts say that the best ideas in brainstorming are often generated after two or three "periods of calm" that give the mind a chance to incubate.

6. Close: Once the flow of ideas has slowed down to a trickle and the previously announced time limit is coming up, give a 3-minute warning. Some of the best ideas are often generated during this extra time at the end. Alternately, you might want to challenge the team to come up with five additional ideas, then don't be surprised if you get twice as many before idea generation comes to a halt.

7. Dismissal: Thank the team members for their participation and let them know what will happen next. Collect all the ideas that were written down, as well as the tape recording, for later processing and evaluation. Encourage the team members to e-mail you additional ideas that might come to them in the next few days.

✓ Three-Minute Activity 8-1: Team Name

Brainstorm a name for your team or class. Jot down all brainstorming ideas on a flip chart, then save the list for Critical Thinking Activity 10-5.

✓ Ten-Minute Exercise 8-2: Creative Thinking Warm-Up

Brainstorm *as many uses as you can think of* for a one-foot square piece of aluminum foil. Then compare your list with the list given at the end of the chapter in Figure 8.2. Now can you think of five additional "crazy" ideas?

✓ Team Activity 8-3: Creative Problem Solving Project

Conduct the idea generation phase for your design project—for the problem you explored and defined in the hands-on activity of Chapter 7. You may use the same team, or a team enhanced by additional members. For example, if you are a team of students in a senior design project, consider including some engineering freshmen or liberal arts students. Do not forget the creative thinking warm-up.

a. If the entire class is using the same design topic, brainstorming can be done in class, either using verbal brainstorming or the panel method. The instructor will collect all ideas.
b. If each team of students will work on its own design project, brainstorming will need to be done as part of a team meeting. Schedule sufficient time. The note taker should be in charge of collecting and safe-keeping all ideas. To keep a record of the brainstorming session, the ideas can be typed up and a copy handed to each team member, to be added to the notebook. Also, the team process observer needs to write a brief summary about the brainstorming experience of the team. These two write-ups become a part of the creative problem solving learning process being documented in the notebook.

Debriefing

What were the results of your brainstorming exercise? How did the process go? Were you pleased with the outcome and the variety of ideas that were generated? Would you have been able to think them all up yourself? Is it necessary to be an expert in the problem area to have creative ideas? Did you and your team members get tired? That should not surprise you—brainstorming is mentally exhausting. This is why it is preferably done in the morning, when people are well rested and have fresh minds. That is also why it is not usually done for more than one hour at a time or for more than two problems per day. Under optimum conditions and with an experienced team, your output will become even more productive and creative.

Did you experience some shortcomings or notice a problem during your verbal brainstorming session? This technique usually works well in all-female or all-male groups of up to a dozen members, especially with people who are comfortable with each other and like to express themselves verbally. It also works well in a collaborative, innovative climate. But what if you do not meet these conditions—what if you have shy (or domineering) team members? What if you have a group of 20, 100, or more people that you want to involve in brainstorming? What if you do not have a leader to keep the group focused? What if there is open conflict between team members who must, for some reason or other, be involved in the brainstorming? To address different circumstances and problems, variations of verbal brainstorming have been developed. We will now briefly look at some of the more popular techniques.

Other brainstorming methods

Written brainstorming has been found to work well for engineers and for mixed-gender groups. It allows for teams larger than a dozen members, and it works well for shy people. The disadvantage of written brainstorming is the lack of direct verbal interaction between the team members; the quantity of ideas may thus be reduced. Some of the written brainstorming methods can be used by individuals working alone on a project, or they can be done sequentially (by letter, on a bulletin board, or by e-mail) by a group of people who cannot meet in the same place at the same time. *For all written brainstorming, make sure you follow the four brainstorming rules.* Quickly write down each idea, just as it comes to mind—don't worry about the ideas being practical, crazy, good, or dumb; also do not be concerned about grammar or spelling.

Pin card method: People sitting around a large table write down ideas on note cards—one idea per card. These cards are then passed to the left around the table, and group members are asked to add their related ideas and improvements to the original idea on the card. Several levels of additions can be made to the original idea in this way. Since this process is somewhat anonymous, it can get people involved who otherwise may feel too intimidated to contribute creative ideas. When the process of sending new cards around has slowed down to a trickle, the session is terminated, and the cards are collected for later evaluation by a different team. One application for this method could be in a family circle with several teenagers, because the people involved can concentrate on ideas and will not be influenced by an argumentative tone of voice.

Brainstorming sessions should "bubble" with laughter. Funny ideas are often stepping stones to the best solutions.

Crawford slip writing: This is used to collect ideas when large groups of people want to be involved in brainstorming. After the problem definition has been presented, each participant is asked to write down 20 to 30 ideas on slips of paper, with each idea on a separate slip. The slips are collected quickly, before the people have time to make corrections or delete ideas. A different task force is then used for sorting the ideas into categories and evaluating them to arrive at a workable solution. This method can be used by large organizations—thousands of people can be involved in this way. Sometimes the number of ideas can be cut down if the people are asked to do a bit of prejudging and only submit their top two or three ideas. But then the most unusual, crazy idea may be thrown out too soon—thus prejudging is not usually a good approach. Group interaction can be inserted into the process by having small groups of two or three people brainstorm ideas for submission.

The "Ringii" process: An interesting method with minimal face-to-face interaction is the Japanese "Ringii" process. Here, an idea is submitted on paper to others in an organization. These people may make any modification or addition to the idea. The original proposer can then use these suggestions to rework the original idea, or a synthesized solution can be worked out by an independent panel. The second approach can be used in cases when the original proposer wants to remain anonymous. This process is beneficial in large and small organizations (including families) when there is some problem with communication, with people being confrontational, or with conflicting schedules.

Panel method: If a large group is present, say from 20 to 30 people, and it is not possible to separate them into smaller groups for brainstorming, the panel format can be used. Seven volunteers are chosen from the group and formed into a panel. The problem definition is presented to the entire group, then the panel verbally brainstorms the problem for 15 to 20 minutes, with these ideas being posted on a flip chart. The other group members write down their own original ideas as they listen to the panel and try to hitchhike on the posted ideas. The posted ideas of the panel are collected for later evaluation, together with the written ideas of the audience. We have found this variation suitable for the classroom, where a second panel of students can get a turn after the first 10 minutes and where students get rewarded for turning in additional unique ideas.

Story board: Here a matrix visually displays ideas in several categories. This method can be used for brainstorming, planning, or idea evaluation. Title cards (headers) are made up for the important factors involved in a problem. (In implementation planning, these could be the words who, what, where, when, why, and how.) These headers are posted across a large bulletin board so a logical relationship exists among the categories. The first category is always "purpose." Then each category is brainstormed, and the ideas are posted on index cards or post-it notes below the appropriate header. Through this visual arrangement of ideas, additional creative ideas and solutions can be triggered, because the items in the different category columns on the board can be combined in different, unexpected ways. After the brainstorming, the group conducts a critical thinking session to eliminate the idea cards that do not meet the objectives. Then the remaining ideas are creatively improved. Many organizations outside advertising and film making use the storyboard for planning, communication, implementation, and follow-up.

Imagination is more important than knowledge, for knowledge is limited while imagination embraces the entire world.

Albert Einstein

Electronic brainstorming or bulletin board: Electronic brainstorming is used by Bill Gates of Microsoft in his company. People are connected via e-mail; when someone has a creative thought, it can be sent to other computers where a signal will flash on. Thus instant feedback and hitchhiking ideas can be obtained. The low-tech equivalent is the bulletin board. The problem definition (with a short briefing about the problem's background) is posted in a prominent place for several weeks; anyone can post new ideas as well as hitchhiking ideas at any time. These ideas are then collected and evaluated by a team or a single judge. The bulletin board is a method that is very appropriate for children.

TRIZ uses three tools to encourage inventive thinking based on science and technical knowledge:

1. A patent search reveals the evolution of technical systems.

2. Contradictory needs must be accommodated with problem solving, not trade-offs.

3. An ideal, imaginary system models how all functions can be met.

Although engineers in general like this method because of its emphasis on science, they need training in the creative aspects of Steps 2 and 3.

The method is taught in the former USSR and other European countries from fifth grade on up.

Other methods: Consult the references at the end of the chapter if you want to find out more about the following techniques. In the *gallery method,* group members work silently on their own flip charts; after inspecting the ideas of the others, they elaborate on their own ideas. The *nominal group technique* is used when time is very short, since it combines idea generation, evaluation, and decision making. *Method 6-3-5* was the first written brainstorming technique (developed in 1970). Six people are instructed to produce three ideas in five minutes; these are then passed to the next person in the circle, and the process is repeated five more times. To relieve the stress of this method, it was modified. Thus, in the *brainwriting pool,* people can work at their own pace. In the *Delphi method,* ideas are collected by questionnaire or on-line, and several rounds are conducted until consensus is achieved. The method requires a judge or "jury" and is often used for planning the future direction of an organization. *TRIZ* was developed by a Russian inventor; its aim is to help engineers who use the method be more inventive in solving technical problems (see sidebar). *Mindmapping* is a great tool for individual brainstorming—see Reference 2.2. *Integrated problem solving* is used when only a few ideas are expected for simple or for very complicated problems. Brainstormed ideas are discussed one at a time and then combined into a compound solution. The *collective notebook, morphological creativity,* and *Synectics* were discussed in Chapter 7, and the *idea trigger method* was demonstrated in Chapter 4.

What to do when you are "stuck"

Several techniques are available to start ideas flowing again when a team is "stuck." These methods "force" the mind to make creative leaps. The next step in creative problem solving—making ideas better and more practical using the "engineer's" mindset—uses force-fit thinking.

Imagine success or imagine the worst: One of the easiest ways to free a mind that is stuck on a problem is to reverse the direction of the problem-solving process in the imagination. Call a time-out and turn the problem around. Instead of focusing on the problem and trying to think of solutions, concentrate on the ideal state or the "what should be." The mind will fill in the steps on how to get to this ideal state. Record all ideas; usually the team will quickly move to verbal brainstorming after discussing this change in viewpoint. If brainstorming has been hampered because of the presence of constraints, mentally remove the constraints for 10 minutes and generate ideas with this new "frame." Another way to turn the problem around is to brainstorm the "worst things to do" to solve the problem. The absurd ideas that will be generated will serve as stepping stones to practical and innovative ideas, or they will generate laughter and loosen up people's thinking to where they are able to continue productive brainstorming.

Force-fitting two unrelated ideas. Activity 8-4 is an illustration of this technique. It can get a sluggish brainstorming session going; it can be used for improving and hitchhiking on ideas that have already been posted. When brainstorming has slowed down, team members select two very different ideas that were generated earlier and attempt to fit them together. This process will often result in additional creative ideas.

✓ Three-Minute Activity 8-4: Force-Fitting Ideas

In a group of five or more people, quickly brainstorm this concept: How could you use the idea of caged white rats to improve the food and atmosphere in a school cafeteria?

Examples: Have a wild animal decorating scheme. Serve pizza in the shape of white rats. Use a squirrel cage for students to let off steam. Have a magician perform in the cafeteria during lunch time. Have students do a research project using white rats to test the nutritional value of typical cafeteria meals. Can you see how different aspects of the two unrelated ideas lead to creative as well as practical ideas? This technique can be used as a creative thinking warm-up for a brainstorming session. Invent additional pairs of unrelated ideas that would make a good warm-up.

Free association: This technique stimulates the imagination. The process is started by jotting down—on the blackboard—a symbol that may or may not be related to the problem. This can be a picture, a word, a sketch, a numeral, or a relationship. The process is continued by jotting down a new symbol suggested by the first. This chain is continued until creative ideas related to the problem emerge. These ideas then become part of the brainstorming process and are recorded. You might already be familiar with this method from children's games and psychology.

Big dream/wishful thinking: Group members think of the biggest, far-out dream solutions to the original problem. Then the big-dream idea is further developed by wishful thinking and by asking related what-if questions. All ideas coming out during this process are recorded. When these ideas begin to be more closely related to the problem at hand, continue with regular verbal brainstorming. You can really have some fun with this approach. This technique helps to "loosen up" a group that is too analytical and practical-minded.

> **REMINDER**
>
> **Be sure to follow the four rules even if you brainstorm alone.**

Forced relationship matrix: This method resembles morphological creativity. From definitions of possible forms and elements of the original problem, relationships are determined between them—such as similarities, differences, causes, and effects. These relationships are recorded and then analyzed to find new ideas and patterns. Especially when opposing and absurd concepts are combined in different ways, creative ideas may suddenly emerge. This technique can be practiced by rearranging the words in a short sentence. For example, different combinations of the two ideas of PAPER and SOAP give us paper soap and soap paper (both nouns), soapy paper and papery soap (adjective/noun combination), or papered soap, soaped paper, soap "wets" paper, or soap "cleans" paper (verb forms). Then each of these concepts is used as a "trigger" for creative ideas, depending on the original problem. In the example, this approach could lead to some good ideas if your goal is to develop washable wallpaper or a new way of packaging soap.

Thought starter tools: Dr. Alex Osborn, the inventor of verbal brainstorming, developed a thought-starter chart (shown in Table 8.2) as a tool for helping people generate creative ideas. The acrostic SCAMPER—substitute, combine, adapt, magnify/modify, put to other uses, eliminate, rearrange/reverse—will remind you of this list. Idea generator tools based on Dr. Osborn's approach have been developed by other inventors and are available commercially, either as small tables, hand-held tools, decks of cards, large wall charts, or software packages. In essence, they are just different ways of asking "what if?" and "what else?"

Table 8.2 Dr. Osborn's Nine Thought-Starter Questions

1. **Substitute?** Who else instead? What else instead? Other place? Other time? Other ingredient? Other material? Other process? Other power source? Other approach? Other tone of voice?
2. **Combine?** How about a blend, assortment, alloy, ensemble? Combine purposes? Combine units? Combine ideas? Combine functions? Combine appeals?
3. **Adapt?** What else is like this? What other idea does this suggest? Any idea in the past that could be copied or adapted?
4. **Magnify?** What to add? Greater frequency? Stronger? Larger? Higher? Longer? Thicker? Extra value? "Plus" ingredient? Multiply? Exaggerate?
5. **Modify?** Change meaning, color, motion, sound, odor, taste, form, shape, or texture? Other changes? New twist?
6. **Put to other uses?** New ways to use object as is? Other uses or purpose if modified?
7. **Eliminate?** What to subtract? Smaller? Lighter? Slower? Split up? Less frequent? Condense? Miniaturize? Minify? Streamline? Understate? Simplify?
8. **Rearrange?** Other layout? Other sequence? Change pace? Other pattern? Change schedule? Transpose cause and effect?
9. **Reverse?** Opposites? Turn it backward? Turn it upside down? Turn it inside out? Mirror-reverse it? Transpose positive and negative?

Attribute listing: This technique is used as a checklist or a matrix. All important attributes, parts, elements, or functions of the problem or object under consideration are listed; then the team focuses on each part in turn for new ideas. The questions to ask are, "Why does it have to be this way?" or, "Could it be done differently?" When an attribute listing is combined with Osborn's thought-starter questions, it is called the sequence-attribute/modifications matrix (SAMM). Although its major application is for identifying promising areas for brainstorming, it also can be used as a tool to get idea generation started in a particular area.

Bionics: This simple technique is useful for starting creative thinking. It employs analogy to living organisms by asking: "How is the problem solved in nature?" Examples: People in a Synectics brainstorming session thought up the idea of using the pressure distribution in a camel's foot on sand to design a new tire for a dune buggy. The wings for a superlight aircraft were designed using the wing of sea gulls as a model to make the aircraft maneuverable yet stable in high winds. The structure of a moth's eye (perhaps the most antireflective surface known) was taken as a basis for improving the performance of optical-disc storage systems. Flight tests with owls in an anechoic chamber yielded insight on how to reduce vortex noise generated by the frame of an aircraft.

The force-fit game: Force-fitting is not always serious business—a game can be used as a creative thinking warm-up. To play the force-fit game, the team is divided into two groups and given the problem definition statement. Group 1 shoots out an idea that is completely unrelated to the problem. Group 2 tries to turn the idea into a practical solution for the original problem. If they succeed, the second group earns a point; if not, the point goes to the first team. The two teams alternate in posing crazy questions and finding good applications. All ideas and solutions are recorded. The game combines imaginative thinking and wild ideas with the process of force-fitting two unrelated ideas and is thus a good creative exercise in its own right. The game is continued for a few rounds, until answers come easily; regular brainstorming is then resumed. Younger students especially like this activity.

A "value added" of brainstorming is building group spirit.

Resources for further learning

Reference books

8.1 Henry G. Altov (Altshuller), *The Art of Inventing: And Suddenly the Inventor Appeared,* translated and adapted by Lev Shulyak, Technical Innovation Center, Worcester, Massachusetts, 1994. This is *the* manual on TRIZ.

8.2 Charles Clark, *How to Brainstorm for Profitable Ideas,* Creative Education Foundation, 1050 Union Road, Buffalo, New York 14224. This is just one of the fine books available from this organization for helping people do brainstorming. This organization was founded by Dr. Alex Osborn in 1945 and teaches workshops and summer programs on creative problem solving.

8.3 V. A. Howard and J. H. Barton, *Thinking on Paper*, Morrow, New York, 1986. This small hardback book teaches how to generate ideas by writing. The focus is on writing as a thinking tool and thus goes beyond the traditional view which considers writing as communication.

8.4 Stanley Krippner and Joseph Dillard, *Dreamworking: How to Use Your Dreams for Creative Problem Solving,* Bearly, Buffalo, New York, 1988. This textbook-workbook combination would be of interest to students wanting to find out more about the value of dreams in creativity and idea generation; it gives many examples from history.

Laughter is the brush that sweeps away the cobwebs of your mind.

Randall Munson, founder and president of Creatively Speaking

8.5 H. A. Linstone, *The Delphi Method: Techniques and Application,* Addison-Wesley, Reading, Massachusetts, 1975. This group idea generation technique is designed for futures forecasting.

8.6 ✓ Alex F. Osborn, *Applied Imagination—The Principles and Problems of Creative Problem-Solving,* third revised edition, Scribner's, New York, 1963. This book by the inventor of brainstorming is well worth reading (especially by team leaders); it explains the technique and its applications.

8.7 George M. Prince, *Practice of Creativity,* Macmillan, New York, 1970. This book is required reading for anyone who wishes to study Synectics. Also, it has useful comments on committee meetings and the briefing document.

8.8 Arthur B. Van Gundy, Jr., *Techniques of Structured Problem Solving,* second edition, Van Nostrand Reinhold, New York, 1988. Over one hundred proven problem-solving techniques are explained and evaluated.

Idea generator tools

Software programs such as IDEA GENERATOR PLUS, IDEA FISHER, MINDLINK PROBLEM SOLVER, BRAINSTORMER (based on morphological creativity), and INSPIRATION have been available in the last few years for brainstorming using a computer. We are not aware of rapid advances in this software. Some people enjoy these tools and find them useful, especially for brainstorming alone; others think they are too complicated and do not match the productivity of a classic team brainstorming session. We recommend that you interview some users and try out a software program before purchase, to see if it would meet your needs and expectations.

Card decks to help generate creative ideas during brainstorming are available commercially, such as the THINKPAK by Michael Michalko, the WHACK PACK by Roger Von Oeach, and BOFF-O!™ (Brain on Fast Forward) by Marilyn Schoeman Dow/ThinkLink. Because these cards are visual, hands-on, and playful, they can be quite effective in quickly generating "wild, wonderful, workable" ideas. These tools, as well as books featuring creativity and innovation, are available from the ACA Bookstore (associated with amazon.com) on the website of the American Creativity Association at **www.BeCreative.org**.

Eighty Uses for a Square of Aluminum Foil

1. Wrap food.
2. Cook (bake) food.
3. Conductor.
4. Ball.
5. Sun reflector.
6. UHF antenna.
7. Frost hair.
8. Christmas decoration.
9. Drip pan liner.
10. Boat for mouse.
11. Wrap package.
12. Shred for tinsel.
13. Scarecrow.
14. Stencil.
15. To make a relief print.
16. Hold hot or sticky pan.
17. Wrap pop for freezer.
18. Wrap sandwich.
19. Use as lid.
20. Mirror.
21. Crinkle and make a texture.
22. Imprint (rubbing).
23. Punch holes for filtering sand.
24. Cover vent.
25. Silver confetti.
26. Distress signal.
27. Start fire.
28. Get rust off other metals.
29. Put in shoe for temporary repair.
30. Demonstrate static electricity.
31. Jewelry.
32. Funnel.
33. Cake decorating tool.
34. Bookmark.
35. Temporary fuse.
36. Mouse suit.
37. Deflector.
38. Angel halo in Christmas pageant.
39. Flag for an alien country.
40. Wrap candies that you can eat in church.
41. Cigarette lighter.
42. Use as fan.
43. Little table mat.
44. Eye mask.
45. Stuffing for drafts and holes.
46. Melt to use as filling.
47. Make little toy animal.
48. Make little toy dishes.
49. Make "emergency" wedding ring.
50. Wrap for a small bouquet of flowers.
51. Beautify a flower pot.
52. Catch water under flower pot.
53. Book covers for "silver" library.
54. Make emergency drinking cup.
55. Make windmill toy.
56. Crease every inch, then use as ruler.
57. Make play money.
58. Make wall decoration to cover defect.
59. Make a butterfly mobile.
60. Roll up, use to blow soap bubbles.
61. Shower cap.
62. Shade for a transplanted plant.
63. Fountain for architectural model.
64. Garbage bag for "yucky" stuff.
65. Recycle.
66. Grill cover.
67. Shelf liner.
68. Candy mold.
69. Gift wrap.
70. Pie pan.
71. Emergency gas cap for car.
72. Picture frame.
73. Bird cage liner.
74. Window shade.
75. Party streamers.
76. Hair "spikes."
77. "Tin man" costume for doll.
78. Emergency purse to carry small stuff.
79. Creative art material.
80. Shoe shield for walking through mud.

Figure 8.2 Answers to Activity 8-2: creative thinking warm-up.

Exercises for "artists"

8.1 Warm-Up Exercise

Find different uses for one of the following "fun" objects as a warm-up exercise for a brainstorming session—a worn sock, a mirror, a feather, a bucket of sawdust, a peanut, an old sneaker, a Frisbee, or a pumpkin.

8.2 ✓ What-if Creative Thinking Warm-Up

Pose a what-if question and play around with it for a while, preferably in a group (but this activity can be done alone also). The what-if questions do not have to be practical; the exercise is even more valuable if you practice it with a wild or impossible idea. If you cannot think of a what-if question, select one from the following list.

a. What if gravity were suspended for 10 minutes each day—how would bedrooms have to be redesigned?

d. What if one country were suddenly occupied by aliens from outer space—how would (or should) people react?

e. What if trash could be made desirable—what would be the effects? How could it be made so?

f. What if people all looked identical—how would one be identified as an individual?

g. What if insects worldwide suddenly quadrupled in size—would this mean a new food supply or a disaster?

h. What if you were stranded on a desert island with the three people you most dislike—what would you do to make this a pleasant experience?

If someone laughs at your idea, it's likely a sign that you've been very creative.

8.3 "Ringii" Process

Find an application for the "Ringii" brainstorming method. Go through the procedure, then write up a summary of your results.

8.4 ★ Crawford Slip Writing ★

If you belong to a club where more than ten people attend a meeting, look for an opportunity to apply the Crawford slip writing method. Are you looking for ideas for some club activity or fund-raiser? Get together with the club officers to make up a problem definition statement and arrange for a brief period of idea generation—with people silently writing their ideas on slips for 10 minutes or so. Then collect all ideas. You may want to enlist the help of a committee to evaluate the ideas and find the best solution.

8.5 ★ Force-Fitting Example from Technology ★

Think about the development of the razor in the course of history. Draw an analogy to the design of a lawn mower. What kind of a mower would you design using each type of razor? Now reverse the process—can you think of an improved shaver by drawing an analogy to advanced lawn mower technology? Now extend the analogy to other types of cutters and hair-grooming tools—can you think up some wild as well as practical modifications?

8.6 ★ SAMM ★

Do a literature search and write a summary report (including an example) that explains the sequence attribute/modification matrix (SAMM). How is it related to Dr. Osborn's nine thought-starter questions and the method of attribute listing? How do these techniques compare to the storyboard?

Chapter 8 — review of key concepts and action checklist

Your role as "artist": For brainstorming, break out of the left-brain mold and use your quadrant C and quadrant D thinking modes. Welcome eccentric ideas! "Artists" take information (from the problem definition phase) and transform it into new ideas.

Verbal (classical) brainstorming: This method for generating creative ideas with a group was developed in 1938 by Alex Osborn. The four rules are: 1. Generate as many solutions as possible—quantity counts. 2. Wild ideas are welcome—be as creative as you can be. 3. "Hitchhiking" is encouraged—build on the ideas of others. 4. No criticism is allowed—defer judgment until later. A number of written brainstorming methods have been developed since the 1970's to address different circumstances and accommodate larger groups. Special techniques can be used to jump-start groups whose creative thinking is "stuck."

Procedure for leading a brainstorming session:

1. Briefing and review of the problem definition statement.
2. Review of the four brainstorming rules.
3. Explanation of the brainstorming procedure that will be used.
4. Creative thinking warm-up exercise (5 minutes).
5. Brainstorming (say 20 minutes, with an extra 5 minutes at the end).
6. Closing and collection of all ideas.
7. Thanking and dismissing the group.

Recipe for a mini-adventure:

Go on an occasional wild goose chase. That's what wild geese are for.

Action checklist

☐ Think about applications for brainstorming in your daily life. Also, just for the fun of it, schedule a brainstorming session as a social activity to create a song, dance, or children's story book.

☐ Try using Post-it notes or note cards (one per idea) when brainstorming a small problem with a small team during a meeting. Shuffle the notes around to generate additional creative ideas through force-fitting.

☐ Use the bulletin board method to brainstorm a problem in a situation where you want to involve a larger group but where it is impossible to get everyone together at the same time. Post the problem definition on the bulletin board and have a stack of blank cards and pens available. Encourage the participants to check the bulletin board frequently to add hitchhiking ideas.

☐ Make it a personal habit to always carry a pack of notecards with you. When an interesting idea comes to you, immediately jot it down or visualize it in a weird image or story.

9 Creative Evaluation

What you can learn from this chapter:
- The goal of this key problem-solving step: developing high-quality, innovative, optimized concepts or solutions through synthesis.
- Traits of the "engineer's" mindset; the four rules and best timing for this second round of brainstorming.
- Three steps for improving ideas: grouping, synthesis, force-fitting.
- Application in engineering design. Process case study.
- Further learning: references, exercises, review, and action checklist.

Look at Figure 3.17 on page 80 and note the position of the "engineer" squarely between the right-brain "artist" and the left-brain "judge." This step in creative problem solving alternates between the two ways of thinking. Creative idea evaluation is primarily a second round of brainstorming, with the goal of developing the "wild and crazy" ideas of the first round into better and more practical concepts for solving the defined problem. This explicit idea improvement phase or directed synthesis is missing in other problem solving schemes. It is a key step for obtaining high-quality, innovative engineering designs. In the knowledge creation cycle, this synthesis activity is part of knowledge conversion from conceptual to systemic knowledge through combination. In the present chapter, we will investigate the mechanism of this thinking process; in Chapter 11, we will see it in action as part of the Pugh method of conceptual idea evaluation.

The role of the "engineer"

Imagineering:
Let your imagination soar and then engineer it down to earth.

Creative idea evaluation is more focused than the divergent thinking of the "artist." We want to add some convergent thinking to clarify concepts and arrive at practical ideas that have the potential for implementation and solving the problem. Although this key process of "engineering" ideas into practical concepts, solutions, and product designs can be applied by any person to improve the quality of the original output of brainstorming ideas, it is ideally suited to be a team activity. In our framework, it is not to be confused with idea judgment which uses explicit criteria and is the next phase in creative problem solving.

Figure 9.1 The "engineer's" mindset.

What do "engineers" do? They design, build, manage, categorize, combine, develop, and synthesize ideas to put them to practical use. Similarly, we will work with our brainstormed ideas as "raw materials" and try to improve them through additional creative thinking. We will question each idea: "How can it be used to construct a superior idea? What is useful or valuable about this? How can this be improved?" In the "engineer's" mindset (illustrated with the team in Figure 9.1), we switch rapidly between quadrant D and quadrant A thinking while keeping a positive, nonjudgmental attitude. Can you think of at least three reasons why green is a good color to represent the "engineer's" mindset?

The four rules of creative idea evaluation

Do you recall the four brainstorming rules? For creative idea evaluation, we also have four rules.

1. Look for quality and "better" ideas.
2. Make "wild" ideas more practical.
3. Synthesize ideas to obtain more complete, optimized solutions.
4. Maintain a positive attitude; continue to defer critical judgment.

You can be wrong, you can commit errors in logic, even record inconsistencies, but I won't care if you can help me to useful new combinations.

J. W. Haefele, Procter & Gamble

Rule 1: Look for quality and "better" ideas. Instead of quantity, we are now aiming for quality. Look for the good in each idea and try to make it even better.

Rule 2: Make "wild" ideas more practical. Use wild ideas as stepping stones or thought starters to generate more practical solutions. This requires iteration between creative and analytical thinking.

Rule 3: Synthesize ideas to obtain more complete solutions. Instead of hitchhiking, we will now try to integrate, synthesize, force-fit, or meld different ideas to develop optimal solutions.

Rule 4: Continue to defer critical judgment. We will continue to abstain from quick judgments and negative comments. A positive attitude is essential during this step—it will help us generate additional creative ideas as we combine intuitive, innovative thinking with analytical, logical, more pragmatic thinking.

A good number of engineers have a double dominance in quadrant A and quadrant D thinking; thus this step should not be difficult to learn and apply. Engineers who are quadrant A and quadrant B dominant must be careful not to be negative and fault-finding during this stage.

Timing and preparation

At every stage, synthesis involves the generation of alternative solutions, that is innovation, evaluation, and decision making.

Commission on Engineering and Technical Systems, National Research Council, 1991

How soon after brainstorming should creative idea evaluation be done? Wait at least one day if the same team will be involved. Brainstorming and creative idea evaluation are both mentally exhausting and are thus more productive if done with fresh minds. By letting the conscious mind rest and the subconscious mind incubate the ideas that were generated, the thoughts that will come up during the evaluation phase will be more creative. This time lag will also give the facilitator a chance to do some preliminary organizing work with the ideas, if desired. In a typical 20- to 45-minute brainstorming sessions, a team may generate from 40 to over 200 ideas. Working with a large pool of ideas can be unwieldy, unless we have a structured approach for making the job easier. The three-step procedure described below is such a tool.

To prepare for idea evaluation, the facilitator must obtain a stack of index cards. We have used three sizes (3x5, 4x6 or 5x7) and think the 4x6 cards work well for most situations. Some teams may prefer to use Post-it notes and work vertically on a wall instead of horizontally on a table. Each brainstormed idea is written down separately with a heavy pen so it can easily be read by a team from a distance. Thus cursive writing is not suitable here. The facilitator, an assistant, or the team members can do this task. Start writing at the top of the card to leave some blank space for notes at the bottom, and include the identification number. When new ideas come to mind during this process (as is quite likely), they are written down on cards, too, and added to the stack.

The facilitator needs to bring the following materials to the evaluation session: the completed idea cards, blank cards, pens in different colors for writing on the cards, paper clips, rubber bands, a flip chart, markers, and masking tape. The meeting room should have a large table or two, as well as empty wall space where flip chart pages can be posted.

The creative idea evaluation process

Creative evaluation is a three-step process as shown symbolically in Figure 9.2 on the following page. It is an open-ended activity involving brainstorming—thus the results are not entirely predictable, even though a structured approach is used.

Synthesis: [syn, together + tithenai, to place]
1. the putting together of parts or elements so as to form a whole.
2. a whole formed in this way.

Task 1—sorting related ideas into categories

The idea cards are randomly spread out over the table. The team gathers around the table to ponder the ideas in silence for a few minutes and let the ideas sink into the subconscious mind. Then it is time to begin looking for similarities and shuffling the cards around. Some ideas seem to

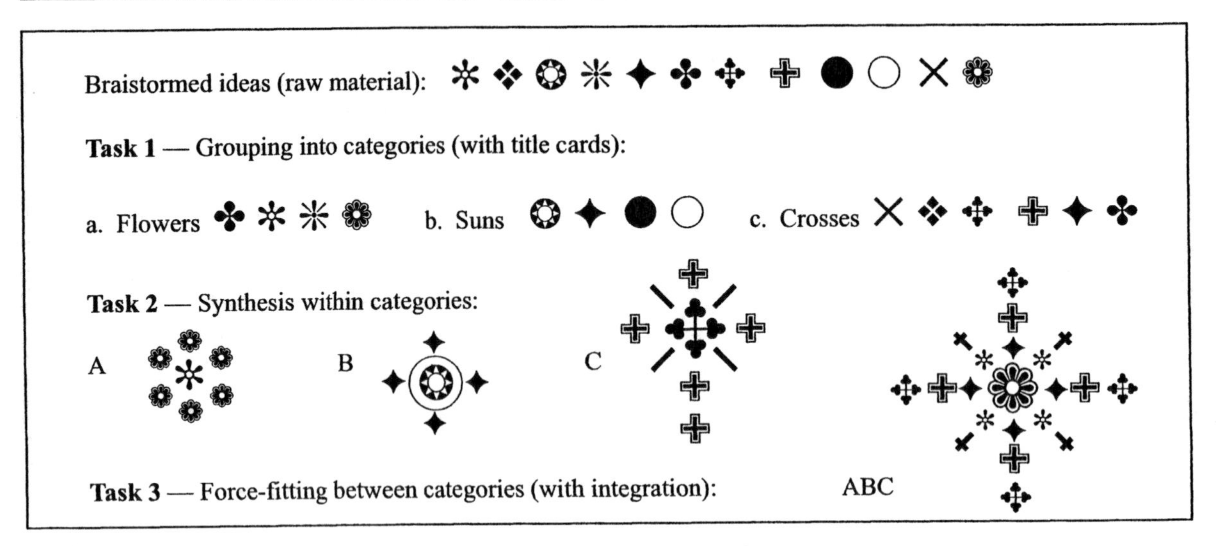

Figure 9.2 Symbolic diagram of the creative idea evaluation process.

naturally want to be together. For these similar ideas, make up a "title" card in a different color, and any ideas that seem to fit can be placed under this category. At this point, do not make these categories too broad and bunch ideas together that do not have much in common. It is quite all right to have many different categories. Team members can have brief discussions about where the ideas should go, but do not get bogged down with quibbling. If an idea seems to fit into more than one category, make up a duplicate card and enter the idea in both.

Again, jot down any new ideas that come to mind (on new cards) and add them to the pool. In our experience, we have found that the sorting process is accomplished rather quickly. Our brain naturally likes to group and categorize ideas. Ideas that do not fit into any obvious category can be placed in the "odd ideas" category. With the title card on top, the idea cards in each category are bundled together with a rubber band. If more than seven categories are present, repeat the process by combining two or more subcategories into a new "umbrella" category. For some topics, it may be difficult to come up with category headings. In this case, ideas can be sorted according to well-known ideas, novel ideas, and wild ideas, or according to the degree of difficulty of implementation—simple (inexpensive) ideas, "meaty" (more challenging) ideas, and difficult ideas (requiring major resources and innovation).

The facilitator can do this organizing work ahead of time. When an entire class is working on the same project (and where only a 45-minute class period may be available to conduct idea evaluation), we recommend that the instructor and a class assistant do this sorting task. They can then assign a category or two to each student team in such a way that each team will have approximately the same number of ideas to work with. The instructor may be able to add some duplicates or new ideas to some of the categories to create a better balance.

Task 2—developing quality ideas within a category

After all ideas have been sorted into categories, each team needs to work with one category at a time. If the team is large, categories may be assigned to heterogeneous subteams of three to five members. Have a breakout room or widely separated tables ready for the subteams to work on their assigned categories. At the start of Task 2, conduct a brief creative thinking warm-up. The objective now is to "engineer" the many ideas or idea fragments within the category down to fewer, but more completely developed, practical, and higher-quality ideas. The team members can discuss the ideas in the category; they can add detail; they can elaborate; they can hitchhike on ideas; they can force-fit and combine ideas. Idea synthesis—combining several concepts or ideas into a new whole—is a key mental process that should be especially encouraged and practiced. Synthesis and integration are illustrated in Table 9.1 with ideas from a brainstorming session by a combined group of high school and first-year college students.

Ten-Minute Team Activity 9-1: Idea Synthesis

a. Starting with the nine ideas given in Table 9.1, use integrated problem solving to synthesize a different solution from the one given in the example. Will your solution solve the original problem?

b. Use the wild idea, "Flood the school's hallways in the winter; keep the doors open at night and create ice tracks for sliding," for insight into underlying needs as well as a stepping stone for further creative idea generation and synthesis. You may want to start the process by expanding the problem through asking a chain of "why" questions. This idea is particularly "wild" if you live in a warm climate but will still serve as a trigger to more practical ideas.

When two ideas are combined, this is considered to be a new idea. To save time, changes and additions to ideas can be made directly on the respective cards. Use paper clips to fasten cards together that have been combined into one idea, with the most developed, synthesized idea placed on top of the stack. Don't be in a hurry to discard "wild" ideas or ideas that do not seem to fit; try to use them as triggers to new creative ideas. It is possible that the most useful and innovative solution to the original problem is hiding out among the wild ideas. Attempt to make well-known ideas better. Examine each novel idea closely. The danger here is that the team may suddenly get carried away with one of the novel ideas. If this happens, do not stop evaluating all other ideas. Continue to look for ways to improve and synthesize ideas to come up with fewer, but higher-quality solutions.

When the team has gone through all of its ideas, the improved ideas for each category can be written on large sheets and posted on a blackboard or wall to facilitate the next step. Alternatively, depending on the types of categories and the original problem, we have seen teams who

Table 9.1 Example of Idea Synthesis Within a Category

Brainstorming topic: How can schools be made better. In all, 262 ideas were generated.

Category: Countywide School System Changes

1. *Specific academies at different schools.*
2. *Skill centers at different high schools.*
3. *Have separate high schools for gifted students.*
4. *Saturday school taught by engineers.*
5. *Create many boarding schools (wild idea).*
6. *More business and trade schools.*
7. *Schools in factories.*
8. *Areas of excellence in all schools.*
9. *Students select the school they want to attend.*

Can #1 and #2 be combined? Yes, and two alternatives come to mind. a. Combine academies and skill centers at each school,, with the same area focus (such as math and science, languages, or the arts and applied arts). b. Have a complementary emphasis. Going with the idea of complementary emphasis seems to lead to a higher-quality school, so let's keep going with this idea. Can it be combined with #3? One of these academies/skill centers could be designated strictly for gifted students. But with several excellent academies/skill centers in a larger community, the gifted would have a challenging environment and would by their very presence help improve the quality of the schools even more. Thus all the academies/skill centers should incorporate programs for the gifted.

What about #4? Saturday school—that's an interesting concept; it could be used to enrich the academic and cultural programs at these academies/skill centers. Yes, let's go with this idea, but let's include other professional people from many walks of life, and also include summer programs. Idea #5 looks especially impractical—but what about creating inviting, homelike areas in existing schools for neighborhood group study under the supervision of parents or older student mentors? With #6, #7, #8, and #9 added, the comprehensive result was:

New Countywide School System Concept:
The school system will be restructured to have diverse, combined academy/skill centers with special programs for the gifted as well as for business and trades (with sponsors from the community); other schools will have special centers of excellence, and all will have Saturday enrichment programs and innovative curricula. Students select the school they want to attend. Schools will be open until 10 p.m. for group study with mentors and as community activity centers—with emphasis on community support for learning and culture by people of all ages.

used tape or tacks to arrange the category and the improved idea cards (or Post-it notes) on a wall as elements of a storyboard. If student teams cannot complete Task 2 during one class period, they can finish the discussion and the posting as homework, but this happens only rarely.

Task 3—force-fitting unrelated ideas between categories

The teams now try to combine the most developed ideas from all categories to come up with superior solutions. This is truly a force-fitting activity because these ideas are very different. Mentally try out different combinations of final ideas (be they simple, meaty, or difficult). Entirely new and interesting ideas may be generated through this process. Again, post the improved, final ideas. However, for some types of problems, it is impossible to distill the large number of original ideas down to a few comprehensive solutions; creative idea evaluation instead results in lists of valuable ideas that, implemented together, will solve the problem. In this case, the entire list is carried forward to idea judgment.

Through this process of examining and discussing ideas, the team gains an understanding of the logic, meaning, and purpose of the ideas, as discussed in more detail in Chapter 11. This is one of the most important benefits of this approach; it enables the team to find high-quality solutions. Thus creative evaluation should not be rushed. This activity can easily take two or three times as long as the original brainstorming, even when the facilitator has done the grouping and Task 2 has been subdivided among several teams. Another benefit of this second round of brainstorming is that completely new ideas may pop up. Quite often, the best idea for solving the problem is generated at this time.

Some teams have trouble stopping at this point, for several reasons. Quadrant B people feel uncomfortable with unfinished business; they want to immediately adopt one of the final "better" ideas as the solution to the problem (or as *the* design concept). Some people want to keep working to exhaustion to find a "perfect" solution—which is impossible. In the next two chapters, we will show how to develop a "best" or optimized solution. Others may drift into a critical mode and begin judging and tearing down the final ideas. To prevent these inappropriate responses, it is necessary to STOP! It will be the responsibility of the "judge" to determine which of the solutions will be best and should be implemented.

✓ Team Activity 9-2: Creative Idea Evaluation

Conduct the creative idea evaluation with the brainstorming ideas from Team Activity 8-3. Start with a creative thinking warm-up. Summarize the improved ideas for each category on a large sheet of paper for later posting on the wall to facilitate idea judgment by the team. If you have a complex problem or a design project, study the Pugh method in Chapter 11 first, before conducting the conceptual idea evaluation with your team.

After you have reduced the number of ideas to fewer but more practical solutions, take the time to analyze the process and the results. Was it easy to avoid negative criticism? Were you able to generate additional creative ideas? Are your synthesized solutions quite different from the original brainstorming ideas? How were the team interactions contributing to the results?

Application in engineering design

We have found in our freshman classes that the creative idea evaluation process for a conceptual design project can have completely different outcomes at the end of Task 3. One result is a list of ideas that can best be described as a specification list for product design. If this happens, this list can be taken by a team to develop at least three different conceptual designs that would solve the original problem. If several teams are working on the same project, each team can generate one concept. These different concepts can later be evaluated and optimized with the Pugh method (see Chapter 11).

If each team in a class worked on different categories for the same topic, Task 2 often results in options for different components or aspects of a product. For example, in a simple project to design an improved luggage carrier, one team looked at handle design, another at the wheels, a third and fourth at two different structural concepts, and a fifth at special features and strapping. In such a case, each team discusses and selects from this "menu" to develop one or two initial conceptual designs for later optimization. In essence what happens here is that the synthesis or force-fitting of Task 3 is postponed and will occur during the Pugh method evaluation and design optimization process.

Whatever form the results of idea evaluation takes, the note taker must keep careful track of all materials that are being generated, and the note taker and process observer should work together to summarize and write up the outcome at the end of each task. An example of "soft" idea evaluation is given below to illustrate the process. Design project examples will be discussed in Chapter 11.

Example of Creative Idea Evaluation

Problem Definition Statement: How can stress be reduced for employees faced with major changes in job status (dismissal, transfer, or plant closing)?

This problem was brainstormed with managers and engineers in a workshop. The first brainstorming session was short and resulted in 29 different ideas that were presorted into five categories by the facilitator, typed up, and handed back to the same group the following day. During the evaluation, the group decided to focus on things that managers can do. With this viewpoint, ideas were then improved and added in each category (shown in boldface lettering below). The teams worked from typed-up idea summaries, not with idea cards, since the idea pool for each category was small. The "better" and "best" ideas were then developed on larger sheets of paper. The team went on to idea judgment; we have marked the ideas they chose for implementation with a star (✱). Note that most of these ideas did not appear in the first brainstorming session but surfaced or were engineered during creative idea evaluation. This is why this second round of creative but more focused thinking in the "engineer's" mindset is very important and should not be skipped.

Category A—Things Management Can Do:

1. Management needs to organize the company to allow for horizontal interaction (information flow and movement of workers).
2. Management needs to be aware of trends in society and the marketplace, watch for new opportunities, prepare new products to meet the new needs. This requires creative thinking. ✱
3. Management needs to understand change and technology, as well as the impact these factors have on the company and the employees. **Manage innovation to provide jobs.** ✱
4. Training for workers must be continuous.
5. Training for managers must be continuous; they must be prepared to deal with change creatively.
6. **Reduce bureaucracy.**

Category B—Education and Training to Give People Options:

1. Set up job rotation, so workers will be more versatile and can move to other positions within the company if their position is abolished.
2. Train workers **continuously** in new technology **and languages** so they are qualified for new jobs (either in the old company or elsewhere).
3. People should be educated in the schools **and in the media** to expect change. With this mindset, it will be accepted practice to always have an alternative option or two to fall back on if necessary.
4. Managers need to be continuously informed about the changes technology brings to their companies.
5. Workers have to be given the time by management to become competent in their new jobs in high tech (this may take as long as a year or more for complicated computers).
6. New employees should be trained for the job that they are expected to do. Also, they need to have a clear job description.
7. **Managers need to talk to school boards, influence media. ✷**
8. **Unions and management need to brainstorm together. ✷**

Category C—Measures for Helping Dismissed Workers:

1. Unemployment insurance.
2. Job placement programs by the company or the government (paid by the company).
3. Counseling to help locate another job (paid by the company).
4. Counseling for employee and family to cope with this change (paid by the company).
5. Assistance with relocation (real estate, finding a job for spouse, etc.).
6. Set up an organization like a personalized chamber of commerce to assist relocating workers.
7. The company should offer comprehensive retraining **(including languages)** or support/sponsor the employee for further education at a college or other school. ✷
8. National **or global** data bank to match workers with job openings in the whole country **or overseas. ✷**
9. **Managers should be on the lookout for networking with other companies that may be able to use workers (make pensions portable). ✷**

Category D—Measures to Avoid or Prevent Dismissal of Workers:

1. All employees agree to voluntary pay cuts to keep workers from being dismissed. **Managers support this plan.**
2. All employees agree to reduced work hours (especially when the company's difficulties are expected to be only temporary); this will avoid laying off people. **Managers support this plan.**
3. Personnel surveys should be taken to match people to jobs for productivity and morale.
4. **Have bonus or profit sharing.**

Category E—Measures That Reduce Stress in the Company:

1. Assess quota levels fairly and adjust them when changes have occurred.
2. Have a mediator to minimize/remedy interpersonal conflicts.
3. Reward company loyalty; give merit recognition.
4. Pair each new person with a mentor. **✷ (top-ranked idea)**
5. Allow for mistakes; look at mistakes constructively (like the Japanese). Mistakes are a learning opportunity. This will avoid a cover-up of mistakes that can be damaging to the company. ✷
6. Arrange for social activities for employees and management together to make people more comfortable with each other.
7. Foster a spirit of cooperation, not competition. Emphasize the benefits of teamwork.
8. **Influence government policies to avoid those that are counterproductive. ✷**

Resources for further learning

9.1 Eugene S. Ferguson, *Engineering and the Mind's Eye,* MIT Press, Cambridge, Massachusetts, 1992. This book examines how engineers lose touch with the real world through too much reliance on computer models and a lack of hands-on experience in their education and workplace.

9.2 Win Wenger, *A Method for Personal Growth and Development,* United Educational Services, 1991. This source book on image streaming gives step-by-step instructions on how to learn this technique as an individual and how to teach it to groups.

Think smart: find the "best" in even the "dumbest" idea!

Adam Macklin, engineering student

Exercises for "engineers"

9.1 Disaster—So What?

a. Suppose that while you are out of town for a relaxing weekend with your family or friends, your car with all your money, luggage, and everything is stolen. Find at least ten ways to turn this apparent disaster into an interesting, positive, or enjoyable experience.
b. Do a creative evaluation—can you engineer and integrate these ideas into one or two practical solutions?
c. Discuss the results and application of this exercise with two or three friends—will the results affect the way you will plan your future vacation trips?

9.2 Sensory Experiences and Sales Ad

First, buy a fruit or a vegetable that you have never eaten before. Examine it, taste it, eat it (if necessary after cooking it). Use all five senses (sight, touch, smell, taste, and hearing) to describe and appreciate this new experience. Write each statement on a separate card. Note the shape, color, flaws, textures, flavor, sound-producing aspects, odor, temperature, possible uses. Draw many analogies as you go along, finding image-filled ways to describe the event. Be wildly poetic! Next, sort the statements with the method of creative idea evaluation. Make up several categories; combine ideas within the categories and then between the categories. Use one of these improved ideas and write a sales ad for this fruit or vegetable. Would you buy this fruit or veggie based on your experience? Would you buy it based on your ad? Test this last question on several of your friends. The exercise illustrates an application of creative idea evaluation to improve writing.

9.3 Brainstorming and "Engineering" Ideas

Brainstorm a problem from the following list, either alone or with a group. Then do a creative evaluation with the brainstormed ideas a day or two later to come up with improved ideas.

a. What can a person do to get more time daily for regular exercise?

b. How can team activities be made more pleasant for people who hate group activities?
c. How would you change the school system so more students would go on to study math, science, and engineering?
d. In what way can paperwork be reduced in your organization?
e. In what way can a particular procedure (specify) be improved?
f. Develop concepts for a child's playground toy that is sturdy, safe, and recyclable (play with ideas; do not do the actual design).
g. How would you improve communications between parents and their teenagers?
h. How can parents teach their children time-management skills?
i. Identify an environmental problem in your community that needs to be addressed (polluted river, full trash dumps, air pollution, etc.). How could such an effort be organized?

9.4 ★ Image Streaming ★

Image streaming is a technique developed by Dr. Win Wenger, president of the Institute for Visual Thinking in Gaithersburg, Maryland. By integrating right-brain and left-brain thinking, it can help improve your mental abilities. To do this exercise, you need another person or a tape recorder. For this exercise to be effective, you must talk out loud, not just think to yourself. You may also want to use a timer set at 20 minutes. Here are the steps:

1. Close your eyes and turn on the tape recorder (or ask your friend to listen attentively).
2. Start describing what you "see" (blotches, patterns, images from your memory, a person, object, or scene from your past). Describe all aspects of the image: smells, sounds, colors, feelings of texture, temperature, whatever sensory information is attached to the image. Visualizing, intellectually analyzing, and speaking aloud all use different parts of the brain. Thus this exercise does what is called "pole bridging" in your brain.
3. Continue to follow your image streams with rapid talk until the time is up. Remember to report everything that comes up, even if you think a particular impression is not important. Look for as much detail as possible.
4. During or at the end of a session, develop a humorous interpretation for the "messages" that have come to mind, if you can.

Integration, or even the word "organic" itself, means that nothing is of value except as it is naturally related to the whole in the direction of some living purpose.

Frank Lloyd Wright, architect

Practice this techniqueregularly, in 10- to 30-minute sessions; this will make it easier for you to access your right brain during problem solving for new and useful creative ideas. Image streaming is closely related to an ancient method of learning. Socrates, through asking questions, would cause his students to examine their inner and external perceptions; they had to describe what they found. Through this technique, they gained understanding and personal growth, the mark of true education.

Chapter 9 — review of key concepts and action checklist

Your role as "engineer": Group, sort, organize, build on, develop, integrate, engineer, and synthesize ideas to create good potential solutions to the original problem. When used with the Pugh method, optimized design concepts result from this process.

The four rules of creative idea evaluation: 1. Look for quality and "better" ideas. 2. Make "wild" ideas more practical. 3. Synthesize ideas to obtain more complete, optimized solutions. 4. Maintain a positive attitude; continue to defer critical judgment.

The three-step process of creative idea evaluation:

1. All ideas generated in the brainstorming phase are written on index cards, one idea per card. These ideas are grouped into categories.
2. After a creative thinking warm-up, the team works with one category at a time and tries to combine, develop, and synthesize the ideas within a category to obtain fewer, but higher-quality solutions. These improved ideas are written on a flip chart and posted on the wall.
3. Finally, the team or teams try to force-fit ideas between the categories. This process can yield some very innovative solutions.

Answers to many single questions can be organized together to form solutions to problem satements which become advanced definitions, which themselves are subject to new and higher questions, and the cyclical power of the creativity and innovation engine continues on...

Bruce LaDuke,
ACA Focus, *Vol. 10, No. 1*

Action checklist

☐ Practice the creative evaluation process by doing a brief exercise before applying the technique to your design project.

☐ Next time you are in a brainstorming session and the ideas that have been generated are rather hum-drum, ask that the organization, the committee, or the team do a second round of brainstorming (on a different day). Facilitate the process by collecting the brainstormed ideas, writing them on note cards (or Post-it notes), and organizing them into categories. Then conduct the creative evaluation during the next meeting. Don't forget the creative thinking warm-up!

☐ When you are tempted to make a negative comment about someone's idea, try to use the imperfect idea as a stepping stone and generate at least three "better" ideas. Or come up with your own idea and then work with the other person to integrate both ideas into one solution.

☐ Remember to maintain a positive attitude during creative idea evaluation. New ideas are fragile; treat them with care during this stage in the creative problem solving process.

☐ Continue to encourage your own creative thinking with humor and other actions that will reduce stress in your life.

10

Idea Judgment

What you can learn from this chapter:

- The goal of idea judgment: finding the best solution for implementation.
- Traits and tasks for the "judge."
- What is good judgment? Examining values, presuppositions, and bias. Thinking about consequences. Ethics in engineering. The costs of "blowing the whistle."
- Attributes of critical thinking; connections to creative problem solving.
- Idea judgment as a two-stage process:
  (a) Ranking different options by using valid criteria and an appropriate judgment technique.
  (b) Making decisions based on analytical and intuitive processes.
- Further learning: references, exercises, review, and action checklist.

In this stage of creative problem solving, we find the ideas or solutions among the "better" ideas from the previous step that will best solve the original problem. During idea judgment, we establish evaluation criteria and then sift and rank the ideas and solutions according to the criteria, before making a decision on which is best. Judgment is part of the analytical activities in Step 3 of the knowledge creation cycle.

The critical power ... tends to make an intellectual situation of which the creative power can profitably avail itself ... to make the best ideas prevail.

Matthew Arnold,
The Function of Criticism at the Present Time, *1864*

A story by Galileo is retold in the November-December 1992 issue of the *American Scientist.* A column was stored horizontally by supporting its ends on piles of timber. But since it was possible that the column could break in the middle under its own weight (as had been observed in various situations in the past), someone suggested that a third support be added at the center. Everyone consulted agreed that this would improve the safety of the column, and the idea was implemented. A few months later, the column broke in two anyway, at the center. The cause of the failure was the new support that failed to settle at the same rate as the end supports—the column broke when too much of its weight was no longer supported at the ends.

How relevant is this story today? It serves as a reminder that solutions to problems can be the direct causes of failures, if our judgment is flawed. Recent notable examples are the collapse of the sky walks in the Kansas City Hyatt Regency Hotel (where changes in the support rods

Figure 10.1 The "judge's" mindset.

weakened the structure) or the space shuttle Challenger (where extra O-rings in the booster rocket were accepted uncritically). Let us now look at the role of the "judge" and the techniques used to examine ideas and render decisions.

The role of the "judge"

In the "judge's" mindset (see Figure 10.1), the critical, conscious mind comes into full action. In some ways, the role of judge seems to be natural since it is easier to criticize than to explore new options or to transform ideas or to do something about them. But if we spend all our time being a "judge," we won't accomplish much. Also, it is important to remain impartial about the ideas we are judging. As a "judge," it is our job to find the best idea and not wait for the perfect idea. "Judges" themselves are not perfect, but they need to make wise decisions based on evidence and principles. Why could purple be a suitable color to represent the "judge's" mindset?

"Judges" need a sense of timing to figure out which decisions should be made quickly and which decisions should be made only after a long, careful study. "Judges" have the responsibility to note flaws and then devise ways of overcoming them with creative thinking. Being a good "judge" takes practice because it is not easy to recognize the shortcomings of the ideas while still keeping an eye on the positive features. As "judges" we want to give the producer a solution worth defending—one that will be as trouble-free as possible. Thus a judge's responsibility is to consider the risks involved in the proposed solutions. All solutions have some possibility of failure—nothing is entirely foolproof. Even carefully planned implementations carry an element of surprise, especially when dealing with an innovative idea, because we cannot predict all the reactions to such a solution.

As "judges," we must be able to see the value of learning and improvement that failures represent. Failures can be starting points and motivation for growth. Also, "judges" have to decide if the timing is right for a new idea—a 1999 idea for the year 2000 marketplace could ruin a company. Thus judges need to use future-oriented quadrant D thinking to balance the quadrant A critical and analytical modes and the quadrant B risk-averse mindset.

Five-Minute Activity 10-1: Failure and Wisdom

In groups of three, discuss the statement, "Experiencing failure makes us better judges." Then share your most interesting insight with a larger group. Especially focus on the aspect of wisdom—how can it be acquired?

Any engineering disaster is made up of three parts: unexpected circumstances, poor design, and ethical failure.

Gary Halala, materials engineering professor, SUNY Stony Brook

Who should be the "judge"? Should a single person (the team leader, facilitator, middle-level manager, or client) be the "judge," or should a team be involved in the judgment process? If much time is available, if a high-quality solution is needed, if there could be a problem with acceptance of the final solution, and if other people need this learning experience or training, then the team approach is best. If none of these factors are present, judgment by an individual will be more expedient. A combined approach is possible. An individual can do the preliminary selection, and a committee (or team) can make the final choice. Or the team can make the preliminary selection, with an individual (or a smaller group) making the final decision. At many universities, faculty members are hired through such a combined process: a search committee made up of faculty, staff, and perhaps student representatives narrows down a long list of candidates to three, and the president, chancellor, or provost makes the final decision.

Before we continue the discussion of the creative problem solving process, we need to pause and think about some important issues that are foundational to idea judgment. We commonly assume that "judges" use good judgment and critical thinking skills. Is this a valid assumption? In this section, we will look at factors—such as values and bias—that we must consider as "judges." In the following section, we will investigate the relationship of critical thinking to creative problem solving.

What is good judgment?

On a product's warranty statement, we saw this warning: "We cannot be responsible for the product used in situations which simply make no sense." Good sense—or good judgment—is difficult to teach, because it is best learned through experience with failure. However, we do not need to experience the failure personally; we can learn from studying the failures of others. This is particularly true in engineering.

Good judgment comes from experience. Experience comes from poor judgment.

Ziggy

With increased use of technology, good judgment becomes critically important. There has been a tendency to rely on computer models in place of hands-on experience, instead of integrating the two. One source of danger is that the user of a complex analytical computer program may not be able to discover all the simplifying assumptions made by the program designer. The "precision" of the computer's numerical output can give a false sense of security as to the validity of the calculations, even when critical factors unique to the particular problem are not included. Designers introduce another level of potential flaws when their high-tech designs do not consider the user interface. Eugene S. Ferguson, history professor emeritus at the University of Delaware, describes an example in "How Engineers Lose Touch," in *Invention & Technology* magazine, Winter 1993, page 24:

> The designers of the Aegis [on the missile cruiser USS Vincennes], which is the prototype system for the Strategic Defense Initiative, greatly underestimated the demands that their designs would place on the operators, who often lack the knowledge of the idiosyncrasies and limitations built into the system. Disastrous errors of judgment are inevitable so long as operator error rather than designer error is routinely considered the cause of disasters. Hubris and an absence of common sense in the design process set the conditions that produce the confusingly overcomplicated tasks that the equipment demands of operators.

Nine out of ten recent failures [of dams] occurred not because of inadequacies in the state of the art, but because of oversights that could and should have been avoided.

Ralph Peck, foundation engineer, 1981

The problem here was that the operators aboard the Vincennes were overwhelmed with more information than they could assimilate in the few seconds before a crucial judgment had to be made about shooting down a plane, with the result that they mistook a commercial airliner for a hostile military aircraft. Human abilities (and limitations) must be considered as part of the context in any design or solution.

A new engineering design (which is the solution to a problem or need) must combine analysis with intuitive, tacit knowledge gained from experience. Even then, the judgment will contain a degree of uncertainty. Thus "judges" have the responsibility to detect errors made at any point during the design or problem-solving process. They must eliminate flaws and evaluate (and document) the risks, consequences, and uncertainties of alternative solutions to the best of their abilities, before making the final decision on which solutions are to be implemented. "Judges" must be able to imagine all the things that could possibly go wrong with the proposed solutions; yet "judges" must also have a flexible mindset that allows them to see uncertainty in a positive light.

Consequences

Thinking about consequences is difficult, especially for young people, but it is so very important, because once we make a choice of one thing, other things will no longer be possible. We cannot have the cake and eat it, too. When people decide to get married, their new commitment separates them from the singles dating scene. If someone chooses to get into drugs, he or she may lose job, health, family, and reputation. Thinking about consequences can help us make decisions on the best timing for implementing a solution. One mother was asked by her son for a loan so he could buy new tires for his car—the old tires were in terrible shape, and he used the car daily to go to work. Since he still owed her money on a previous car maintenance loan, she was at first reluctant to advance still more. When she thought about the consequences of postponing the tire replacement—the increased danger of driving during the approaching winter season—she decided the risk was not worth it. She loaned him the money immediately instead of waiting a few weeks.

Bad design results from errors of engineering judgment, which is not reducible to science or mathematics.

Eugene S. Ferguson, Engineering and the Mind's Eye

Here is another example. During the 1989 TECHNORAMA at the University of Toledo, Molly Brennan, a young engineer from General

Motors, was invited to the campus as the main speaker. She was one of the drivers of the *Sunraycer,* a solar car that won the race across Australia a year earlier. During a conversation, she explained that her sister had come home from school one day with her planned schedule of classes. Their mother—an English teacher—noticed that her daughter had not signed up for physics. When asked for an explanation, the girl replied that she did not want to take physics because none of her girlfriends were taking the class. The mother was insistent that her daughter not shortchange her future options. So the girl talked her friends into taking physics (even though the school counselor tried to dissuade them). Do you know what happened to the four girls? One is now a researcher in science, one is a medical doctor, and two are engineers!

A poignant illustration of the lack of critical thinking and its consequences is provided by an old legend retold by Iron Eyes Cody in *Guideposts* in July 1988—we leave it up to you to think of situations where the moral of this story would be relevant today:

Whatsoever a man soweth, that shall he also reap.

Letter to the Galatians 6:7,
New Testament
(King James Version)

> Many years ago, Indian youths would go away in solitude to prepare for manhood. One such youth hiked into a beautiful valley, green with trees, bright with flowers. There he fasted. But on the third day, as he looked up at the surrounding mountains, he noticed one tall rugged peak, capped with dazzling snow.
>
> I will test myself against that mountain, he thought. He put on his buffalo-hide shirt, threw his blanket over his shoulders and set off to climb the peak.
>
> When he reached the top he stood on the rim of the world. He could see forever, and his heart swelled with pride. Then he heard a rustle at his feet, and looking down, he saw a snake. Before he could move, the snake spoke:
>
> "I am about to die," said the snake. "It is too cold for me up here and I am freezing. There is no food and I am starving. Put me under your shirt and take me down to the valley."
>
> "No," said the youth. "I am forewarned. I know your kind. You are a rattlesnake. If I pick you up, you will bite, and your bite will kill me."
>
> "Not so," said the snake. "I will treat you differently. If you do this for me, you will be special. I will not harm you."
>
> The youth resisted awhile, but this was a very persuasive snake with beautiful markings. At last the youth tucked it under his shirt and carried it down to the valley. There he laid it gently on the grass, when suddenly the snake coiled, rattled and leapt, biting him on the leg.
>
> "But you promised—" cried the youth.
>
> "You knew what I was when you picked me up," said the snake as it slithered away.

⌛ Five-Minute Team Activity 10-2: Consequences

With two other people, discuss examples of consequences that are reversible with hard work and some that are not. How are value systems related to dealing with consequences?

Values, presuppositions, and bias

The only tyrant I accept in this world is the "still small voice" within me.

Mahatma Gandhi

Judges are required to do much critical thinking. However, critical thinking is not taught well in schools, where only the analytical aspects of logical reasoning may be introduced. But effective critical thinking and decision making are whole-brain processes, as we shall see in this chapter. Experience also enters the picture. How do we know that we have made a good judgment? Judgments need the test of time—we learn from the outcomes, from our failures, from our experiences with judgment. Besides critical thinking skills and experience, a third factor influences our ability to render a good judgment, and that is our personal belief system of values, principles, and moral standards with its presuppositions and biases. Presuppositions are strongly held, implicit paradigms that influence thinking and can prevent a "judge" from seeing the merits (or flaws) of particular solutions. Thus one advantage of having a whole-brain team involved in the judgment process is to keep inappropriate paradigms from dominating the judgment. To be good "judges," we not only have to be cognizant of our own personal biases, we must be aware of cultural bias, prejudice, and false assumptions that could influence the decision-making process. Bigotry and a "politically correct" view inhibit reasoning. As a "judge," we must allow dialogue and explore beyond the limits of our own tribe and preferred ways of thinking.

Ethics is the code of morals or standard of conduct of a particular person, religion, group, or profession.

Principles and moral standards guide human behavior and thus are linked to survival. Many educators are troubled because such virtues as honesty, loyalty, discipline, responsibility, and accountability—all aspects of personal integrity—are increasingly being lost in our society, even though we seem to demand them in our political leaders. Why do we expect leaders, judges, and professional people to follow high ethical standards? Since morality involves the principles of right and wrong in conduct and character, who determines what is "right" or "wrong"? Where do personal and cultural values come from? How is conscience developed? Who has the responsibility for teaching moral values? These are difficult but very important questions that are often neglected, as is the related teaching about personal responsibility—knowing what is right and accepting the consequences for one's decisions and actions.

Team Activity 10-3a: Cultural Values

In a group of five to seven made up of people from at least three different cultural backgrounds, discuss the following hypothetical situation without making a judgment as to which solution is "better"; instead, look at the different outcomes in terms of the underlying cultural values.

A man is in a building with his mother, wife, and child. Suddenly, there is an explosion, followed by a rapidly spreading fire. The man has to make a quick decision: which person in his family should he carry to safety first, since all three have suffered injuries that keep them from walking away on their own?

Team Activity 10-3b: Cultural Values (continued)

If you cannot find a multicultural group, discuss and develop reasons and explanations for saving the mother, the wife, or the child. What would be some of the values underlying each case? Look at the list of values that have been brought out in your discussion. Can you make a distinction between personal values and societal/cultural values? What happens when personal values conflict with societal values? Would the outcome of the discussion be different if a woman had to decide which family member to save: husband, father, or son?

Ethics in engineering

The following two examples are based on cases discussed in *Engineering Times,* the monthly paper published by the National Society of Professional Engineers (NSPE). In "You Be the Judge" this statement is given preceding each ethics case:

> Although engineering is a profession of precise answers based on scientific principles, engineers work in the real world where business, ethical, or even human-relations questions have no easy answers. Many engineers have turned through the years to NSPE's Board of Ethical Review for impartial help in making ethics judgment calls. Do you want to try your hand at deciding a case? Below are situations posed to the board. Note: It should be understood that each ethics case has its own answer; each case is unique. The general response in one case may not fit what appears to be a similar problem.

Engineers shall hold paramount the safety, health, and welfare of the public... If engineers' judgment is overruled under circumstances that endanger life or prperty, they shall notify their employer or client and such other authority as may be appropriate.

NSPE Code of Ethics, Sect. 11, Rules of Practice.

Case 1—Utility Cost Consultant

The situation: N. R. Gee, P.E., a specialist in utility systems, offers industrial clients the following service package: a technical evaluation of the client's use of utility services (electricity, gas, telephone, etc.); recommendations, where appropriate, for changes in the utility facilities and systems; methods for how to pay for such utilities; a study of pertinent rating schedules; discussions with utility suppliers on rate charges; and renegotiation of rate schedules. Gee is compensated for those services solely on how much money the client saves on utility costs.

What do you think? Is it ethical for Gee to be compensated this way?

What the board said: Gee is acting ethically in accepting such a contingent contract arrangement. *See the February 1989 issue, page 3.*

Case 2—Expert Witness

The situation: X. Burt, P.E., was retained by the federal government to study the causes of a dam failure. Later, Burt was retained as an expert witness by a contractor who filed a claim against the government demanding additional compensation for work performed on the dam.

What do you think? Was it ethical for Burt to be retained as an expert witness under these circumstances?

What the board said: The Board of Ethical Review found that Burt's actions were unethical. *See the April 1989 issue, page 3.*

Whistleblowers

Taking an ethical stand in today's materialistic world can be very costly. Two out of three whistleblowers in the past lost their jobs in the organization whose wrongdoing they exposed. The results are economic hardship, anger, depression, persecution, and isolation for the "ethical resisters" and their families. Blacklisting makes it almost impossible to find a job in a similar field. Why do these courageous men and women stand up for what they believe despite the high cost? Legislation was passed under the Bush administration that offers some protection to whistleblowers both in the private and public sector. Roger Boisjoly, the engineer at Morton Thiokol who tried to prevent the launch of the Challenger space shuttle in January 1986, now frequently speaks on university campuses. He is a champion for training professionals in ethical sensitivity. From painful personal experience, he knows that technical education is not enough to meet the ethical challenges of the workplace.

When you have God, the law, the press, and the facts on your side, you have a fifty-fifty chance of defeating the bureaucracy.

Hugh Kaufman, EPA whistleblower

Only when scientific (left-brain) and spiritual (right-brain) reasoning are integrated will society's problems be solvable and solved. This requires thinking along new paths. It has been said that the technological development and achievements of *Homo sapiens* have far outstripped moral and ethical development. Can you support this opinion with concrete evidence? Can you cite evidence supporting an opposing view?

Critical thinking

When we have learned to think critically, we should exhibit certain attributes shown in Table 10.1. Critical thinking, as taught in secondary schools and at the undergraduate college level, aims to develop the characteristics and skills listed in Table 10.2. The problem is that this teaching is often narrowly focused and mostly in the area of literary or artistic criticism. Connections to everyday life—the workplace, human relationships, the media, and responsible citizenship—are missing.

Table 10.1 Attributes of Critical Thinking

- We are aware of the potential for distortion in the way the world is presented by the media.
- We are aware of physical limitations in the perception of reality and its interpretation by our mind (i.e., we know how the lack of sleep or the consumption of certain substances can affect judgment).
- We are able to recognize mental blocks, overgeneralization, and false rationalization.
- We are able to assess the "language" and thinking preferences involved in the verbal description of problems and ideas.
- We are honest with ourselves.
- We recognize and value evidence and feelings.
- We resist manipulation, we overcome confusion, we ask the right questions, and we seek connections to make a balanced judgment independent of peer pressure.

Table 10.2 Outcome Objectives of Teaching Critical Thinking

- Critical thinkers decide on what they think and why they think it.
- Critical thinkers seek other views and evidence beyond their own knowledge.
- Critical thinkers decide which view is the most reasonable, based on all the evidence.
- Critical thinkers make sure that they use reliable facts and sources of information; when they state a fact that is not common knowledge, they will briefly say where they have obtained the information.
- When critical thinkers state an opinion, they anticipate questions others might ask and thus have thoughtful answers ready to support their opinion.

Current models of critical thinking

Three views are prominent in current literature on teaching critical thinking at the college level: argument skills, cognitive processes, and intellectual development; they are summarized below. In contrast, we will introduce the Brookfield model that sees critical thinking as a whole-brain activity employing the creative problem solving mindsets.

Argument skills—Students are taught the skills of analyzing and constructing arguments based on informal logic. This emphasis on analytical skills may improve the students' ability to justify beliefs they already hold. But it has been found that students are unable to translate this learning to everyday issues.

Cognitive processes—Here students interpret problems or phenomena based on what they already know or believe. They construct a mental model of the problem or situation around a claim or hypothesis that is supported by reasoning and evidence. Three kinds of knowledge contribute to the model: the facts involved in the particular discipline, knowing the procedures on how to reason in the discipline, and metacognition, which means evaluating the goals, the context, the cause-and-effect relationships, and the progress of inquiry or problem solving. However, new learning is not stored as a collection of isolated facts, but as meaning constructed into patterns or scripts as understood by the student. Professors rarely teach the strategies, procedures, and metacognition explicitly; thus students are not learning how to apply knowledge and critical thinking in unfamiliar situations.

Perry's Four-Stage Model

1—Authorities have "the answer."
2—One's own opinion is valuable.
3—Knowledge and people are connected.
4—A commitment is made to contextually appropriate decisions.

Intellectual development—this approach examines students' relationship to belief and knowledge. The best-known model has been developed by William Perry (Ref. 10.8) and expanded by Mary Belenky and associates (Ref. 10.1). This model forms the basis of much research in the area of critical thinking done today at the college level. It has been found that the majority of students do not progress beyond the second stage of Perry's model. We believe that students who are taught to apply creative problem solving will learn critical thinking skills and progress to higher stages in the Perry model.

Thinking Activity 10-4: Perception

Alone, or with one other person, look at the shapes below. Can you see patterns in the individual shapes? Can you discern meaning from the sequence and repetition? If your paradigms and usual habits of perception keep you from recognizing the message in the shapes, read the next paragraph.

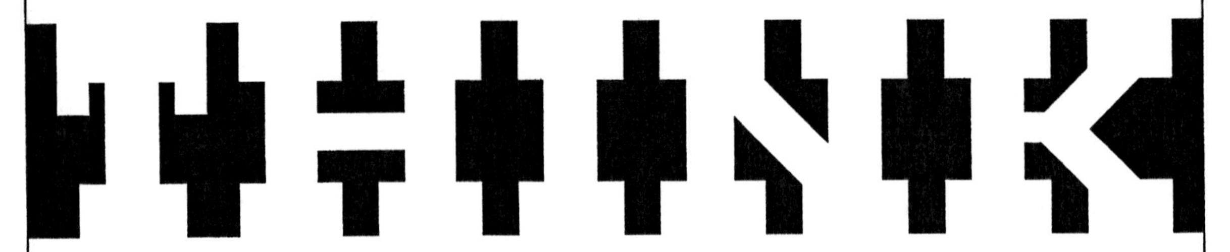

Instead of analyzing the individual symbols, look at the sequence of shapes differently—what if the meaning were hidden in the empty spaces between the shapes? Why do you think it is difficult for some people to immediately "see" the answer? What were your presuppositions about the nature of the assignment and how to solve it? Drawing this figure was at first a very frustrating task in the word processing program, since the usual approach was to draw the outline of the shapes. When the problem was turned around and perceived as solids (not lines), it was a breeze to build the individual (white) shapes with a series of rectangles.

Critical thinking and creative problem solving

The current view appears to consider critical thinking as a form of analytical problem solving. However, the goal of critical thinking is not necessarily to find a solution, but to construct a logical representation of a situation or position based on plausible arguments and evidence. This is a major difference from the creative problem solving approach. Let's illustrate this difference in an example. In a court of law, a couple is involved in a custody case, in which the opposing lawyers are trying to build the strongest case (by argument and supporting evidence) for their client's position. If creative problem solving were used, the estranged parents would try to define the real problem and work together to develop a solution that would be acceptable to all, but would above all consider the needs of their child. What if creative problem solving had been used at an earlier time in the marriage—would it have helped to build a strong family instead of an adversarial relationship?

To criticize is to appreciate, to appropriate, to take intellectual possession, to establish [in essence] a relation with the criticized thing and to make it one's own.

Henry James

In professional practice such as engineering, business, teaching, or architecture, reasoning combines aspects of critical thinking and problem solving. In the current (traditional) view of critical thinking, the central element is the ability to raise relevant questions and critique solutions without necessarily posing alternatives. But critical thinking needs a more broadly defined concept that includes playing with ideas and creatively developing analogies and metaphors, not just logical reasoning. In creative problem solving, ideas, solutions, designs, and products

There is no evidence that any of the skills of critical thinking learned in schools and colleges have much transferability to the contexts of adult life.

Stephen D. Brookfield

are critiqued, but the process is taken an additional step in that ideas for making improvements must be proposed or considered. Creating these ideas requires right-brain thinking and a positive attitude to balance a critical mindset that looks only for flaws.

Becoming critical thinkers (Brookfield model)

In this whole-brain model, people break out of the analytical pattern of critical thinking that they have been taught in school through some trigger event (a positive happening or a tragedy). They go through a period of self-examination; they explore and test alternatives and new paradigms. Through this process, they develop new perspectives; they then try to choose the "best" and integrate it into their life. This results in changed attitudes, confirmed beliefs, and altered subconscious feelings. Others (including instructors) can assist in the process by affirming self-worth, listening attentively, and showing support for the effort. They can provide motivation and encouragement for risk taking, evaluate progress, and supply a contextual network and resources.

This model, summarized in Table 10.3, considers critical thinking by teams as a key to maintaining a healthy democracy. Pressure against critical thinking is seen as coming from people in power who want to preserve the status quo, especially if it is inequitable, such as from political dictators, labor leaders, employers, teachers, family members, professional groups—anyone with a vested interest in continuing a paradigm and hierarchy. Critical thinking is needed to counter the bias and influence of the media in the news and in entertainment.

Table 10.3 Characteristics of Critical Thinking—Brookfield Model

- Critical thinking is a process, not a result; it includes the continuous questioning of assumptions. It is important to understand the context of problems, the underlying assumptions, and social value system.
- Critical thinking is a productive and positive activity: it includes creativity and innovation. Imagination is practiced; possibilities and alternatives are explored. This leads to reflective skepticism—change is not simply accepted because it is new. Consequences of actions are anticipated.
- Critical thinking is emotional as well as rational—it is whole-brain thinking where we recognize our assumptions within the framework of our personal beliefs and commitments as well as within the context of the world around us. Criteria are not strictly objective but subjective. Role playing, decision simulation, and preferred scenarios and futures are valid creative thinking strategies. Poetry, fantasy, drawing and painting, songs, and drama are means to release creative imagination and thus help in developing critical thinking.
- Critical thinkers are curious, flexible, honest, and skeptical—they can distinguish bias from reason, facts from opinion. They can use thought rationally and purposefully together with feelings and intuition to move toward a future goal.

Let us now continue with the task of judging ideas. Idea judgment is a two-step process: first, ideas and solutions are ranked by using carefully developed evaluation criteria (quadrant A analytical thinking with quadrant D brainstorming). Then the final decision is made on which idea will best solve the problem and should be recommended for implementation. Needs, values, social context, benefits, and risks are considered (mainly quadrant B and some quadrant C thinking). Although left-brain aspects dominate, "judges" must be able to iteratively cycle through the whole brain to develop the best solutions and make wise decisions.

Ranking different options

In this section, we will discuss the importance of having a list of evaluation criteria—many judgment techniques work best when they are supported with a good list of criteria. We will also examine some specific judgment techniques, and we will look at the question of what to do when solutions have flaws.

The list of criteria

A good list of criteria includes all factors that influence a problem or decision. Let's say you have been looking for a new job and are fortunate to get four different offers. Which one should you take, when the fifth option is to remain in your current position? How do you make the best decision? What would be some of the important factors (and feelings) that should be considered? It takes time to make up a valid list of criteria. The list can be developed through regular brainstorming—the more criteria, the better! Through creative evaluation, the criteria are further refined, and the most useful and important criteria are selected. Make sure the evaluation is balanced between analytical and intuitive criteria, between quantitative and qualitative factors. Sometimes a weighting system is used. It can simply be based on rank, for instance from 1 to 5, with the highest number assigned to the most important criteria. Or the weighting factor for each criterion can be voted on by the team members (or some other qualified panel) and be assigned this averaged value.

I don't think that you can make change in an area this important unless you also know what has to be maintained, unless you have people of real seasoning and judgment.

President-elect Bill Clinton, December 22, 1992, defending his choices for cabinet posts.

Criteria can also be thought of as the boundaries, limits, or specifications that the solution must fit to solve the problem. For example, government laws and regulations must be observed. A component of a larger system may be constrained by size, weight, and other physical limitations. However, if time permits, limits should be questioned. Are they merely arbitrary conventions? Why do they exist? Could the limits be overcome through creative thinking, the development of a new paradigm, or the application of new technology? Think of specifications not as chains but as challenges! As a "judge," you need to also pay attention to intuition—what attributes do you "feel" the ideal solution should have?

Why does a certain criterion or solution seem "right" and another "wrong"? When we develop a list of criteria, we have an opportunity to put the solution into a larger context. We need to look to the future and consider the factors that will make implementation easier and more successful. A good list of criteria will help us understand all the important factors that are involved. In particular, we should look for criteria that will answer some of concerns listed in Table 10.4.

Table 10.4 Factors Involved in Successful Implementation

1. **Motivation:** Why would people want to accept the solution—what are their motives? How can we motivate them to buy the product? Does the idea, service, or product meet customer needs?
2. **People:** How will people be affected by the solution? Will it be difficult to use the solution? Will they need to make changes in their lives? Is the product marketable? Who are the customers?
3. **Cost:** What will be the costs to you, to others? Will the solution be affordable? Will the product be easy to manufacture? Will it be easy to service or maintain? Can it be reused or recycled? Does the idea have other applications? Is the idea feasible? Does implementation require new technology?
4. **Support:** What support is available for implementation? What resources—such as materials, equipment, information, training, or people—will be needed to successfully implement the solution?
5. **Values:** What social values are involved? What will be the benefits to people? What are the safety issues? What are the dangers to the environment?
6. **Time:** Will the solution take a long time to implement? Will there be a short-term or long-term application for the solution?
7. **Effects:** What will be the consequences of the solution? What effects will it have on other activities in your organization, in your life, in your community?

A list of criteria is very useful for analyzing the quality of different ideas and solutions and their capability for solving the original problem. Criteria can point out areas of weakness and can identify ideas that have too many shortcomings and should thus be dropped. But very rarely will an idea emerge as a clear "winner" that will satisfy all the criteria. Thus some additional evaluation techniques need to be employed to further sift and rank the ideas.

Techniques for idea judgment

We will look at three types of judgment techniques: quick methods, advantage-disadvantage comparisons, and specialized techniques.

Quick procedures

When we do not have time to develop a good list of criteria that will allow us to rank ideas, we can use some type of **judgment by vote**. Voting can be done in a number of ways, openly or preferably by secret ballot. A major disadvantage of quick voting is a lack of explicit criteria.

Each person makes decisions based on his or her own values or prejudices; there is no common, agreed-upon standard by which the judgment is made. Sometimes people make judgments without knowing why they make the choice, or they are swayed by peer pressure or "groupthink." Quick votes tend to discourage the discussion of flaws in the ideas under consideration. If an idea is voted "the best," this seems to tacitly imply that it cannot have any faults since it has been accepted by the majority.

The only thing necessary for the triumph of evil is for good men to do nothing.

Attributed to Edmund Burke, 1729-1797

A large number of ideas can be quickly reduced to a more manageable level by an individual or a group through a **single criterion**, such as cost. Caution is in order because the limits of a single criterion can often be overcome with additional brainstorming; thus a hasty decision here could eliminate potentially good choices. For example, let us assume that in the search for hiring a research engineer the single criterion used to cut down the list of applicants is having a Ph.D. degree. It could happen that the best candidate could be someone only two months away from getting the degree, or an experienced person who made important discoveries in a new field and never found the time to complete the academic work for the doctorate. Thus the best judgment is rendered when a list of carefully thought-out criteria is used to measure the worth of the ideas that have made it to this creative problem solving step.

Advantage/disadvantage techniques

The simplest approach with this type of judgment tool is to make a **separate listing of advantages and disadvantages** for each idea, with one column for all its advantages (positive marks or pros) and one column for all its disadvantages (negative marks or cons). The idea with the most advantages and least disadvantages "wins." This method has a major weakness because one negative can be so important that it could outweigh several or even all positives. Let's say we are developing an inexpensive consumer product. If one idea has all positive responses to the list of criteria (or a long list of advantages) except for high manufacturing costs, this one negative mark will be a serious barrier to our ultimate success. Thus we must not add up the positive and negative marks and take the arithmetic results without some critical thinking about what each negative mark implies. If sufficient time is available, the negatives can prompt another round of creative problem solving to eliminate them.

When we add a third column to this evaluation to take the long-range potential of each idea into account, wfà∞ave the **advantages, limitations, and potential (ALP) method**. This method makes it somewhat easier to give a fair evaluation to untried, creative ideas that depend on their potential benefits for acceptance. When we are interested in setting priorities or weighting factors among the criteria, we can conduct a paired **comparison analysis** to rank the criteria relative to each other. This method works best when the number of criteria and the number of solution alternatives are relatively small.

If we construct an **advantage-disadvantage matrix** with the list of criteria in a column to the left and the ideas to be evaluated across the top toward the right, we have a method that compares each idea with all the others for each criterion. To illustrate, let's take another look at the job selection problem. As shown in Table 10.5, each of the five job options has advantages (+) and disadvantages (0). So, which option should you choose? Let's say you brainstormed with your family and came up with the criteria listed in the left-hand column. The job options are arranged across the top of the matrix, and each job is evaluated against the criteria. For Job 1, the salary offer is very good, and this advantage receives a plus mark. For Job 2, the pay is low (a disadvantage) and this is scored a zero. This process is repeated for each criterion and each job.

Table 10.5
Example of an Advantage/Disadvantage Matrix

List of Criteria		Job Options 1	2	3	4	5
Pay		+	0	0	0	+
Other benefits		+	+	0	0	+
Personal growth		0	+	+	0	0
Good for the family		+	+	0	+	0
Independence		0	+	+	+	0
Status		0	+	0	0	+
Excitement/adventure		+	+	+	0	0
Quality coworkers		+	0	+	+	+
Supportive boss		+	+	0	+	0
Fits with life goals		+	+	0	0	+
Total	+	7	8	4	4	5
	0	3	2	6	6	5

When the matrix is completed, the scores are added separately for checks and zeroes. In this example, Options 1 and 2 are fairly close, with the next three (including #5, the present job) separated from the top two by a larger gap. Small differences in points are not important; thus the two top options must be considered further. Can the negatives be removed through negotiation, such as the salary offer in Job 2? Perhaps the lack of quality coworkers is only temporary and can be expected to improve. Or the negatives in Option #1 are only short-term, not long term, and can thus be tolerated.

The criteria need to be reviewed carefully and perhaps supplemented with with a weighting system for the entire list. In the example, is pay more important, or should the potential for personal growth receive priority? When weighting factors are used, the final results will probably have a much larger spread, and it will be easier to select the best solution. This example illustrates that selecting the right criteria is crucial. We must include all important parameters if we want a true indication of the best options. The advantage-disadvantage matrix is useful for ranking ideas and making decisions, because people working out the matrix will understand why ideas are ranked high, since they have an opportunity to extensively discuss and modify the criteria. The QFD House of Quality is an example of a **matrix employing weighting factors** (Appendix A).

When the advantage-disadvantage matrix employs an existing idea or a benchmark product, process, or service as the standard against which the new concepts are compared on a three-way scale, the technique is

The Internet is loaded with information, much of it extraordinarily useful and valuable, and much of it absolutely false.

George F. Huhn, Managing Consultant, ACA Focus, *Vol. 9, No. 2.*

known as the **Pugh method**. It will be discussed in more detail in Chapter 11. The Pugh method is a team approach to creative design concept evaluation, but it can be used for all kinds of ideas, not just design concepts. An existing product or idea is used as the datum. Each new idea or design concept is compared to the datum for each criterion and judged to be substantially better (+), essentially the same (S), or considerably worse (–). The Pugh method is an iterative technique; it goes through many cycles. The highest-scoring idea or concept of each round is chosen as the datum for the subsequent round, and ideas are continuously being improved with further creative thinking and synthesis among the ideas evaluated on the matrix. This process finally results in a consensus on an idea or concept that cannot be improved any further. Weighting factors may be added in the last round to confirm the best solution.

Other judgment techniques

When we have only a small number of ideas, the **advocacy method** can be used. Here the group members are assigned one or two ideas each and have the task of defending them to the group. They take turns emphasizing the positive aspects of their ideas (and how these ideas meet the criteria). This method has one disadvantage in that some serious weaknesses of the ideas may be overlooked, especially if the process is not accompanied by a good list of criteria. However, the procedure is valuable in that it gets excitement and intuition about innovative ideas back into the judgment process.

Reverse brainstorming is the opposite of the advocacy method. Here the group members criticize the weaknesses and flaws of each idea. This approach is an advantage for successful implementation because this knowledge will enable you to plan to overcome the weaknesses. However, this technique must be used with other, more positive methods to overcome this negative thinking mode, and a strong effort must be made to develop "cures" for the weaknesses. Here, too, having a list of criteria will provide guidance to critical thinking. Reverse brainstorming tends to be used for evaluating ideas that are "not invented here," and care should be taken not to use a double standard and misjudge valuable ideas just because someone else thought of them.

When more data are required to make a judgment, **experimentation** may be the best tool. If only a few solutions have to be evaluated, we can choose Edison's **trial-and-error method**. Techniques based on a statistical approach, such as the **Taguchi method of designed experiments**, can be used for evaluating a large number of options and interdependent parameters.

Which one of the many available techniques should be chosen for a specific application? Use a technique that you are comfortable with and that matches the level of sophistication and complexity of your problem.

Your choice will also depend on the time that you have availab
ing the judgment discussion, good communication and interperso
are very important. Typically, if an idea is 90 percent right and 10 percent inadequate, people will jump on the flaw and imply that anyone who put this kind of faulty idea forward must be an idiot. As "judges" we must guard against such an attitude, or valuable ideas and solutions will be discarded needlessly. We can continue to maintain a safe environment for expressing ideas with sensitivity and wisdom, so the criticism of an idea will not be taken as a personal attack but as an incentive for collaboration to creatively improve the idea.

You also need to bring a dose of skeptical thinking to the process. With critical thinking, we ask, "What is wrong with this idea?" With skeptical thinking, we dig deeper and ask,, "Is this true? What evidence do we have to support this?" George F. Huhn, a board member of the American Creativity Association, has this list of questions to ask when judging ideas and solutions to problems:

__ Do I have enough information (facts) to know this is true or correct? Am I rushing to judgment?
__ What are the consequences if this is not true or correct?
__ Do the proponents have a selfish interest in persuading me to believe this?
__ Do I need to believe this? Why? Am I deliberately overlooking or ignoring weaknesses in the arguments?
— What would be the consequences if I did not believe this?
__ How would I feel if I found out with incontrovertible truth that this was not true? Would I be disappointed? Would I look for something else?

⌛ Critical Thinking Activity 10-5: Peer Pressure

If you have a group of fifteen or more young people, you can conduct an interesting demonstration of peer pressure using the results of Team Activity 8-1. Post the list of brainstormed names (there should be at least 10 choices). Each person is given five sticky dots to vote for the preferred names—they can distribute their "votes" in any way they wish, including giving multiple votes to one or more ideas. Before the voting starts, ask two people (if possible one male, one female) to vote a secret ballot instead of using the sticky dots (also with five votes) and collect their ballot. After the group has finished voting with the dots, use a differently colored marker (or dots) to show the vote of the secret ballots. Then have a discussion of the results—how did the group's vote show peer pressure compared to the secret ballot vote?

⌛ ✓ Team Activity 10-6: Project Criteria and Judgment

Using the results of Team Activity 9-2, brainstorm a list of criteria with your team. Don't forget the creative thinking warm-up. Then engineer these criteria to obtain an improved list of valid criteria. Use your list with an appropriate judgment technique (such as the Pugh method) to find the top-ranking ideas. Discuss flaws and how they can be improved by combining the best features of different ideas or through additional creative thinking. Also consider the potential risks of implementation. If there is disagreement, examine the underlying values and thinking preferences.

Decision making

The methods we have just discussed result in ranked ideas, but they do not make the final decision. Criteria clarify priorities and may give a good indication of which may be the best solution for implementation, but decision making is a separate judgment activity. Our cultural heritage influences our decision making. Western civilization is in philosophy and practice a competitive system in which opposing views are argued and fought out in court, in politics, in business, and in everyday living. But through this process of attack and defense, positions become more rigid and thus creative solutions more difficult. Also, losers will not be in a mood to support winning ideas. All participants lose credibility.—just look at a recent political campaign for an illustration. In contrast, other societies are known for cultural traditions of cooperation.

Decision making has been defined as selecting a course of action to achieve a desired purpose. As a "judge," how can we be sure to make good decisions? We will need to appraise the situation and decide which form of decision making is most appropriate for the problem at hand. Important decisions with long-term effects and strong organizational impact require more thought, care, and time, whereas decisions on minor issues can be made quickly and routinely. Established procedures, standards, and policies in an organization are useful since they form a framework for decision making that can reduce time and error. As a "judge," we must also realize that it is impossible to please everyone.

Common decision making approaches

Let us look at ways of how people make decisions. Some of these techniques are used by teams, some by individuals. The selection of the most appropriate method depends on the particular circumstances.

Engineers shall acknowledge their errors and shall not distort or alter the facts. ... Engineers shall advise their clients or employers when they believe a project will not be successful.

NSPE Professional Obligations Section

Coin toss: When we have two options that are equally good, a toss of the coin can help us decide which one to pick, since either choice will give a good result. Sometimes it helps to focus on the consequences—make the decision based on the least troublesome or risky alternative.

Easy way out: If we have a number of equally good solutions, the easy way out will lead to the quickest and least painful resolution to a problem. Care should be taken not to ignore the long-term implications.

Checklist: We can make up a checklist that needs to be satisfied by the best solution. The quality of the solution will depend on the quality of the checklist. This list of minimum criteria is useful if all important points that the solution must meet are included. The list helps to pinpoint solutions that will really solve the original problem.

Advantage/disadvantage matrix: For best results, the matrix should come with a weighting system and include much thought about the nonremovable disadvantages and the validity of the criteria. It is a useful tool for identifying the most promising options or solutions. We can make a decision to select the highest-ranking option, or we can have a team vote. The team members should understand why they are casting their votes for particular ideas. If we feel that we want to vote for an idea even if it is ranked low in an evaluation matrix, we need to explore this prompting of our intuition—it is probable that we have used a very analytical approach when developing the criteria which may have led us to leave out important values that need to be brought out into the open.

You need to develop a careful balance between making judgments based on past experiences and keeping your mind open to new possibilities.

Mark Von Wodtke, Mind Over Media

Common consensus: This is the lowest level of group decision making. A decision that is reached quickly by common consensus is usually a mediocre solution because only what the majority likes and agrees with is being incorporated in the solution and thus implemented. Creative solutions have a knack of stirring things up—thus they are not easily accepted. When a quick decision has to be made, people tend to throw out ideas they don't like, ideas that make them uncomfortable initially, or ideas that would require change. It takes time to make creative ideas understood and accepted. Common consensus may be expedient to quickly solve an urgent problem, because delay has serious consequences. For the long term, a better-quality solution should be sought.

Compromise: People with widely differing views may choose solutions through compromise—a second level of group decision making. Compromise is a trade-off; some good parts are given up by both parties to gain acceptance of part of the solution. This approach is used frequently in government. Although a compromise may make the solution acceptable to a wider constituency, it may not be the best solution for the entire organization or community, because some good features have to be traded off to make the compromise acceptable.

Compound team decision: This process—the highest level of group decision making—can give a superior solution because the team concentrates on making the solution incorporate the best features of several ideas, to where everyone agrees that no further improvement is possible. This is the approach used in the Pugh method because all objections and weak points have been overcome with additional creative thinking. This is defined as a compound decision, although some people use the word consensus here, too. When a group does a careful job during the entire evaluation process and takes time to champion new ideas, an excellent compound decision and problem solving will be the result. In this process, what people don't like gets improved, not thrown out.

Delay: This is a decision alternative that may have its place. It may give you time to get more data and find a better solution. Perhaps the extra time will make the problem go away. Also, for political reasons, you may want to avoid making a decision. By delaying the decision past a specified deadline, you can exercise what is known as the "pocket veto."

"No" decision: Sometimes, a "no" is appropriate or wise. Perhaps the original problem no longer exists, or the problem and its context have changed so much that implementing any of the proposed solutions would make the problem worse. Changed circumstances demand a new cycle of creative problem solving.

Intuitive decision: Some people make decisions intuitively, without consciously reasoning through the process or working out an explicit set of criteria. Then, to explain their decision to others, they may "invent" rational reasons for their choice. This right-brain approach works quite well with people who have learned to trust their intuition and its reliability in making good judgment in particular situations.

Creative decision making

Many of the decision-making approaches that we have discussed so far are primarily analytical procedures involving left-brain thinking. But because we live in changing times, where the future is unpredictable, we need to use decision making tactics that involve both left-brain and right-brain thinking processes. Dr. H. B. Gelatt, an educator and psychologist, career consultant, author, and trainer, has written the book *Creative Decision Making: Using Positive Uncertainty,* which uses both rational and intuitive techniques for making the best decisions. Because this small workbook provides interesting insights into whole-brain decision making and is drawing parallels to the creative problem-solving process, we want to give a brief summary here.

There seems to be no invention, no matter how sophisticated, that can equal the power, flexibility, and user-friendliness of the whole human mind. We all possess the world's finest multisensory decision-making machine right in our heads. All we have to do is to learn how to use it.

H.B. Gelatt

Uncertainty is present in problem solving when we have too little information. It is equally present when we have too much information, especially when this data is irrelevant, conflicting, incomplete, unconnected, or even wrong. What we know is not the only basis for decision making—both what we want and what we believe strongly influence what we decide to do. This viewpoint is expressed in a four-step framework, and it is based on an attitude that sees uncertainty as positive!

1. Goal: Be focused (left-brain) and flexible (right-brain) about what you want. Goals are not fixed in concrete; they are just guides. Thus be open to change in response to changing conditions and changing expectations. As you achieve your goals, be open to unexpected discoveries.

2. Knowledge: Be wary (left-brain) and aware (right-brain) about what you know. Knowledge is power. But ignorance can be bliss: you haven't learned yet what doesn't work; imagination is valuable, and memory can't always be trusted.

It is impossible to go through life without making judgments about people. How well you make those judgments is critical to the quality of your life. Before you judge someone else, you should judge yourself.

M. Scott Peck, M.D., psychiatrist and author

3. Belief: Be objective (left-brain) and optimistic (right-brain) about what you believe. Wishful thinking has a rightful place. What is reality? Positive thinking can be a self-fulfilling prophesy.

4. Action: Be practical (left-brain) and magical (right-brain) about what you do. Trust intuition. Respond to change, but also cause change. Planning leads to learning and vice versa. What are your personal paradigm shifts? Be playful, not fearful, when making decisions. Play with the limitations of your logic and the bounties of your intuition.

The ideal situation is to develop a balanced approach. Evaluate the actions you could take and the possible results and uncertainties involved—the options, consequences, and probability of success. Have an attitude that asks: "What else?" Use different mindsets: be positive ➛ "explorer," objective ➛ "detective," emotional ➛ "artist," creative-integrating ➛ "engineer," negative ➛ "judge," and controlling ➛ "producer"-implementer. Decision making is at its best when it employs all six mindsets of the creative problem-solving process. Decisions here are seen as having four outcomes: Either they result in a plus or a negative for the self, and a plus or negative for others. When options are being evaluated, they can be passed through this **outcome matrix:**

Self		
Others		
	Plus	Negative

Practice imagining or "inventing" the future. How can old knowledge block new thinking? Get advice; collect different opinions. Use the process of internal debate: Have your left brain supply rational arguments; have your right brain comment on how you feel about each argument. Consider other people—who is on your "left" and who is on your "right"? What do their positions tell you? Who do you want to be like? Why? Whose opinion do you trust? Be optimistic about what might happen—you can change and influence what will happen. Your beliefs can determine what you do, what you want, and what you know. Be versatile; adapt to change. As an idea "judge" and decision maker, be sure to review the available information, options, beliefs, and goals.

This approach was developed to give "decision advice that is more closely related to what people actually do than to what experts say they should do." Positive uncertainty paradoxically combines intellectual/objective techniques and imaginative/subjective techniques into an unconventional wisdom for future planning and decision making. Does using this process make decision making easier? To answer this, we must first ask: "When is decision making easy?" It is easy when we have developed good solutions through the creative problem-solving process; it is difficult when none of the available options really solve the problem. The problem is that creating an effective solution usually takes longer than the time allowed for making the decision. But to create an effective solution to an unstructured problem requires that we employ the capabilities of the whole brain. We can use a decision cycle: 1. As quadrant D "explorers" ask, "What are the possible outcomes? Can we imagine what might happen? Does the decision fit in with our vision of the future?" As quadrant A "detectives" ask, "What is the evidence? Can it be trusted? What does an analysis of the facts show? What is the bottom line?" As "producers" ask in quadrant B mode, "What must we do with what we know? How do we take action? Does the planned decision leave us with sufficient control?" and in quadrant C mode, "Do we value the outcome? How are others affected?"

Final selection

What do you do if you still have more than one best solution at this stage in the judgment process? One or more of the steps in Table 10.6 may help you make the final selection.

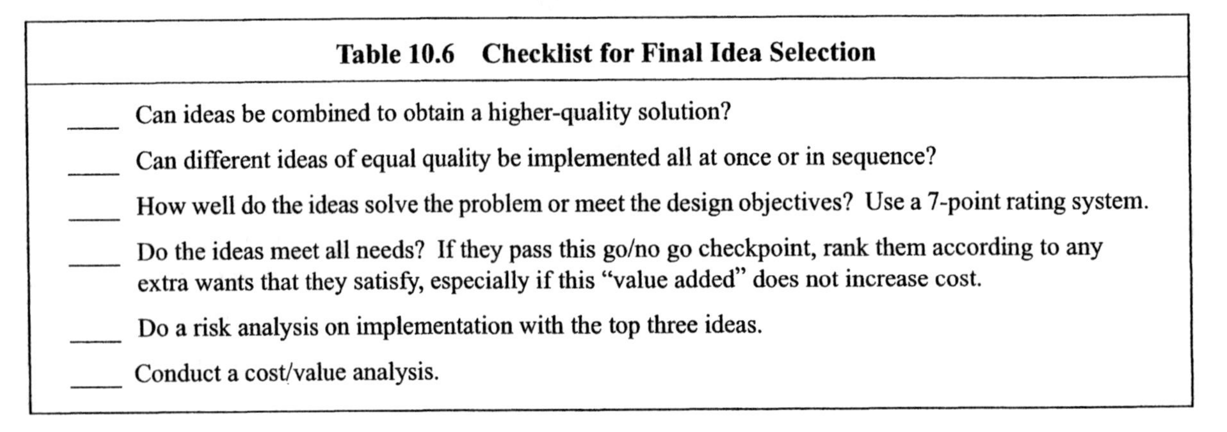

Table 10.6 Checklist for Final Idea Selection

____	Can ideas be combined to obtain a higher-quality solution?
____	Can different ideas of equal quality be implemented all at once or in sequence?
____	How well do the ideas solve the problem or meet the design objectives? Use a 7-point rating system.
____	Do the ideas meet all needs? If they pass this go/no go checkpoint, rank them according to any extra wants that they satisfy, especially if this "value added" does not increase cost.
____	Do a risk analysis on implementation with the top three ideas.
____	Conduct a cost/value analysis.

These final steps in the decision-making process can be done by the team or by management. Select the steps that are most appropriate to solve the problem. For example, a risk analysis (perhaps using Kepner-Tregoe) is only cost-effective for complicated, expensive solutions. The final decision should not be made strictly on the basis of return on investment (ROI) because intrinsic values or benefits to society can rarely be assigned a precise dollar figure. It may help to compare the long-term as well as the short-term costs and implications of implementation (see Chapter 16 for doing a life-cycle cost analysis).

The checklist has two purposes: It provides a last opportunity for improving the final ideas, and the results of this analysis facilitate the final selection. An example is given in Table 10.7. This last judgment activity is immediately followed by decisions on what actions to take to implement the chosen solution and assigning this responsibility to specific people. Implementation will be the topic of the next chapter.

Table 10.7 Case Study of Final Idea Selection

The problem of "How can high schools make learning more relevant?" was brainstormed by a class of 22 honor students in engineering divided into four teams for creative idea evaluation. Each team chose a small list of criteria and ranked the final ideas based on these criteria. The teams and their top-ranked ideas were:

Team 1—Curriculum: Strengthen the curriculum with special academic programs, including a new creative thinking class to increase practical applications and use of problem-solving methods.

Team 2—Teaching: Teachers should be tested before they are hired, not only to determine their amount of knowledge in their field but also to judge their ability to teach and convey this knowledge to others.

Team 3—Environment: Set up "career visits" to businesses and industry.

Team 4—Structure: (a) Restructure high schools to follow a flexible, college-type class schedule and atmosphere, (b) coupled with a positive grading system. *Note that (b) was added to improve (a).*

The final ideas were evaluated with an advantage/disadvantage matrix to determine their weak points with respect to implementation. Next, these ideas were examined in light of the final idea selection questions:

Evaluation Criteria	Ideas: + / o	1	2	3	4a	4a+b
1. Will the solution improve student learning?	yes/no	+	+	+	o	+
2. What will the implementation process be like?	easy/difficult	+	o	o	o	+
3. Will implementation lead to change/innovation?	yes/no	+	+	+	+	+
4. What will be the costs of implementation?	low/high	+	+	+	+	+
5. How many students will be served?	many/few	+	+	+	+	+
6. Will the solution decrease the dropout rate?	yes/no	+	+	+	o	+
7. Will the solution increase college-bound students?	yes/no	+	+	+	+	+
8. Will the solution impact the disadvantaged?	much/little	+	+	+	o	+
9. Can the solution get community support?	yes/no	+	+	+	o	+
10. What is the degree of risk?	low/high	+	+	+	o	+
11. What is its effect on school morale?	up/down	+	+	+	+	+
12. Will teachers accept the solution easily?	yes/no	+	o	+	o	+

Can these four ideas be combined to obtain a higher-quality solution? Career visits and a college-type schedule address different students—thus both of these ideas are ranked of equal importance. Quality teachers are also essential, as are the stated curriculum improvements: stronger academics together with a creative thinking/problem-solving class for more relevant learning.

Can these ideas be implemented all at once or in sequence? The career visits will be easiest to implement. The creative thinking/problem-solving class will require teacher training, different room layouts, and some adjustments in scheduling, all of which will take time to implement. Implementation of the other two ideas will be more difficult. The college-type schedule could be tried in smaller schools first, but encouraging teachers to use a more positive grading system should be fairly easy to do. Implementing teacher testing is a rather thorny issue but is being addressed by some school boards in an effort to increase the competence and quality of teachers.

How well do the ideas solve the problem? The creative thinking/problem-solving class will benefit all students. Quality teachers are also an essential prerequisite. The career visits will serve noncollege-bound students especially, whereas the college-type schedule of course will benefit the college-bound students the most. All four ideas in combination provide a good climate and a more complete solution.

Continued on the next page

Table 10.7 continued

Do the ideas meet all needs? Each idea does not meet all needs—the list of required improvements in public education is simply too long. (Many of the original brainstorming ideas addressed specific needs and could be implemented later in stages.) Implementing the four top-ranked ideas will form a foundation to build on, with further improvements possible as needed in both the academic and vocational areas.

What are the costs and risks involved? The schedule change would place more responsibility on the students; thus it comes with increased risk. Creative thinking requires more flexibility from the teachers. But the overall risk for the country as a whole is much larger if nothing is done to improve schools—thus we cannot afford to wait. A firm determination is required to do whatever is necessary to make the schools better—there is no other priority that is more urgent. The costs of implementing these four ideas will be quite reasonable, and it should be possible to develop strong support from taxpayers, parents, and businesses.

Application to engineering design projects

Engineers have two additional "screens" when evaluating conceptual designs. One is feasibility, the other is technological readiness. The judgment involving feasibility is usually made intuitively: 1. The design concept will never work. 2. The design can possibly be made to work (with a lot of development effort). 3. The design idea looks feasible. Concepts should not be rejected out of hand if they do not appear feasible. The ideas can serve as stepping stones, or they can be improved by combining with other concepts; this will be demonstrated with the Pugh method in Chapter 11. As will be shown in Part 3, the feasibility of a design concept will need to be confirmed through solid modeling, analysis, and perhaps prototyping during the design process.

Decision making based on technological readiness must ask:

__ Are manufacturing processes for the technology available and proven?
__ Have critical parameters been identified and quantified for optimum performance (for example through a QFD House of Quality)?
__ Have the failure modes been analyzed (FMEA and/or FTA)?
__ Is the planned technology mature?
__ What product life-cycle problems might be associated with the use of the technology?

Making decisions on technology is not easy in today's competitive environment. State-of-the-art technology may be quickly outdated with innovation and continuous improvement. On the other hand, product quality (and market share) can suffer if a new technology is used prematurely.

The project team needs to keep documentation on the judgment phase: a summary of the conceptual designs and results of the Pugh method, the criteria for making the decisions, and the results of any analysis conducted for comparing the feasibility of different designs. In industry, the project would now be ready for submission to management for the go/no go design review. Student teams may need to prepare their optimized design concept for a "sales" presentation and report to the sponsor.

⌛ **Ten-Minute Thinking Activity 10-7: Analyzing a Decision**
Alone, or with a small group, analyze how you made a recent decision. Could the process have been improved if you used the whole-brain approach? What were the underlying values?

⌛ ✓ **Decision-Making Activity 10-8: Final Project Decisions**
Using the results of Team Activity 10-6 and the discussion in this chapter, make your decision on which of the final ideas or solutions to implement. If you employed the Pugh method, explicitly make the decision to implement the top-ranking design or the highest-ranking solution (or combination of solutions). Include a brief explanation of how you made the final decision and why the selected solution is best.

Resources for further learning

10.1 Mary F. Belenky et al., *Women's Ways of Knowing: The Development of Self, Voice, and Mind,* BasicBooks, New York, 1986. These researchers found that Perry's model may not be valid for female students.

10.2 ✓ Stephen D. Brookfield, *Developing Critical Thinkers—Challenging Adults to Explore Alternative Ways of Thinking and Acting,* Jossey-Bass, San Francisco, 1988. This book shows that critical thinking is a productive process enabling people to be more effective and innovative. Available by the same author and publisher is an audio tape, *Becoming Critical Thinkers: Learning to Recognize Assumptions That Shape Ideas and Actions,* 1991.

10.3 Michael J. French, *Invention and Evolution: Design in Nature and Engineering,* Cambridge University Press, New York, 1988. This paperback book contains many examples of designs and products. Judgment and design decision making are taught implicitly by example.

10.4 ✓ H. B. Gelatt, *Creative Decision Making: Using Positive Uncertainty,* Crisp Publications, Los Altos, California, 1991. This workbook encourages exploration of rational and intuitive techniques to make the best decisions.

10.5 Myron Peretz Glazer and Penina Migdal Glazer, *The Whistleblowers: Exposing Corruption in Government and Industry,* BasicBooks, New York, 1989. This paperback summarizes the values and experiences (and the price paid) of sixty-four courageous ethical resisters and their spouses.

10.6 Spencer Johnson, *"Yes" or "No": The Guide to Better Decisions,* Harper Business, New York, 1992. The fictional story of a businessman's hike up a mountain teaches important decision-making concepts.

10.7 Joanne G. Kurfiss, *Critical Thinking: Theory, Research, Practice, and Possibilities,* ASHE-ERIC Higher Education Report 2, Clearing House on Higher Education, George Washington University, 1988. This report surveys theories and research into current college practices of teaching critical thinking as argument skills, cognitive processes, and intellectual development.

10.8 William Perry, *Forms of Intellectual and Ethical Development in the College Years: A Scheme,* Holt, Austin, Texas, 1970. The Perry model detailed here forms the basis of much current research in critical thinking.

10.9 ✓ Henry Petroski, *Design Paradigms: Case Histories of Error and Judgment in Engineering,* Cambridge University Press, New York, 1994. This book presents studies of famous engineering failures.

To attain knowledge,
add things every day.
To attain wisdom,
remove things every day.

Lao-Tsu,
in the Tao Te Ching

Exercises for "judges"

10.1 ✓ Check Your Assumptions
Ann and Barnaby are found dead on the living room floor in the middle of a pool of water and broken glass. Write a story of what happened.

10.2 Fable
First, examine one of Aesop's fables and analyze the moral value that is being taught. Next, write your own fable.

10.3 ✓ Failure
Imagine that you are a senior citizen giving a talk to a group of high school students. What would you tell them about the value of failure? Include a funny story (true or invented) about your personal failures in school (or in life) and how this helped you develop good judgment.

10.4 ★ The Power of the Telephone ★
Do not answer the phone the next two times it rings. How does this make you feel? Was it difficult to do? How much power does the phone have over you? How would you reason with someone who thinks that an emergency call might be missed if the phone is not answered? What strategies would let *you* decide when you want to talk on the phone?

10.5 ★ Judging a Television Program ★

a. Alone or with a group, watch a television program. Record it on a VCR for a later rerun. After the program ends, judge it quickly on the basis of positives and negatives. Write down these judgments.
b. Develop a set of thoughtful criteria for judging TV programs (including the news). Also consider your values—on what are you basing your criticism or choice of criteria? If you can, involve people of different age groups when you make up the list of criteria.
c. Now run the taped program again and judge it using the list of criteria. Is your judgment different this time? From working out the list of criteria, did you gain some insight into what makes a "good" program and which programs are just a waste of time (or, even worse, garbage for your subconscious mind)? Are there differences in the criteria based on age, or are there universal criteria?
d. Why do you like your favorite programs? Do you feel guilty when your viewing does not include many educational programs?

10.6 ✓ How to Criticize

Make up a scenario in which you have to criticize someone. Write it in such a way that you start out with two positive statements. Then make a wishful statement about the item you want to change, followed by another positive statement about the other person. Then conclude with a hopeful, cooperative, positive statement. Here is an example:

Critical statement: Ugh, you smell awful; why can't you quit smoking!

Better way: I appreciate your visits—you have a way of cheering me up. And it is so thoughtful of you to take off your sneakers before walking across my nicely buffed floor. I wish you could take the same care with your health and quit smoking. I bet this could even increase your endurance—you might win the marathon next time! Let's make a pact for mutual support and encouragement—I'm willing to give up snacking on junk food; this way we'll both be winners.

10.7 ✓ How to Accept Criticism

It is easy to feel dejected. It is normal to be put on the defensive when receiving criticism. It is abnormal to look at the criticism as an opportunity for self-improvement. Be abnormal! Think of a situation when you were criticized. But instead of thinking of defenses or feeling hurt, place yourself "outside" the situation. Analyze the criticism. Was there a basis for it? What situation brought it about? What should you change to avoid this situation in the future? If the criticism is unjustified, mentally write it on a piece of paper, then imagine throwing it in the trash (or down the toilet). Then let the matter rest.

Factors involved in critical thinking:

- ***Learning how to learn—preparing for change.***
- ***Learning from visual representations.***
- ***Lifelong learning—a necessity.***
- ***Metacognition—how to assess one's own progress, be aware of what one does or does not understand, and pursuing strategies for filling the identified gaps.***

Susan M. Brookhart, National Forum, *Vol. 78, No. 4, Fall 1998*

10.8 Authoritarian Environment (continued from Exercise 6.7)

Now that you have learned several steps in the creative problem solving process, do you have some new ideas about overcoming the "follow the rules" barrier? Divide into two groups of three people each, with one group representing authority, the other the creative problem solvers. Make up a scenario where a creative idea is "sold" through negotiation and compromise. Note that "breaking the rules" does not mean breaking the law. The new idea must be legal, moral, and ethical—it simply does not follow the traditional way of doing things.

10.9 ★ Critical Thinking and Democracy ★

In a democratic society, is freedom always coupled with responsibility? In what way? In a group, discuss and answer questions such as: Does everyone have the right to drive a car, or is it a privilege? Who decides what is right—is there such a thing as absolute truth? Do "pro-life" and "anti-abortion" have the same meaning? Is "pro-choice" a good word to use for someone talking about abortion rights, or does it disclose a fallacy in thinking about choice in the larger context of life and values? Do the ways the questions in this problem were posed reveal the underlying values and bias of the authors? Is it possible to do value-neutral teaching? If yes, would it be desirable or undesirable? Why?

10.10 Criteria and Voting

a. In groups of three people, share and discuss an experience with voting in which the lack of specific criteria caused problems. How were the problems resolved? What was learned from the experience?

b. Alternatively—especially in an election year—discuss some of the reasons mentioned in the media that people use to judge a candidate's suitability for political office.

c. In a group of six, brainstorm a list of ten important criteria for one of the following: Supreme Court Justice; President of the United States, mayor or manager of your city, member of the local school board, U.S. ambassador to the United Nations. Was it easy or difficult to reach a consensus on the ten most important criteria?

10.11 Personal Values

a. How would you explain to a 5-year-old child that taking a candy bar in a store without paying for it is wrong? Would your explanation be different if the child were your brother, a friend of your brother's, or a stranger? Why or why not? What if the child were a teenager?

b. How valuable is a good name? Brainstorm this question with a group of your peers and with a multigenerational group. Do the answers come out differently, or is there a common ground? What values are being expressed by the participants in the discussion?

c. Discuss who gets hurt when students cheat on an exam because they did not make the effort to thoroughly learn the material. What are the consequences to learning, personal relationships, the future?

d. Find examples of people who have overcome handicaps or personal tragedies. How did they do it? What inner resources do they have? How did their beliefs change because of these experiences?

To make better decsions,
I use my head and
I consult my heart.

Spencer Johnson, M.D.

10.12 ★ Cultural Values ★

Surveys have found a conflict between personal and cultural values. Trial by a jury of one's peers is considered to be an important value in our democratic culture, yet people are increasingly unwilling to serve on a jury. What personal values do you think these people have that conflict with the cultural value? What values must a democratic society have to survive? How important are hard work, discipline, respect for law and order, service, tolerance, and honesty to the survival of democracy? How prevalent and respected are these values in our society today? How are freedom and personal commitment related? What values undergird a caring community? What are some important values in a society dominated by scarcity? What are important values in an affluent society?

10.13 Application of Skeptical Thinking

Use skeptical thinking when evaluating television and other commercials, printed ads, music, TV programs, magazines, newspaper articles, political speeches, and "buzz" words. In a group, select three different items, discuss, and write a brief analysis.

10.14 Gender Differences
Look at Figure 4.4. Would the message be different if the two people bailing were women? Why? What values, feelings, and bias are coming into play? What if the people at both ends were heterogeneous couples?

10.15 ★ Evaluation of Conflicting Opinions ★
Find newspaper or journal articles that give two opposing points of view on a certain subject. For example, *USA Today* carries a daily feature that presents two views on a current issue. Give a brief summary of each; then indicate your agreement or disagreement with the expressed views. Support your viewpoint with additional facts or point out where the writers should have supplied more information.

10.16 ★ Detecting Fraud and Swindlers ★
Many people from all walks of life become victims of swindlers every year, but self-confident individuals seem to be particularly vulnerable, as is the secret desire of getting something for nothing. Other clues are strong pressure to act and glowing testimonies from "satisfied customers." Research the topic and prepare a list of strategies that can help your team develop good judgment and protect from falling prey to fraud.

Critical thinking is the intellectually disciplined process of actively and skillfully conceptualizing, applying, analyzing, synthesizing, and/or evaluating information gathered from or generated by observation, experience, reflection, reasoning, or communication, as a guide to belief and action.

Michael Scriven and Richard Paul

Chapter 10 — review of key concepts and action checklist

Your role as "judge": Use an analytical, critical mindset together with positive, creative thinking to decide which ideas are best and to find or develop the best solutions for implementation. Look ahead and consider the impact of the solution; assess values and bias. It takes experience with failure to develop good judgment. "Judges" also have to deal with uncertainty, risk, and ethics.

What is good judgment? It is an ability to detect errors made at any point during the design or problem-solving process, to eliminate flaws, and to evaluate the risks, consequences, and uncertainties of alternative solutions. It involves an awareness of bias, underlying values, and presuppositions that can influence judgment. Good judgment is ethical.

Critical thinking and creative problem solving: The current view considers critical thinking as a form of analytical problem solving; the goal is not necessarily to find a solution but to construct a logical representation of a situation or position based on evidence and plausible arguments. Critical thinking as taught in schools is often very narrowly focused on developing argument skills and cognitive processes based on facts and procedures in a particular discipline. The Perry model sees critical thinking as a development in four stages, where the majority of students do not progress beyond the second level by the time they graduate. The Brookfield model considers critical thinking to be a whole-brain process which includes skeptical thinking; it is rational as well as intuitive, creative, and contextual. This type of critical thinking in teams is essential for maintaining a healthy democracy.

Idea judgment ➛ ranking ideas: Develop a list of criteria. Criteria are standards used for judging; they are best developed through brainstorming with a team and should consider such factors as motivation, people, cost, support, values, time, and consequences. Techniques are available to help evaluate and rank ideas.

Table 10.8 Idea Judgment in a Nutshell

1. Objective	___	What is the current problem situation?
	___	What is the idea trying to do?
2. Positives	___	What is worth building on?
3. Negatives	___	What are the drawbacks?
	___	What is the worst thing that could happen?
4. Probability	___	What are the chances of success?
	___	If the idea fails, what can be learned?
5. Timing	___	Is the timing right for this idea?
	___	How long do you have to make your decision?
6. Bias	___	What assumptions are you making?
	___	Are these assumptions still valid?
	___	Do you have some blind spots?
7. The Verdict	___	What is your decision?
	___	How will it affect people?
	___	What is to be done next?

Idea judgment ➛ decision making: Traditional forms are mostly analytical. Cultural values influence attitudes in decision making (cooperative versus adversarial). Creative decision making sees uncertainty as positive. Features are: 1. Be focused and flexible about goals. 2. Be wary and aware about knowledge. 3. Be objective and optimistic about beliefs. 4. Be practical and imaginative about actions. Use all six creative problem-solving mindsets to make the best decisions! In summary, Table 10.8 is a simple checklist for a "judge" that you can apply as an individual when evaluating ideas.

Action checklist

☐ Ethics in engineering is important. Find out where you can learn more: a course, books, professional societies? Then take time to become better informed on some of the current ethics issues in engineering.

☐ When making your next important decision in a team, make it a whole-brain process; check that your list of criteria includes items from all four quadrants. Consider facts as well as values.

☐ Copy the "Idea Judgment in a Nutshell" table and place it in your wallet or purse; refer to it when you need some guidance when making a quick decision as an individual.

☐ Analyze a case in the past where you made what you feel was a "wrong" decision. What aspect of the judgment process do you need to improve to prevent this from happening again?

11

The Pugh Method

What you can learn from this chapter:
- The place of conceptual design in product development.
- Economic benefits of the Pugh method.
- Phase I: development of criteria and conceptual designs.
- Phase II: convergence to a superior concept.
- Application: examples and hints for conducting a concept evaluation exercise with your team.
- Further learning: references and review of key concepts.

The Pugh method of creative design concept evaluation was developed by Stuart Pugh, a design and project engineer with many years of practice in industry. He later became professor and head of the design division at the University of Strathclyde in Scotland. He came to recognize that designs done purely by analysis were "somewhat less than adequate" because it took a long cycle of modifications to satisfy the customer. He realized that engineers need to see the whole picture in product design and development; they need an integrated approach to be competitive. Although the Pugh method has its most direct application in product design, the procedure and thinking skills used can be applied to many other situations where different ideas and options have to be evaluated to find an optimum solution.

We will discuss the principles of the Pugh method in the context of conceptual design and product development. We will say a few words about its economic benefits and will then present the features of the Pugh method and the results of the Phase I and Phase II evaluations. We will briefly examine three examples and then give some hints for conducting your own Pugh method exercise for design concept evaluation.

Improving the practice of engineering design in U.S. firms is essential to industrial excellence and national competitiveness.

National Research Council, 1991

The product development process

In Figure 11.1, product development is compared to the creative problem-solving process. Each step in creative problem solving results in an output, which is indicated as a boxed item on the right-hand side of the

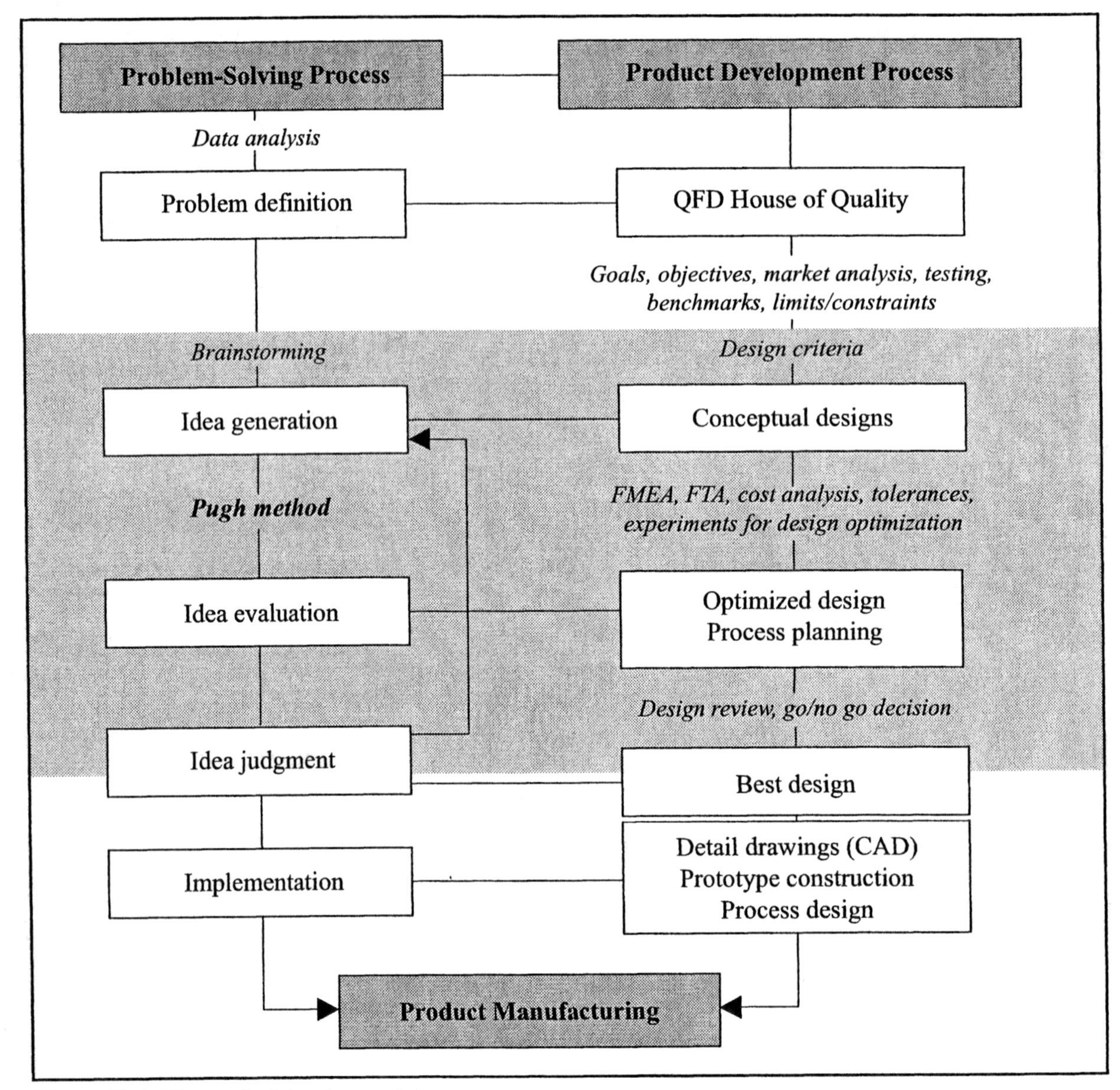

Figure 11.1 The product development process.

flow chart. We can thus visualize how the Pugh method fits within the context of product development. The goal of problem definition is to come up with a comprehensive list of design criteria. Brainstorming with team members from several departments (including R&D, design, manufacturing, sales, and service) is done during this stage. The customer's voice is critical in this data analysis process, and it must be part of the evaluation criteria. Benchmarking is often used here: the best competing product in its class is taken as the standard—and then goals are set for doing better for each important product feature involved in satisfying the customer. The QFD House of Quality (see Appendix A) is an excellent tool for identifying critical product features.

When a team uses the Pugh method, it makes a conscious effort to overcome the negative features and shortcomings of proposed design concepts. Because the ideas and concepts are placed together on an advantage/disadvantage type matrix, the process encourages force-fitting and synthesis at a conscious as well as at a subconscious level. The conceptual designs are based on the criteria and are generated in the "artist's"mindset. Here, the product designers and engineers can really play around with many different ideas and approaches on how to satisfy the criteria. At this stage, the emphasis is more on exploring alternatives than on finding one adequate solution. Only after a number of very different concepts are on hand should the process move on to evaluation in the "engineer's" mindset. The ability to obtain insight into the problem, to think of many different approaches, and to use stepping stones to solve the problem—these skills can be improved with a number of techniques as part of creative problem solving (see especially the different methods described in Chapters 7 and 8). Dynamic computer models or physical models of the proposed concepts can yield useful information at this stage that cannot be obtained with a static model. Most of all, the team members must continue to ask questions with an open mind.

Optimization only happens if you want it that way. It takes extra effort to get that extra plus.

Sidney F. Love

After the first phase of the Pugh method has reduced the number of concepts and has produced superior designs, additional studies can be made, such as cost analysis, FMEA (Appendix D), and FTA (Appendix E), with the goal of optimizing the most promising designs while simultaneously initiating process planning with feedback to product design through revised criteria. This is known as "total design" or "concurrent engineering." At the close of Phase II, one or at most two superior concepts will emerge. The design and evaluation process is then followed by a design review and a go/no go decision conducted by management. If the design is approved, detail drawings and prototype construction (if needed) are authorized, and manufacturing process design is completed.

Economic benefits of the Pugh method

A major portion of the cost of a product is determined during the design concept phase (from 70 to 85 percent); thus it is imperative that the best thinking and design tools be used to develop a "best" design at lowest cost. The Pugh method is a key tool for achieving this objective. The Pugh method eliminates engineering changes late in the product development process or, what is even more costly, after the start of production. Through intensive discussion and analysis, no flaws are overlooked. The Pugh method is an effective communications tool, and the team members gain a common understanding of the problem and the different options and solutions. When the best solution emerges, every person on the team understands why this solution is best; each person is ready to champion this concept.

Table 11.1 Ten Cash Drains in Product Development

1. **Technology push—but where's the pull?** Americans love technology for its own sake, and major resources are spent on developing technology even where no discernible market needs have been identified. Conversely, if we have a strong market need without proven technology, a new concept should not be selected during product or system design unless prior research activities have developed the concept to a sufficient level of maturity. It is an advantage to have one member of the conceptual design team from the advanced technology or R&D department.

2. **Disregarding the voice of the customer.** The voice of the customer must drive the activities of the entire organization. Chapters 5 and 7 have addressed this topic.

3. **The eureka concept.** This is when someone has this "great new idea" that becomes the only concept that is considered seriously. Such concepts are often proven very vulnerable by the time they reach the market. The Pugh method objectively assures that only well thought-out concepts are developed, not ideas that people "fall in love with."

4. **Pretend designs.** Here the emphasis is simply on being new, not necessarily better. When we compare the new design against a competing benchmark, we will be able to avoid developing an inferior product just for the sake of novelty.

5. **The pampered product.** We can prevent pampered products (who will only perform at their best when given a lot of maintenance) when we do optimization studies to make the product robust and reliable by reducing variance. The product must be able to stand up to use (and abuse) by all kinds of customers and under varying operating conditions.

6. **Hardware (and data) swamps.** This phenomenon occurs when we have so much prototype iteration that the entire team becomes swamped with the chore of debugging and maintaining the experimental hardware, leaving no time to evaluate data or improve designs. With the Pugh method, two to four iterations at most of the best design will be needed for fine-tuning and verification.

7. **Here's the product; where's the factory?** Manufacturability is an important criterion that is introduced very early in the design concept phase of concurrent engineering.

8. **We've always made it this way.** This cash drain not only holds for product design, it especially applies to process planning and design when the operating points of the manufacturing processes are being specified. The values of the process parameters are often chosen rather arbitrarily by experience. In product design, the group discussions and the creative thinking involved in the Pugh method create an environment where "the old way of doing things" will be questioned and improved if possible.

9. **The need for inspection.** The need for extensive inspection is minimized through the use of on-line quality control. The control chart (one of the SPC tools) has been used to good benefit. When production approaches zero defects, frequent inspection obviously will become superfluous.

10. **Give me my targets; let me do my thing.** When people in organizations work in isolation—without looking at the context of the entire process or product—the result is a subsystem that cannot be integrated or a product design that cannot be produced. Successful manufacturing requires teamwork and cooperation between horizontal and vertical levels throughout the organization, together with an appropriate level of management.

Here are some additional reasons why the Pugh method should be used. It prevents a company from making costly mistakes in the choice of products. If you have to convince managers to use the Pugh method, you can emphasize the cost savings. The material in this section is an excerpt from a lecture that Dr. Don P. Clausing has developed. Formerly vice president at Xerox Corporation, he is now a professor at MIT. He points out ten areas (shown in Table 11.1) where companies waste money during the product development cycle; using the Pugh method will help avoid these pitfalls. Money is wasted primarily in two ways: in the expense of developing the product, and in the lost opportunity of having a better product that could maintain or gain market share.

The Pugh evaluation process

Table 11.2 outlines the key steps in Phase I of the Pugh method. The matrix is basically an advantage/disadvantage evaluation scheme. During the early parts of Phase I, the number of concepts under consideration increases because each alteration to a concept is considered to be a new concept. In later rounds and in Phase II, the number of concepts carried forward to another round of improvements decreases, since ideas are merged and synthesized. Weaker concepts drop out as the quality of the remaining concepts increases. Developing and understanding the list of criteria is a key task for the evaluation team.

Table 11.2 The Pugh Evaluation Process

Phase I

1. **Criteria:** The list of evaluation criteria is developed through team discussion. A benchmark or datum is selected, usually the "best" existing product. If no comparable product exists, one of the new concepts (selected at random) can serve as datum.

2. **Design concepts**: Original design concepts are brainstormed by individuals or small teams.

3. **Evaluation matrix:** Each design concept is discussed and evaluated against the datum. Through the discussion, new concepts emerge; they are added to the matrix and evaluated.

4. **Round 1 results:** The results of the first round are evaluated, and the top-ranking concept is selected as the datum for the next round. During an incubation period, the teams improve the original design concepts by borrowing ideas and components from each other, as well as through additional creative thinking. Then Steps 3 and 4 are repeated with these improved, synthesized designs (further rounds).

Phase II

5. **Better designs:** The weakest designs are dropped; the improvement process is continued for additional rounds with fewer but increasingly better concepts. During the process, the strong, surviving concepts are engineered to more detail; the criteria are expanded and further refined. The weak points of the concepts are being eliminated. The team gains insight into the entire problem and solution.

6. **Superior concept:** The process converges to a strong consensus concept that cannot be overturned by a "better idea." The team is committed to this superior design and wants to see it succeed.

1. The list of criteria

The criteria used in the evaluation matrix are comprehensive, relevant, and explicit; they must incorporate the objectives of the planned product, its purpose, and its targeted market. Performance specifications have been established through testing or a benchmarking analysis of the competition (see Appendix B). Constraints (such as cost ceilings and government regulations) are identified. For components, specifications or tolerance limits are set to make the part fit into the context of the whole product. When evaluating nondesign ideas, the criteria may not be as technical but they still must include all important aspects of what the solution needs to accomplish to solve the problem.

These deficiencies [of not considering people's aspirations and needs] have to be recognized, otherwise misdirected engineering rigor will always give rise to bad total design. This implies that design teams should always include non-engineers.

Stuart Pugh

In the early stages of idea generation and development, it is better not to have too many constraining or detailed criteria. Did you know that the first successful modern airliner was built within two years from a one-page list of performance specifications? This was the DC-3, perhaps the most successful airplane ever. In comparison, the specs for a new aircraft today might fill several trucks. Different organizations, especially large companies such as Ford or General Motors, usually have established procedures to come up with the design criteria. These design criteria traditionally have reflected the "voice of the boss" or the "voice of the engineers." The Japanese have perfected the art of collecting the voice of the customer through such methods as quality function deployment (QFD), see Appendix A. This technique efficiently culminates in a list of criteria which assures that the designed product is responsive to the market needs and customer wants, not engineering or technology requirements. It assures that customer needs are not lost somewhere between the design shop and the factory floor. Each worker understands how this or her job contributes to meeting the customer needs.

Four criteria should always be included: quality, low cost, manufacturability, and environmental impact (in manufacturing, product life cycle, and final disposal). U.S. engineers have learned from the Japanese that low cost means product excellence which is designed into the product from the start. We usually add one more criterion to student projects: the concept must be creative. Otherwise, students will go back to the "tried-and-true" and will not take a risk with innovative ideas.

2. Conceptual designs

After the design criteria have been obtained by the design team, conceptual ideas are worked out over a period of days (for students) or weeks (in industry). These conceptual designs are outline solutions to the design problem, where the rough sizes and structural relationships among the major parts have been determined. Also, decisions have been made on how each major function will be performed. A conceptual design is worked out in sufficient detail to allow estimates of cost, weight, and overall dimensions to assure at least a reasonable probability that the product is feasible. Chapter 16 will provide information on how to make economic decisions during conceptual design. During Phase I, major points in the product features need to be decided as well as the rationale for making these decisions. An arbitrary decision, especially early in the design process, means a wasted opportunity for increasing quality and decreasing cost. When we use an established parameter without questioning, we are in fact making an arbitrary decision. During the conceptual design phase, we must question the "accepted" or conventional way of doing things and look for alternatives.

Broad solutions to the defined problem are developed in the form of design options. This phase places the greatest demands on the designers in terms of creative thinking, since innovation can originate here, not only in the product but also in manufacturing process design. When the designers choose a shape for a part, they must also think about how this part, this shape, is going to be manufactured. Is there an easier way to achieve the same purpose with a different material, shape, or way of making it?

3. The evaluation matrix and process

Then the conceptual designs are submitted for evaluation. A group meeting is called for all those involved in product development. The meeting is held in an ample conference room with a large board covering an entire long wall. An evaluation matrix is set up on the board, with the design criteria listed in the left-hand column of the matrix. The large drawings of the design concepts are posted across the top of the evaluation matrix. Depending on the number of design concepts submitted, the matrix may take up the entire wall of the room. The main features of each concept are explained by its developers to the whole evaluation team. Immediately after the presentation, the concept is evaluated against the datum using the three-way rating scale given in the sidebar.

Pugh Method Evaluation Scale:

+ **means substantially better,**

— **means clearly worse**
→ **this item needs attention,**

S **means more or less the same.**

The first concept entered on the matrix is the datum. Its features are explained first (whether it is an existing product or a new concept). It is used as the standard of comparison for the first round. A "plus" mark is given when a new concept is markedly better, a "minus" mark when it is definitely worse, or an "S" when it is about the same as the datum. The three-way evaluation may appear rather primitive, but it is easy to do with a team. The results are effective, because the objective is not quantitative, precise information but a movement toward increasing quality and superior satisfaction of all criteria. Inexperienced group members may be very defensive and protective of their design and will argue about every minus mark. They need to remember that this

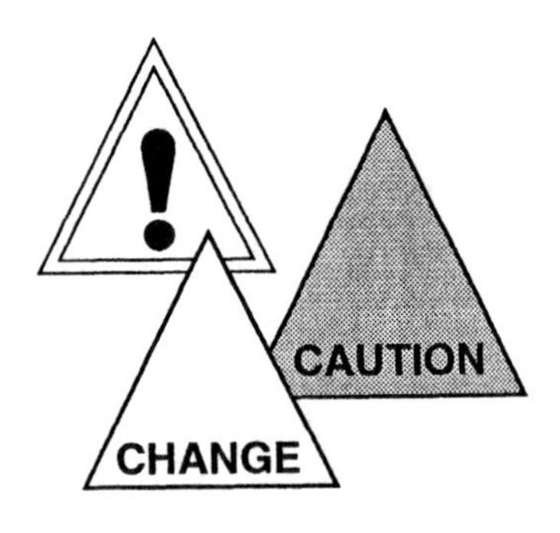

evaluation serves to point out weaknesses or potential problems in the design that must be overcome for a product to be competitive (and the company to survive). The judgment only determines as objectively as possible if the concept is better or worse than the datum or benchmark. But because the negative sign has an emotional impact on students (since it is seen as criticism), we propose using a different symbol for items that fall short and thus need attention—the delta (Δ) which gives the more positive message of caution or change needed!

So that the evaluation team can judge each concept carefully, each concept is drawn large enough to be visible to everyone in the room. At this early stage, the conceptual designs are outline solutions only. To allow for a fair comparison, the drawings should be carried out to a similar level of detail and follow the same format. It is difficult to compare concepts if one resembles a sketch on the back of an envelope and another is a beautiful artist's perspective or a detailed CAD drawing. Questions and disagreements about the criteria often occur during this comparison process, due to differing interpretation. The open discussion during the creative evaluation helps clarify and resolve these ambiguities, and the criteria become increasingly better defined and useful.

4. Evaluation of the first round results

When the first-round matrix has been completed, the results are critically examined. Is there a criterion that received no plus signs all across the matrix? This indicates that none of the new concepts considered an important customer need, and the teams must address this. If a criterion received all positives, it will need to be refined and made more specific. Criteria that are least important can be dropped in future rounds to simplify the evaluation. What new criteria were brought out in the discussions? If they are important or clarify an ambiguity, they must be added to the list. Typically, the customers become more precisely identified. For example, if we are working on ideas to improve a curriculum, the customers are not only the students, but the parents, the teachers, the profession (business, law, engineering, medicine, etc.), the future employers of the students, the taxpayers, and perhaps even society as a whole, with each customer group having different requirements.

The difficulty in concept evaluation is that we must choose which concepts to spend time developing when we still have very limited data on which to base the selection.

David G. Ullman

The scores in each column are now added separately for the positives and the negatives (or the deltas). Positives never cancel out shortcomings—thus do not add them together mathematically to obtain an overall ranking of the concepts. The ultimate goal of the process is to obtain concepts whose shortcomings have all been eliminated with improved ideas. With the results of the matrix, the design teams can take their concepts back to the drawing board and target their improvements to the identified weaknesses in preparation for the next round of evaluation. The concept that had the highest number of positives becomes the datum for the next round. Its creators in the meantime will try to improve this design even further by eliminating the identified shortcomings. Did you notice that the design concepts now have to try to meet or beat a higher standard to remain in the running? And the datum concept must be improved as well, or it will be surpassed by the others.

"SILENT" EVALUATION

For very argumentative teams, the evaluation can be performed by each member independently on a sheet of paper; the results are then compiled.

5. Fewer but better designs

In Phase II, the emphasis changes from conceiving additional concepts to synthesizing higher-quality designs by combining ideas and dropping the weaker concepts. During the process, the strong, surviving concepts are engineered to more detail; the criteria are expanded and further refined. The weak points of the concepts are being eliminated. The team gains insight into the entire problem and solution. The top concepts are analyzed in more detail. For example, recall the process for the design of the new Oakland Bay Bridge. At this stage, the original design concepts were narrowed down two basic designs with two options each. More studies and design details were then authorized to facilitate the final decision among these four concepts.

What if a negative cannot be eliminated? This could be the case for high manufacturing costs. If a competitive price is very important, other concepts that do not have this "flaw" will have to be pursued. This could be an area where the development of new technology is needed or a new paradigm must be discovered. Teams must not discard low-scoring concepts too quickly—they may contain valuable stepping stone ideas that can be merged with some of the other concepts.

In Phase II, the competition sharpens since weaker concepts are dropped and only the strongest are carried forward for further development, again with the highest-scoring design from each round becoming the datum for the next round. Phase II is continued through several iterations and incubation periods over the span of weeks or months, depending on the complexity of the product that is being developed. (Introductory student projects rarely have time to go beyond one Phase II round at most.) The list of criteria undergoes continuous refinement. After each evaluation session, the weakest designs are eliminated; the remaining designs are improved through further creative thinking and engineered in more detail. It is through this process of discussion, review, and evaluation for the purpose of improvement that the team members grow to understand why the solution that finally emerges is best: all good points of the design have been defended and all the negative points eliminated. Only at the very end may a weighting system be used on the criteria for confirmation. With the absence of negative points in the top designs, weighting factors will usually not be able to add additional insight to the evaluation, except perhaps in the case when two very different designs are emerging that appear to be almost equally strong.

6. Convergence to a superior concept

Phase II ends when it is no longer possible to improve the best concept. This means the best solution has been found! The team has a strong commitment to this design and is confident that it will succeed. The team knows why this is the best solution—it will have no negatives that may have been overlooked during a traditional, more cursory review.

Before the first prototype can be constructed, the superior concept is thoroughly analyzed with one or more of the following tools: FMEA, FTA, engineering analysis, and cost analysis. The drawings are now in the piece-part design phase, where suppliers are consulted. And finally, a go/no go design review is conducted to answer three vital questions: Is the design inherently superior to competitive benchmarks? Does the design meet consumer requirements? Will the new product be timely, or is it already outdated? Only if all three questions are answered in the affirmative are detail drawings and prototype construction authorized. These drawings are production-intent and precise since all foreseeable problems have been solved, thus eliminating the need for engineering changes after the start of production. The purpose of the prototype is to confirm the design, not to identify and correct problems.

Why should the Pugh method be used? If a company wants to produce a product that is best, it has to start by selecting the best design concept. This statement assumes that companies want to produce a "best" product. However, as pointed out by Jim Hibbits, former president of Monarch Analytical Laboratories in Toledo, Ohio, such an assumption may not be valid. From his experience, he has found that engineers generally are not searching for the "best"—which is relatively easy. They are encouraged by their management to find something that is better than the competition, that can be made cheaper, and that will yield a profit and result in greater market share. There is rarely a reward for manufacturing the "best." Yet it is our conviction that U.S. manufacturing can regain a leading position if we widely adopt the goal of meeting the customer's needs with "best" products, "best" service, and extra value. When successful Japanese products are critically examined, it can be seen that they are by no means the best possible designs—they are just better than what we are accustomed to making. In the service sector, many businesses could improve with staff providing "best" service. The Ford 2000 goal on which the entire global enterprise is being restructured simply says: "We want to be the best automaker in the world."

Synergy is a key ingredient in the creative mental process. By synergy, I mean the mental result of interaction between different specialized parts of the interconnected brain— The creative ideas that can result from the interaction between the differing modes of analysis and synthesis, between rational processing and intuitive processing, between facts and feelings, between linear processing modes and global thinking.

Ned Herrmann

What makes the Japanese difficult to surpass in excellence is that they have introduced the concept of quality as an expandable commodity, not a fixed standard. When we talk about a "best" product, design or service, this means only with presently available, cost-effective technology and methods. A technological breakthrough or paradigm shift can quickly increase customer expectations and expand the requirements that constitute acceptable quality. Getting the Pugh method accepted routinely into organizational procedures will not only involve training for the employees, it will require support and commitment from management and an attitude that keeps working for continuous improvement. See the related discussion on total quality management in Appendix F.

Synergy happens two additional ways in the Pugh evaluation: at the team level through the discussion process, and in the emergence of a superior design concept from many different stepping stone ideas.

Applications

To show how the Pugh method actually works, we will now present some simple examples. An advanced illustration is the classic "teaching" example used by Professor Pugh—the design of an automobile horn; it can be found in his book (Ref. 11.3). But to really experience the process and experience the power of this technique, you must make up a team and go through several rounds in your own evaluation exercise. Such an exercise may take half a day or more, depending on how much time is spent in thinking up and sketching improved concepts, as well as on the total number of concepts that are being evaluated and the number of criteria used. A good starting point is having four different concepts with ten important criteria. Senior design project teams will have more time to do the Pugh method evaluation than a one-term freshman class doing a simple conceptual design.

Example 1: Design of a better school locker

We did the exercise summarized in Table 11.4 in 1991 with secondary school students in a Math/Science Saturday Academy program which focused on creative problem solving. The parents who sat in the back of the room as observers were at first rather skeptical when the students selected the locker topic for the exercise—they thought it would be too difficult since these students did not know anything about design. The amazing outcome proved that creative thinking and teamwork are more important for developing conceptual designs than technical drawing skills.

This class had 30 students, and it was important for the development of their communication skills and self-confidence that we allowed them enough time to explain their design in each round, to critique each other's ideas, and to defend the different concepts. Unfortunately, our schedule did not allow us to continue the project beyond the second round. Each team made additional improvements to their concept for their final group presentation that involved "selling" their locker idea to school administrators or taxpayers. The Pugh method helped them identify potential problems with implementation and acceptance; it also gave them a clear understanding of the benefits. Both the students and the parents enjoyed this conceptual design activity which was also a good teambuilding and networking tool uniting participants from very diverse backgrounds.

Table 11.3 Designing a Better School Locker

Problem briefing: The most important problems with lockers were determined through a survey and Pareto analysis. The students who came from city, suburban, and rural schools interviewed their classmates with a questionnaire to find the real problems with lockers. The problems were ranked as follows: 1. Open door blocks access to neighboring locker. 2. Not enough space. 3. Not enough shelves. 4. Noise. 5. Ugly looks. 6. Trouble with lock operation. 7. Damage to books, clothes, or students' skin. The students made up teams of four or five, brainstormed a list of design criteria, developed their design concepts, and then presented these designs to the entire class for evaluation.

Design criteria: 1. Efficient arrangement of groups of lockers. 2. Shape and size of individual locker for retrofit if possible. 3. Interior space divisions. 4. Door redesign. 5. Improved lock. 6. Recycled or recyclable materials; easy to manufacture. 7. Acceptable because of good looks, low noise, and easy use.

Advantages and features of Round 1 locker design concepts:

1a Shelves and drawers; foam edge; I.D. card or fingerprint lock.
1b Desks with individual lockers; separate small locker for coats; card lock.
2 Double width; many drawers + shelves, roll top door; button combination lock.
3 Door opens 90°, then slides straight back; extra shelves; ABC lock.
4 Retrofitable to present lockers; stopper at 90° for door; extra shelves.
5 Foam rubber door gasket; floor drain; "laser beam" door.

Results from the first round of evaluation (Phase I):

Evaluation Criteria	Design:	Old	1a	1b	2	3	4	5
1. Easy to use (door)			S	S	S	+	S	+
2. Nice looking			+	+	+	S	S	S
3. Reasonable cost		D	–	–	–	S	+	S
4. Acceptable to administrators		A	+	–	–	S	+	–
5. Better storage (students)		T	+	+	+	+	S	S
6. Improved materials		U	S	S	+	+	S	S
7. Locker arrangement in hall		M	S	+	–	S	S	S
8. Lock design			+	+	+	+	S	+
9. Convenience			S	–	+	S	S	+
Total	+		4	4	5	4	2	3
	–		1	3	3	0	0	1

This was a nice selection of different concepts, from retrofit improvement of the present "old" locker to futuristic innovation. At this point, no single concept was "the winner"—the design teams had to return to the drawing board to see how they could improve their designs by borrowing ideas from each other and adding new ideas. Concept 3 was designated as the datum for the second round, since it had the most improvements. Concept 1b was set aside since it involved redesign of desks (a different problem).

Round 1 brought out some changes for the criteria as well. Lock design was now eliminated as a criterion since it was taken for granted that all concepts would incorporate an advanced lock. Designs for Round 2 were engineered to more detail (for example, rough dimensions were required.)

Results from the second round of evaluation:

Evaluation Criteria	Design: Datum	1	2	3	4	5	6
1. Easy to use (door)		S	S	S	S	S/–	–
2. Nice looking		S	+	S	+	S	S
3. Reasonable cost		S	–	S	S	+	–
4. Acceptable to administrators		S	–	S	S	+	–
5. Better storage (students)		S	S	S	S	S	S
6. Improved materials		S	S	S	S	S/–	S
7. Locker arrangement in hall		S	S	S	–	–/S	S
8. Cleaning		+	S	S	+	+	–
9. Convenience		S	S	S	+	S	–
10. Safety		S	–	S	+	S	–
11. Long-term development/innovation		S	S	S	+	–	S
12. Short-term improvement/retrofit		–	–	–	–	+	–
13. Space required (floor area)		S	–	–	–	S	S
Total	+	1	1	0	5	4	0
	–	1	5	2	3	4	7

Advantages and features of Round 2 improved locker designs:

1. "The Slider": Door opens 90°, slides back; lunch compartment; bottom grid shelf for boots; umbrella hook; 2 top shelves; pencil/pen holder; recycled plastic in various colors; mirror-reverse units of two; standard size, floor mat for muddy boots.

2. "Roll-Away Locker:" Roll top door; recycled plastic; lunch shelf at top, book drawer with lower front edge (for visibility) next, pencil drawer in middle, vents at bottom section; each compartment pulls out separately with release button to give access to space in rear (for coat and umbrella), thus locker is deeper than standard size.

3. "The LETLOCK": Design like datum, except available in many color combinations; door opens 90°, then slides back; shelf and drawer at top, shelf and drawer at bottom; vent at top; standard-size interior (door "pocket" requires 1" extra for storage), thus locker requires more space for installation; with "alphabet" lock.

4. "The Wider Locker": Standard width extended by 8"; increased coat hook size; foam door gasket, extra shelf and drawer at bottom; floor mat for muddy boots; door hinges let door open to only 90°; assorted colors in recycled material; integral lock sunk into door—no protruding parts; door will spring open when lock is released.

5. "The Stopper Locker": Foam gasket around door; top compartments for lunch and miscellaneous, followed by book shelf, 3 hooks, pocket for pens and pencils, bottom shelf above removable perforated boot shelf; recycled plastic; all doors open to right to 90° only; width 20" (or 12" for retrofit); recycled plastic replacement door available.

6. "The Convenience Master": Standard size for retrofit; top shelf adjustable 6" up or down, with hook on shelf; 8" shelf and drawer below, with 12" book drawer, and drain at the bottom; recycled plastic in standard blue or choice of colors; "garage-type" push-up door with vents. *This new concept— received favorably by the audience—would need more work to overcome its low evaluation.*

Example 2: Design of an improved lamp

This team of high school honor students in an engineering summer program selected the problem of inconvenient lamp switches as their design project. During problem definition, they realized that the problem had to be expanded to encompass the design of the whole lamp. The students conducted a customer survey, did a Pareto analysis, brainstormed ideas for solving the identified problems, and thus developed a list of design criteria. The students then formed small teams to come up with different design concepts for the Pugh method exercise. Basically, three categories of designs appeared: (1) improvements over existing table lamps; (2) innovative—completely new—concepts; and (3) novelty lamps. The novelty lamps did not score well in the first-round evaluation since they cost more than most people would be willing to pay; also, they did not address some other important identified customer needs.

For the second round, a traditional lamp with many improvements (built-in timer switch in the base, flexible shade, retractable cord, and fluorescent bulb) was the datum. An innovative design scored very high: it could be used as a table lamp or a pole lamp, and it had an interesting shade that could be fixed for up-down indirect or task lighting or expanded to expose a lighted column for room lighting. The novelty lamp in the shape of a tree with jade leaves and little lighted apples received almost all negative marks, yet the team remained steadfast in not wanting to change its concept: it had "fallen in love" with its design. This is one of the fatal dangers in product design and thus demonstrated the utility of the Pugh method for identifying a flawed design. Lack of time and resources prevented the high-scoring team from pursuing its innovative design. Several people who saw the team's presentation on this design felt that it had potential to be patented—it was a real invention.

Invention, its development, testing, and implementation, always involved adventures. To have a victory over a technical problem takes flexibility of the brain and bravery. If you are looking today for adventures that are useful for the human race, invent! For technical creativity, you have to start to prepare yourself from an early age.

Henry Altov (Altshuller), president of the Inventor's Association of the former Soviet Union

Example 3: Application to course syllabus design

With teamwork skills being demanded by industry and with team projects taking up extra class time, how can instructors make room in an already crowded class syllabus for new activities? It is clearly impossible to cover everything—thus how do you determine what to cover and what to leave out? Students must still be taught the fundamentals, but they do not thrive when we "load dump." We must incorporate time to reflect and make connections to the content of other courses. Also, which topics should be covered in class, and which topics can be assigned for self-study or computer-based instruction? The Pugh method is an excellent tool for identifying topics that can be dropped and for ranking those that need to be emphasized. An example is given in Table 11.4 (first published by Edward Lumsdaine and Jennifer Voitle in a paper entitled "Contextual problem solving in heat transfer and fluid mechanics," *AIChE Symposium Series,* Vol. 89, No. 295, 1993, pp. 540-548).

Table 11.4 Pugh Method Evaluation for Heat Transfer Course Syllabus (Excerpt)

Criteria	Topic:	a	b	c	d	e	f	g	h
Relevance to subject		+	S	+	+	+	+	+	+
Usefulness		+	–	+	+	+	S	+	+
Teachability		+	–	+	+	+	S	+	+
Duplication		S	S	S	S	S	S	+	+
Fit with context		S	S	S	S	+	S	+	+
Need in subsequent courses		+	–	+	+	+	–	S	S
Need in industry		+	–	+	+	+	–	+	+
Need in design		+	–	+	S	+	S	+	+
Integration with other courses		+	–	S	+	S	S	S	S
Integration with computers		S	–	–	S	+	–	+	S
TOTAL	+	7+	0+	5+	6+	8+	1+	8+	7+
	–	0–	7–	1–	0–	0–	3–	0–	0–

Topics in course syllabus:

a = energy balance
b = product solution, T(x,y) = X”Y”
c = use of temperature charts
d = dimensionless parameters
e = finite difference methods
f = boundary layer theory
g = heat exchangers
h = convection correlations

Evaluation key: **S = satisfactory** **+ = advantage** **– = disadvantage**

The Pugh evaluation matrix of Table 11.5 is an excerpt from a much more extensive list since a valid course evaluation must include all the course topics. Instructors can brainstorm their own list of criteria and priorities. The two heat transfer instructors (who team-taught the course) were surprised at some of the results of their analysis. For example, they had always included the product solution method; yet undergraduate engineering students have little opportunity to apply the method and gain proficiency. The method solves a very limited class of problems encountered only in textbooks, not out in the field. Being able to use the method does not increase understanding of heat transfer principles or phenomena. But because of its mathematical elegance (and because the instructors understood the method so well), “it was one of our favorite topics to teach, even though it left many students frustrated and confused.” A survey of the entire curriculum showed that only two elective undergraduate courses taught the method, and previous knowledge was not needed. Cutting this topic from the syllabus freed up two weeks.

The Pugh method can be used to evaluate an entire curriculum. Each topic in a course can be judged against a list of criteria based on requirements from different customer sectors—students, alumni, professors teaching follow-on or prerequisite courses, accrediting agencies, and employers in industry. This process is changing the focus of these courses since it critically examines what we teach. The Herrmann brain dominance model can then be used on each topic to assure that different thinking and learning modes are addressed. Student learning improves considerably as the courses are streamlined, fundamentals are emphasized, different hands-on activities are added, and realistic team projects are required for applied creative problem solving and computer use.

The question we need to ask is if we, as a nation, can afford to keep the teaching of creativity a seldomly used option in our design courses.

Edward Lumsdaine, 1991

The Pugh method exercise

Hints for leaders

Doing a Pugh method exercise with a group of students will generate a lot of discussion. It helps to divide the class into teams (with one design per team) to limit the number of design concepts that will be submitted. Even then, some teams come up with multiple ideas which they want to have evaluated. Students get upset if the evaluation process is hurried: the facilitator must remain strictly neutral and not make unilateral decisions about criteria and the evaluation marks. Ask for a vote on a particular evaluation when there is no consensus. We use large sheets of paper to keep a record of the evaluation. It is also a good idea to ask each team to submit a one-paragraph description of their design for each round (as this will make it easier to write up a project summary later). Refer to Figure 17.11 for an example.

Most of our student teams have come up with very interesting concepts to an assigned or self-selected problem. They are disappointed that our introductory 30-hour course does not provide for building a prototype or for implementing the superior solution. With the new ABET emphasis, many engineering students can now look forward to a complete project in their senior capstone course. Since students of all ages want to keep their conceptual sketches, it is a good idea to ask them to use CAD for the drawings if possible so they can print out a copy for the instructors as well as having a record for their project reports (if required). We have found in our classes and workshops that teams will typically keep working with their designs to have a strong product to "sell" in their team presentation as part of their final exam—they gain experience with an attitude of continuous improvement.

Homework assignment

1. Select a topic from the numbered assignments (pages 282-283).
2. In teams of three to five members, define the problem, collect data, prepare a briefing, and brainstorm design criteria.
3. Brainstorm conceptual ideas based on the criteria.
4. Evaluate these ideas using a creative idea evaluation technique; then develop the most meritorious ideas into a conceptual design.
5. Brainstorm a list of evaluation criteria.

If the class has less than four teams, each team should generate two very different concepts. Since the drawing must be recognizable from a good distance, it need to be at least the size of half a flip-chart page. Specify only nontoxic product materials; give rough dimensions.

Preparation for conducting the evaluation

Select a large room with a long blackboard or blank wall. Have flip charts, masking tape, and water-based markers available. Write the initial list of evaluation criteria on a separate sheet and post as the left edge of the matrix. Then post the conceptual designs (including the datum) in a row above the evaluation matrix. Invite non-engineers to participate as evaluators. Have an assistant to help keep track of the results if possible. Have a table with extra sheets of paper and markers ready so new design ideas can be sketched and added to the matrix immediately.

Evaluation activity

1. Briefly review the features and problems of the datum.
2. A representative of the team explains the advantages and features of the first new design concept.
3. Then the concept is evaluated through group consensus against the datum. Through a process of lively discussion, the criteria become better defined. Team members must not become defensive of their concepts but look at the negative marks as opportunities for improvement. They can examine the other concepts for ideas to borrow.
4. Repeat the explanation and evaluation for all the concepts posted. As the concepts are discussed, the team members will think of new ideas. These should be sketched, added to the matrix, and evaluated.
5. Encourage positive thinking—put-downs and sarcasm are out. Look at all ideas as stepping stones for further creative thinking!

When teams comprise people with various intellectual foundations and approaches to work— that is, different expertise and creative thinking styles— ideas often combine and combust in exciting and useful ways.

Teresa M. Amabile, Harvard professor of business administration

Results and further activities

1. Idea "engineers" must not feel proprietary about their designs; the objective is to find the best solution for the entire team or company.
2. Add up the totals for the positives and negatives (or inverted deltas).
3. The concept with the highest number of positives becomes the datum for the next round.
4. The teams now have a new assignment: to improve the designs based on the evaluation results and using additional creative thinking.
5. Repeat the process for one or two more rounds. Then have the teams prepare a final evaluation with the superior concepts.

Comments and hints

1. Leave plenty of time for the teams to thoroughly discuss the concepts and criteria. This activity cannot be rushed.
2. Remember that this exercise teaches the process—the end product and its quality are incidental the first time around. But in an invention project or upper-division design course where the focus is on

producing a truly superior solution, the result becomes important.

3. In subsequent rounds, add more detail to the drawings. Also, obtain cost estimates; do a risk assessment; do other design analyses as required to evaluate and improve the design.
4. Finally, write a brief summary about your experience with the Pugh method—what have you learned?

To our amazement, years after having gone through a Pugh evaluation exercise, students still remember every detail.

They remember their team members fondly, the project topic, how much fun they had, and most importantly, what they learned.

Design topics for the Pugh method assignment

11.1 Creative Problem Solving Project

Use the "better ideas" from the "engineer's" mindset in your team exercise project to conduct the Pugh method evaluation and optimize a "best" solution. The topic need not be a conceptual design.

11.2 Traveler's Spill-Proof Cup

Examine a plastic traveler's coffee cup. Fill it with warm water and experiment with the cup. Can you identify some problems? Does it spill easily? Does it splatter when being filled? Is it comfortable to hold and to drink from? Does it keep its contents warm? Is it spill-proof? Can you improve on the design using the list of criteria?

Criteria: 1. Easy to manufacture. 2. Recyclable material. 3. Easy to open and close (if applicable). 4. Will keep contents warm or cold. 5. Equally suitable for left- and right-handers. 6. Stable in moving vehicle. 7. Won't burn fingers. 8. Easy to clean. 9. Comfortable. 10. Nice appearance. 11. Spill-proof during fill. 12. Spill-proof during use. 13. Competitive cost. 14. Special selling points (such as space for advertisement or logo).

11.3 Toy for a Small Child

Make a conceptual design for a child age 3-5, following the list of criteria. The conceptual design does not have to be a completely novel idea; it may be an improvement of an existing product. The datum is a nice, cuddly teddy bear.

Criteria: 1. The toy encourages creative play in teams. 2. The toy encourages each child's imagination. 3. Selling cost is less than $20. 4. The toy must be robust. 5. The toy must be cleanable. 6. The toy must be compact for storage. 7. The toy must be easy to manufacture.

11.4 Educational Toy for Grade-School Kids

Brainstorm a list of criteria and develop conceptual designs for an educational toy suitable for children from kindergarten through grade six. Important criteria are that the toy encourage imaginative play and cooperation among students.

11.5 Mosquito Trap

Design a quiet suction device that will trap and destroy mosquitoes and similar flying insects.

11.6 Toboggans and Sleds with Brakes

Each winter season, people are killed in toboggan or sledding accidents. Design a braking system for sleds and toboggans. Also incorporate a steering mechanism if possible.

11.7 Visual Aid for an Engineering or Science Concept

A science principle can be selected (by vote). Develop a list of criteria, then discuss and modify the list before doing conceptual designs for a visual aid or hand-on apparatus to help teach the principle.

11.8 Redesign of Hour Glass

Design a simple hour glass that makes a sound when time is up. Or, design a simple hour glass (to be used when playing Pictionary®, for example) that can be "dumped" instantly, so people do not have to wait for the glass to empty itself before starting a new play.

11.9 Improve a Business Procedure

Think of a business procedure, paperwork, or way of doing something that could be simplified or made more efficient through creative thinking. Brainstorm a list of criteria and develop a best concept.

11.10 The Problem with Crutches

People on crutches cannot carry a beverage or a plate with food. That's just one of the problems. Do a customer survey; then come up with ideas and design concepts that might solve the problem.

11.11 Reusable Packaging

The goal is to reduce trash. Design a breakfast cereal box that has other uses so that the packaging will not be thrown away. As a datum, take a box of Corn Flakes. The assignment is wide open and will allow you to think of some very innovative approaches to the problem.

With products changing more often than ever before, time-to-market pressures have become as profound as material costs, market share, and overhead. The whole process—design, analysis, prototyping, testing, and manufacturing—has got to go much faster.

Mark Clarkson

Resources for further learning

11.1 Michael J. French, *Conceptual Design for Engineers,* Springer-Verlag, New York, 1985. This book, by a British engineering educator, links the creative design function with analytical engineering by emphasizing synthesis. It is written for engineering sophomores; the vocabulary is interesting but does differ from American usage.

11.2 Sidney F. Love, *Planning and Creating Successful Engineered Designs: Managing the Design Process,* revised edition, Advanced Professional Development, Los Angeles, 1986. This easy-to-read book focuses on the iterative principle of design. It makes both the systematic procedure and the creativity of the engineering design process understandable to managers and engineers as well as the management of engineering design. It includes a chapter on computer-aided design.

11.3 Stuart Pugh, *Total Design: Integrated Methods for Successful Product Engineering,* Addison-Wesley, New York, 1991. This book provides the framework of a disciplined design and evaluation method for creating products that satisfy the needs of the customer. It includes examples from many fields and a wide selection of design exercises.

11.4 David G. Ullman, *The Mechanical Design Process,* McGraw-Hill, New York, 1992. This book discusses the human interface with mechanical products; it includes the application of quality function deployment, concurrent design, robust design (Taguchi methods), function mapping, and the Pugh method.

Chapter 11 — review of key concepts

Purpose: The Pugh method of creative design concept evalution was originally developed to design better, more competitive products that truly met customer needs. Teams make a conscious effort to overcome all weak points in a design. The advantage/disadvantage type matrix encourages force-fitting and synthesis of design concepts. The discussion of the design concepts and criteria helps the team understand why the superior design that emerges is best. The Pugh method helps prevent the manufacture and marketing of flawed concepts and products.

When analysis becomes as natural a part of the design process as breathing, the synergy between design and analysis helps create a greater understanding of how the world behaves— from the view right at your desktop.

Rich Gallagher, former head of interactive software development for ANSYS, Inc.

Phase I steps:
1. The list of evaluation criteria is developed and the datum (benchmark) is selected, usually the best existing product.
2. Original concepts are brainstormed and conceptual sketches made.
3. Each concept is discussed and evaluated against the datum. New concepts that emerge are added to the matrix.
4. The first-round results are evaluated; the top-ranked concept becomes the new datum, and teams improve their designs for additional rounds.

Phase II steps:
5. Weaker designs are dropped. The evaluation is continued with incresingly stronger concepts (which are engineered to more detail). Criteria are expanded and refined. The teams gain valuable insight.
6. The process converges to a strong consensus that cannot be improved further. The team is committed to this superior concept.

Application: The Pugh method can be used to evaluate and optimize ideas, concepts, solutions in many areas; they need not be design projects. Anyone can learn the process, let's say from sixth grade on up. It is a useful tool for evaluating and redesigning course syllabi and curricula.

The best way to learn the principles of the Pugh method is to see it in action. Conduct a Pugh evaluation exercise!

12 Solution Implementation

What you can learn from this chapter:

- Putting your idea into action—the role of the "producer."
- Step 1—selling your idea: strategic planning and working for acceptance.
- Step 2—the tactical work plan: who does what, when, and why; schedules and budgets; risk assessment.
- Step 3—implementation monitoring and follow-up; final evaluation of lessons learned.
- Student example from the workplace.
- Time management—a cure for procrastination and other time wasters.
- Further learning: references, exercises; review, and action checklist. Review of Part 2: critical judgment and personal application.

There once was a farm family. In early spring, its members toiled hard to prepare the ground. Then they planted the seeds. All through the summer they hoed, they watered, they pruned, they staked, they weeded, they fertilized—they did everything to ensure getting a good crop. Came fall and harvest time, the family sat on the porch in rocking chairs. The neighbors were astonished. "Why aren't you out in the field harvesting your bountiful crop?" The family members smiled and said: "It is reward enough for us to just look at all this beautiful produce."

The most difficult thing in the world is to put your ideas into action.

Johann Wolfgang von Goethe, German philosopher and poet

Let's compare the story to the creative problem solving process. As "explorers" and "detectives" we defined a problem. As "artist" we generated interesting ideas on how to solve the problem; as "engineers" we then worked to make these ideas more practical. As "judges" we examined and ranked ideas and then made decisions on which solutions would solve the original problem best. We revisited the "engineer" and "judge" as we synthesized an optimum solution using the Pugh method. If we stopped now, we would be like the farm family sitting on the porch. We need one more step, so we can reap the payoff for all our work. This last step is usually the hardest—this is where you become the "producer"; this is where you need your energy, your persistence, your careful planning and self-discipline as well as your interpersonal skills.

The role of the "producer"

In the "producer's" mindset (Figure 12.1) we take action—well-planned action. What are some characteristics of "producers"? They are managers; they have something to fight for. They are good communicators and follow a strategy. As paradigm pioneers, they have much courage. This word is related to the word heart. "Producers" take heart, are optimistic, do not give up. We will see how this attitude will help us implement our ideas. Implementation is time-consuming and involves several key steps. Although quadrant B and C thinking are emphasized, "producers" use the entire creative problem-solving process—the whole brain—since implementation is a new, unstructured problem. Why could red be an appropriate color for the "producer's" mindset?

Figure 12.1 The "producer's" mindset.

Dr. Robert Warner, Dean of the School of Information and Library Studies at the University of Michigan and formerly in charge of the National Archives in Washington, D.C., said this in a speech to honor students on the Dearborn campus:

> Dream the impossible dream; if something is important enough to you, take risks and get involved, even if the odds are against you. Pick a good cause, work hard with the right people, do the things that will get you growing support, and with a bit of luck, your impossible dream will become reality.

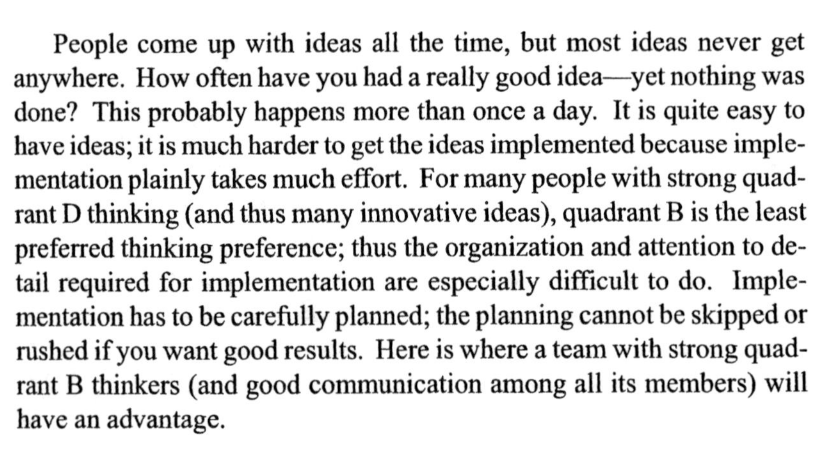

People come up with ideas all the time, but most ideas never get anywhere. How often have you had a really good idea—yet nothing was done? This probably happens more than once a day. It is quite easy to have ideas; it is much harder to get the ideas implemented because implementation plainly takes much effort. For many people with strong quadrant D thinking (and thus many innovative ideas), quadrant B is the least preferred thinking preference; thus the organization and attention to detail required for implementation are especially difficult to do. Implementation has to be carefully planned; the planning cannot be skipped or rushed if you want good results. Here is where a team with strong quadrant B thinkers (and good communication among all its members) will have an advantage.

Dealing with risk

People seldom hit what they do not aim at.

Henry David Thoreau, writer, engineer, naturalist

Even with the most careful planning, do not expect everything to go perfectly and smoothly. You will be dealing with people and change, risk and uncertainty, and thus you can expect opposition. We will look at some strategies we can use to overcome opposition and work for acceptance. Opposition is natural, because when you implement an idea—especially if it is an innovative idea—you are asking people to change.

People often see change as threatening. Idea implementation is difficult also because by this time, you (and the other people closely involved in the process) may be sick and tired of the whole problem; you are exhausted, you have run out of physical and mental energy. Conversely, you may be so excited and eager about implementing your solution that you may skip the required care in tactical planning and support building.

Although "producers" work to minimize risk, they are committed to seeing their ideas implemented; they make great personal sacrifices and shoulder the risks involved when doing something new. In Chapter 6, we met Art Fry and his invention of the Post-it™ notes. Here is the story of how he got 3M to adopt and implement his idea:

> Post-it notes were by no means an instant success. First, Art Fry had to sell the idea to his boss. Geoffrey Nicholson initially was rather skeptical because he foresaw that these notes would be expensive. But he agreed that the idea was worth testing. The two of them distributed samples throughout the company. When secretaries began using the notes, their bosses wanted to get in on this product, too. Soon, the 3M people were sold on the idea. Also, they suggested various improvements to make the notes practical in many different ways.
>
> The next problem was—how to produce these pads, since 3M was geared to manufacture many different types of tapes and other products in rolls only. Here, the personal commitment really came in. Art Fry finally had to invent and construct an assembly machine in the basement of his home. When it was complete, one of his basement walls had to be torn down to get the machine out.
>
> In 1977, three years after Art Fry first thought of the idea, 3M was ready to test-market the product in four cities with eye-catching, fancy displays and large newspaper ads. The tests were an absolute failure—so much so that the company decided to kill the product.
>
> But Art Fry and his boss persuaded the commercial office supply people to try another approach—they needed to talk to the customers to find out what the problems were. They found out that people who used the product loved it and wanted more, but the others had no idea what the notes were. This was one product that had to be "experienced" before it would be bought. Thus for the next marketing test, thousands of samples were given away. The company figured that a 50-percent repurchase would indicate a wild success. However, to their astonishment, the test resulted in a 94-percent repurchase, and sales took off. In two years, distribution was nationwide and across Canada; the following year the product was marketed throughout Europe. Post-it notes became one of 3M's most successful office products.

When selling something, present both sides and act modestly and a bit doubtful.

Benjamin Franklin

The moral of the story is—don't give up too soon. Maintain your personal commitment and get the support of others. Many creative ideas can and do fail in the beginning. Just keep working to overcome objections and barriers with creative problem solving; persist, and you will succeed. Also note that Step 4 in the knowledge creation cycle was crucial to the "sale"—people needed to see for themselves how useful the notes were. Once they were "hooked" on their convenience, they were more than willing to buy them, even at the high price. It is interesting that 3M did not get a patent on the original concept of the notes.

The first task that you need to do to implement your solution is to analyze which portions of the process will be your responsibility and which parts will be done by other people. In other words, you have to plan your strategy. Will you have to convince others of the benefits of your ideas, so they will accept, fund, and implement your solution? Will you and your team be in charge of the entire implementation process? In this case, you will have to check that the solution is working as designed. Also evaluate the creative problem-solving process and the lessons learned. We will discuss each of the three responsibilities in turn.

There is only one way under high Heaven to get anybody to do anything. And that is by making the other person want to do it.

Dale Carnegie

Selling your idea

Gaining acceptance for your idea involves careful planning. Analyze your targeted audience; prepare a list of benefits; develop a strategy on how to make an effective sales presentation. Use the whole brain, but concentrate on quadrant C as you develop your selling strategy.

Analyze the context and plan your approach: Your plan of attack for implementing your solution will depend on your situation. Do you have an innovative idea that could help your organization solve a problem? In this case, your primary objective is to sell your idea to management. If your team worked on an assigned problem, some team members (to provide continuity and tacit knowledge) may become part of the implementation team, and the biggest problem may be to get people to accept the solution. Are you an inventor of a product who has to persuade someone to bankroll the new venture or help you get at patent? Perhaps you have to talk to a family member to get your idea implemented. In all these cases, your primary goal is to convince your audience of the benefits of your idea. Determine who your audience is and what these people may want from you. Use the tips in Chapter 5 to plan an effective approach.

Why do you need a selling plan? The selling plan will prepare you to make a sale. You can have the best idea in the world, but if it is poorly presented, the result could be "no sale." A good presentation often sells an idea that is only fair. We have all seen mediocre products that are well presented and are thus finding ready acceptance. Thus you need to know some selling strategies, and you need to employ techniques that make for an effective presentation.

The list of benefits: The central part of your selling presentation is the list of benefits. What benefits can be reaped by implementing your idea or solution? What unsatisfactory situation will be remedied? How will these advantages directly benefit your audience and motivate them to accept your plan? Brainstorm as many direct (and indirect) benefits as you can think of. Look back to the list of judgment criteria and the results of the final evaluation matrix (if you used one); these are good starting points for bringing out the strengths of your idea. If your team went through several rounds with the Pugh method, you will all know why your idea is the best possible solution. Cast these benefits into words that will be easily understood by your audience. Having a list of benefits can also serve for self-motivation, to encourage you not to give up in case implementation runs into snags. Try to imagine and address possible arguments that may be brought up by people opposing your idea. What are their perceptions about your idea—will you need to correct misunderstandings or can you build on existing expectations?

Principles of selling

Let's look at some general principles of selling. Keep in mind that selling is not a one-shot deal—consider it in the whole context of gaining acceptance and overcoming opposition. Your success will also depend on the kind of person you are—your character, your reputation, your integrity, your good name. Never jeopardize these to get a quick sale. Do not look at "making the sale" as winning a battle of wits; think of it in terms of building long-term relationships. Contextual thinking is important. Whether you are selling an idea or a product, the "buyers" are very much interested in the associated service that they expect you to provide. Table 12.1 summarizes some selling principles to keep in mind.

When the police department in Bowling Green used an undercover narcotics agent in the disguise of a Santa Claus to nab a drug dealer, this creative (and very successful) approach caused quite an uproar in this Ohio community. Training in creative thinking is needed by everyone, not only for problem solving but also for building understanding and broader support for innovative solutions.

Dealing with opposition and working for acceptance

As you develop your selling plan, be prepared to deal with opposition diplomatically. Don't be surprised if you run into opposition at some point in the implementation process—you can expect it, especially if you are working with a very creative idea! For an actual example, see the story in the sidebar. A person opposing your idea can have any number of reasons. Table 12.2 summarizes possible reasons for opposition and presents ideas on how to develop acceptance.

The following points also help build easier acceptance of new ideas. They relate directly to the knowledge-creating step of acquiring tacit knowledge:

- Ideas are "tryable" — The implementation and benefits can be demonstrated with a small pilot program or in a test market.
- The innovation is reversible — If the idea doesn't work, is it possible to go back to the way things were done before, without a lot of hassle?
- The idea is divisible — The idea can be implemented in easy steps or

Table 12.1 Principles of Idea Selling

1. **Don't oversell.** Be moderate. Don't be arrogant; don't exaggerate. Listen carefully to what your audience has to say. As soon as you have made the sale, stop. If you keep talking, the "buyers" may change their minds!
2. **Don't give up too soon.** Selling may take longer than expected—thus don't give up too soon. It took Christopher Columbus two years to persuade Ferdinand and Isabella to let him have his ships. It took Chester Carlson seven years to sell his idea of the Xerox process. If you have a worthwhile idea, persist! Try a different approach—it just may succeed.
3. **Watch your timing.** Don't be impatient; watch for the right moment. Don't try to sell an idea to people when they are tired, annoyed, or unwell. Don't approach the boss just after he or she had some bad financial reports (unless of course the idea happens to be the perfect solution to that problem—then that would indeed be auspicious timing). Also be sure *you* are feeling well. At work, midweek is usually better than Monday or Friday; mid-morning or mid afternoon is better than close to lunch time or just before it is time to go home.
4. **Be brief and to the point.** Chapter 5 gives an outline on how to prepare an effective 30-second message. People are busy—30 seconds may be all you have to capture their interest and sell your idea. Boring your audience may be fatal to your sale.
5. **Plan your presentation carefully—use visual aids.** This way, people will be able to visualize and remember your ideas better. Check that the visual aids are of good quality and easy to understand and that all your equipment is in working order (see Table 5.10 and page 146).
6. **Make your ideas easy to accept.** Emphasize the benefits that your idea and the accompanying changes will bring to the people in your audience. If appropriate, describe in some detail how the implementation could be done. Have someone who is supportive of your idea in the audience.
7. **Avoid confrontation and controversy.** Expect opposition. Don't argue. It is better to stop the discussion on good terms and come back again another day. Also consider whether you would be willing to compromise to get at least some part of the idea accepted.

stages that require only small changes each time.

- You can build on what is already there — The idea represents an improvement, not a radical change; it maximizes previous investment and relates to things people are already familiar with. It fits in with the culture and explicit long-term goals of the organization.
- The idea is concrete — The idea is something that is real, that can be visualized and felt. The idea will have tangible results, not vague benefits, and it will have strong publicity value and good potential for increased status for the people who will be involved.

If none of these favorable conditions and approaches apply, investigate whether you can do the project on your own. Do it on the side, unobtrusively. Don't seek the limelight until you have your project running well. Many successful small companies have taken this approach; they were able to develop a sizable market share before they were noticed by their large, powerful competitors.

Table 12.2 Reasons for Opposition and How to Gain Acceptance for Ideas

1. **Lack of understanding.** Criticism may merely be someone saying, "I don't understand what you are doing and why you are doing it." Make sure you explain your ideas in terms your audience is able to understand. Be empathetic.

2. **No direct benefits.** Your idea might not have much in it that directly benefits your opposition. Many people first ask, "What's in it for me?" Consider your audience's priorities.

3. **You are not following the rules.** People—especially quadrant B thinkers and organizations—like things to go on in their customary ways. Have an attitude of sympathetically agreeing with your audience and their legitimate right to voice their concerns. Listen carefully to their arguments; they may bring out ideas that you can combine with yours to make it "their" project. Present your implementation plan and budget in a structured format. Get others involved in the implementation; do not try to do everything yourself. Emphasize the mutual benefits. By all means avoid being patronizing. Positive thinking can be contagious!

4. **Prejudice and other loyalties.** People may not have anything against your ideas; they just don't care about you, perhaps because of subconscious prejudices. It's simply a fact of life, an extra barrier to overcome. If you are young, old, or otherwise an outsider, you can count on this type of opposition. Can you find common ground? Can you find champions in the "in" group, one at a time? Also, people may have other loyalties or commitments. You may have a better idea than your colleague; but if the boss is his grandmother, she may support his idea in preference to yours. People skills are important; sometimes it takes years to build a foundation of trust, as U.S. companies that want to do business in the Far East have discovered. Do not take the opposition personally; answer objectively ; do not talk back; remain calm and friendly. Above all, do not bully, counterattack, or ridicule the opposition. Good-natured humor (at your own expense) can defuse hostile moments.

5. **Implied criticism.** People may take your idea as criticism that their ways are no good. This is a very important stumbling block in an organization and requires sensitivity and diplomacy. An established policy that encourages an attitude of continuous improvement helps.

6. **Fear of change.** People may fear your idea will bring change and a loss of status, influence, or position. Do not underestimate the force of this factor. Try to anticipate these concerns in your planning and presentation. Acknowledge that these points are valid and deserve to be taken seriously. Concentrate on the positive aspects of your idea; emphasize how the idea will benefit the person as well as your company, even though there may be a short time span when things may be harder, until everyone has become accustomed to the improved conditions.

⌛ Team Activity 12-1: Overcoming Opposition

In a team of four, select one of the reasons for opposition listed in Table 12.2. Pick a good idea (preferably from an earlier brainstorming and creative idea evaluation exercise). Then prepare a small skit in which two team members are trying to sell the idea to the other two who adopt the selected opposing mindset. Then switch roles and repeat the exercise, with the same or a different reason for opposition. Finally, evaluate the results of the exercise. How easy was it to put yourself into an opposing point of view and come up with positive arguments and responses to sell the idea?

Self-motivation

Sometimes the most important person who has to accept change is you! Or you may be in a position where you are the lone champion of a novel idea. Use the list in Table 12.3 for encouragement and self-motivation. Then give it your best shot. Get on your feet—you are the "producer" who gets the action rolling! If you have a habit of procrastination, study the section on time management in this chapter. For encouragement, read about people who are champions. Look around and note what is good in your community. How did those things get established? Many times the answer will be "because someone had a good idea and sold the idea to others who had the means to get it implemented." You can find examples all around you. Here is an illustration from our personal experience.

Table 12.3 Checklist for Self-Motivation

___ I have a good plan of attack.
___ I can be proud of my past successes.
___ There is a big potential payoff.
___ I'm getting encouragement from one or more supporters.
___ I believe in myself.
___ I have faith in my idea—it's a great idea!
___ I have an alternate plan B to fall back on.
___ I have to succeed—there is simply no acceptable alternative.

Case Study: Exercise Trail

As a family with four young children we were vacationing in Switzerland and really enjoyed the exercise trails that were the rage over there thirty years ago. Each community had its own trail, and people were out at all hours of the day and night exercising. When we returned home to Brookings, South Dakota, one of our children asked: "What if our town had one of these trails?" Since these trails were usually laid out in a forested area, shade would be a problem: summers in South Dakota are much hotter than in Switzerland. Also, a major life insurance company was sponsoring the trails there by furnishing signs and equipment. We assembled the snapshots we had taken of our family exercising at various "exercise stations" and took them to the local parks and recreation department. To our considerable surprise, the idea was very well received. We then obtained the materials about the trails from the sponsoring insurance company in Switzerland. Because the instructions for each station's exercises were copyrighted (and in German), we translated the material and made new sketches to go with the instructions. A local artist improved these sketches, the 3M company donated a strip of land that included a wide belt of trees for shading, and the National Guard did the actual construction of the trail and exercise stations. Thus the community had its exercise trail—one of the first in the United States—at little cost.

There is a sequel to this story. A few months later, just two weeks after we moved to Tennessee, we read an article in the daily paper by a columnist reporting a conversation with a local resident about the Swiss exercise trails. We contacted the columnist and were put in touch with a delightful middle-aged Swiss couple. Since the men were rather busy, Frida and Monika decided to work together to try to sell the idea to the city of Knoxville. Well, there it got hung up in red tape, but people in a community nearby heard about it. So one evening, the two housewives (with shaking knees and pounding hearts) gave a presentation about these trails to the Oak Ridge City Council. Thus it came about that a year later that community had its own safe trail located in a lovely park next to the police department. And the two families became lifelong friends in the process.

If your problem-solving assignment was to develop a conceptual engineering design for a client, the only implementation activity your team may have to do is to prepare an effective final presentation. You may want to study the following section about implementation, so you can give your client some "value added" by including aspects in your design (and presentation) that will help in getting your project results accepted and implemented.

⌛ ✓ **Team Activity 12-2: Selling Plan**

For the final or "best" solution from the judgment phase, prepare a list of benefits and a selling strategy. Think about your targeted audience and the presentation format that will be required. What potential opposition will you have to address? What "value added" aspects will you be able to highlight?

The work plan and implementation

As we have seen in the preceding section, selling a creative idea or innovative solution or making a final project presentation may be the only task that you need to do as a "producer." In this case, the job of setting up a work plan and doing the actual implementation will be someone else's responsibility. But what if you are that person who is given the task of implementing the solution? How do you go about the task of preparing for an actual implementation? The predominant thinking preference here will be quadrant B, because the work plan maps out the exact steps needed for implementation—who does what, when, where, and why. As a "producer," you also address the prevention of possible failure; you prepare time schedules and cost budgets. Since implementation is an unstructured problem, you must be prepared to use the other five creative problem solving mindsets to consider alternative ideas, the context, and the people interface.

We will discuss a variety of work plan models in a moment, but first we want you to examine a simple example given in Table 12.4. Let's say you seriously want to change a bad habit. You will not accomplish your goal unless you adopt a "battle plan." This is an illustration that we have adapted from a plan to stop smoking recommended by the American Cancer Society. How does this plan address the problem of change and strengthen motivation?

The biggest room in your house is the room for improvement; the largest window is the window of opportunity.

Roger Milliken, CEO of a Fortune 500 company.

Table 12.4 Example of a Work Plan: "How to Stop Smoking"

1. Why do you want to quit smoking? List as many reasons as you can think of in support of a smoke-free "you." What are the obstacles—what are some reasons for smoking that you will need to overcome? To sell the plan to yourself, you need a list of persuasive benefits and an analysis of the opposition you will have to overcome.
2. Change to a low-tar, low-nicotine cigarette. Select a Quit Day two or three weeks in the future. Mark the day on your calendar—you are preparing your mind for action!
3. Chart your smoking habits for at least two weeks. Write down how many cigarettes you smoke a day and when you smoke them. Go over this list and rank which cigarette you think is the most important or desirable to you, such as the one with morning coffee; the next most important one; and so on, down to the least important. Habits are easier to change when you bring them from the subconscious mind to undergo critical judgment.
4. Eliminate one of the cigarettes you routinely smoke. It may be the most important one, or the one in the middle of your list, or the least important one. Secure a supply of "oral substitutes"—mints, gum, ginger root, dried fruits and nuts, raw vegetable sticks, popcorn, or even mouthwash—and use them instead of reaching for a cigarette. You are beginning to put your plan into action with an easy first step.
5. Repeat each night, at least ten times, one of your reasons for not smoking, from your "list of benefits." Visualize a healthier, more attractive "you." You are maintaining your motivation and programming your subconscious mind to support your efforts.
6. Quit on Quit Day. Try different substitutes as the urge to smoke recurs. Enlist a friend or family member in a series of busy events, such as going to the movies, playing tennis, or taking long walks. Keep reminding yourself of the shocking risks of cigarette smoking to yourself and to those you love. Strengthen your motivation by looking at the long-term benefits and consequences.
7. Set up a daily or weekly reward structure for success—treat yourself to something special that you can now afford to buy with the money you are saving by not smoking, or make a contribution to a worthy cause. If you give in to temptation, it is no reason to give up your plan. Analyze the cause of the failure; learn from it; eliminate the contributing factors; forgive yourself, and start again. Can you strengthen your support system? Can you recruit others (including medical help) to encourage you? Can you make this a true team effort with family members and friends? Believe in yourself—you have a good plan and you will succeed!

Implementation strategies

The purpose of a work plan is to make sure that the idea or solution will be put to work—that it will work right and be on time as well as within budget. Several different procedures are available for the work plan; the complexity of the problem will determine which approach should be taken. For example, a very complicated implementation involving many people and tasks should employ a PERT chart for planning and progress monitoring; moderately complex problems can be handled with a flow chart, and simple projects will need nothing more than a time/task analysis (Gantt chart) or answering the 5-W questions. When you are dealing with a complicated, risky, expensive implementation, you may want to include a risk analysis as part of your work plan. Consider the resources available to help make implementation successful. Following are brief descriptions of some of these strategies. The use of Microsoft Project planning templates for engineering design is described in Chapter 15.

Copycat: If your idea is similar to one that has been successfully implemented before, you will save much time and trouble if you can just copy the procedure, maybe with some minor adjustments.

The 5-W method: It asks the questions "who, what, where, when, and why." Begin by brainstorming and listing the required implementation tasks. Then ask a series of questions about each task: Who will do this task? What will they do? Where and when will they do it? Specifically answer each of the questions. Next, ask "why" for each of the questions. Why should these people do this particular task? By asking "why," you provide a reason or justification for each action to ensure that no major activity is overlooked. Finally, give or obtain the go-ahead for the implementation and set the plan in motion. The storyboard format can be employed very nicely with this method. Note that the "how" is not specified. If you feel uncomfortable about letting people make their own decisions on how they will do their assigned tasks, try to restrict your directions to broad outlines only. Most likely you will get much better performance and cooperation when you let people do their own thinking and decision making about their jobs—avoid "micromanaging" the project. However, when it is critical to the project that certain procedures be followed, be sure to specify these and explain why.

Flow charts: They visually present all the activities that must be performed sequentially during implementation. The basic elements of the flow chart are activities (designated by rectangles), decision points (indicated by diamonds), and arrows. The flow chart is then used as a guide for implementing the idea. Time estimates for each activity may be added to the flow chart. Flow charts are useful for showing simultaneous activities (through different branches) and prerequisites. Figure 14.1 is a simple flow chart of the engineering design process and documentation.

In a well-organized system, all of the components work together to support each other.

W. Edwards Deming, quality expert

PERT (program evaluation and review technique): This method is a planning tool as well as a progress monitoring tool for large, complicated engineering projects. It consists of events (start or completion of a task) connected by activities (actual performance of the task). Time estimates are given for each activity. The completed chart represents a network of interconnected activities and can identify bottlenecks and critical paths to the successful completion of the project. Complex networks are handled with computer programs that can track thousands of activities and events. The most difficult part of setting up a PERT network is identifying all the activities that precede the completion of specified events. Although this type of detailed planning takes a considerable effort, it leads to implementation that is quite routine.

Time/task analysis (Gantt chart): This is one of the simplest work plan formats and visually presents the time requirements of each implementation task. Typically, every task that must be completed to implement the idea is listed in the left-hand column of a lined chart, with the time scale across the top. Then the time required to complete each task is estimated as accurately as possible, as well as the target date for completion. These estimates must be realistic; it is a common mistake to underestimate the time required to complete each task. For each task, a time line is drawn from starting date to the projected completion date. The chart clearly shows simultaneous or overlapping activities. An example used in a research project proposed to the Department of Energy is shown in Table 12.5. Gantt charts for student design project plans are included in Figures 15.1 and 15.2.

Table 12.5 Time/Task Analysis for the Development of a Solar Concentrator

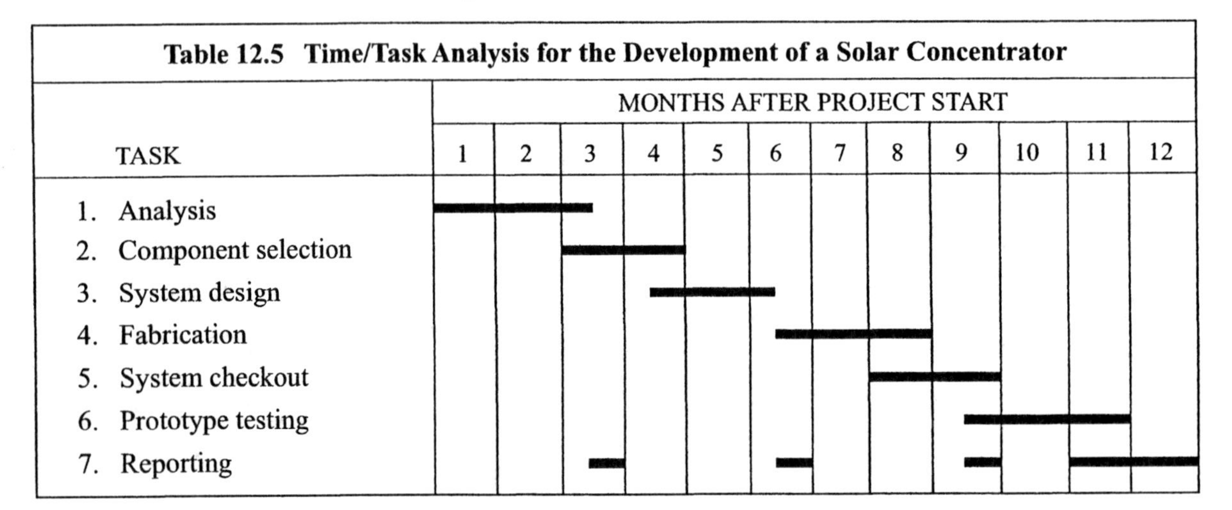

TASK	MONTHS AFTER PROJECT START											
	1	2	3	4	5	6	7	8	9	10	11	12
1. Analysis												
2. Component selection												
3. System design												
4. Fabrication												
5. System checkout												
6. Prototype testing												
7. Reporting												

Even if you are the only person scheduled to do the work, you still need to make a work plan or list of what needs to be done and when. Estimate how much time it will take and what support items you need to get each task done. Also find out where you need more information or training or where you could or should get other people involved. Do any of these people need special training? Be sure to include these activities in the time schedule for a successful implementation. If you are a procrastinator, you have to make up your mind to do each day what you have scheduled to do. Set a starting date and reasonable deadline—add intermediate dates as milestones for longer projects, so you can gauge if the implementation is on track.

Cost budget: Information from the work plan is useful for preparing the implementation cost budget. When you know who does what and for how long, you will have a good grasp of the labor costs. Usually, it is easiest if the costs for each task are estimated first for various cost categories such as salaries and wages (including fringe benefits), equipment, materials and supplies, travel, and communications (computers, telephone, etc.). Then the totals for each category can be calculated. Are the required funds available? For projects that are funded from an outside source, you may be able to include a certain percentage for overhead costs. If a ceiling for the total budget exists, various adjustments and trade-offs may need to be made between the different tasks and their estimated costs. If the budget cannot be met without seriously jeopardizing the implementation, fund raising may have to be added as one of the primary activities in the work plan. Fund raising is in itself a new problem that will require creative problem solving. The task of making up a budget is much easier if you can follow a specified format and examples of budgets for similar projects and if you can work with a person who has the required accounting information.

Risk analysis: A number of techniques are available for risk analysis. The Kepner-Tregoe problem-solving method uses a special method during the implementation process to anticipate possible difficulties and develop countermeasures. This technique, known as **potential problem analysis**, should be applied to very important projects where major obstacles to implementation are anticipated. This procedure requires a considerable investment in time and effort and thus would not be cost-effective on a routine basis. Alternatively, an FMEA (see Appendix D) can be conducted *on the implementation process* in order to identify areas that have high probability scores for possible failures. This will allow you to develop contingency plans for the most critical areas. Other risks may have been identified during reverse brainstorming and other judgment activities; appropriate measures to deal with these risks should be incorporated into the work plan.

⌛ **✓ Team Activity 12-3: Work Plan**

For the implementation problem in Activity 12-2, prepare a work plan. This may involve an actual implementation, or it may be a specific outline for preparing the final project presentation. Determine who will need to do what, when, why, and where. Prepare relevant charts, schedules, budgets, and visual aids.

Implementation monitoring and final project evaluation

Your idea has found acceptance; work plans have been prepared and approved; your project has been funded; and the work has been executed by those assigned to carry it out. Thus you now have only one remaining responsibility—to make sure that the solution actually works. In a small project, it may be possible to make arrangements to personally check up on the success of the implementation. Depending on the type of problem and organization, you may need to work out an implementation monitoring plan as part of the work plan. There is nothing complicated about a monitoring plan; you are probably using one for keeping track of your car's maintenance or for doing seasonal jobs around your home. A monitoring plan provides for periodic checks to make sure things keep on working right as part of preventive maintenance and follow-up. Finally, reflect on the results as well as on the entire creative problem-solving process as a learning experience.

Whether you think you can or can't, you're right.

Henry Ford

Implementation monitoring and follow-up

Occasionally, a creative problem-solving team has the responsibility for long-term monitoring of the implementation. Procedures able to confirm that the solution has been implemented correctly are needed. Has the problem been solved without causing other problems? Has any resistance to the change appeared, and has this been handled properly? Have modifications been made to the solution to make it "fit" better? A first review should be scheduled within two weeks after implementation has been completed, with a follow-up in six to twelve months.

The people involved in problem definition are probably best suited for monitoring the implementation of the solution and can be assigned the responsibility for follow-up. In a manufacturing plant, SPC methods (see Appendix C) can yield quantitative data to verify the improvements that have been achieved with the solution. These diagnostic tools should be employed on a specified schedule, not only to provide continuous monitoring over the long term but also to pinpoint new problems that may arise due to the implementation or to identify modifications made (purposely or inadvertently) during implementation. Specific testing may also be required to confirm the expected results.

The following tasks are usually part of the confirmation process:
- Make all required changes in the appropriate drawings.
- Make all required changes in the affected processes and procedures.
- Dispose of all "old" forms, drawings, equipment, and materials.
- Notify all individuals at all levels that are impacted by the change.
- Set up audit functions for easily-reversible situations.

The monitoring team may also have responsibility for these tasks:
- Conduct a midcourse review and make appropriate corrections if needed to stay on target.
- Obtain customer feedback on the improvement.
- Investigate wider applications of the successful solution or change.
- Obtain employee feedback (acceptance or resistance) to the change.
- Monitor if longer-term associated requirements have been instituted, such as training for employees, development of new instructions manuals, computer and software updates, and management procedures.
- Evaluate and improve the creative climate in the organization.

Do you have a favorable climate, a mechanism, and a team responsible for receiving and evaluating ideas for continuous improvement and innovation within the organization? See Chapter 18 for a discussion of the organizational climate needed for innovation.

Final evaluation of learning

It is very helpful if you keep a journal or take notes during the entire creative problem solving process. When the project has been completed, write a brief summary of the results and your achievements. Then sit back and review what the process has done for you. What have you learned? Did it help you grow? Can you use this idea somewhere else? Did you achieve all or some of the goals? How did the process help you communicate with people? Did the process open future opportunities for you? If the solution did not work out, you can still ask what you can learn from the experience. Sometimes we learn more from our so-called failures than from things that are too easy. As a "producer," reflect on where you could have avoided needless battles. If you were involved in the project as a supervisor, facilitator, or instructor, identify areas where you can give your people positive feedback. Continue to help them focus on accomplishments as well as learning; encourage them to be supportive of each other by recognizing the team's collective contributions.

Keep your summary and conclusions in a file. As you become more practiced in creative problem solving, this projects file will grow into a valuable data base. It will help you become more efficient, since you can reuse "ideas that worked" and avoid (or improve) ideas that did not work. As part of honing your "producer's" mindset, review the time management skills discussed in the next section.

⌛ ✓ **Team Activity 12-4: Presentation of Project Results**

To conclude your problem-solving project as well as to give you another opportunity to practice your skills, prepare a final team presentation. Use the thinking preferences of your team members to advantage; different people can take on different roles and leadership at various stages in the process; all must be involved and make a substantial contribution to the final presentation. Depending on the subject of your team project and the setting (formal class, workshop, self-study, assigned problem by a client), you can choose the appropriate format—oral presentation, skit, written report—possibly including visual aids and a "selling" video.

The presentation should include: 1. Summary of the problem and goals. 2. Description of the creative problem-solving process. 3. The optimized solution. 4. The list of benefits. 5. A work plan (if appropriate). 6. Risk prevention and how to motivate people to accept the solution and change. 7. Description of implementation results (if available from a pilot project). 8. Evaluation of what was learned (or how the success of the solution could be measured). In the presentation, be enthusiastic; be positive! Finally, evaluate the presentation—what ideas were especially effective and creative? Jot down ideas used by others that you may want to try yourself sometime.

Case Study: Creative Problem Solving Applied at Work

This simple case study was written by a student in an extension class taught at Delco-Chassis in Sandusky, Ohio, during the fall of 1991 as part of a pilot program in engineering in which nontraditional students in technical fields can earn a B.S. degree while continuing to work. Ellen chose this case study as the subject of her "thinking report" (which was a class requirement with the objective of making connections between classroom learning and outside application or context).

"The Room Temperature Compromise"
Ellen Stanton

Problem Background: With a goal of increasing the productivity of the workers, we are concerned with the office environment. My job title is designer, and I work with computer-aided design drafting software (CADD). Most of my time is spent at the CADD workstation or at my desk next to the workstation. My work area is located in a room 16 ft x 16 ft which contains two additional workstations. Their operators are the other members of our problem-solving team. Our problem is the air conditioner. It was installed in the ceiling to cool the space occupied by the computer terminals. The temperature is set by the thermostat inside the room, so that the room environment can be controlled separately from the larger office area outside.

Problem Definition: First, we surveyed the group members. The unit blows too much cold air for me. It's not cold enough for Mr. Polar Bear during the OFF cycle. It causes sinus aggravation for the third person. As "detectives" we attempted to describe the parameters of the problem. By inviting a cigar-smoking co-worker in, we observed that during the ON cycle the air blew out from a directional finned vent; the adjacent vent was an intake.

By experiment, using a thermometer and a stopwatch, we found that at a specified temperature setting there is a consistent ON and OFF cycle period of time. This varies day to day as the outer office temperature varies and mixes with our room temperature through the open door. On a specific day, readings were recorded for a thermostat setting of 74 degrees. Cooling started at 75 degrees and ran for 12 minutes; it stopped at 73 degrees and stayed off for 12 minutes. This cycle repeats predictably. For a setting of 76 degrees, cooling started at 77 degrees and ran for 10 minutes; it stopped at 75 degrees and stayed off for 28 minutes.

Idea Generation: After we gave up on changing the thermostat settings when no one was looking, our group began verbal brainstorming. One condition we accepted was that we would not replace the unit, but we would attempt to modify its effects. We came up with the following ideas:

- Remove the walls.
- Shut the system off.
- Redirect the flow with cardboard.
- Diffuse the flow with a box cover with holes in the box.
- Put pinwheels up, covering the vent area, to diffuse the flow.
- Turn the temperature up so the ON cycle is short and OFF cycle long.
- Change the direction of the vent grate fins to point in the opposite direction.

Creative Idea Evaluation: We began to "engineer" the ideas by grouping them:

Simple:
- Shut system off.
- Turn temperature up.
- Change direction of grate.

Meaty:
- Redirect flow with cardboard.
- Diffuse flow with cardboard box full of holes.
- Diffuse flow with pinwheels.

Difficult:
- Remove the walls.

We decided not to combine or modify the different ideas within the categories.

Idea Judgment: As "judges" we defined and selected our evaluation criteria to be: optimal people comfort, low cost, quick implementation, easy operation, simplicity of solution, and effectiveness of solution. We ranked the ideas as follows:

1. Change direction of grate (simple).
2. Redirect flow with cardboard (tricky to implement but reduces wind chill).
3. Turn temperature up (reduces cycle ON time; makes Mr . Polar Bear uncomfortable).
4. Cardboard box with holes (bulky and unaesthetic).
5. Pinwheels (aesthetic but distracting).
6. Remove walls (involves other workers' time and cost).
7. Turn system off (impractical because computer systems generate enough heat to make it uncomfortable for the workers in the room).

Solution Implementation: We selected and implemented items 1 through 3. We reviEwed the results at each step, before proceeding with the next one.

1. Change directional vent. ➛ Results are limited; air deflects off the opposite wall and results are unsatisfactory; the workers still have wind chill.
2. Redirect flow by taping place mat onto the grate. ➛ This makes the air flow toward the open door. The room conditions are improved, but now the worker outside the door is complaining of wind chill.
3. Turn the temperature up, leaving the place mat in position. ➛ This reduces the ON cycle time and wind chill, but Mr. Polar Bear is melting during the OFF cycle.

Final Evaluation: We had to go back and do some additional "engineering." By Edison's trial-and-error technique we arrived at an improved solution. Mr. Polar Bear introduced a personal-sized fan that keeps him cooler during the OFF cycle and allows a higher temperature setting to decrease the ON cycle time. This solution is not perfect, but it is a noteworthy exercise in compromise for everyone concerned. The solution was accepted since the three persons affected the most were part of the problem-solving team. The solution was "no cost" to the organization and thus did not require the approval of management.

Comment: This was a fine, practical application by a student who was learning the creative problem-solving process. The coworkers had no previous creative problem-solving instruction or experience. By spending additional time in the "engineer's" mindset, the team might have come up with a higher-quality solution earlier.

Time management

If you don't take time to plan, you are planning to waste time.

Poor time management is a serious hindrance to being a good "producer" and getting things accomplished. Many students (including those in honors classes) have told us that procrastination has a negative impact on their success in college and achieving goals.

Planning

Collecting data: Planning must begin with an analysis of your life goals and priorities. People usually take the time to do the things they really want to do. The question is—are the things that you are doing in tune with your priorities and goals? You have to take a careful look at your long-range goals, not just your short-term goals. Make sure that your short-term goals do not prevent you from reaching your long-term goals. What are your priorities? How are you actually spending your time now?

Assignment 12-5: Time Use Data

1. Make a list of your long-term goals (beyond one year to about ten years).
2. Make a list of your short-term goals (things you want to accomplish within the current year).
3. Make a list of priorities.
4. Complete a daily time log for two consecutive weekdays, in 15-minute increments.
5. Complete a weekly work schedule (or a class and study schedule if you are a student).

As you are preparing your weekly schedule, take your daily energy cycle into account. Do you do your best work when you get up at 5:00 a.m. or do you slowly come alive in mid-afternoon, reaching peak performance when most people have gone to bed? If you have a choice, schedule your most important work or classes during your peak times for better alertness and learning. Do your routine tasks in your "low" periods. For many people, one hour of mental activity during daylight hours is worth one-and-a-half hours of night-time learning and studying.

Spiritual, Finance, Career, Family, Social, Self, Health, and Leisure

Priorities: What are some of the things that are important in life? Starting from a young age, education takes a considerable time commitment. Later, the emphasis shifts to work and career, although learning activities of one sort or another should never be discontinued to keep our minds functioning well into old age. To have a balanced life, we must make time for all the aspects listed in the sidebar.

Your weekly schedule: Include all the routine tasks you normally do in a week: work, commuting, shopping, meetings, learning, social time, recreation and exercise, hobbies, meal preparation, household maintenance, bookkeeping, correspondence, eating, sleeping, health, personal care, etc. If you are a student, each college course on the average requires two hours of study time for each hour in class (or for each credit hour for lab courses). It is important that new college students set up this type of discipline at the very beginning of the first term. If you slack off for three weeks, you have set up a habit that will take an extra effort to correct. If you made top grades in high school without doing much studying, you may be in for a shock if you do not establish good study habits when starting college.

As a transition to the lifelong learning mode, class time is for learning "what" to learn and for evaluation of learning. The bulk of the learning occurs during individual or group study time.

Now look at your list of priorities. What other things besides working or learning are important aspects in your life right now? Do you have family responsibilities? Mark time for these on your schedule. What about your spiritual life? Mark time for your religious commitments. What about your health? Fill in time on your schedule for regular meals, sleep, and exercise—an aerobic workout at least a half hour three times a week as a minimum. What about social activities (including helping others) and leisure? Are you running out of uncommitted time slots on your schedule? Can your social activities be combined with other items? Could you exercise with family and friends? Could mealtimes be used for maintaining your social relationships? Can you and your "significant other" be together during religious and hobby activities? Some creative thinking is required here to come up with solutions that will help you get all your priorities into your weekly schedule.

What about finances? If you are working and studying, you must schedule both, with adequate commuting time. If you must work while in college, you will have to reduce your class load. Here is a formula that can give you a starting point:

Credit-hour load = 1/3 [48 – (number of hours of work per week)]

You can adjust your course load as you go along. If you are continuing your education while working, some other areas may have to be temporarily curtailed, such as recreation. Television watching is often an area where substantial cuts can be made. Are you merely working to finance a car? Calculate the financial gain possible by giving up the car and graduating a year earlier. On the other hand, are loans (instead of work) a good option when you consider the long-term burden of a heavy debt that may seriously limit future job and education choices?

Last, but not least, you must schedule time for yourself—for "personal care and maintenance." This includes personal grooming, household chores such as doing the laundry and keeping your abode in reasonable order, taking care of correspondence (including paying bills), as well as planning and evaluation. Also schedule one hour a week just for yourself—to daydream, think, reflect, give yourself a special treat, indulge in a hobby, or soak in the bathtub. To prevent running out of energy and affecting your creative thinking ability, leave yourself some breathing space between various activities while you experiment with different arrangements in scheduling your week. You will encounter

Murphy's law, "If anything can go wrong, it will." Life will be less stressful if you have some built-in flexibility. Try the "best" version of your schedule for a week. Basic flaws will quickly become apparent. Make changes as needed. You may also have to learn some new attitudes about keeping to a schedule if you are too rigid or too casual.

Progress toward your goals: Compare your list of long-term goals with the completed weekly schedule. Circle those items on the schedule that have anything at all to do with achieving your top three goals. If you have many circles—good for you. If you don't have any circles, you need to evaluate your goals and daily activities. If you are not incorporating actions into your daily schedule that move you toward your goals, you will not reach those goals; you will be under stress and dissatisfied. If you want to be an engineer but your class schedule is made up entirely of music and social studies, your activities will not move you toward your goal. Why did you pick these courses? Why do you want to be an engineer? Is this your personal goal, or is someone pressuring you? Perhaps you need to change your goal! If you want to be a writer but your present responsibilities do not leave you with a free minute during the week, you must restructure your life to make time for writing. Thinking, planning, and managing your time may not be exciting, but they are the way to fulfilling your ambitions. Do this type of goal evaluation at least once a year, perhaps at year's end or on your birthday.

My father can climb the highest mountain,
My father can swim the deepest ocean,
My father can fly the fastest plane,
My father can fight the toughest tiger,
But most of the time he just takes out the garbage.

Author unknown

Personality and modes of thinking

If you have trouble with time management, this may be part of your personality style and preferred modes of thinking. Table 12.6 summarizes time management problems and solutions. Note that it is possible to have traits of more than one style. The functions and interactions of the different "talents" can be enhanced with good time management. The goal is to find those time management strategies that work for you!

Daily priorities: separating the important from the urgent

As your last activity each evening (or as the first thing in the morning), get into a habit of spending a few minutes making a daily "list of things to do." This list starts with the information from your weekly schedule. Many things will be added to the list, such as appointments, things that did not get done the day before, or unexpected items. These items can be identified according to three different priority categories:

A = First priority ➛ things that absolutely have to be done.
B = Second priority ➛ things that should be done.
C = Third priority ➛ things that can be put off if necessary.

Table 12.6 Time Management Solutions for Different Personality Styles

Personality A	is a perfectionist who does everything right the first time; organizes everything logically; can concentrate well, and works unusual hours to get the job done.
Problems:	Not delegating work; underestimating the time needs for doing the tasks.
Solutions:	Set priorities and review often. Don't get lost in details. Learn to delegate tasks without being too critical. Be realistic in your time estimates.
Personality B	is steadfast, dependable, and punctual, yet can tolerate the lack of time sense in others.
Problems:	Getting upset with rescheduling; inflexible; not deviating from schedules, which can result in unfinished work.
Solutions:	Learn to be more flexible. Analyze your time use. Work unusual hours when needed. Adopt new ways of doing things; avoid ruts.
Personality C	is a "people person," a team worker aware of time problems and accepting suggestions.
Problems:	Letting others impose. Prone to procrastination; wasting time by being very talkative.
Solutions:	Learn to use calendars and schedules to plan and identify priorities. Make "to do" lists and obey them; follow written instructions, even if boring. Develop listening skills to focus attention. Use a time log to analyze where time is being wasted.
Personality D	is a go-getter, someone who sees the big picture, who works long hours, who has abundant energy, who barges right in when a job needs doing.
Problems:	Impatient with tactical planning and time use analysis. Underestimating time needs which leads to schedule conflicts and pressure.
Solutions:	Learn to plan carefully. Be aware of the serious consequences of avoiding quadrant B thinking. Establish a follow-up evaluation routine. Learn from the analysis; accept the advice of others.

How do you determine which things should be done and which can be neglected? It is frustrating when we have so many urgent things to do in a day that we never get around to doing the really important things. Here is a strategy that allows you to do more of the important things and less of the urgent—it can help you choose priorities for your daily list.

	URGENT	**Not urgent**
IMPORTANT	I	II
Not important	III	IV

On a piece of scratch paper, first list all the things you think you have to do that day. Then draw a large rectangle (as indicated above) on a blank piece of paper and divide the box into four quadrants. Label the axes and quadrants as indicated. Let's discuss each quadrant in turn, so you can decide into which category (or quadrant) a particular task or item on your scratch list should be placed.

Category I — important and urgent. These tasks must be done, the sooner the better. These will be your priority A activities. Mark them as such on your list! Once you have developed good time management habits, this category will not be a problem since crises and deadline-driven projects will be prevented with creative thinking in Category II.

A stitch in time saves nine.

Proverb

Category III — urgent but unimportant. These can be real time killers. If you spend time doing these, you will not accomplish anything of substance. Why? Because these truly are unimportant! People will try to make you believe that these are important, but you will have to use your own judgment and wisdom in making this determination. Don't let others rob you of time and put you under pressure. Watch advertisers. They are masters at this game, "Buy this article today or you'll lose a once-in-a-lifetime opportunity." Use critical thinking to determine if this is really true. What should you do with the items that fit into this category? Most can be tossed into the wastepaper basket or crossed off your agenda. If you are frequently interrupted by people and phone calls, devise a system that will limit these to less than 20 percent of your time.

An ounce of prevention is worth a pound of cure.

Proverb

Category IV — neither important nor urgent. Here we find trivia, busy work, and fun activities. If you have time for relaxation, turn to items in this category, but on most days, these can be safely neglected.

Category II — important but not yet urgent. Items here are crucial time management targets and include planning, prevention, preparation, building relationships, exploring opportunities, recreation, and large projects that cannot be done in one day (term papers, home remodeling, Christmas shopping, whatever). Items here get put off precisely because they are not urgent. If these items are postponed, they will grow to be urgent, and you will not be able to do them well or creatively. Take care of as many of the Category II things as possible in your daily activities under priority A and B. If you can't get them done during the week, the weekend might be less busy. Delegate—see if you can find someone else to do them. If you begin working on these activities early, in smaller "bites," you will be in control. Also, your subconscious mind can incubate ideas and thus help you with good solutions as you progress through the project. Many Category II items may be connected to your achieving your long-term goals; thus you must schedule them into your weekly schedule! Try to spend at least half your daily time in Category II.

Frequently ask: What is the best use of my time right now?

Time wasters and dealing with procrastination

We have just seen that lack of planning—being ruled by the urgent but trivial—can be a big time waster. What are some other time wasters that we need to look out for? How do we deal with the biggest time waster of all—procrastination? Table 12.7 is a list of time wasters—see if any of these items apply to you. Table 12.8 provides scheduling tips.

Table 12.7 Time Wasters

Attitude: This can be seen as a lack of interest, poor listening habits, or unnecessary perfectionism and is largely a sign of imbalance, with too much focus on some all-consuming activity and too much attention to detail in quadrant B, while neglecting personal relationships and consideration for others in quadrant C.

Lack of self-discipline: This can result in temper, impatience, inability to say no, and carelessness. It may involve avoidance of quadrant B as well as quadrant C thinking. Some of these items may have a cause-and-effect relationship: a person who can't say no can get overcommitted with various responsibilities which can lead to stress which in turn is expressed in temper, when the lack of planning is the root of the problem.

Interruptions: Here we have gossip, visitors, phone, junk mail, as well as music, TV, parties, friends. These items have to do with priorities as well as with wisdom and judgment in making choices. Fun, relaxation, social interactions are desirable—within limits. The focus on quadrant C must be balanced with other activities that will lead to the attainment of short-term and long-term goals.

Lack of planning: This results in waiting, unproductive meetings, unnecessary errands. The combined deficiency of quadrant B and D thinking leads to inefficiency in work and other areas. A few minutes of thinking ahead can prevent wasted hours. Get into the habit of making shopping lists and agendas.

Unused information and lack of communication: Good communication is needed when living and working with other people. A team functions well only when each person knows what is expected. We can also waste time when we do not have enough information or when we have so much clutter that we do not even know where to start looking for a specific item. At least a rudimentary system for organizing one's life and maintaining communication (quadrant B and quadrant C skills) will go a long way to prevent lost time.

Disorganization: This can include a cluttered work space, misplaced items, poor memorization skills, and poor scheduling habits—all symptoms of a general lack of attention to quadrant B thinking. As a first step, set up a simple filing system for important records: health (including medical test results), finances, taxes, and insurance, school records and job information for updating your résumé. Also have a "tickler file" to remind you of monthly tasks. File important paperwork with its computer file backed up on a floppy disk.

Procrastination: This is an attitude that says: "Do it later." It is a major reason for wasted time and overloading the "urgent and important" category. Procrastinators have developed a habit of putting off doing things that they see as unpleasant or difficult. They may have indecisive personalities; they may have a fear of failure. They may have a problem with authority—they just don't want to do things someone is telling them to do. How can you overcome such a habit? Do a time study to find how long it really takes to do a job if no interruptions occur? It usually comes as a big surprise to find that it takes less than half the estimated time. Give yourself a reward for completing an unpleasant task quickly. As you learn to be more creative, it will become easier to overcome the fear of failure. Try breaking big jobs down into smaller tasks—then do these smaller things one at a time. Pat yourself on the back as you see your accomplishments grow.

Table 12.8 Scheduling Tips

1. In your daily, weekly, and monthly schedules, include time for planning.
2. Do not over-schedule; leave flex time—as much as 25 percent.
3. If you have too much to do, get help. Learn to delegate; don't volunteer for too many tasks. If you accept a new task, drop one of your other responsibilities.
4. Handle each piece of junk mail only once—make instant decisions.
5. Take a break after each 2-hour period of study or working at a computer.
6. Organize your work space efficiently. Learn to take shortcuts; eliminate the trivial.

Resources for further learning

12.1 Mortimer V. Adler, *How to Speak, How to Listen,* Macmillan, New York, 1983. This book talks about two essential business skills—sales talk (including establishing your credibility, the use of persuasion, and giving effective lectures) and active listening.

12.2 Marion E. Haynes, *Personal Time Management,* Crisp Publications, Los Altos, California, 1987. This book is recommended for college students.

12.3 Victor Kiam, *Going for It! How to Succeed as an Entrepreneur,* Morrow, New York, 1986. Here you can find very good insight into the art of selling—which is not seen as a one-shot activity but something that needs to be done continuously.

12.4 Alfie Kohn, *No Contest—The Case against Competition: Why We Lose in Our Race to Win,* Houghton Mifflin, Boston, 1986. The author shows that gaining success by making others fail is an unproductive way to work or learn, whereas cooperation makes people happier, better communicators, more secure, and more productive.

12.5 Sunny Schlengler and Roberta Roesch, *How to Be Organized in Spite of Yourself: Time and Space Management that Works with Your Personal Style,* New American Library, New York, 1989. This book offers practical time and space management hints for ten types of people: perfectionist, hopper, allergic to detail, fence-sitter, cliff-hanger, everything out, nothing out, right angler, pack rat, or total slob.

12.6 ✓ Anthony M. Starfield, Karl A. Smith, and Andrew L. Bleloch, *How to Model It—Problem Solving for the Computer Age,* McGraw-Hill, New York, 1990. This book contains a nice chapter on how to build a PERT chart and identify critical paths. This small softcover book also exemplifies thinking processes and decision points in problem solving.

Exercises for "producers"

12.1 Time Management

From the discussion of time management, select a tip that you think would benefit you most in terms of immediate results. Make up an implementation plan and carry it out over the span of three weeks. Write a brief evaluation of your results and what you have learned.

12.2 Body Maintenance

Are you in good physical shape? Are you eating a healthy diet? Do you exercise on a regular basis? Now is the time for developing good health habits (or break a bad habit). Pick an item you would like to change, and write out an implementation plan. Make a schedule, a list of benefits, set goals, and then do it. As an example, review the "stop smoking" plan.

Idea implementation takes careful preparation and hard work.

The only person who likes a change is a wet baby.

12.3 ✓ Role Model

History as well as society today is filled with people who are "producers"—people who are willing to stand up (and even die) for their beliefs and ideas. Find a role model and describe what qualities you admire in this person. What are your criteria—on what are you basing your judgment? Now repeat this activity with another person—someone who is very different from the first model in ethnic background, beliefs, education, sex, age, etc. Despite the differences, what "producer" traits do they have in common? What are the most striking aspects of their battles? Select one attribute of these "producers" that you think is especially important; describe how you could possibly make it a part of your life.

12.4 ✓ Selling Technique Analyses

a. From the marketplace or business world, identify some selling techniques. What makes them effective (or not effective)? What are the objectives? How well do they address customer needs? How could you resist this type of approach? Write a brief essay.

b. Scan your junk mail for a couple of weeks and collect as many letters from charitable organizations asking for support as you can. Then analyze the content. What approaches do you find appealing and persuasive—what approaches do you find distasteful and negative? Select the best and the worst and prepare a 2-minute presentation.

c. How do you educate children to become discerning consumers able to resist hype in advertising? Research the topic; write a summary.

d. What selling techniques can you discern in a successful résumé?

12.5 ★ Selling Yourself ★

How do you come across to other people? Are you projecting a positive self-image? Observe how you interact with people. Do you make derogatory remarks about yourself (especially in response to a compliment)? Make a pact with a friend to monitor and encourage each other. Write a brief summary on how you will do this.

12.6 ★ Exodus Continued ★

This is a follow-on to Exercise Problem 3.4. In the Bible (Old Testament), analyze Exodus Chapter 4, Verses 1-17. This is a fascinating example of God teaching Moses creative problem solving. Can you identify problem definition, idea generation, creative idea evaluation and judgment, "selling" the ideas, and implementation? Make a sketch of the process using the four brain quadrants of thinking preferences.

12.7 Selling Plan

Take one of the top ideas from an earlier creative evaluation or judgment exercise and develop a selling plan. Indicate the benefits and what you would do to build acceptance and overcome opposition. Determine who your audience or customers are and address their concerns.

Chapter 12 — review of key concepts and action checklist

The "producer's" mindset: To reap the problem-solving payoff, plan carefully; have courage; be persistent; maintain good communication. Putting an idea into action is hard work.

Selling your idea: Develop your selling strategy; use a list of benefits geared to your audience. Apply effective selling techniques. Idea selling requires quadrant C thinking. Overcome reasons for opposition (perceived criticism and fear of change) and work for acceptance. Motivate yourself—be a champion of your idea.

Without planning, all you have is a party.

Jim Pierce

The work plan: It specifies who does what, when, and why and can include risk analysis. Budgeting and scheduling are key items. Use time/task analyses (Gantt charts) to visualize overlapping activities.

Implementation monitoring and final evaluation: To check that the solution is working right, plan a first review within two weeks after implementation, with a follow-up six to twelve months later. Also, write a brief summary of the results and your experience with creative problem solving. What have you learned? Inform the team of results.

Action checklist

- ☐ Hugs are good for you emotionally; hugs help your mind think more creatively. Have you hugged someone today? Three hugs a day for a child are not too much, and we never outgrow this need.
- ☐ If you have a problem with time management, make an action plan and immediately start developing better habits.
- ☐ Are you nurturing innovation champions and paradigm pioneers to develop a strong support base for positive change in your organization?
- ☐ Ask or observe a successful "sales person" for tips you can adopt.
- ☐ Monitor the voice "within." Is it telling you that you cannot learn something new, or that a task is too difficult, or that you may be ridiculed, or that what you're doing is not good enough? Redirect these negative thoughts into positive images and commands to yourself.

It is well to think well; it is divine to act well.

Horace Mann

- ☐ Use the creative problem solving process, not just in a team design project, but anytime you face a baffling problem on the job or at home.
- ☐ Read the biography of a "producer."
- ☐ Make a plan to accomplish an important six-month goal. Incorporate the steps into your daily "to do" list as priority items.

Review of Part 2

The questions and problems below will help you apply and practice what you have learned about creative problem solving and critical thinking. If you are in a formal class, selected problems from this review may be part of your final exam.

Personal Application

It has been said that people can be divided into four categories: those who make things happen, those who watch things happen, those who wonder what happened, and those who are unaware that anything happened. Your studying this book should prevent you from ever belonging to the third and fourth group—you now have an increased awareness about problems and how they are solved using different parts of the brain. The question now is: Will you be a bystander, or will you be a "producer," a person of action who applies and benefits from what has been learned?

1. Describe three areas where you see a potential for applying creative problem solving in your personal life. If possible, express these as positive goals.

2. Indicate three areas where you could possibly apply creative thinking skills in your workplace or in your education and training activities.

3. Develop a concise work plan for accomplishing one of the targets from Problems 1 or 2 above. Include when, where, why, how, and what you are planning to do, and who else might be involved in helping you reach the stated goal. Also brainstorm a list of benefits.

4. Let's think about accountability. You have had the opportunity to learn some valuable thinking skills. You can now make a choice: you can quickly forget what you have learned; you can make an effort to apply your new knowledge in many different situations. Also, you can pass on this knowledge and teach some of the skills to others. You can continue to improve your skills. How will you review your accomplishments and chart your progress over the next year? Make up a monitoring and evaluation plan.

5. This is a project for a family or a circle of friends. Brainstorm a list of problems that would be fun to solve creatively on a weekend. Then choose one of the projects, using these criteria:
   a. If you are a tense person, pick a project that will teach you to be more relaxed. Take the project through all five steps, including implementation.
   b. If you are inclined to be lazy, select a small project that will require you to be organized and energetic. Take the project through all five steps, including a work plan that has a reward structure for progress and successful completion. Then carry out the project.
   c. If you are in a rut or depressed, go for a project that will involve a high degree of creativity, physical activity, and helping someone else. Determine that you will carry your full load, as you and your team take the project through all steps in creative problem solving. Then have a well-deserved celebration!

   After your project has been put into action, write a brief summary of the results from the point of view of all four quadrants of thinking preferences.

Review of Part 2 continued

Critical Evaluation

All your life you will be asked to evaluate many things: college courses and instructors, conference presentations, the job performance of coworkers, project proposals, written materials, etc. Many forms use a range of statements, for example from "poor" to "excellent," or "strongly agree, agree, neutral, disagree, strongly disagree." These types of evaluations have serious shortcomings because the evaluator is not being asked to use explicit criteria or explain the answer. Evaluations are helpful when they show why an item has been rated high or low. People will be encouraged by knowing what has been most helpful to you and why—and how they can become even better. Those who are not rated very high will learn about possible causes of difficulties—and they will be given positive ideas and concrete suggestions on how these areas can be improved. To make good judgments, use your whole brain—evaluation is a serious responsibility.

7. Describe your interest and involvement in this textbook so far. In what way has the material met (or not met) your expectations? How successful were the authors in accomplishing the goals they set in the preface of the book? Also evaluate the amount of work required to learn each topic. Would you have liked more or fewer explanations, activities, examples, questions, group discussions, or applications? Which topics did you skip? Why?

8. Describe in a one-paragraph summary the highlights of what you have learned. What type learner/thinker are you? Which activities taught you the most? How has creative problem solving improved your thinking skills?

9. In a group of three people, brainstorm and optimize a list of criteria that can be used to evaluate a book, a class, a workshop, or a conference.

10. Review all chapters that you have studied in this book. For each chapter, select something that you particularly appreciated and something that you would score low—using your list of criteria from Problem #9. Next, look over your selections and choose three high-ranking and three low-ranking items to be used in the following two problems.

11. For each of the low-ranking items, explain why you rated the material or presentation low. Indicate the main criteria you used to make your judgment. Then describe how you would improve the situation, taking the context of the problem into account. Also think about what might be possible causes of the deficiency. How do you think your solutions would be received by others with different thinking preferences from yours. What do you think about this piece of advice: Critics are not allowed to "rewrite the play"?

12. For each of the high-ranking items, explain the reasons for your favorable rating. You may offer suggestions and ideas on how these items could be made even better. How was your evaluation influenced by your own thinking preferences?

13. Should an evaluation be submitted anonymously or with the evaluator's name? Make a case with supporting arguments for both views.

14. Evaluate the value of two textbooks—this and another (long-term, short-term, economic, non-economic). How do the results influence your decision about keeping or selling the books?

Part 3

Application in Engineering Design

LOOK WILLIAMS, I TOLD YOU THE CUSTOMER WANTS A BETTER CHEESE SLICER, NOT THAT CONTRAPTION!

I'M SORRY SIR, IT'LL BE ANOTHER WEEK. YOU KNOW DESIGN, THEY ALWAYS WANT TO BUILD BETTER MOUSETRAPS.
DESIGN
D.K.

13

What Is Design?

What you can learn from this chapter:

- The definition of engineering design and its broad implications.
- How to use this book in different contexts: as a freshman in engineering or technology; as a senior in a team capstone design project; as a professional in the workplace.
- Further learning: references, exercises, review, and action checklist.

Introduction to Part 3

We devote the last part of this book to applying the techniques of creative problem solving to the engineering design processes. First, in this chapter, we consider the question of exactly what is meant by *engineering design,* and we define it in a way that leads to implications for all stages and aspects of the design process. Because this book addresses the needs of three different levels of learners in design—freshmen being introduced to conceptual design, engineering seniors in capstone design courses, and teams learning to apply creative problem solving in multidisciplinary design projects in industry—Chapter 13 also constitutes a curriculum guide. It shows how Chapters 13 through 17 provide the basis of assignments appropriate for the respective learners.

The scientist explores what is — the engineer creates what has never been.

Theodore von Kármán, Hungarian-born American physicist and aeronautical engineer

In Chapter 14, we present a sequence of twelve steps leading to a successful design. The following two chapters introduce two practical software tools—one for project planning to help designers and students stay on track (in Chapter 15) and the other for economic analysis (in Chapter 16), because economic decisions in design now have to be made early in the process. Chapter 17 contains a complete set of design communication formats. For the practicing design professional, these chapters can provide a structure and rationale for organizing a design project. For engineering students, the assignment sets provide a series of steps that build upon one another to lead to a successful one- or-two-semester team design project. Finally, we will examine the question of what happens when professionals want to apply creative problem solving in the workplace. Chapter 18 will address the problem of change in an organization and will look at the factors that can help make innovation happen.

What is engineering design?

Design has three activities: **understand, observe, *and* visualize.** ***Remember, design is messy; designers try to understand the mess. They observe how their products will be used; design is about users and use. They visualize, which is the act of deciding what [the product] is.***

David Kelley, from an interview with Bradley Hartfield in The Designer's Stance

You probably already have an idea of what constitutes "the design" of something. Perhaps you think of a design as

- __ a set of drawings showing how something is put together,
- __ a sheet of specifications of how a product is to be built,
- __ a list of components that a device is made of,
- __ a rendering (drawing) of the appearance of an object,
- __ a description of the purpose and operation of an apparatus,
- __ a scale model of the product or a building.

An engineering design can be all of these—and more. Implicit in virtually all definitions of engineering design is the idea that a person, or more likely, a team of people, conceived of a device, a product, an apparatus, a program, a chemical compound, or a construction as the *solution to a problem*. The items in the list above represent the *results* of a problem solving process. The designer(s) first had to imagine one or more ways to solve the problem before the product or device could be put on paper or in computer memory. Typically, such problems can be solved in many ways—there is no one right answer, as we have seen in Part 1 and Part 2 of this book. The designer(s) must choose from a wide array of possibilities to create the design solution that satisfies the objectives of the people who will use the product. Every aspect of a product, a device, or a construction, represents such choices. Designers must decide on shape, material, size, assembly method, color, and so on, to fully specify the design. These decisions, then, are the *essence* of design.

Once all the decisions are made, then the design is ready to be communicated. The quality of the design is reflected in how well these decisions are made and how well they are communicated:

- __ Do the decisions result in a solution that satisfies the ultimate users?
- __ Do the decisions represent the best choice among all reasonable alternatives?
- __ Do the decisions consider *all* of the objectives of the users?
- __ Do the decisions follow accepted engineering practice?
- __ Do the decisions satisfy all applicable codes and regulations?

Of course, there are many additional criteria specific to any practical engineering design by which to judge its quality and effectiveness. The point here is that engineering design is all about decision making. Therefore, we make the following definition of engineering design:

Engineering design is the communication of a set of rational decisions obtained with creative problem solving for accomplishing certain stated objectives within prescribed constraints.

Implications of our definition of design

The definition of design in the given terms and framework has implications for the design process. In all that follows we refer to design results as a design *product*, meaning any kind of design output, including plans for consumer products, computer programs, buildings, bridges and roads, chemical compounds, military hardware, or production machinery. This methodology thus applies to all engineering disciplines.

Implications for the user. The user provides, directly or indirectly, the objectives for the design product to satisfy his or her needs. The designer's *customer* is sometimes this user of the design product. However, for many systems, the user is represented by an agent, such as the city planner who hires a civil engineering firm to design a building for the city. In that case, the city planner represents the citizens, and he or she would specify the building to satisfy their needs. In the case of a consumer product, the manufacturer's marketing group assumes the role of "agent" for the intended buyer, specifying features that consumers would presumably favor. In either case, the objectives for the design come either directly or indirectly from the users of the design product.

Implications for the designer's customer. The persons who bring the design project to the designer are thereby the designer's direct *customers*. A design customer typically sets the context for the design, spells out what is expected of the designer, and sets any constraints on the design. We define a *design constraint* as a requirement of the design within which the designer must work. There are, of course, other constraints not specified by the customer, such as those placed by environmental conditions, industry standards, or safety codes. The designer assumes the responsibility for determining the product objectives that the user (through any agent) would like in the product. Sometimes these are clearly spelled out to the designer by the customer, but frequently they are not. Sometimes the designer discovers a conflict between his perceptions of what the user needs and what his customer specifies. Resolution of such conflicts can lead to superior products that satisfy user needs better than competing designs, or they can lead to disaster if the user's needs and desires are subverted.

Activity 13-1: Quality Characteristics

The Japanese quality leader Kaoru Ishikawa defines a "substitute quality characteristic" as an indirect measure of quality derived from the customer's "true" quality characteristic. An example is a user who wants a house paint which "lasts for 15 years" (the true quality characteristic), and the paint manufacturer who substitutes the measure of no chipping, cracking, fading, or peeling during an accelerated life test under very extreme conditions for two years (the substitute quality characteristic). Brainstorm (in a group if possible) three other true and substitute quality characteristics. Then discuss the question of whether substitute quality characteristics are necessary.

Implications for products. A design product is often the result of a series of trade-offs between competing objectives. For example, the user of a lawnmower may want low cost, reliability, and low maintenance. Yet the latter two objectives usually work against low cost, so some trade-offs to reach the most attractive balance between features and cost is necessary. Thus the weighting of relative importance of competing objectives is an important aspect of engineering design. Competing objectives can inspire invention and new technology to avoid trade-offs.

Implications for designed processes. A process, not just a product, can be a design result. A process is a series of tasks or activities designed to accomplish some desired result. Thus process design easily fits our definition of design. The objectives of process design could be characteristics such as high production rate, low defect rate, short cycle time, low cost per unit, low down time, and low maintenance. In this case, the user of the device would be the people in the production unit, and the customer is likely to be the production department supervisor.

Implications for civil and building systems. A civil engineer in a design firm typically receives information on a project from an agent of the users. Thus, a design firm contracted to design a bridge for a state highway would have a state official in a state office of construction or department of transportation as a customer. The state official is an agent of the citizens of the state, representing their interests. In civil and building systems in particular, engineering designers assume a much broader stewardship role for the needs and desires of the users, since the users typically cannot be expected to know the complexities which would bear on making the best designed product for their use.

Implications for ethics and stewardship. For any product to be used by the public, the designer assumes the role of steward of resources and public safety. The public vests trust in the designer that the product will exhibit good engineering practice and that all reasonable and prudent safety precautions will be taken. At times this can mean that some of the users' wishes may not be fulfilled. It is the mark of a responsible designer that he or she is willing to counter user or customer demands for features in the design which are unsafe or wasteful of public resources. Of course, if it is possible, a good designer will find inventive ways to satisfy the users without compromising safety or wasting resources.

Activity 13-2: Public Interest

In a group, brainstorm user desires or demands associated with the following design projects which may require that the designer examine and possibly oppose or seek other alternatives. State what public interests may be at risk. a. Dam to create a recreational lake. b. Highway through a national park. c. Electric car. d. Back-up electrical power supply for the homeowner. e. Riding lawnmower. f. Computer program which can unlock security-coded systems.

Ethics implications for other-than-intended use. The designer assumes a legal liability for the safety of the product, even when it is used in ways that the designer never intended. The now infamous example of a person using a lawnmower as a hedge-trimmer illustrates the point well. In this case, the designer could have designed-in a disabling mechanism to stop the mower in case it is tilted over a certain amount or when the operator's hands are removed from the handles except for a short period for starting. Of course there will be ways to misuse a product that the designer simply could not have foreseen. A prudent designer will pause to imagine typical ways that people could misuse the product and include safety mechanisms to thwart the misapplication or protect the user from harm when attempts for misuse are made.

A common mistake that people make when trying to design something completely foolproof is to underestimate the ingenuity of complete fools.

Douglas Adams,
Mostly Harmless

Implications for communication. Defining design as the *communication* of a set of decisions may seem surprising to you. Yet consider what a design would be without this communication— a set of decisions in a designer's head. The act of communicating the decisions gives the design its reality. Typically this communication includes the result—the designer describes the chosen alternative. However, a complete package of exemplary design communications always includes a description of the set of alternatives considered for any decision, along with the rationale for making the decision. A thorough job of design communication serves to validate the credibility of designers and their decisions. Because of the critical importance of design communications, a sequence of assignments is given in Chapter 14. Refer to Chapter 5 as the primary resource of design communications descriptions.

Using Part 3 of this book as a curriculum guide

In addition to serving as a resource for design professionals, this book is a source of assignments for a variety of engineering design courses. As examples we present assignment sets for conceptual design in freshman engineering courses, for disciplinary capstone design projects, and for multidisciplinary design teams that may be vertically integrated across the curriculum. The listed hours required are estimates based on individual projects. For teams of three to five persons, project person-hours will be approximately the listed hours multiplied by the number of persons involved. For projects of similar scope, the resulting output for teams will not necessarily be increased by the same factor, due to overhead for team operation and logistics. However, the quality of the results—the creativity, and innovation displayed—can be significantly better in well-functioning teams. Assignments that are too time-consuming for individuals or more logical for a group effort should be done by teams. The numbered assignments are described in detail in Chapter 14.

Assignment set for conceptual design in freshman engineering or technology courses

The assignment set in Table 13.1 would be appropriate for a one-semester freshman course involving a semester project stressing conceptual design and multidisciplinary teamwork. The instructor can augment this assignment set with other assignments directed toward particular course goals. Assignment numbers refer to assignments in Chapter 14, with formats given in Chapter 17. Experiences with conceptual design among freshman teams have been very successful (see Reference 13.1). A course in a quarter system would expect a time investment in the range of 50 hours, with a minimum of additional assignments and a somewhat lower quality of output expected (because the main emphasis will be on the learning process, not the project results due to the time constraints).

Table 13.1
Assignment Set for a One-Semester Freshman Course with a Conceptual Design Project

Assignment #*	Est. Hours	Description
DP-1	2	Project concept statement
Table 4.5	2	Team ground rules charter (teams only)
DP-3 (optional)	4 – 18	Survey of users to determine desired features
DP-6	4 – 8	Project plan and Gantt chart
DP-6A	12 – 18	Written proposal for design project
DP-6C	3 – 6	Oral presentation of project proposal
DP-7A, DP-7B	2 – 4	Sketches and brief descriptions of conceptual designs
DP-7	10 – 20	Evaluation of alternative conceptual designs (Pugh I)
DP-12B	3 – 6	Final design evaluation by design team
Ch. 4, p. 115	1 – 3	Peer evaluation of team members (teams only)
DP-12	8 – 12	Final written report
DP-12A	6 – 9	Final oral presentation
Total	50 – 106	

* Assignment numbers refer to assignments described in Chapter 14 (using formats described in Chapter 5 and Chapter 17 under the respective DP numbers) or material included in Chapter 4.

Assignment set for disciplinary capstone design projects

The assignment set in Table 13.2 would be appropriate for a one-semester course involving a semester design project. For a two-semester (or a three-quarters) course, the optional assignments would be included, and the assignments would require an investment in hours toward the upper end of the range given. Again, assignment numbers refer to assignments in Chapter 14, using formats compiled in Chapter 17. The quality of the project results would be as important as the learning process.

Table 13.2 Assignment Set for a One- or Two-Semester Capstone Design Project

Assignment #*	Est. Hours	Description
Table 4.5	2	Team ground rules charter (teams only)
DP-1	6	Project concept statement
DP-2	3	Table of design constraints
DP-3 (optional)	4 – 18	Survey of users to determine desired features
DP-4	6	Table of design objectives
DP-5	4	Design problem analysis and statement
DP-6	6	Design project plan and Gantt chart
DP-6A,B	12 – 18	Written proposal for design project and executive summary
DP-6C	3 – 6	Oral presentation of project proposal
DP-7	10 – 20	Evaluation of alternative conceptual designs (Pugh I)
DP-7 (repeat)	6 – 12	Evaluation of alternative conceptual designs (Pugh II)
DP-7A, DP-7C	9 – 14	Overall concept drawing (and revisions)
DP-7B	1 – 2	Brief descriptions of conceptual designs
DP-8	6 – 8	Written mid-project progress report
DP-8A	24 – 36	Design decisions
DP-8B	8 – 24	Assembly drawings
DP-8C	2 – 12	Bill of material
DP-6 (repeat)	1 – 2	Updated project plan and Gantt chart
DP-8D	4 – 12	Oral presentation of progress report
DP-9 (optional)	2 – 8	Detail drawings
DP-10 (optional)	6 – 12	Test plan
DP-11	12 – 42	Evaluation results report (report of design review) Note: hours include modest prototype testing.
DP-12	8 – 12	Final written report
DP-12A	6 – 9	Final oral presentation
DP-12B	2	Final design evaluation by design team
Ch. 4, p. 115	1 – 3	Peer evaluation of team members (teams only)
Total	121 – 295	(for one or two semesters, respectively; some hours are split among team members; optional assignments included only in the two-semester total.)

* Assignment numbers refer to assignments described in Chapter 14 (using formats described in Chapter 5 and Chapter 17 under the respective DP numbers) or material included in Chapter 4.

Assignment set for vertically integrated design projects for multidisciplinary teams

Table 13.3 would be appropriate for a one- or two-semester project involving a multidisciplinary team that may attract team members from various disciplines and various levels of study (freshmen through graduate students). Such courses can have powerful pedagogical advantages, both for the more experienced student and for novices. The experienced members learn from organizing and leading the project, and, notably, from mentoring the less experienced members. A key to setting up such

Table 13.3 Assignment Set for a Vertically Integrated Multidisciplinary Team Design Project

Assignment #*	Est. Hours	Description
Chapter 14	6 – 8	Ad/recruiting multidisciplinary team members & leaders
Chapter 14	3 – 6	Multidisciplinary team member contract
Table 4.5	2	Team ground rules charter (teams only)
DP-1	6	Project concept statement
DP-2	3	Table of design constraints
DP-3 (optional)	4 – 18	Survey of users to determine desired features
DP-4	6	Table of design objectives
DP-5	4	Design problem analysis and statement
DP-6	6	Design project plan and Gantt chart
DP-6A,B	12 – 18	Written proposal for design project and executive summary
DP-6C	3 – 6	Oral presentation of project proposal
DP-7	10 – 20	Evaluation of alternative conceptual designs (Pugh I)
DP-7 (repeat)	6 – 12	Evaluation of alternative conceptual designs (Pugh II)
DP-7A, DP-7C	9 – 14	Overall concept drawing (and revisions)
DP-7B	1 – 2	Brief descriptions of conceptual designs (and revisions)
DP-8	6 – 8	Written mid-project progress report
DP-8A	24 – 36	Design decisions
DP-8B	8 – 24	Assembly drawings
DP-8C	2 – 12	Bill of material
DP-6 (repeat)	1 – 2	Updated project plan and Gantt chart
DP-8D	4 – 12	Oral presentation of progress report
DP-9 (optional)	2 – 8	Detail drawings
DP-10 (optional)	6 – 12	Test plan
DP-11	12 – 42	Evaluation results report (report of design review) Note: hours include modest prototype testing.
DP-12	8 – 12	Final written report
DP-12A	6 – 9	Final oral presentation
DP-12B	2	Final design evaluation by design team
Ch. 4, p. 115	1 – 3	Peer evaluation of team members (teams only)
Total	130 – 313	(for one or two semesters respectively; some hours are split among team members; optional assignments included only in the two-semester total.)

* Assignment numbers refer to assignments described in Chapter 14 (using formats described in Chapter 5 and Chapter 17 under the respective DP numbers) or material included in Chapter 4.

design projects is defining expectations so that team members with varying levels of experience in any needed discipline can contribute, succeed, and be recognized (and graded) based on reasonable expectations. Typically, a student team leader or organizer advertises on the Internet for team members with needed expertise, and the team members contract with the leader and with appropriate faculty for their expected contributions to the team project. Such projects can be industrially sponsored or based on interests (i.e., team competitions) of the organizers.

Resources for further learning

13.1 William Shelnutt et al., "A Multidisciplinary Course Sequence Stressing Team Skills, Conceptual Design, Creative Problem Solving, Professional Practice, and Computing Skills for Students Entering the William States Lee College of Engineering," *ASEE Annual Conference Proceedings*, June 1997, Milwaukee, WI.

Exercises

13.1 Decisions Hidden in Design
Place the following objects before a group:

> Small plastic container used to protect a roll of 33mm film, a paper clip, and a spiral-bound notebook.

Now, brainstorm a list of decisions which had to be made in the design of each object. Then, on a separate sheet, brainstorm on the objectives or design criteria (such as low cost or long durability) which might have been considered in making each of the decisions for each object. Evaluation: How do the objectives for all three products compare? Which of these objectives, or criteria, would be applied in the design of most consumer products?

13.2 Design Credibility
Suppose a designer seeking funding for her invention brought you a copy of plans for a new vacuum cleaner. She is very excited and seems to have a thorough design. List any major questions you might want answered before you would be willing to invest your savings in this venture. What form of verification for the inventor's answers would you be willing to accept?

Engineering problems are under-defined; there are many solutions, good, bad and indifferent. The art is to arrive at a good solution. This is a creative activity, involving imagination, intuition, and deliberate choice.

Ove Arup

13.3 Resource and Environmental Ethics (group discussion)
Suppose you are designing a new cellular telephone for extended talk and standby time. Your position and promotion depend on the success of this design. You discover an inventor's new battery which will triple the talk and standby time while occupying less space than conventional batteries. On further investigation you find that the battery uses very corrosive and toxic chemicals, although the inventor assures you that the battery cladding protects from leakage for many times the battery's life. Discuss and list on a flipchart or blackboard the conditions under which you might specify the battery for your new cellular phone design.

13.4 Safety Ethics (written assignment)
Suppose you have just been promoted to the position of Chief Engineer for a newly successful company producing lawn care products. The president of the company, Mr. Alfred Sloan, attributes its rapid success to

"giving the customers what they want." New federal government guidelines for lawnmower safety have required the additions of several safety features such as "dead-man" kill switches, foot protectors, object deflectors, and others. The director of marketing has just brought the president a survey of new customers, which shows that they dislike the new features because they see them as cumbersome, and some tend to clog the mower more frequently. Some customers report that they have removed the devices, even though it was a difficult and time-consuming task. The president asks you to look into making the devices modular and easily removable, with perhaps some subtle hints as to how to remove them in the owner's manual. Write a response memo to Mr. Sloan, outlining your approach to the problem. You are free to discuss the problem with others to determine what you should do.

The means by which we live has outdistanced the ends for which we live. Our scientific power has outrun our spiritual power. We have guided missiles and misguided men.

Martin Luther King, Jr.
Strength to Love, *1963*

13.5 Environmental Ethics (written assignment)

Suppose you are a newly graduated civil engineer just hired by a land development company to lay out the streets and drainage paths for a large suburban development. You are told by the Vice President for Planning, Mr. Patron, that in this area a developer's responsibility is simply to provide for any drainage not entering directly into storm sewers so that it follows the prior contour of the land, using the same drainage paths as before the development was created. You design the development layout in this way, providing catch basins to prevent washing into adjacent property. However, in touring the development under construction, you are confronted by an angry homeowner who complains that the drainage into his property has been increased several-fold, pointing out his washed-out driveway. You respond that the drainage path is the same as it was before the development, but concede that there does seem to be some erosion now. You go back to the office and study the problem, concluding that either the flow into the catch basin was much more than you had calculated, or the construction had inadvertently diverted additional drainage to this side of the development. Decide what you should do, if anything, and explain your decision and recommendations in a memo to Mr . Patron.

13.6 Energy and Ecology Ethics (group discussion or essay)

Suppose you are a mechanical engineer in a large architectural and engineering firm. You are part of a team designing a new classroom building for a public university. Your responsibility is the design of the heating/ventilating/air conditioning (HVAC) systems for the building. The university committee charged with providing input to the design simply tells you to design the systems for comfort, accurate and flexible control, and lowest capital cost. The committee is particularly interested in lowest capital cost, because any funds left over from those allocated by the legislature for constructing the building can be used to buy equipment and furnishings. In considering alternative conceptual designs for the HVAC systems, you discover that System A provides comfort with very good control, can be installed for approximately $1,500,00 in capital costs, and has a predicted annual energy consumption cost of $300,000. System B, however, does an equally good job at comfort and control with a capital cost of $2,000,000 and annual energy costs of $150,000. The architect in charge of the project tells you that the committee won't care about the annual energy cost savings of System B because the cost of energy comes from another budget over which they have no control, but that they would like to spend the half million dollars in savings of capital costs of System A on equipment and furnishings. Identify your direct customer and the ultimate customers (users), and elaborate on their interests. Also identify any public interest involved. Outline your approach and comments to the building committee.

13.7 Trade-offs in Design (group discussion or essay)
Exercise 13.6 involved a trade-off between capital costs and annual energy costs in building HVAC design. Similarly, discuss any three of the trade-offs below, identifying (a) competing interests of users or their agents, and (b) any public interests involved.

__ Bodies of water for recreation or for drinking water.
__ Power versus fuel economy for automobiles.
__ Battery energy versus weight in laptop computers.
__ Tread life versus skid resistance in automobile tires.
__ Tuning sensitivity versus squelch in radio receivers.
__ Features and capabilities versus ease-of-use in computer programs.
__ Suction power versus noise in vacuum cleaners.
__ Strength versus weight in lightweight concrete mixtures.
__ Effectiveness versus toxicity in chemical insect sprays.
__ Speed versus maneuverability in fighter aircraft.
__ Reliability and validity versus cost and time in design of a customer survey.
__ Capital and operating costs versus waiting time in design of a customer checkout area in a grocery store.

Engineers... are not superhuman. They make mistakes in their assumptions, in their calculations, in their conclusions. That they make mistakes is forgivable; that they catch them is imperative. Thus it is the essence of modern engineering not only to be able to check one's own work but also to have one's work checked and to be able to check the work of others.

Henry Petroski,
To Engineer Is Human

13.8 Who Is Your Customer?
For this activity, you need two very different containers of dental floss. Take them apart. Count the parts and also consider their complexity. Which design do you think is better? Why? Do you think the designs could be improved? If you designed a better dental floss dispenser, who would be your customer? Where would you have to go to "sell" your improved design? What factors would you have to consider?

13.9 Spoon
Think about the knowledge embodied in this simple utensil: evolution of form, function, materials, manufacturing processes and equipment, packaging, marketing, quality standards, energy sources. Brainstorm options for each of these items. What problem(s) is a spoon designed to solve?

Chapter 13 — review of key concepts and action checklist

Purpose of Part 3 of this book: (1) Application of the principles of creative problem solving to the engineering design process. (2) Resource for professionals in the workplace and for students at all levels: in freshman engineering courses, in senior capstone design courses, and on multidisciplinary teams in vertically integrated design courses.

What is engineering design? Engineering design is a product or process as the solution to a problem satisfying constraints and objectives (based on customer needs).

> **Engineering design is the communication of a set of rational decisions obtained with creative problem solving for accomplishing certain stated objectives within prescribed constraints.**

Implications: This definition applies to all engineering disciplines. The customer is not always the user—the person dealing directly with the designer may be an agent of the user (which could be society as a whole). Designers must resolve conflicts between user needs and customer specifications and between competing design objectives. Designers have a role of steward of resources and public safety. Communcation is central to engineering design as discussed in detail in Chapter 5.

Many people think the creative process only involves the initial schematic design and design development phases of a project. Actually, the creative process involves every phase of a project.

Mark von Wodtke
Mind over Media

Action checklist

- ☐ Check the time requirements of your assignments. Plan the pace; start early; leave time for the unexpected and for checking quality.
- ☐ Look at the assignment list. What are the team competencies; where is extra time needed to develop team and individual competencies?
- ☐ A frog jumping into a pot of hot water will immediately jump back out. A frog sitting in a pot that is being slowly heated will die because he does not notice the change in environment. Where can you learn more about ethics and increase your awareness of changing societal values?
- ☐ With our definition of design, does the division between "hard" and "soft" still exist and make sense?
- ☐ Remember the two levels of learning, not just knowing design technology but also the thinking needed to use the technology well.
- ☐ Look to the future: How will the skills you are learning be useful in your career—as an engineer in a big company, as a self-employed entrepreneur, as a manager of an engineering enterprise?

The Engineering Design Process

What you can find in this chapter:
- The twelve steps to quality by design:
  - ➛ Guidelines for the design problem analysis and planning stages.
  - ➛ Guidelines for system and parameter level design.
  - ➛ Guidelines for tolerance/detail design and for design evaluation.
- A complete set of student assignments using sample formats.

Engineering design processes employed by design companies, manufacturers, and individuals vary greatly. Even within companies, the design process varies with the scope of the design problem and the driving forces behind the design project. Even so, we can identify similar steps in virtually all design processes—*because design is essentially a creative problem solving process of decision and communication*. We present here a description of steps in a generic design process that can be applied quite universally. Although manufacturers or design companies will have particular variations to suit their industry and market orientation, these design steps will likely exist in one form or another, expanded or contracted, or perhaps in a modified sequence. The documentation attached to each step serves as a guide to performing that step and forms a record of the process. We regard such documentation as essential, since we have defined design as the *communication* of a set of rational decisions for accomplishing certain stated objectives within prescribed constraints.

Part 2
Part 3
12 Steps to Quality by Design
Part 1

Overview: 12 steps to quality by design

Table 14.1 characterizes the engineering design process as twelve steps with associated documentation (the DP-numbers, referring to formats detailed in Chapter 17) for each step. These steps may be categorized in eight stages:

- **Steps 1–5, design problem analysis stage** ➛ resulting in a design problem analysis and statement document.

Table 14.1 Steps and Documentation in the Engineering Design Process

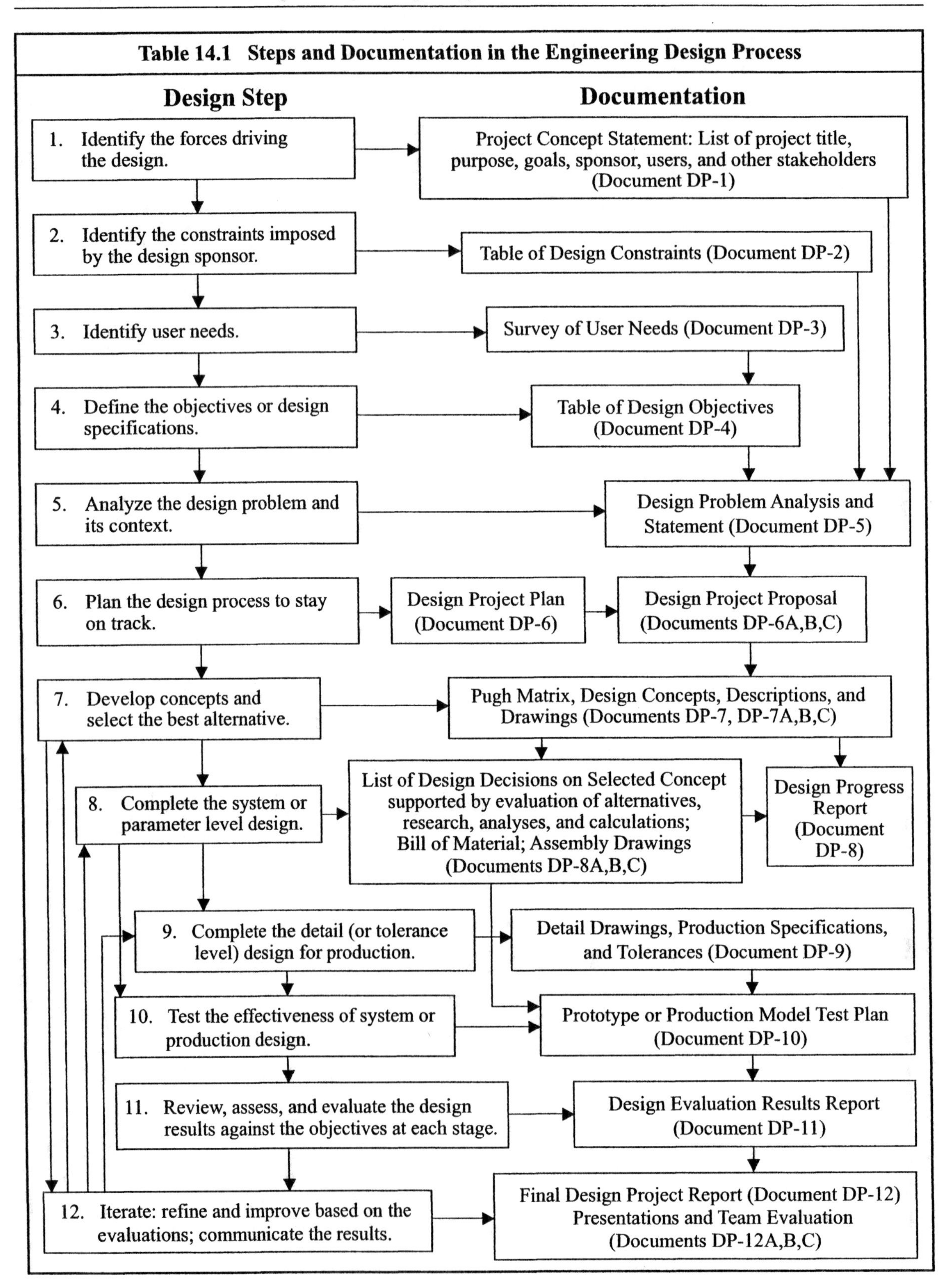

- **Step 6, planning stage** ➛ resulting in a project plan and design proposal document.
- **Step 7, conceptual or system design stage** ➛ resulting in a design concept description (and drawings) document.
- **Step 8, parameter level design stage** ➛ resulting in a set of completed design decisions embodied in assembly drawings and specifications and summarized in a design progress report.
- **Step 9, detailed or tolerance level design stage** ➛ resulting in a set of detailed working drawings, specifications, and tolerances for production.
- **Step 10, planning for and testing the effectiveness of system or tolerance design stages** (may be applied after Step 8 or Step 9, or both) ➛ resulting in a prototype or production model test plan and test results.
- **Step 11, design review and assessment** (this step should be applied after Steps 7 and 8 (preliminary assessment) and Step 9) ➛ resulting in a design evaluation results report.
- **Step 12, refinement and improvement based on design evaluations** ➛ resulting in design improvement iterations and eventually a final design project report.

Note that iterative refining and improvement of the design are built into every stage from concept through prototype testing (Steps 7 through 12).

Design problem analysis stage

Although the major thrust of engineering design is the *synthesis* of design solution alternatives, it must begin with an *analysis* of the context of the design problem. This involves the first five steps of the design process of Table 14.1. You will be using mainly the "detective's" mindset, with input from the "explorer" mainly in Steps 1 and 3 (see Chapter 7).

Step 1 ➛ Identifying the forces driving the design

The designer must be aware of the driving forces behind the design project, since these influence both the definition of the project and the design product. Typically a design project receives its impetus from the design sponsor, a changing market, changing user or stakeholder needs, or technological opportunities, or from a combination of these factors.

Impetus from the design sponsor. The designer's customer, or design sponsor, may conceive of the need for the design, perhaps resulting from an appraisal of company opportunities or from the expressed needs of their customers. This design sponsor may assume some or all of the responsibility for creating a design problem analysis statement (but this does not relieve the designer from the responsibility for checking it for internal consistency). For example, the sales division of a consumer electronics manufacturer (the design sponsor) may solicit a design for a

Project Concept Statement

Design of a Radiofrequency Surgical Probe for Soft Palate Tissue Reduction to Control Snoring and Sleep Apnea

The purpose of this project is to design a surgical probe and the associated low-level radiofrequency generator and controller which would be used by surgeons for shrinking the redundant tissue of an enlarged soft palate and uvula as a treatment for habitual snoring and sleep apnea. With local anaesthetic, the surgeon inserts the insulated probe with a disposable, very-small-diameter, uninsulated needle tip into the soft palate and administers low levels of radiofrequency monitored by thermocouples in the probe. This causes very localized, painless heating damage to the sub-mucosal layers, resulting in a coagulation and absorption of damaged tissue and a stiffening of the palate. Somnion Associates sponsors the design project and a consortium of physicians from Somnion will coordinate the design requirements and specifications, Other stakeholders include the stockholders of Somnion and the health care profession. Major goals of the design include controlled studies within 18 months showing (a) very significant reduction in snoring in patients treated by this procedure, (b) significant reduction of sleep apnea, (c) a virtually pain-free procedure, and (d) virtually no side effects or complications. A multidisciplinary team of two electrical engineers (Jason Metcalf and Rene Farthing), two mechanical engineers (Jennifer Stanley and Fowler Blanch), a biomedical engineer (Colby Suther), an engineer-physician (Michelle Farmer), and an industrial designer of medical devices (Edward Sofill), designated Team Somniprobe, will conduct the project, with Colby Suther serving as the project leader.

Figure 14.1 Example of a project concept statement—using Format DP-1 in Chapter 17.

duplicating VCR to fill out its product portfolio, perhaps at the behest of its larger customers. Or, a utility department of a city may ask for the design of an expanded sewer system in anticipation of population growth. Figure 14.1 above is an example of such a design project concept.

Impetus from a changing market. The design sponsor may point to a changing market as the reason for the design project. Perhaps the market is being influenced by competing or alternative products or by fluctuating price structures. The sponsor may have done some or all of the market research required to determine the causes. If not, the designer must arrange for this research to determine what new product attributes or current product attribute changes are needed to react to the changing market. For example, the marketing department (design sponsor) of a wristwatch manufacturer may ask for new color-coordinated wristwatches to be marketed to young teens as fashion accessories to respond to an emerging fashion trend.

Impetus from changing customer or stakeholder needs. Changing user lifestyles may drive designs for new or improved products. The designer needs to know what factors influence the user's decision to buy and use the product, whether from market research or from current customer/user profiles. The needs of other stakeholders, such as support and service people, can also drive the need for new designs. For example, rising costs of repair services may drive the need to design products with higher reliability, fewer repairs, and reduced disassembly and repair time.

Impetus from a technological breakthrough. Emerging technology may drive new designs by making possible products with significant competitive advantages over existing products. The potential for such new products is obvious only in retrospect. Customers will not be clamoring for the new products until someone recognizes the potential and creates a market. Imagination and creative problem solving play a pivotal role in taking advantage of such technological breakthroughs. The Japanese electronics industry (initially Seiko) recognized the potential for electronic wristwatches and created a market that eventually all but displaced the entire Swiss mechanical watch industry—this despite the fact that the idea was invented by the Swiss (as described in Chapter 1).

Five-Minute Class Exercise 14-1: Technological Development

In small groups, brainstorm on ways that one of the following developments might influence existing products. Identify the product and the influence of the new technological development.

- A new rechargeable battery with 100% greater energy storage per unit weight and volume, moldable into any shape;
- A fiber, made from recycled paper and plastic, that feels like silk and has the strength and resilience of nylon;
- A super-fine-finish stainless steel cladding that can be applied to virtually any shape and size part and which is harder and more durable at temperatures up to 1000°C than any known substance except diamonds;
- A micro-sensor that can reliably detect and identify odors with a sensitivity 100 times greater than the best bloodhound.

Step 2 → Identify constraints imposed by the sponsor

The sponsor of the design project may constrain the design to fit company manufacturing capabilities or a market niche. A civil engineering project sponsor may list cost constraints, municipal or state codes together with environmental considerations. These constraints can form a design envelope within which the design must fit. A design constraint differs from a design objective in that a constraint imposes a specific limitation on the design, but a design objective becomes a target for aiming design improvements. For example, a cost limitation for a municipal park of $1,000,000 in approved tax funds is a *constraint*, but a low cost criterion which the park designer weighs along with other criteria to choose the best park design concept is a design *objective*.

Since the set of constraints will completely limit the design, it is important to define a *minimum* set of *real* constraints—minimum in the sense that only essential constraints are included, and real in that if the product could be viable without a constraint, then it should be omitted. In addition to the constraints imposed by the design sponsor, constraints might be imposed by such factors as:

- Convention or code (such as building codes).
- Compatibility with existing equipment (such as standard tire sizes).
- Compatibility with mating parts of a designed assembly.
- Available power requirements (i.e., 115 V AC current at less than 15 amps).
- Regulatory requirements (such as EPA auto emissions guidelines).
- Industry standards (such as the thread on a light bulb base).
- Product environment (such as temperature, pressure, or humidity).

Quality by design is driven by the customer's wishes at all stages of the design.

Constraints should be expressed in measurable, quantitative terms, such as dollars, pounds, cubic feet, psi, etc., whenever possible. Table 14.2 illustrates the constraints imposed on the design of a municipal park by a county commission (see Format DP-2 in Chapter 17). The designers must choose land and design a park within these constraints. Sometimes the designer must ferret out the constraint from a very general description by the sponsors. In the example of Table 14.2, the county commissioners may have said that the park must include a stream, but not how big a stream. The designers may have had to seek clarification from the sponsors by suggesting a certain minimum size in terms they can visualize and then translate those ideas into a quantitative constraint.

Table 14.2 Constraints Imposed by County Commission on the Design of a Municipal Park

Constraint	Measurement Method	Quantitative Limitation
Size of cleared area, C	Gross Survey	$5 < C < 10$ acres
Wooded area for hiking trails, W	Gross Survey	$20 < W < 50$ acres
Total cost, including land	Verified cost estimate	< $750,000
Included stream flow rate	Weir flow rate	> 5 cubic feet/second

Identify user needs

Either the sponsor or the designers must arrange to determine the needs of the intended users of a design product. Even if you think you know the needs of a typical user, some minimal survey is usually very useful. In particular, if you are designing a product that you would like to use, considering yourself as typical of other users in the market invites gross misjudgments. Typically questionnaires, either written or from telephone interviews or focus groups, provide this information (see page 139 in Chapter 5 for guidance). Format DP-3 in Chapter 17 gives further information for developing user surveys for designed products. The goal of this step is to produce a list of the desired features of a design product and to weight them according to user preferences. *These are the users' definition of quality for the product.* Questions should therefore include opportunities for respondents or focus group members to contribute features they think of and to rank or weight them along with any features that the designers wish to have considered. Table 14.3 is a hypothetical example of a weighted feature list derived from a survey of potential users of a proposed new general-purpose toolbox.

Table 14.3 User Quality Characteristics for a General-Purpose Toolbox
(derived from a survey of potential users; hypothetical data)

Feature		Weight
Lightweight		23%
Easy to carry		17%
Reliable lid clasp, hinges, and handle		11%
Durable over years of continued use		9%
Has many compartments for screws, nails, nuts, and bolts		7%
Relatively low cost (in range of $15–$30)		6%
Rust free		5%
Attractive appearance		4%
Flat top for stacking		4%
Other miscellaneous attributes		14%
	Total	100%

⌛ **Exercise 4-2: Constraints**

Briefly list methods of measurement and arbitrary examples of quantitative limitations for the following constraints:

- The noise emitted by a portable room dehumidifier.
- The size of a laptop computer.
- The concentration of toxic chemicals in the effluent from an industrial plant into a river.
- The thermal pollution (temperature and energy per unit effluent) of cooling water from a nuclear reactor into a lake.

Step 3A ➤ Perform an analysis of the competition or of present methods

With the product users' desires as a context, much can be gained by a thorough analysis of the competition or currently used methods. At a minimum, this analysis should answer the following questions:

- How do the potential design product users handle the problem now?
- If there is no competitive product, how does the problem get solved, if at all?
- If there are competitive products, how do the products stack up against each other in sales and popularity? What are the best features of each product? What are the weakest features of each product? See Appendix B for a discussion on the method of benchmarking.
- What product opportunities does this analysis suggest?

Although such an analysis seems to apply more directly to consumer or industrial products, it has value for civil or municipal projects as well. Taxpayers want to pay only for needed projects that solve real problems they face or that provide services they see as desirable and valuable.

Step  ➛ Define quantitative objectives (or specifications) for the design

Once the wishes, dreams, and desires of the potential product users have been obtained, the designer must translate these expressions into quantitative terms that can be measured. This translation, referred to by Kaoru Ishikawa (Ref. C.4) as the *substitute quality characteristic,* becomes the basis for quantitative *design objectives* or *design specifications*. At this stage of design, these two terms are virtually synonymous, but as the design is refined and enters the tolerance level of design, the objectives become precisely defined as *engineering specifications*. Table 14.4 uses the general purpose toolbox example of Step 3 to illustrate translation of user desires into quantitative objectives with measurable targets (see Format DP-4 in Chapter 17).

Major methods for measuring achievement of objectives during the design process:
- **analytical methods, such as calculation of stresses in support members of a truss,**
- **experimental methods, such as scale model tests in wind tunnels,**
- **simulation methods, such as Monte Carlo computer simulations,**
- **evaluation by experts in the field of concern.**

Table 14.4 Example of a List of Design Objectives for a General-Purpose Toolbox
Weights have been adjusted to account for omission of "other miscellaneous attributes" from Table 14.3.

Design objective	Weight	Method of Measurement or Design Estimatation	Target
Low weight W	27%	Estimate W by volume of material in each part	W < 5 lbs
Ergonomic handle grip	10%	Comfortable to X % of users with 20 pound gross load	X > 90%
Balanced pickup with typical tool load	9%	Feels adequately balanced to X % of users with center of gravity displaced by 1.5 inches front to back and 3 inches side to side	X > 90%
Reliable lid clasp, hinges, handle, and box body with fully loaded box (20 lbs)	13%	Estimated mean number of operation cycles N before failure (verified upon prototype test) of load, shut, pickup, set down, open, unload, and repeat	N > 100,000 cycles
Durable over years of continued use	10%	Accelerated exposure test of box material to UV equivalent of Y years of southwest U.S. sunlight with no discernible discoloration (estimate based on material properties, later verified by prototype test)	Y > 5
Many small parts compartments	8%	At least S compartments of 2 x 3 x 1 inches and at least L compartments of 4 x 8 x 1 inches	S > 4 L > 1
Low cost	7%	Estimated total retail price P	P < $25
Non-corrosive material for all parts	6%	Corrosion resistance R of any metal parts as percent of corrosion resistance of surgical stainless steel	R > 95%
Attractive appearance	5%	Rated attractive by A (number) of potential users	A > 90%
Handle folds into flat surface	5% 100%	F % of footprint as stackable flat top surface	F > 90%

Fifteen Minute Team Exercise 14-3: Design Objectives

In teams, develop examples of specific design objectives, methods of measurement, and targets for each of the following user quality characteristics:

- Comfort of a bicycle seat for mountain trail biking.
- Low maintenance for a riding lawnmower.
- Ease of hearing from a digital cellular phone.
- Convenience for riders of free downtown buses.
- Attractiveness of sidewalks, curbs, and plantings for a downtown plaza.
- Maintainability of a storm sewer system for a subdivision.

Write each team's results on a blackboard or flip chart in the form of Table 14.4. Compare results from the teams and choose the most effective, measurable objectives for each quality characteristic.

Step 5 ➤ Develop an effective design problem analysis statement

Steps 1 through 4 provide most of the information necessary to proceed to the conceptual design stage. However, the information is typically not organized into a succinct statement of the thrust of the design project, one that can convey to all stakeholders the purpose, objectives, and constraints of the project with a view toward the potential for solving the problem. As noted in Chapter 7 (see Table 7.1), we alternate between the "detective's" and "explorer's" mindsets to define the problem. Organization of the information from Steps 1 through 4 into a concise statement that will guide the design effort through to completion is a very worthwhile effort requiring careful, deliberate thought as well as imaginative thinking. The design problem analysis document format (DP-5) in Chapter 17 shows how the material from the design project concept, the table of design constraints, user profiles and needs, and table of design objectives, are integrated and summarized into one document, which culminates in a short statement of the design problem. An up-to-date design problem analysis statement can serve as an evolving master guide for everyone working on the project, as shown in Table 14.5.

Table 14.5 Example of a Design Problem Analysis Statement

Design of a Compact Modular Playground System

Design for Acme Playground Systems, Inc., a modular, cost-effective, outdoor playground system for municipal parks and school playgrounds,

- with a retail price less than $25,000,
- with at least 6 different safety-tested activities including tunnel slides and swings,
- that appeals to children of ages 4 through 12,
- that appears attractive, sturdy, safe, and durable to adults,
- that requires no maintenance for a life of at least 15 years, and
- that can be erected on a footprint of less than 1000 sq. ft in less than 3 days by a crew of four installers.

Step 6 ➤ Plan the design process to stay on track

Once the design problem is laid out, the next step is to plan the project to produce the design. Major industrial or commercial design projects can consume significant human resources and span several years. Thorough planning in such projects produces a better design product and avoids costly overruns and missed deadlines. The same applies to student projects, which perhaps benefit even more from the discipline of visualizing the process before embarking on it. For these reasons we devote all of Chapter 15 to project planning. Here we simply point out the rudiments of the planning process. This detailed planning requires primarily quadrant B thinking skills.

Steps 1 through 6 in engineering design correspond to the problem definition step in the creative problem solving process.

In planning any project, the designer focuses on decision making. Every activity in the plan should be tied to one or more design decisions, either helping to make the decision or validating it. Effective plans answer the following questions:

- What decisions are likely to constitute the design?
- How will the design team gather the data relevant to the decisions?
- How will the design team process and analyze the data into meaningful information helpful in decision making?
- What specific tasks are required to gather the data, process it, make effective design decisions, and test or validate the decisions?
- In what sequence must these tasks be done?
- Which design team member will accomplish each task, how much time will it take, and when will it be done?
- What tasks, materials, services are needed of others not on the design team?
- What contingencies or uncertainties must be provided for in the plan?
- What are the resulting deliverables and milestones characterizing the plan and what dates are associated with each?

Typically, the answers to these questions are summarized in tabular form (see Format DP-6 of Figure 17.6) or graphical forms such as time-line or Gantt charts (see Chapter 15 for examples and guidance). Frequently, designers use software such as Microsoft Project to complete the plan.

Step 6A ➤ Prepare a design project proposal

The first five steps in the design process constitute an analysis and exploration of the context, purpose, and goals of the design. Step 6 sets out a plan for accomplishing these goals in a design project. At this point, the designers most likely need to obtain permission to proceed with the project, gaining consent from the design project sponsor (the designer's customer) to expend the effort and funds necessary. The purpose of the design project proposal (Format DP-6A in Chapter 17) is to obtain approval for carrying out the plans using the design approach outlined in the design problem analysis statement. Many other stakeholders may be interested in the potential of the project and the approach to be taken, so the proposal must be written for a wide variety of technical levels of understanding and expertise. The proposal is the first stand-alone formal document to be submitted to the design sponsors, and it includes an executive summary (Format DP-6B). In student projects, it can be used as a starting model for the two other formal project documents, the design project progress report (DP-8) and the final design project report (Format DP-12). Feedback from the instructor on the proposal can allow the student to refine the material from Steps 1 through 6 before incorporating it into the design progress report. Again, feedback on Steps 7 and 8 in the progress report can be used to refine this material before inclusion in the final design project report. This sequence of feedback and improvement can produce excellent final reports that students can submit with pride to the project sponsors or potential employers.

System level design stage

We will now discuss the first of three design levels. Notice from the flow chart of Table 14.1 that after each level of design, designers refine and improve the decisions in an iterative process (Step 12), *although we discuss this process only once as Step 12.*

Step 7 →

Develop several system level design concepts and select the best alternative

When the project approach and plan have been approved by the design sponsors, the designers may proceed to invent alternate design concepts and select the most promising one. Selection of the design concept is the first and most significant decision (or set of decisions) in the design. It is a highly important design stage, and, unfortunately one given short shrift by novice designers. Too often, designers have a concept in mind from the beginning of the project, and they do not pause to consider other alternatives. Usually too late in the project they discover another concept that would have been better, simpler, easier to manufacture, or less costly than the one they are stuck with. Sometimes designers are not even aware that they are wedded to a design concept—like the fish not recognizing he is in water—until someone points out an obviously better design solution. This is an example of the "Eureka" phenomenon discussed in Table 11.2 (page 268); also see the kitchen design on page 160.

THREE LEVELS OF DESIGN

Step 7:
System or concept design.

Step 8:
Parameter design.

Step 9:
Tolerance/detail design.

To avoid these embarrassments and dilemmas, effective designers follow a disciplined procedure for creating and evaluating design alternatives. This process follows the brainstorming of Chapter 8 and the evaluation and judgment of Chapters 9 and 10. Tools such as the Pugh method of Chapter 11 facilitate comparison of alternatives early in the design process. At first, this may seem like a constricting and time consuming procedure. Paradoxically, such a disciplined procedure leads to *more creative* solutions and *savings in time (and cost)* in the long run since better design concepts are identified earlier in the project.

Some novice designers wish to rush this process, wanting to get to the details of parameter and tolerance design more quickly, since that appeals to engineering problem solvers. Yet, more time spent on fleshing out alternative design concepts early in the design process is time well spent. A change in a concept on paper is inexpensive, one during prototype stage is more costly, and a change during production ramp-up is typically very expensive. Changes necessary in the field, such as recalls of products for safety or warranty repairs, are extremely expensive.

Steps 7 through 12 in engineering design correspond to the idea generation, creative evaluation, and judgment steps in the creative problem solving process.

Genichi Taguchi, the Japanese statistician and quality leader who popularized the concept of robust product design, refers to this stage of design as the *system level design* for imparting quality to products (Ref. F.13). At the system, or functional, level of design the designer views the intended product the way the user sees it and the way the producer may intend to build it. The designer must consider product and process technologies (including emerging technologies) which will bear on the design decisions. For example, the choice of whether to base a design concept for a cellular telephone on analog or on digital technology will profoundly influence the course of the design process, the manufacturing processes, and the introduction of the product to consumers.

Pugh Method Round 1. An additional investment of time for team discussion and brainstorming of concepts can reap the marvelous power of team synergy in this conceptual stage of design. A particularly powerful way to do this is to assign each team member the task of inventing a *complete* design concept that embodies all the best design features he or she can imagine. It is important that each team member produce a complete design concept, with all major decisions and features clearly noted (see Format DP-7B and DP-7C in Chapter 17), since that is the only way that the concepts can be compared to each other. Each team member then presents his or her concept to the rest of the team, taking care to elaborate on how well (or not so well) the concept addresses each of the design constraints and objectives by using a modified Pugh Matrix. (See Chapter 10 on critical thinking, as well as Chapter 11 and the example of a modified Pugh matrix in Format DP-7 in Chapter 17.)

Each team member should score his or her design against the best current design concept and explain the rationale for the scores. A spreadsheet (EXCEL or Lotus 123, for example) facilitates an easy scoring summary if the scores are –1, 0, and +1, meaning worse than the reference, the same as the reference, or better than the reference, respectively, for each of the criteria. Each score is multiplied by the importance weight for each criterion, and the scores are summed for each concept. If no current design exists, then the team selects one of the team member's design concepts as the reference (it doesn't matter which one) and scores the remaining concepts against it. Invariably there will be features in each of the various designs that all team members would like to adopt.

Pugh Method Round 2. A second round of the modified Pugh matrix method is now in order. The team selects the best of the concepts put forth in the first round as the new reference design. Then each team member develops another complete concept for the design, using the best of the features of the designs from Round 1 and any additional ideas that come up from the discussion. Again, team members present their designs and the scoring of each against the design criteria or objectives to the entire team. The best design is then chosen as the team's approved concept, unless clear benefits can be gained from another round of the Pugh method. Efforts should be made to overcome all identified flaws in the chosen design.

In creating the alternative concepts for considerations, documentation in the form of sketches can serve both to communicate ideas among the team members and as documentation of the process to establish credibility in the design product. Every concept proposed by team members for the Pugh matrix comparison should include one or more sketches of the basic features of the concept. For mechanical products, these sketches will likely illustrate the structure or operating mechanism of the concept. For electrical products, the sketches could include tentative schematic drawings and parts layout sketches. For civil engineering projects, layouts of plan views, sketches of structure alternatives, or configurations are in order. Multidisciplinary designs may involve several types of sketches and initial drawings. Concept sketches should be clearly drawn and unambiguous, but at this stage formal drawings are not warranted. The purpose of the conceptual design stage is to create, evolve, and evaluate as many design concepts as possible, and excessive details or complex drawings thwart this purpose. Also, each design concept in each of the Pugh matrices should be succinctly summarized in words (see p. 277). A one- or two-line description of each concept as a key on the Pugh matrices is sufficient if accompanied by a more detailed description (Document DP-7B) in the progress report (see Assignment 14.8).

Quality loss is defined as the financial loss to society after a product has been manufactured.

Also, it has been estimated that the cost of customer dissatisfaction and loss of market share can be as high as four times the cost of warranty claims and repairs.

Parameter level design stage

The next level of decisions are what Genichi Taguchi refers to as the parameter level of design (Ref. F.13). The purpose of this design stage is to optimize the design with respect to performance (measured against objectives based on user needs) and cost. In Taguchi's construct, this means making all decisions so that the design product will exhibit "robust" performance, i.e., it will perform well in the intended environment with relative insensitivity to variation in environmental conditions or user profiles. For example, a portable radio designed for indoor personal use would exhibit robust performance if it operates satisfactorily under extreme heat or cold conditions, is unaffected by moderate moisture exposure (rain), and tolerates being dropped occasionally.

Step 8 → Complete parameter level design

Parameter design involves all decisions required for complete specification of the design product—materials, purchased components, sizes, capacities, components, assembly, finishes, etc. These decisions are typically given as examples in engineering analysis courses for each engineering discipline. Although the particular design analysis technique for determining acceptable parameter levels depends on the engineering discipline involved, the design process is virtually the same for all disciplines. It typically involves the following steps.

- List the major decisions to be made to complete the parameter level design, such as, for a computer cooling system, "calculate the heat removal rate based on temperatures allowable, size the fan blades, size the motor, select the switch and fuse, choose the cooling path, select the filter, select the wiring, and route the wiring."

- For each decision, identify several viable alternatives. If the decision is to determine the level of a parameter on a continuous or discrete scale, the "alternatives" are the *range* of possible choices for that parameter. For a discrete choice example, the decision might be to size a conveyor motor, and alternatives include 3/4 and 1.0 Horsepower motors with acceptable power and torque. For a continuous scale example, consider the decision of the capacity of the windshield washer fluid tank in an automobile when the space allows capacities of up to 5.3 liters.

- Perform analyses, simulations, or model tests to determine the appropriate values of the parameters for each decision. For example, a structural loads analysis yields data on stress for sizing the support beams.

- Perform trade-off analyses for each variable that affects two or more performance objectives in opposing fashion. For example, we may trade off performance on the design objectives for an HVAC system of low initial cost versus low operating costs in selecting a supply fan. A less expensive constant-speed fan will incur higher operating and energy cost than the more expensive variable-speed fan. The choice is a trade-off.

The "parameters" referred to by Taguchi are the specification choices, such as the size, shape and material of a support beam, the power and torque of a drive motor, the output voltage and current of a power supply, or the capacity and flow rate of an aeration system for sewage treatment.

- Make each decision so as to optimize the performance of the entire design against all the performance objectives. One must be careful here not to *sub-optimize* an aspect of the design. Such sub-optimization happens when one achieves a measure of good performance on one or more objectives at the expense of other, perhaps more important, performance objectives. For example, one could satisfy the low-pressure-drop criterion for a hydraulic valve by choosing an expensive, smoothly contoured valve, but at the risk of blowing the project budget. In another example, the choice of tires for an automobile might be made on the basis of low cost and tread life, but the performance objectives of cornering stability, wet-stopping distance, and ride quality may suffer. The safeguard against such sub-optimization is always to consider *all* performance objectives for each decision. This holds even when the choice seems to hinge on only one or two objectives. If others are not affected, the analysis will simply show their nil effects. Frequently, however, such an analysis uncovers conflicting performance among the objectives.

■ Complete documentation of the decisions made, including:

- An annotated list of major decisions made, with rationale for each (see Format DP-8A in Chapter 17).
- Construction, layout, or assembly drawings (Format DP-8B).
- Bills of material, showing every component to be built or purchased along with complete specifications (Format DP-8C). Production specifications, tolerances, and manufacturing or building plans are typically done in Step 9.

In a designed experiment, design parameters are referred to as independent *factors*, their particular setting are referred to as design *levels*, and the quality characteristics are the dependent variables being optimized.

Designed Experimentation. An understanding of the complex interplay among various design alternatives or parameter levels is sometimes not available with the simple tools of trade-off analysis and Pugh method matrices. At the parameter design level, designed experimentation can provide insights on optimization of a complete set of design parameters with respect to one or more measures of performance. For example, if a designer wants to minimize harmonic distortion in an amplifier, he or she may identify several design parameters (e.g., values of components such as resistors, capacitors, and inductors) that are known to influence the distortion level. However, the best combination of parameter levels to accomplish lowest distortion is usually unknown. Typically such effects are not revealed through straightforward calculations. The problem, then, is to find levels of the factors that optimize the quality characteristics (such as distortion). Techniques for conducting such experiments (called variously factorial design, designed experimentation, design of experiments, or planned experimentation) are described in textbooks on quality control (see for example Ref. 5.6). Typically such experiments are conducted via electronic simulation or actual physical test of breadboard or prototype apparatus. A series of such experiments in a sequential, adaptive approach can lead to optimization on a much higher level than that available through conventional one-change-at-a-time experimentation. The sequential, designed experiment approach can also identify the influential factors and any interactions among them.

Robust Design. Taguchi has popularized an extension of designed experimentation for the purpose of creating robust product designs at the parameter or tolerance levels of design. The goal in such experimentation is to design products that are *on target* (i.e., the quality characteristics are centered closely to their nominal values) and *tolerant to variation* from the environment and the user. Taguchi's methodology involves some conventional methods of designed experimentation along with a personalized set of statistical tools that have generated considerable controversy among Western engineers and statisticians. In particular, his introduction of the *signal-to-noise ratio* and the *quadratic loss function* concept have spawned criticism that they are misleading and fail to take

A first measure of robust quality of designed products comes simply from designer awareness of product susceptibility to variation in environment and user patterns.

advantage of modern graphical, data-analytic approaches (Ref. 14.2). The practice of designing robust products using designed experiments, however, remains a very worthwhile approach to improve product quality. Presentation of the techniques of designed experimentation is beyond the scope of this text. Readers are referred to Kolarik's comprehensive introduction to the subject (Ref. 5.6).

A robust design starts with questions such as:

- What can be done to reduce variation of the quality characteristics of the product as it ages? For example, how does the designer of an electronic product provide automatic compensation for changes in resistive components over product life?
- What design practices minimize the effects on performance characteristics of wear over the life of the product? For example, how does an automotive tire designer avoid degradation of handling and braking characteristics of automobile tires as the tread wears?
- What design provisions can reduce the negative effects of overload and outright misuse of products? For example, what design provisions can minimize rutting of asphalt pavement or cracking of concrete pavement subject to occasional grossly overloaded trucks?
- What design provisions can minimize any performance changes caused by occasional environmental conditions outside the expected range? For example, how may a designer of audio tapes provide design features minimizing any loss of mechanical performance due to exposure to high temperatures in a car left out in the sun?

Tolerance or detail design stage

Detailed design for production involves decisions about how the design product is to be made. Although production and industrial engineers should participate in the design process at all stages, their degree of involvement increases substantially during this detailed or tolerance level of design. Production process designers work with product designers to make decisions concerning precision capabilities of production equipment in relation to precision needs of the design product. Both for purchased components and designed components, increased requirements for precision lead to significantly increased costs. Here the term "robust design" takes on a new significance. A design that requires high precision of all components in order to function is not robust. This tolerance level of design involves the following types of questions and decisions.

■ ***How much variation around nominal specifications is tolerable by the design while maintaining acceptable quality characteristics?*** For example, how much ripple in a DC power supply subject to typical input voltage fluctuations is tolerable before the system it is supplying becomes unacceptably unstable?

■ ***What is the least precise tolerance that can be applied to a purchased or built component specification while maintaining high quality characteristics?*** If a design for an amplifier is tolerant of a resistor specified as 100 ohms ± 5% almost as well as one specified as 100 ohms ± 1%, then the design is both more robust and more cost effective.

■ ***How may assembly or construction tolerances be relaxed by a "fault-tolerant" design?*** For example, a construction support system for a suspended catwalk that depends on precise placement of support connectors to avoid overstressing the cables might be made more tolerant of errors in support alignment with a different type of connector, resulting in a safer and less expensive system.

■ ***How may tolerances be specified to avoid tolerance stack-up and ambiguous or conflicting specifications?*** In mechanical systems design, for instance, GD&T (geometric dimensioning and tolerancing) methodology references all dimensions to datum features in the product and avoids redundant or ambiguous directions on parts drawings by using a special set of specification conventions. These standards are based on standard ASME Y14.5M. (See Ref. 14.3 for an introduction to GD&T using these standards; see Ref. 14.1 for locating engineering standards.)

Step 9 ➝ Complete detail or tolerance level of design for production

Steps in completing the tolerance level of the design process includ:

- Defining production (or construction) specifications for each component in the product assembly;
- Defining tolerances for each specification that call for minimum precision consistent with high quality performance;
- Developing documents communicating these specifications and tolerances clearly to production personnel and component suppliers, including detailed drawings (see Format DP-9 in Chapter 17) and specifications sheets or summaries.

Design evaluation stage

Typically, all or parts of the design require testing to validate the decisions. If the design will depend on a new technology or application of existing technology in new ways, tests may be required to confirm the effectiveness of the technology application in the design. Sometimes a test is needed to confirm estimated performance simply because the complex interplay of environmental and operating variables cannot be modeled mathematically. Construction of test models, prototypes, and test apparatus can consume inordinate amounts of time. And yet, all too frequently, the tests do not provide the validation or answer the questions needed by the designers because the test was not well planned.

Step 10 ➤ Test the effectiveness of system or tolerance level designs

In general, such tests are designed to answer specific questions not amenable to calculation or simulation. Tests of a prototype serve to confirm the validity of the entire design. Frequently, tests of some subset of the design decisions are needed, perhaps in an early phase of the design process. For example, a new model of farm tractor may use the drive train of a previous model; thus it only requires tests of a new power-take-off and hydraulic system. Tests of prototypes or components in student projects are routinely sources of frustration, primarily through underestimation of the time required and lack of contingency planning.

For some designs, prototypes are not feasible, such as for ships, buildings, and bridges.

In these cases, a combination of tests for feasibility of aspects of new technology use, computer simulation, and testing of physical scale models can provide valuable guidance or confirmation of design decisions.

Elements of an effective, carefully considered test plan include the following (usually limited to one page, especially for student projects):

- A statement of the specific purposes of the tests. Which decisions will the test validate or illuminate? Vague purposes can lead to disorganized tests with no useful result—if the test does not bear on a design decision, then there may be no need to run the test.
- Specific test objectives, i.e., exactly what needs to be measured during the test, including equipment requirements and specifications.
- A step-by-step procedure for conducting the test, with attention to any variables that need to be controlled or monitored.
- An outline of expected results, preferably in the form of a data sheet with predicted data outcomes—if one does not know roughly what will happen, data collection and analysis are threatened.

Format DP-10 in Chapter 17 provides general guidance for developing a test plan for prototype, component, or production models. As noted in this format, technical societies publish test plans or codes for testing various products, and government agencies publish test protocols for regulated tests (such as the Environmental Protection Agency's guidelines for emissions testing of new vehicles). Even if such test protocols are not required for particular products, designers can gain valuable insight into test planning considerations by studying these plans, especially on how to control for typical confounding variables in classes of tests for particular purposes. A well-designed test plan can pay significant dividends in time saved and in useful results.

Step 11 ➤ Review, assess, and evaluate design results (at each design stage)

Good design promotes an iterative process of assessment and improvement at each stage. Design reviews, test results, and assessments provide opportunities for periodic improvement and redirection of the design at the earliest possible points. Time for such improvements must be built into the design process in order to be effective. The form and nature of such improvements are dictated by the specifics of the project, but documentation of the process should not be neglected. Every design review or test report should include the following types of information:

- A summary of any testing (purpose, objectives, procedures, results).
- A summary of any review or evaluation (by whom, criteria used, methodology, and results).
- Conclusions drawn from testing or design review.
- Recommendations for design changes.
- Plans to implement the changes.

As with the test plan, this review need not be a lengthy document. The results should be succinctly summarized so that the basis and rationale for design changes are clear (see Format DP-11 in Chapter 17).

Step 12 ➛

Iterate: refine and improve based on evaluation, then communicate the results

The initial design products at each stage should virtually always be seen as works in progress—not as finished products. Taking this design iteration process as a separate and distinct step separates product and process and allows a more objective view of the design's successes and deficiencies. Design iteration consists of three parts:

- Review of design evaluation results (from tests or independent review).
- Identification, ranking (and also possible implementation) of potential improvements.
- Communicating the results (in an appropriate format for the particular stage of design).

Student assignments

These assignments should be regarded as the *substance* of the design project, not as "extraneous paperwork." The assignments are designed to coach students as they complete each step in the design process. Each document leads naturally into the next. Some of the tables and figures assigned are to be used as specific parts of the three formal reports: the design proposal, the progress report, and the final design project report. Feedback from the instructor on each assignment can help students stay on track as well as help improve the coherence of the final report.

Student Assignment 14.1 Complete three project **concept statements** for possible design projects, using Format DP-1 in Chapter 17. Be sure to identify the driving forces behind each project. Devise titles that succinctly convey the thrust of the projects. If you do not have external sponsors (customers), list yourself or your team as sponsors. Carefully define the users of the design product. Also list other stakeholders, those people who would be influenced by or interested in the design product. Submit all three project concept statements to your instructor for approval, and select one of the approved concepts for your class project.

Student Assignment 14.2 Make a **table of constraints** similar to Table 14.2 for the class project begun in Assignment 14.1. Use Format DP-2 in Chapter 17 to organize your table. Note the source of each constraint in the table or in an attached discussion.

Optional Student Assignment 14.3 If assigned by your instructor to do so, conduct a **survey of potential users** of your design product according to the guidelines of Format DP-3 in Chapter 17. Limit the scope of your questions in a survey or a focus group to those that define desired features for your design product. Aim for 30 or so respondents, unless you have statistical reasons to choose a particular sample size.

Student Assignment 14.3A Using data from a customer survey, focus group, published studies, or other sources, create a **table of user quality characteristics** for your class project similar to that of Table 14.3. If you create the list from your own preferences, or that of the team, clearly indicate yourself or your team as the source.

Student Assignment 14.4 Create a **table of design objectives** for your class project similar to that of Table 14.4 or Format DP-4 in Chapter 17. Develop quantitative, measurable objectives based on user preferences. Consider targeting specific weaknesses in competing designs. Include objectives representing the best features of competitive products.

Quality means knowing or anticipating what customers want, then translating this vision into practical, innovative, dependable products in a system that can produce these at low cost while generating profits or benefits.

Student Assignment 14.5 Write a **design problem analysis document** for your class project using the example in this chapter as well as Format DP-5 in Chapter 17. In the statement itself, don't try to include everything from your tables of design constraints and objectives—just summarize the major points of sponsor, users, purpose, constraints, and objectives culminating in one concise statement describing the design problem and your approach to solving it.

Student Assignment 14.6 Prepare a **plan for your class project**. Working with your team, do the following steps:

- Brainstorm and list the major design decisions which must be made.
- Complete the task list of Table 14.6.
- Resolve to update the plan at least every 2 or 3 weeks by conferring with each team member on status of tasks, new tasks to be added, milestones, and changes in estimated time required.

Student Assignment 14.6A,B Write a **design project proposal** for your class project using Format DP-6A in Chapter 17. Include an executive summary as in Format DP-6B. Refer to the material in Chapter 5 on design communication as an additional guide.

Student Assignment 14.6C Prepare and deliver a formal technical design project **proposal presentation** using the guidelines for oral technical presentations in Chapter 5 and Format DP-6C in Chapter 17.

Table 14.6 Project Planning Tasks (for Assignment 14.6, Step 2)

❑ Brainstorm and list the tasks required to make these decisions. Combine tasks likely to take less than 2 hours, and split up tasks likely to take more than 6 hours. As a rough guide, a one-semester project should include 40 to 50 tasks and a two-semester project should include 60 to 80 tasks for teams with 3-5 members. Include tasks such as

- Data gathering.
- Data processing.
- Analysis of the data, with graphical and tabular summaries helpful to decision making.
- Creation of alternative concepts for the design product.
- Development of each of the documents in the 12 design steps (except those which have already been done to this point).
- Simulations, analyses, model or prototype construction.
- Testing.
- Evaluation.
- Preparation for oral presentations.

❑ Use a computer application such as Microsoft Project to accomplish the following tasks. In using a computer program for planning, don't get caught up in too much detail. Use the program as a planning tool to help you in managing your project, not as a project itself.

- List and number each task and assign a team member to be responsible for it.
- List the estimate of the time required for each task and the completion date.
- List the predecessors for each task (those task numbers which must be completed before the particular task can be completed).
- Add project milestones (tasks of zero duration indicating a stage of completion in the project).
- Create and print a Gantt chart for the project. Printing Gantt charts can be an exercise in creative problem solving: adjust the scale and print area for each page to avoid yard-long printouts which are virtually unreadable and thereby useless.
- Print a list of the tasks, persons responsible, and beginning and end dates for each.

Student Assignment 14.7 Using the procedure discussed in this chapter and following the format illustrated in Format DP-7 in Chapter 17, construct a **modified Pugh matrix** for your class project. Assign each team member the task of creating a complete, stand-alone concept for your design and evaluating it against the design criteria or objectives for your project. A concept sketch and a two- or three-line description of its features should accompany each concept (Document DP-7B). If no current or competing design exists, arbitrarily choose one of the team member's concepts as the reference for comparison (by drawing lots). In this **Round 1** comparison, discuss the results among the team members and use the Pugh matrix results to choose the best concept.

Then, using this best concept from Round 1 as the reference concept, complete a Pugh matrix **Round 2**. Again, each team member should develop a complete, stand-alone concept for the design using the best combination of features from any of the previous concepts and any other ideas generated during discussion. Then team members present their concepts to each other for comparison and selection of the best final concept (DP-7C). The team members then critique and improve the final concept sketch and description before submitting it to the instructor.

Alternative approach for Assignment 14.7:

Instead of using the modified Pugh matrix, follow the standard Pugh matrix and the procedure outlined in Chapter 7. Verify with your instructor which approach is to be used by your team.

The function of the concept drawing is to present the salient features of the concept in an easily understood format.

Student Assignment 14.7A Once a design concept is selected, the features should be clearly illustrated in an overall design concept drawing. The **concept drawing** is much more than a sketch. The example of the Budweiser beer can of Format DP-7A in Chapter 17 shows the added detail and explanatory power of such a concept drawing over a simple sketch. Such drawings use call-outs, expanded views, section views of various parts, pointers to features, and concise descriptions of operation or manufacture that call attention to the decisions made in selecting the concept. This overall concept drawing may evolve through several stages, but it is important to retain the drawings at each stage to document the process and the rationale for the changes. For this assignment, create an overall concept drawing for the selected concept of your class project, using the drawing of the Budweiser beer can as a benchmark. Concept drawings for projects from other engineering disciplines may require different forms, as outlined in Table 14.7. The concept drawing differs from the assembly drawing (Document DP-8C) in that the concept drawing stresses the user-driven features of the design whereas the assembly drawing shows how the design product will be constructed or assembled.

Table 14.7 Alternate Design Concept Drawings

- Concept drawings for electronic products may require both a schematic drawing of circuits and a component layout sketch.
- Concept drawing for some civil engineering projects may require contour maps of excavation sites, plan-view layouts of structures, or schematic diagrams of building water and waste systems.
- Concept drawings for industrial engineering projects may require assembly line routing schematics and as-is or could-be process maps for industrial process design.
- Concept drawing for chemical engineering projects may require annotated piping and equipment schematics for chemical reactors.

Use the type of concept drawing suitable for the discipline of your project. Remember that the purpose of the concept drawing is to convey the selected overall concept for the design, and especially the features that address user needs, not to show the details which may not yet be available.

Student Assignment 14.8A Prepare an annotated **list of design decisions** for your class project using Format DP-8A in Chapter 17 and the discussion of this chapter as guides. List each major decision requiring 8 hours or more of calculation, analysis, or evaluation. Make sure you describe alternatives for each decision and discuss how the decision is to be made. If the decision has been made, describe it in the past tense—if not, describe what will be done. This constitutes the primary record of your design, and it will provide the rationale to establish credibility in what you will have done.

Student Assignment 14.8B Prepare an **assembly, construction, or layout drawing** for your class project using as an example the drawing DP-8B in Chapter 17 for a mechanical product. There should be only one overall assembly drawing for the product. The assembly drawing refers to all components used to make up the product, subassemblies, purchased components, or built components. Use the style of drawing appropriate for the engineering disciplines involved, as given in Table 14.8. Remember to use the drawing conventions of the disciplines for all notations—this task requires mostly quadrant B thinking, not quadrant D creativity. Multi-disciplinary projects call for multiple subassembly drawings representing the contributions from the various disciplines.

Table 14.8 Alternate Assembly Drawing Styles (with Bill of Material)

- A schematic diagram, or wiring diagram, or integrated circuit diagram for electronic products.
- A piping and reactor vessel layout for chemical process design.
- A production process equipment layout for production engineering plant design.
- As-is, should-be, and could-be process maps for redesign of industrial engineering processes.
- Construction drawings, structural plans, and appropriate section views for civil and building structures.

Bill of Material Specifications

- The call-out number of each item from the assembly or layout.
- The quantity of each item.
- A notation of whether the item is to be built (B) or purchased from suppliers (P).
- A complete description of each item, including engineering specifications or vendor part number if the item is purchased, or reference to a detail drawing if the item is to be built.

Student Assignment 14.8C Prepare a **bill of material** to accompany your class project assembly drawing. This document will list the tentative specifications for each item called out in the assembly drawing, as indicated in Table 14.8. For some projects this bill of material amounts to a list of raw materials needed to construct the designed product. For others it becomes a list of vendor-supplied components making up an assembly. Still others are a combination of the two. In any case, the combination of the assembly drawing (or layout, or construction drawing) and the bill of material should completely specify all materials and components required to construct or assemble the design product. At this point the specifications may be generic or tentative, but can be converted into final engineering specifications, including tolerances, at the tolerance level of design of Step 9.

Student Assignment 14.9 Develop **detail drawings** of each manufacturing component of the designed product using Format DP-9 in Chapter 17 as a guide (for mechanical parts). Include complete **specifications** of all purchased parts. For manufactured parts specify **tolerances** consistent with high-quality robust design, using standards appropriate to the discipline, such as ASME Y14.5M for mechanical components. Include a detail drawing for each manufactured item on the assembly.

Student Assignment 14.10 Prepare a **test plan** for your class project. You may wish to test an entire **prototype** or only a **component** of the system. Use Format DP-9 in Chapter 17 as a guide. For this student project, the plan should be very brief, perhaps one or two pages. Do not lose the benefits of planning by including only generalities, however. Pay particular attention to the test's purpose and objectives in quantitative terms. A good job of documentation here will pay time-saving dividends in writing the evaluation results report and the final project report.

Student Assignment 14.11 Prepare a **design evaluation results report** for your class project. Use Format DP-11 in Chapter 17 as a guide. This report will also be used as part of the final design project report.

Student Assignment 14.12B Have each team member complete an **evaluation of the final quality of the design** using Format DP-12B in Chapter 17. Team members should complete these evaluations independently of each other, then summarize them for the final report.

Student Assignment 14.12 Prepare a formal, comprehensive, **final design project report** for your class project, using Format DP-12 as a guide. Weave together all previous assignments in this series, beginning with an update of your progress report (DP-8). Take into account particularly any feedback from your instructor on these assignments. Have as an overall goal the production of an excellent design report document that you can present with confidence to potential employers.

Student Assignment 14.12A Prepare and deliver a **final, comprehensive oral presentation** of your class project using Format DP-12A in Chapter 17. Stress the validity of the decisions making up your design and your rationale for each. Since guests from outside your discipline and outside your institution may attend, you will want to gear the presentation to a more general audience. Keep in mind their lack of knowledge of the context of your project, and relate the purpose, scope, and results in terms they are likely to understand. Assign parts of the presentation to various team members, but avoid switching speakers too often.

The designer has a passion for doing something that fits somebody's needs, but that is not just a simple fix. The designer has a dream that goes beyond what exists, rather than fixing what exists.

David Kelley
high tech product designer

References

14.1 William H. Middendorf and Richard H. Engelmann, *Design of Devices and Systems,* third edition, Marcel Dekker, Inc., New York, 1998. This book covers product and system design from conceptual, economic, and ethical aspects to modeling, decision making, and accelerated life testing. It includes discussions on product liability and reliability, locating engineering standards, legal, medical and sociological implications of system design in a multidisciplinary approach for upper-level undergraduate engineering students.

14.2 J.J. Pignatiello and J.S. Ramberg," Top Ten Triumphs and Tragedies of Genichi Taguchi," *Quality Engineering,* Vol. 4, No. 2, pp. 221-225, 1992.

14.3 William Tandler, *An Introduction to Geometric Dimensioning and Tolerancing Based on ASME Y14.5M 1994,* Multi Metrics, Inc., 865 Lemon Street, Menlo Park, CA 94025-6110, 1996.

15

Project Planning

What you can learn from this chapter:

- Project planning terminology.
- An introduction to using Microsoft Project 98.
- Rules of thumb on scope and time required for student projects.
- Use of project plan templates for student projects.
- Rules of thumb for defining tasks in student projects.
- Personalizing—to a specific team project—the generic project plan templates based on Microsoft Project 98 which are included on the CD-ROM attached inside the backcover of this book.

Project planning, like most organizational tasks, must tread the fine line between too little or too much detail—it must strike a good balance. If the team creates a plan with too little detail, it cannot guide task planning day to day. If, on the other hand, there is too much detail, the plan becomes ponderous and requires too much time to maintain. Our goal here is to provide some guidance on how much and what kind of details are useful for promoting effective and efficient projects.

Typically, two organizational tasks dominate during the early planning phase of a design project: team organization and project organization. Tools for organizing and managing teams for productivity were presented in Chapter 4—review the checked items if necessary:

- ☑ A well defined team charge or mission statement
- ☐ **A timetable or project plan**
- ☑ Team ground rules
- ☑ Team member roles
- ☑ Meeting agendas
- ☑ Meeting minutes
- ☑ Running task lists

No matter how high one's aspiration may be, it must be attained step by step.

From the teachings of Buddha

Our purpose in the present chapter is to elaborate on the second item—project timetables and plans—and to present templates for project planning that may be used for various classes of projects. The templates are provided on the CD-ROM packaged with this textbook; for use, they require that Microsoft® Project 98 be installed in your computer. This software is commonly available in engineering computer labs.

Organizing and planning design projects

Although modern project planning tools (such as Microsoft Project 98) can be used to plan and monitor many details of projects, our purpose in this chapter is to only introduce the concepts of task planning and resource allocation. The addition of other details, including costs, is a natural extension of these primary concepts but is beyond our scope here.

Terminology

In the following discussion we use certain terms common in planning design projects. The definitions of these terms are listed in Table 15.1.

Table 15.1 Glossary of Project Planning Terms

Critical path — the series of linked tasks requiring the greatest amount of time for completion of the project (the path constraining the earliest completion date).

Delay time — an extra amount of time added after commencing one task before starting another task to avoid over-allocation of a resource (e.g., waiting until a stereo lithographic machine has produced one model before starting another since the machine will produce only one at a time).

Deliverable — an item produced by the project (such as a product, a service, a plan, or a report).

Gantt chart — a graphical representation of tasks in a vertical list, with linked horizontal time bars for each listed task.

Lag time — extra time added between tasks made necessary by the process (e.g., time for weather aging of a prototype exterior finish before testing).

Lead time — overlapping time before the completion of one task and the initiation of a second (e.g., if after a five-day computer data reduction is underway for three days, a formal engineering analysis of the data can begin, the lead time is two days).

Milestone — an important state of accomplishment in the project, usually represented by a task of zero duration. A milestone might be finishing the specifications for a product.

Over-allocated resource — a resource that has more call for use during a particular time period than that resource has available (e.g., an engineer who can only spend five hours per day on a project being scheduled for eight hours).

PERT chart — a graphical presentation of tasks as linked boxes with each box typically showing the task name, ID number, start and finish dates, and task duration (the PERT acronym originally stood for Program Evaluation and Review Technique).

Predecessor — a task that must be completed before another task can begin.

Resource — a qualified person or item of equipment that is required to accomplish a task.

Task — a well-defined portion of the work to be accomplished during the project. A task requires a certain amount of time for accomplishment by resources, including both personnel and equipment.

Task duration — elapsed time (usually in working days) required for the team to accomplish the task.

Negotiating project schedule constraints

If the team sponsor has charged the team in sufficient detail to proceed, the team would have been given an idea of the time frame it has to do the work. Such time-frame constraints are typically derived totally from external factors, such as competitive forces, market trends, business cycles, and funding opportunities. Only when subjected to realistic planning based on the human and physical resources available does the time frame achieve reality. Usually this requires a series of negotiations between the project sponsor and the design team, as the plans become more and more detailed. If the timetable is inflexible and the available resources (see Table 15.2) are inadequate to complete the project in the required time, then it may be necessary to reduce the scope of the project. Sometimes a scale-up in resources and funding may allow acceleration of the project timetable. The interplay of external timetable pressures, available resources, the project tasks, and provision for unknown contingencies is a dynamic planning challenge which underscores two basic criteria for an effective plan: flexibility and adaptability. A useful plan provides for continuous updating to meet these challenges.

Identifying internal and external resources

Early in the planning process the people, physical resources, and funding available to the project dictate its pace. Although these may be subject to negotiation with the project sponsor, availability of such resources becomes a driving constraint to the project schedule. Sometimes the procurement of external assistance provides a valuable alternative to inadequate internal resources. A component of the design may be "farmed-out" to an external design firm, either to save time or to supplement internal expertise, or both. For example, a firm designing an aircraft may contract with a vendor firm to design the engines, the avionics, or other subsystems. Such contracted services are designated as external resources in the project plan. Depending on the relationship with the vendor design firm, the tasks involved in the contracted services may be presented in less detail than those of internal resources. In any case, the plan should contain periodic design reviews and progress checks to make sure that the designs of major systems and subsystems mesh with each other, whether accomplished internally or externally.

Table 15.2 List of Resources that May Be Available to a Design Project

- The design personnel assigned to the project (the design team) with any specific expertise noted.
- Office space and equipment.
- Prototype construction facility (if applicable) with necessary tools and equipment.
- Prototype testing facility (if applicable) with necessary instrumentation.
- External design support under contract (may include a list of personnel and resources).

Using Microsoft® Project 98 for Student Design Projects

Although the commercially available planning program, Microsoft Project 98, is a powerful program with a commensurate learning curve to take full advantage of its features, it is possible to use the program for relatively simple plans. A few hours of familiarization should suffice to take advantage of basic features. For students, we recommend that both the costing and the resource (people) assignments be omitted. Then, by entering a list of tasks and milestones with estimated task durations, a table of completion dates and accompanying Gantt charts may be generated. The tool allows easy modification of the plan as the project progresses—an important benefit.

Creating the project task framework

A good way to begin planning a design project is to construct a framework for containing all the design tasks. This framework will include the overall timeframe as well as all reporting and design review tasks imposed by the sponsor at the agreed-upon milestones. For example, the sponsor may want formal design reviews at critical decision points (such as completion of the conceptual and parametric designs stages) and formal written reports at specific points. By starting with these milestones, the project director can begin to see the effects of time and resource constraints. Figures 15.1 and 15.2 represent task frameworks for one-semester and two-semester design project courses, respectively. Note that milestones are represented as zero-duration tasks signaling completion of a project phase. Your instructor may give you the data required for creating a similar framework for your course. At this point, we are interested in task durations, not work required per task.

Assignment 15-1: Project Task Framework

Modify the Task Framework of Figure 15.1 (or Figure 15.2 if you have a one-semester project timeframe) to represent the milestones, reports, and other details of your design project requirements. Do not add additional tasks specific to your project at this stage. Be sure to modify the date scale to represent your time period.

Project Scope

Two projects with identical titles can have very different levels of effort. For example, a project entitled *Design of a Children's Playground Set* could represent a one-time construction of a playset for a particular playground based on a single sponsor, or it could represent a comprehensive effort to research playground needs across the country, to assess children's likes and dislikes, to assess nationally competitive designs, to select from a wide variety of low-cost, high-durability alternatives, and to integrate manufacturing considerations into every design decision. Clearly the latter would require an effort several times greater. If it were to be completed in the same amount of time, it would require a much larger team.

Effort-Driven Versus Task-Duration Scheduling

Effort-driven scheduling is based on the idea that the more resources put on a project task, the sooner that task will be completed (i.e., it will have a shorter duration). If a task requires 16 hours of work, and two people are assigned to it, then it can be completed in 8 hours. Task-duration scheduling assumes a fixed period of time to do a task, regardless of the amount of resources assigned to it. Microsoft Project 98 defaults to effort-driven scheduling. In the templates for student projects, however, task durations were input without specific task assignments to team members, so the actual durations are not known.

In student projects, the scope of the project should be clarified in the Project Proposal (Document DP-6A in Chapter 17) and then reflected in the project plan. Project instructors or advisors will necessarily expect a more ambitious scope from a larger team than from a smaller team. This scope must be seen in the number and size of tasks included in the plan.

Student Project Scope Rule of Thumb

Per team member, the project should require approximately
40 – 50 task hours/credit hour (for semester hours)
25 – 35 task hours/credit hour (for quarter hours).

The first indication of the project scope should appear in the Project Concept Statement—Document DP-1 in Chapter 17. If the project begins with specific deadlines and limitations on the scope, these should be specified.

Identifying project tasks and responsibilities

The most important activity in planning a project is developing a list of tasks to be performed. This list may range in size from 50 or so tasks for a one-semester engineering senior design project to tens of thousands of tasks for the design of a jetliner. Typically, tasks are arranged around steps of design, such as the 12 Steps of the Engineering Design Process outlined in Chapter 14 (see Table 14.1 on page 328). To begin, list each of the 12 steps, then beneath each step, list everything that has to be done to complete that step. The following rule of thumb will help to get a list of tasks averaging about four person-hours to complete.

Task Definition Rule of Thumb (for student projects):

Combine tasks requiring less than about two hours of work.
Separate tasks taking more than about six hours of work.

For a particular project, the number of tasks is a measure of the level of work. More tasks indicate a greater level of work, providing the above rule of thumb or something similar is used. The templates of Figure 15.1 and Figure 15.2 could represent projects varying considerably in size, depending on the number of team members and other resources assigned to the projects.

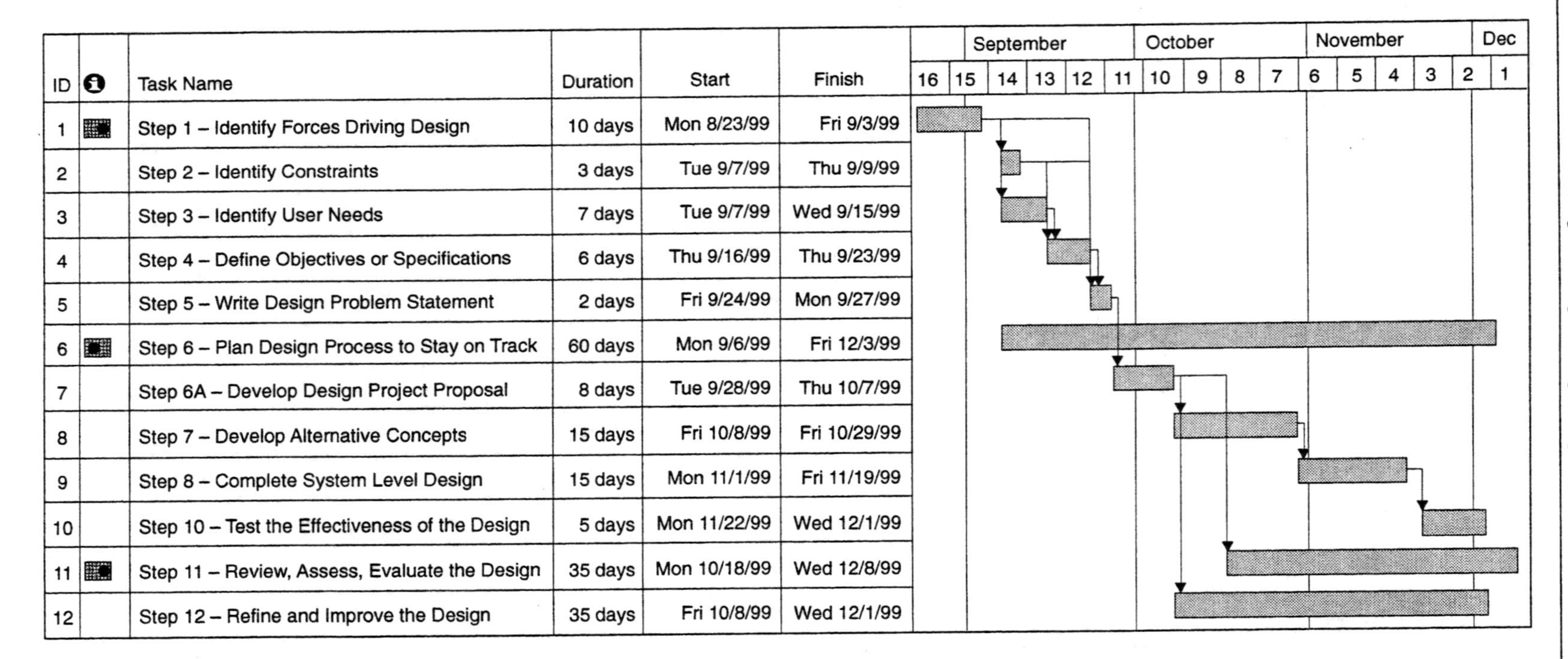

ID		Task Name	Duration	Start	Finish
1		Step 1 – Identify Forces Driving Design	10 days	Mon 8/23/99	Fri 9/3/99
2		Step 2 – Identify Constraints	3 days	Tue 9/7/99	Thu 9/9/99
3		Step 3 – Identify User Needs	7 days	Tue 9/7/99	Wed 9/15/99
4		Step 4 – Define Objectives or Specifications	6 days	Thu 9/16/99	Thu 9/23/99
5		Step 5 – Write Design Problem Statement	2 days	Fri 9/24/99	Mon 9/27/99
6		Step 6 – Plan Design Process to Stay on Track	60 days	Mon 9/6/99	Fri 12/3/99
7		Step 6A – Develop Design Project Proposal	8 days	Tue 9/28/99	Thu 10/7/99
8		Step 7 – Develop Alternative Concepts	15 days	Fri 10/8/99	Fri 10/29/99
9		Step 8 – Complete System Level Design	15 days	Mon 11/1/99	Fri 11/19/99
10		Step 10 – Test the Effectiveness of the Design	5 days	Mon 11/22/99	Wed 12/1/99
11		Step 11 – Review, Assess, Evaluate the Design	35 days	Mon 10/18/99	Wed 12/8/99
12		Step 12 – Refine and Improve the Design	35 days	Fri 10/8/99	Wed 12/1/99

Figure 15.1 Project template for a one-semester design project course (included as Microsoft Project file "1-Sem Templ" in the CD-ROM provided at the back of the textbook). The task/time diagram to the right of the task descriptions is referred to as a Gantt chart. This template uses the 12-step design process of Chapter 14. Note that the task duration column shows elapsed time, not the amount of work in each task. To be useful in project planning and project management, this template should be expanded by identifying (sub)tasks under each step. Note the differences between this template and that of Figure 15.2 for two semesters: Step 9 on detail or tolerance design is omitted in the one-semester template, and Step 10 on testing is significantly shorter. The task durations shown are working days, not including weekends, semester breaks, or holidays.

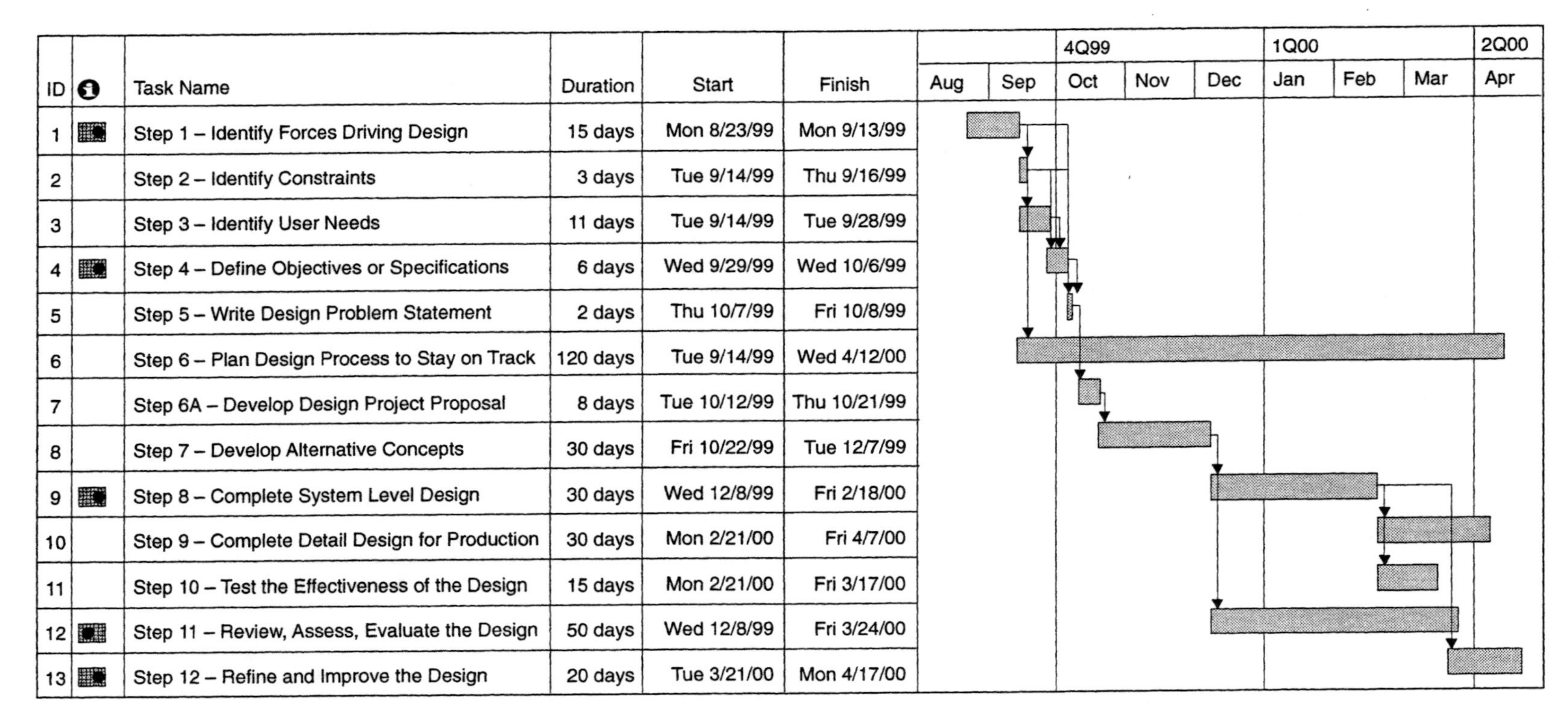

ID		Task Name	Duration	Start	Finish
1		Step 1 – Identify Forces Driving Design	15 days	Mon 8/23/99	Mon 9/13/99
2		Step 2 – Identify Constraints	3 days	Tue 9/14/99	Thu 9/16/99
3		Step 3 – Identify User Needs	11 days	Tue 9/14/99	Tue 9/28/99
4		Step 4 – Define Objectives or Specifications	6 days	Wed 9/29/99	Wed 10/6/99
5		Step 5 – Write Design Problem Statement	2 days	Thu 10/7/99	Fri 10/8/99
6		Step 6 – Plan Design Process to Stay on Track	120 days	Tue 9/14/99	Wed 4/12/00
7		Step 6A – Develop Design Project Proposal	8 days	Tue 10/12/99	Thu 10/21/99
8		Step 7 – Develop Alternative Concepts	30 days	Fri 10/22/99	Tue 12/7/99
9		Step 8 – Complete System Level Design	30 days	Wed 12/8/99	Fri 2/18/00
10		Step 9 – Complete Detail Design for Production	30 days	Mon 2/21/00	Fri 4/7/00
11		Step 10 – Test the Effectiveness of the Design	15 days	Mon 2/21/00	Fri 3/17/00
12		Step 11 – Review, Assess, Evaluate the Design	50 days	Wed 12/8/99	Fri 3/24/00
13		Step 12 – Refine and Improve the Design	20 days	Tue 3/21/00	Mon 4/17/00

Figure 15.2 Project template for a two-semester design project course (included as a Microsoft Project file "2-Sem Templ" in the CD-ROM provided at the back of the textbook). Contrast this template with the template of Figure 15.1 on the opposite page for the one-semester format. Note that the task durations are working days, not including weekends, semester breaks, or any holidays.

Creating a detailed project plan

The time spent in making a detailed list of tasks should be viewed as much more than "paperwork." The thinking processes, the questions asked, the conferral among team members, the realization of effects of deadlines, and the breakdown of a formidable project to doable steps are the inherent benefits of creating a plan with worthwhile detail. To make the process less burdensome, we have created the generic detailed task lists in Microsoft Project 98 for one- and two-semester projects shown in Tables 15.4 and 15.5. These tables are based on the same 12-step plans as outlined in Figures 15.1 and Figure 15.2.

Personalizing the generic plans

In order to personalize Table 15.4 or Table 15.5 shown on pages 360-363 for your particular design project, take the steps outlined in Table 15.3.

Table 15.3 Steps for Customizing the Planning Template with Your Project Data

1. Load Microsoft Project 98 into your computer (if you do not have it already available).
2. Insert the CD-ROM disk from the back of this book into your drive and load the plan into Microsoft Project (click file, open, then select file "1 sem templ date" or "2 sem templ date," and click OK) depending on whether you have a one-semester or a two-semester project.
3. Click file, save as, and save a copy of the original plan template to your computer hard disk.
4. Click file, save as, and save a copy of the template named as your project plan.
5. Change the start date for Task 2, "Identify tentative design topic" to the year, month, and day of the start of your project.
   - Double-click on the "start" column for Task 2 ➛ a monthly calendar will appear.
   - Click on the right or left arrows until you get the year and month of the start of your project.
6. Click on a cell anywhere outside the calendar; the dates of all tasks will automatically change to accommodate your new start date. Change the calendar to show the specific start and end dates and holidays for your semester(s):
   - Click on "tools" in the toolbar at the top of the screen.
   - In the drop-down selection box, click on "change working time."
   - In the resulting calendar, change working time and non-working time to fit your particular calendar by selecting any day(s) and clicking the working or non-working time buttons. Note that weekends are shown as non-working (gray shaded) by default, but you may change that if you wish. Be sure to modify spring and fall break times and holidays if necessary. Click OK when finished.
7. Change, modify, delete, or add tasks as necessary to personalize the plan to your project.

Time Saving Hint

Do one change at a time, and review the results of your change carefully, then save the file if OK. This way you can always go back if the change causes unwanted results by exiting without saving the file, then getting the "good" version back. You may also practice and see the results of changes without saving the file. Changing predecessors through deleting a task or changing listed predecessor tasks sometimes causes unexpected results.

Setting predecessors for tasks

In Tables 15.4 and 15.5, the rightmost column lists for each task the predecessor tasks required to be completed before that task may begin. For example in Table 15.4, a predecessor for Task 14—Prepare Survey of Potential Users—is Task 11—Prepare Table of Constraints. Therefore, the numeral 11 appears in the predecessor column for Task 14. Microsoft Project uses this approach to keep track of the required sequencing of tasks. Clearly then, if it were possible to complete tasks simultaneously, these tasks must not be identified as being predecessors for tasks in this group. Listing unnecessary predecessors prolongs the project unnecessarily. List a predecessor for a task if and only if the task cannot be begun without completing the predecessor.

Student Exercise 15-2: Predecessors

Note the key dates for task completion. Remove the predecessor for Task 14 in the template for Table 15.4. What effect does this have? Did the project final report date change, and if so, by how much?

Obtaining approval for the initial plan

Once you have created a draft of the detailed plan, you should coordinate it with the project team members and the project sponsors. This will probably lead to quite a few useful suggestions for improvement. Now that you have a plan in a computer format, take advantage of its greatest strength—ease of change!

Monitoring, updating, and revising the plan

A useful project plan always answers the question, "How will we accomplish the remaining work in the remaining time?" This means that the plan must be updated at specific points or in step with major shifts in direction of the project. Never present a plan at a meeting with the comment that the project is "behind schedule" or that the plan is "out of date"! A current project plan (or schedule) should always answer the question above, so it shows what will be done in the remaining time and is therefore never out of date. Set specific dates in the plan for reviewing

Table 15.4 Detailed Task List Template for a One-Semester Student Project

ID	ℹ	Task Name	Duration	Start	Finish	Predecessors
1		**Step 1 – Identify Forces Driving Design (DP-1)**	**12 days**	**Mon 8/23/99**	**Wed 9/8/99**	
2	■	Identify tentative design topic	5 days	Mon 8/23/99	Fri 8/27/99	
3		Advertise for and recruit team members	1 day	Mon 8/30/99	Mon 8/30/99	2
4		Write team member "contracts"	1 day	Tue 8/31/99	Tue 8/31/99	3
5		Develop consensus on team ground rules	1 day	Wed 9/1/99	Wed 9/1/99	4
6		Set meeting schedules and guidelines	1 day	Thu 9/2/99	Thu 9/2/99	4,5
7		Develop Project Concept Statement (DP-1)	2 days	Fri 9/3/99	Tue 9/7/99	6
8		Submit project concept statement	1 day	Wed 9/8/99	Wed 9/8/99	7
9		**Step 2 – Identify Constraints (DP-2)**	**4 days**	**Wed 9/8/99**	**Mon 9/13/99**	**7**
10		Query project sponsors on constraints	1 day	Wed 9/8/99	Wed 9/8/99	7
11		Prepare Table of Constraints (DP-2)	2 days	Thu 9/9/99	Fri 9/10/99	10
12		Submit table of constraints	1 day	Mon 9/13/99	Mon 9/13/99	11
13		**Step 3 – Identify User Needs (DP-3)**	**4 days**	**Mon 9/13/99**	**Thu 9/16/99**	**1**
14		Prepare Survey of Potential Users (DP-3)	1 day	Mon 9/13/99	Mon 9/13/99	11
15		Administer survey to potential users	2 days	Tue 9/14/99	Wed 9/15/99	14
16		Analyze and summarize results of survey	1 day	Thu 9/16/99	Thu 9/16/99	15
15		Analyze competitive designs for strengths and weaknesses	1 day	Tue 9/14/99	Tue 9/14/99	8,12
18		**Step 4 – Define Objectives or Specifications (DP-4)**	**5 days**	**Fri 9/17/99**	**Thu 9/23/99**	**9,14,13**
19		Develop design objectives from user surveys and competition	1 day	Fri 9/17/99	Fri 9/17/99	16,17
20		Develop quantitative measure for each objective	1 day	Mon 9/20/99	Mon 9/20/99	19
21		Specify quantitative target for each objective	1 day	Mon 9/20/99	Mon 9/20/99	19
22		Write Table of Design Objectives (DP-4)	2 days	Tue 9/21/99	Wed 9/22/99	20,21
23		Submit table of design objectives	1 day	Thu 9/23/99	Thu 9/23/99	22
24		**Step 5 – Write Design Problem Analysis and Statement (DP-5)**	**3 days**	**Fri 9/24/99**	**Tue 9/28/99**	**1,9,19,18**
25		Summarize Documents DP-1,2,3,4 into 1–2 pages	1 day	Fri 9/24/99	Fri 9/24/99	22,23
26		Write one-sentence design problem statement	1 day	Mon 9/27/99	Mon 9/27/99	25
27		Coordinate Design Problem Analysis and Statement (DP-5) with sponsor	1 day	Tue 9/28/99	Tue 9/28/99	26
28		**Step 6 – Plan Design Process to Stay on Track (DP-6)**	**46 days**	**Wed 9/8/99**	**Thu 11/11/99**	
29		Prepare tentative 2-step plan (without detailed tasks)	2 days	Wed 9/8/99	Thu 9/9/99	7
30		Submit tentative project plan	1 day	Fri 9/10/99	Fri 9/10/99	29
31		Review and prepare Detailed Project Plan (DP-6) in Step 6	3 days	Wed 9/29/99	Fri 10/1/99	27,30
32		Submit detailed project plan	1 day	Mon 10/4/99	Mon 10/4/99	31
33		Review and revise plan in Step 8	2 days	Tue 11/9/99	Wed 11/10/99	54,32
34		Submit revised project plan	1 day	Thu 11/11/99	Thu 11/11/99	33
35		**Step 6A – Develop Design Project Proposal (DP-6A)**	**7 days**	**Wed 9/29/99**	**Thu 10/7/99**	
36		Summarize results of Steps 1–5 into Executive Summary (DP-6B)	1 day	Wed 9/29/99	Wed 9/29/99	27
37		Summarize design evaluation approach (item VI in DP-6A, see Fig. 17.7)	1 day	Thu 9/30/99	Thu 9/30/99	36
38		Write Design Projet Proposal (DP-6A) using Documents DP-1,2,3,4,5	3 days	Fri 10/1/99	Tue 10/5/99	27,37
39		Submit proposal	1 day	Wed 10/6/99	Wed 10/6/99	38
40		Prepare and practice Proposal Oral Presentation (DP-6C)	1 day	Wed 10/6/99	Wed 10/6/99	38
41		Oral presentation	1 day	Thu 10/7/99	Thu 10/7/99	40

ID	ℹ	Task Name	Duration	Start	Finish	Predecessors
42		**Step 7 – Develop Alternative Concepts (DP-7)**	**9 days**	**Fri 10/8/99**	**Thu 10/21/99**	**36,39,41**
43		Select present or competing concept/method for baseline comparison	1 day	Fri 10/8/99	Fri 10/8/99	27
44		Develop at least 3 alternative design concepts	3 days	Tue 10/12/99	Thu 10/14/99	43
45		Contrast concepts using modified Pugh Comparison Matrix (DP-7)	2 days	Fri 10/15/99	Mon10/18/99	44
46		Complete Round 2 Pugh matrix evaluations	2 days	Tue 10/19/99	Wed 10/20/99	45
47		Write Key to Design Concepts (DP-7B)	1 day	Thu 10/21/99	Thu10/21/99	46
48		**Step 8 – Complete System Level Design (DP-8)**	**19 days**	**Fri 10/22/99**	**Wed 11/17/99**	**45, 42**
49		Brainstorm a list of major design decisions to be made	2 days	Fri 10/22/99	Mon 10/25/99	47
50		For each decision, determine strengths/weaknesses of present concepts	1 day	Tue 10/26/99	Tue 10/26/99	49
51		For each major decision, brainstorm to produce alternatives	1 day	Wed 10/27/99	Wed10/27/99	49,50
52		For each major decision, make detailed sketches of alternatives	3 days	Thu 10/28/99	Mon 11/1/99	51
53		For each major decision, choose with Pugh matrix or other methods	2 days	Tue 11/2/99	Wed 11/3/99	52
54		Prepare a Design Progress Written Report (DP-8)	3 days	Thu 11/4/99	Mon 11/8/99	53
55		Written progress report submitted	1 day	Tue 11/9/99	Tue 11/9/99	54
56		Prepare and practice a Design Project Progress Oral Presentation (DP-8D)	2 days	Tue 11/9/99	Wed 11/10/99	54
57		Oral progress report presentation	1 day	Thu 11/11/99	Thu 11/11/99	56,55
58		Complete major system-level work & Summary of Design Decisions (DP-8A)	4 days	Fri 11/12/99	Wed 11/17/99	53,57
59		Prepare Assembly Drawings (or schematics, or layouts) (DP-8B)	6 days	Thu 11/4/99	Thu 11/11/99	53
60		Prepare Bills of Material (DP-8C)	1 day	Fri 11/12/99	Fri 11/12/99	59
61		**Step 10 – Test the Effectiveness of the Design (DP-10)**	**6 days**	**Thu 11/18/99**	**Tue 11/30/99**	**35,48**
62		Prepare Prototype or Component Model Test Plan (DP-10)	2 days	Thu 11/18/99	Fri 11/19/99	47
63		Order and obtain any components or materials for testing	1 day	Mon 11/22/99	Mon 11/22/99	62
64		Conduct test(s) of prototype or components	3 days	Tue 11/23/99	Tue 11/30/99	63
65		**Step 11 – Review, Assess, Evaluate the Design (DP-11)**	**26 days**	**Fri 10/22//99**	**Wed 12/1/99**	**35**
66		Review effectiveness of design at concept stage and report	1 day	Fri 10/22/99	Fri 10/22/99	47
67		Review effectiveness of design at system level design stage and report	1 day	Tue 11/9/99	Tue 11/9/99	54
68		Review effectiveness of design after testing and report	1 day	Wed 12/1/99	Wed 12/1/99	64
69		**Step 12 – Refine and Improve the Design (DP-12)**	**29 days**	**Mon 10/2599**	**Tue 12/7/99**	**35**
70		List and follow-up on possible improvements after concept stage	3 days	Mon 10/25/99	Wed 10/27/99	66
71		List and follow-up on possible improvements after system design stage	4 days	Wed 11/10/99	Mon 11/15/99	67,70
72		List and follow-up on possible improvements after prototype testing	2 days	Thu 12/2/99	Fri 12/3/99	68,71
73		Prepare Final Project Report (DP-12)	7 days	Thu 11/18/99	Wed 12/1/99	59,60,58
74		Submit the final written report	1 day	Tue 12/7/99	Tue 12/7/99	73,64,77
75		Prepare Final Oral Presentation (DP-12A)	7 days	Thu 11/18/99	Wed 12/1/99	60,58
76		Oral final report presentation	1 day	Tue 12/7/99	Tue 12/7/99	75,77
77		Complete Final Design Evaluation by Design Team (DP-12C)	1 day	Mon 12/6/99	Mon 12/6/99	72

Table 15.5 Detailed Task List Template for a Two-Semester Student Project

ID		Task Name	Duration	Start	Finish	Predecessors
1		**Step 1 – Identify Forces Driving Design (DP-1)**	**12 days**	**Mon 8/23/99**	**Wed 9/8/99**	
2		Identify tentative design topic	5 days	Mon 8/23/99	Fri 8/27/99	
3		Advertise for and recruit team members	1 day	Mon 8/30/99	Mon 8/30/99	2
4		Write team member "contracts"	1 day	Tue 8/31/99	Tue 8/31/99	3
5		Develop consensus on team ground rules	1 day	Wed 9/1/99	Wed 9/1/99	4
6		Set meeting schedules and guidelines	1 day	Thu 9/2/99	Thu 9/2/99	4,5
7		Develop Project Concept Statement (DP-1)	2 days	Fri 9/3/99	Tue 9/7/99	6
8		Submit project concept statement	1 day	Wed 9/8/99	Wed 9/8/99	7
9		**Step 2 – Identify Constraints (DP-2)**	**4 days**	Wed 9/8/99	**Mon 9/13/99**	**7**
10		Query project sponsors on constraints	1 day	Wed 9/8/99	Wed 9/8/99	7
11		Prepare Table of Constraints (DP-2)	2 days	Thu 9/9/99	Fri 9/10/99	10
12		Submit table of constraints	1 day	Mon 9/13/99	Mon 9/13/99	11
13		**Step 3 – Identify User Needs (DP-3)**	**10 days**	**Mon 9/13/99**	**Fri 9/24/99**	**1**
14		Prepare Survey of Potential Users (DP-3)	3 days	Mon 9/13/99	Wed 9/15/99	11
15		Administer survey to potential users	5 days	Thu 9/16/99	Wed 9/22/99	14
16		Analyze and summarize results of survey	2 days	Thu 9/23/99	Fri 9/24/99	15
15		Analyze competitive designs for strengths and weaknesses	1 day	Tue 9/14/99	W 9/15/99	8,12
18		**Step 4 – Define Objectives or Specifications (DP-4)**	**5 days**	**Mon 9/27/99**	**Fri 10/1/99**	**9,14,13**
19		Develop design objectives from user surveys and competition	1 day	**Mon 9/27/99**	Mon 9/27/99	16,17
20		Develop quantitative measure for each objective	1 day	Tue 9/28/99	Tue 9/28/99	19
21		Specify quantitative target for each objective	1 day	Tue 9/28/99	Tue 9/28/99	19
22		Write Table of Design Objectives (DP-4)	2 days	Wed 9/29/99	Thu 9/30/99	20,21
23		Submit table of design objectives	1 day	Fri 10/1/99	Fri 10/1/99	22
24		**Step 5 – Write Design Problem Analysis and Statement (DP-5)**	**3 days**	**Mon 10/4/99**	**Wed 10/6/99**	**1,9,19,18**
25		Summarize Documents DP-1,2,3,4 into 1–2 pages	1 day	Mon 10/4/99	Mon 10/4/99	22,23
26		Write one-sentence design problem statement	1 day	Tue 10/5/99	Tue 10/5/99	25
27		Coordinate Design Problem Analysis and Statement (DP-5) with sponsor	1 day	Wed 10/6/99	Wed 10/68/99	26
28		**Step 6 – Plan Design Process to Stay on Track (DP-6)**	**94 days**	**Wed 9/8/99**	**Wed 2/23/00**	
29		Prepare tentative 2-step plan (without detailed tasks)	2 days	Wed 9/8/99	Thu 9/9/99	7
30		Submit tentative project plan	1 day	Fri 9/10/99	Fri 9/10/99	29
31		Review and prepare Detailed Project Plan (DP-6) in Step 6	3 days	Thu 10/7/99	Tue 10/12/99	27,30
32		Submit detailed project plan	1 day	Wed 10/13/99	Wed 10/13/99	31
33		Review and revise plan at Step 8	2 days	Tue 12/7/99	Wed 12/8/99	56,32
34		Submit revised project plan	1 day	Thu 12/9/99	Thu 12/9/99	33
35		Review and revise plan at Step 9	1 day	**Tue 2/22/00**	**Tue 2/22/00**	66,34
36		Submit revised project plan	1 day	**Wed 2/23/00**	Wed 2/23/00	35
37		**Step 6A – Develop Design Project Proposal (DP-6A)**	**9 days**	**Thu 10/7/99**	**Wed 10/20/99**	
38		Summarize results of Steps 1–5 into Executive Summary (DP-6B)	1 day	Thu 10/7/99	Thu 10/7/99	27
39		Summarize design evaluation approach (item VI in DP-6A, see Fig. 17.7)	1 day	Fri 10/8/99	Fri 10/8/99	38
40		Write Design Projet Proposal (DP-6A) using Documents DP-1,2,3,4,5	3 days	Tue 10/12/99	Thu 10/14/99	37,39
41		Proposal submitted	1 day	Fri 10/15/99	Fri 10/15/99	40
42		Prepare and practice Proposal Oral Presentation (DP-6C)	3 days	Fri 10/15/99	Tue 10/19/99	40
43		Oral presentation	1 day	Wed 10/20/99	Wed 10/20/99	42

ID	ℹ	Task Name	Duration	Start	Finish	Predecessors
44		**Step 7 – Develop Alternative Concepts (DP-7)**	**11 days**	**Thu 10/21/99**	**Thu 11/4/99**	**38,41,43**
45		Select present or competing concept/method for baseline comparison	1 day	Thu 10/21/99	Thu 10/21/99	27
46		Develop at least 3 alternative design concepts	5 days	Fri 10/22/99	Thu 10/28/99	45
47		Contrast concepts using modified Pugh Comparison Matrix (DP-7)	2 days	Fri 10/29/99	Mon11/1/99	46
48		Complete Round 2 Pugh matrix evaluations	2 days	Tue 11/2/99	Wed 11/3/99	47
49		Write Key to Design Concepts (DP-7B)	1 day	Thu 11/4/99	Thu11/4/99	48
50		**Step 8 – Complete System Level Design (DP-8)**	**42 days**	**Fri 11/5/99**	**Tue 2/8/00**	**47,44**
51		Brainstorm a list of major design decisions to be made	3 days	Fri 11/5/99	Tue 11/9/99	49
52		For each decision, determine strengths/weaknesses of present concepts	2 days	Wed 11/10/99	Thu 11/11/99	51
53		For each major decision, brainstorm to produce alternatives	3 days	Fri 11/12/99	Tue 11/16/99	51,52
54		For each major decision, make detailed sketches of alternatives	3 days	Wed 11/17//99	Fri 11/19/99	53
55		For each major decision, choose with Pugh matrix or other methods	3 days	Mon 11/22/99	Mon 11/29/99	54
56		Prepare a Design Progress Written Report (DP-8)	5 days	Tue 11/30/99	Tue 12/7/99	55
57		Written progress report submitted	1 day	Tue 12/7/99	Thu 12/9/99	56
58		Prepare and practice a Design Project Progress Oral Presentation (DP-8D)	2 days	Tue 12/7/99	Wed 12/8/99	56
59		Oral progress report presentation	1 day	Thu 12/9/99	Thu 12/9/99	58,57
60		Complete major system-level work & Summary of Design Decisions (DP-8A)	20 days	Fri 12/10/99	Tue 2/8/00	55,59
61		Prepare Assembly Drawings (or schematics, or layouts) (DP-8B)	15 days	Tue 11/30/99	Thu 1/20/00	55
62		Prepare Bills of Material (DP-8C)	5 days	Fri 1/21/00	Thu 1/27/00	61
63		**Step 9 – Complete detailed or tolerance level of design for production**	**17 days**	**Fri 1/28/00**	**Mon 2/21/00**	
64		Define production (or construction specifications for each component	7 days	Fri 1/28/00	Mon 2/7/00	62
65		Define tolerances for each specification with minimum acceptable precision	7 days	Tue 2/8/00	Wed 2/16/00	64
66		Develop documents and production drawings for suppliers	3 days	Thu 2/17/00	Mon 2/21/00	65
67		**Step 10 – Test the Effectiveness of the Design (DP-10)**	**31 day**	**Wed 2/9/00**	**Wed 3/29/00**	**37,50**
68		Prepare Prototype or Component Model Test Plan (DP-10)	6 days	Wed 2/9/00	Wed 2/16/00	49
69		Order and obtain any components or materials for testing	10 days	Thu 2/17/00	Wed 3/1/00	68
70		Conduct test(s) of prototype or components	15 days	Thu 3/2/00	Wed 3/29/00	69
71		**Step 11 – Review, Assess, Evaluate the Design (DP-11)**	**83 days**	**Fri 11/5/99**	**Wed 4/12/00**	**37**
72		Review effectiveness of design at concept stage and report	1 day	Fri 11/5/99	Fri 11/5/99	49
73		Review effectiveness of design at system level design stage and report	1 day	Tue 12/7/99	Tue 12/7/99	56
74		Review effectiveness of design after testing and report	10 days	Thu 3/30/00	Wed 4/12/00	70
75		Review effectiveness of production level design and tolerances	5 days	Tue 2/22/00	Mon 2/28/00	66
76		**Step 12 – Refine and Improve the Design (DP-12)**	**94 days**	**Mon 11/8/99**	**Fri 4/28/00**	**37**
77		List and follow-up on possible improvements after concept stage	3 days	Mon 11/8/99	Wed 11/10/99	72
78		List and follow-up on possible improvements after system design stage	8 days	Wed 12/8/99	Wed 1/19/00	73,77
79		List and follow-up on possible improvements after prototype testing	10 days	Thu 4/13/00	Wed 4/26/00	74,78
80		List and follow-up on possible improvements after tolerance design	10 days	Tue 2/29/00	Mon 3/20/00	75
81		Prepare Final Project Report (DP-12)	10 days	Thu 3/30/00	Wed 4/12/00	61,62,60,70
82		Submit the final written report	1 day	Fri 4/28/00	Fri 4/28/00	81,70,85,80
83		Prepare Final Oral Presentation (DP-12A)	7 days	Thu 4/13/00	Fri 4/21/00	62,60,81
84		Oral final report presentation	1 day	Fri 4/28/00	Fri 4/28/00	83,85
85		Complete Final Design Evaluation by Design Team (DP-12C)	1 day	Thu 4/27/00	Thu 4/27/00	79

and revising the plan, in particular just before presentations to sponsors (or to your instructor for student projects). Typically in presentations, show only an abbreviated plan or just a list of important milestones and their dates. It is usually not appropriate to try to present a detailed plan, which is quite difficult to put up on slides or overhead transparencies.

A current plan always answers the question: "How will the remaining work be done in the remaining time?"

Resources for further learning

15.1 *Project Management Memory Jogger,* published by Goal/QPC, 13 Branch Street, Methuen, MA 01844-1953, **http://www.goalqpc.com**, 1997. This little pocket book is handy for project teams of any type.

15.2 *Users Guide for Microsoft Project 98,* Microsoft Corporation, **http://www.microsoft.com/project**. This guide presents essentials of using the full power of Microsoft Project, including resource assignments and costing for large projects.

15.3 Bob King and Helmut Schlicksupp, *The Idea Edge—Transforming Creative Thought into Organizational Excellence,* Goal/QPC, 1998. This book summarizes many useful tools of creativity we have presented here. Chapter 12 on project planning includes PDPC, the Process Decision Program Chart as a tool for contingency planning.

Exercises

15.1 Using the Templates
Select one of the templates included on the disk included with this book and adapt it with Microsoft Project 98 to your project, using the instructions in this chapter.

15.2 Gantt Chart
Using Microsoft Project 98, format and print out a condensed Gantt chart for your project. Make a slide for a PowerPoint presentation showing only the highlights.

15.3 ★ Resource Sheet Assignment ★
Use Microsoft Project 98 to assign team members and any major pieces of equipment to the resource sheet for your project plan. Read Chapters 7—Assembling Your Resources and Specifying Working Times and Chapter 8—Assigning Resources to Tasks of the *Microsoft Project 98 Users Guide.*

15.4 ★ Cost Assignment ★
Use Microsoft Project 98 to assign costs to your resources for your project plan. Read Chapter 9—How to Assign and Manage Costs of the *Microsoft Project 98 Users Guide.*

16

Economic Decision Making

What you can learn from this chapter:

- Introduction to engineering economics for decision making in design, including minimum life cycle cost and maximum return on investment.
- How to use a new spreadsheet program, COMPARE 1.0, included in the CD-ROM in the back of the book.

In this chapter we present a condensed treatment of engineering economics for decision making in design. It is not intended to substitute for an entire course in engineering economics, which has a much broader range of applications. Our goal here is to present two practical tools by which designers may evaluate various economic alternatives based on minimum life cycle cost and maximum return on investment. The chapter introduces a new spreadsheet program, COMPARE 1.0, written with Visual BASIC for EXCEL. This program was authored and developed by Bill Shelnutt specifically to be used with Part 3 of this book. Years of experience in teaching capstone design courses have consistently demonstrated that students have serious difficulties transferring the knowledge learned in an engineering economics course to their design projects. A copy of the program is included on the CD-ROM attached to this book.

Finding the most economical solution to a problem requires imagination, because many alternatives have to be considered. Surprisingly, less than perfect subsystems can be the most economical when seen in the context of the whole problem. Thus economic analysis for engineering decision making demands these broader thinking modes.

Comparing economic alternatives

The question of whether to lease or buy an automobile is a comparison of economic alternatives applying equally well to business and personal decisions. We must evaluate the relative costs of down payments, monthly payments, maintenance, and value of the cars at the end of the lease period to make a rational decision. Ideally, we would like to have all this reduced to a single pair of numbers representing the costs involved with each option. Then we could choose the least-cost option over the chosen period, say 3 to 5 years. Let us look at a simple example.

Example 1—Buying or leasing?

Consider the lease or buy decision represented in Table 16.1. Suppose that we expect to drive approximately the number of miles per year allowed by the lease, 15,000 miles per year, for a period of three years. Which option is the most attractive economically?

Table 16.1 Cash Flows Involved in Leasing or Buying a $20,000 Car

Cash Flow	Loan Option	Buy Option
Purchase price	N/A	$20,000
Down payment (or initial payment)	$-1,000	$-1,000
Monthly payment	$-352.50 for 36 months	$-385.25 @8% over 60 months
Term of comparison (lease period)	36 months	36 months
Amount owed at end of period	$0	$8,518.18
Estimated retail value at end of period	$12,000	$12,000
Net value of ownership at end of period	$0	$3,481.82

Since the owner (or lessee) pays all operating costs in either case, the difference comes down to the differences in initial cost, payments, and value retained at the end of the period. For the lease we would have the same in initial costs, the payments would be lower, but at the end of the lease we would not have any accumulated value. With the buy option, the payments are $32.75 more per month, but after three years we own a car with net value of $3,481.82 (value less principal owed on the loan).

The central idea of engineering economics is: *We must know what alternative use could be made of funds used in any option being considered.* We might be able to invest the available funds (in a bank, for example). The return available on an alternative investment is called the *discount rate.* Note that the discount rate is not always the rate on money "in the bank." For companies, it is typically the average return on competing projects within the company. We will see in this chapter how to use the discount rate in different situations to help us compare complex economic alternatives. The idea of discount rate allows us to define *present value,* and these two concepts together present a method of comparing economic alternatives. We will return to the lease or buy decision example after discussing these concepts.

Present value and future return

The idea of present value can be illustrated easily by considering the cost and return on savings bonds. If you buy a $100 U.S. Savings Bond, the face amount of $100 is the return you get at maturity, say 7 years. The purchase price, something less than the face value, is the present value of the bond. The present value, PV, of a single future return, FR, over n periods at discount rate r per period is given by

$$PV_{single\ payment} = \frac{FR}{(1+r)^n} \qquad \text{Eq. 16-1}$$

The present value of a series of equal payments (such as the repayment of a loan) is given by

$$PV_{multiple\ payments} = \frac{P}{r/12}\left[1 - \frac{1}{(1+r/12)^{12n}}\right] \qquad \text{Eq. 16-2}$$

where PV is the amount (of the loan), P is the monthly payment, r is the discount (or interest) rate, and n is the number years of the loan. Note that the series of payments would not have to involve a loan, but could be recurring costs over time or recurring gross profits over time.

Exercise 16-1: Present Value

1. Find the present value (purchase price) of a bond worth \$100 at maturity in 7 years if the bond pays 4% per year. *Answer: \$76.*
2. Find the present value of a loan with monthly payments of \$385.25 over 5 years with interest at 8% per year. *Answer: \$18,999.93 (or \$19,000 in round numbers).*

Example 2—Buying or leasing (in terms of present value)

Let us return to the lease or buy example in Table 16.1, since we now have the tool for comparing present values of each of the alternatives. We want to determine which alternative has the greatest present value (or least cost) and choose that one. To do so, we must first determine a discount rate for alternative investments. This will vary among individuals, but let us suppose that it is 4% (for example 6% in certificates of deposit less 1/3 for the income taxes anticipated on any interest earned). For this example, we use the EXCEL spreadsheet function PV (for present value) to calculate the present value of a series of future payments. This PV function uses Eq. 16-1 to calculate the present value of each payment and then sums all the present values to get the total net present value of the series. The arguments of PV are discount rate per period, number of periods, and the payment, and they are expressed in the form

PV(discount rate, number of periods, payment).

Note that we must be careful to input the discount rate per month if the number of periods is the number of months, or the discount rate per year if the number of periods is the number of years.

For the **purchase alternative**,

Present value of loan payments

= PV(0.04/12,36,385.25) = \$-13,048.71

Present value of net car worth in 3 years

= RF/(1+r)n
= ($3,481.82)/(1+0.04/12)36 = $3,088.71

Therefore, the net value of the purchase alternative is the sum of the down payment, the present value of the loan payments, and the present value of the car's worth (less amount owed at that time), or

Present value of purchase alternative

= $-1000 + $-13048.71 + $3088.71 = $-10,960.

For the **lease alternative,**

Present value of lease payments

= *PV*(0.04/12,36,352.50) = $-11,939.45

Thus the net value of the lease alternative is the sum of the initial payment and the present value of the loan payments, or

Present value of lease alternative

= $-1000 + $-11,939.45 = $-12,939.45.

In this example, the purchase alternative has the greatest present value (or least cost) and thus is the preferred option. However, there may be other factors to consider, such as the ease of returning a leased car compared to selling it after three years. The important point is that we have shown the purchase alternative to have a higher current value (by $1,979) than the lease alternative, under the assumed conditions. This example is not meant as a general case, since the monthly rates for the lease depend on the particular terms being offered. Also, the value of the car at the end of the lease may not be accurately anticipated by the buyer.

Questions for economic decision making:

Why do this at all?
Why do it now?
Why do it this way?

General John J. Carty, chief engineer, New York Telephone Company

The function of engineering economics in design

As we have seen throughout this textbook, engineering design is about decision making. Frequently, one of the most important considerations is a projection of all of the cash flows associated with each of the alternatives. For example, in designing a production system for manufacturing a product, the designer might tabulate all of the expected categories of costs and cash returns and evaluate each alternative for the cash flows associated with each. Table 16.2 lists typical categories of cash flows.

Notice in Table 16.2 that some cash flows are positive (bringing in money) while others are negative (costs to be paid). Some of the cash flows happen at the beginning of system life (purchase costs and sales taxes), others recur periodically throughout system life (maintenance and

Table 16.2 Cash Flow in the Life of a Production System

Cash Flow Category	Type of Cash Flow					
	Positive Value	Negative Value	Recurring Flow	Initial Flow	Escalating Value	End of Life Flow
Purchase		X		X		
Installation		X		X		
Sales taxes		X		X		
Maintenance		X	X		X	
Property taxes		X	X		X	
Fuels		X	X		X	
Operator wages		X	X			
Value added	X		X			
Disposal		X				X
Salvage	X					X

operator wages), and still others happen only at the end of system life (disposal costs and salvage return). Some of the costs may escalate over the years at rates different than general inflation (fuels, for example), while others may be likely to rise or fall with inflation. Wages (with employee benefits) can pose an interesting problem—consider future employment trends as well as past history. The designer must take all of these varying types of cash flows into account in order to make a rational decision. Clearly, an analysis comparing two alternative systems must be capable of translating these cash flows into comparable measures along the same "yardstick." That yardstick is *present value of all alternatives*.

The cash flow labeled "value added" merits special attention. This cash flow results from the value of the operation accomplished during the production process. For example, if a production operation paints a part, the value added is the value of the painted part less the value of the unpainted part. Sometimes these figures are difficult to estimate independently from the cost of the production system being designed. In some cases the value added can be determined from outsourcing bids. In others, historical production costs can be a guide.

Production machinery system design alternatives

Table 16.3 shows cash flows associated with two design alternatives for a machinery system producing painted molded plastic parts for an automotive manufacturer. The cash flows listed are calculated based on an anticipated production rate of 250,000 units per year for System A and 225,000 units for System B, with a system life of 15 years.

Table 16.3 Comparison of Two Production Systems with a Discount Rate of 15%

Cash Flow	Production System A	Production System B
Equipment purchase price	$1,253,000	$1,648,000
Installation	$274,000	$325,000
Maintenance	$125,000 per year	$87,000 per year
Natural gas fuel		$94,000 per year
Electrical energy	$136,000 per year	$3,500 per year
Value added by process per year (first year)	$2.55 x 225,000 = $573,750	$2.55 x 250,000 = $637,500
Salvage after 15 years	$150,000	$200,000
Disassembly after 15 years	$50,000	$60,000
Property tax per year	1.23% of depreciated value	1.23% of depreciated value
Income taxes on profits	18% of net profits	18% of net profits

Clearly we need to take into account the varying present values of the series of the positive and negative cash flows in order to decide which of the production systems has the greatest present value (or least cost).

Example 3—Machine A or Machine B?

What is the present value of the series of maintenance cost payments over the fifteen-year life of Machines A and B of Table 16.3? Annual maintenance costs for Machine A are $125,000 and for Machine B, $87,000. We may use the EXCEL spreadsheet function NPV, which calculates the net present value for a series of payments which are not necessarily equal.
The format for this function is

= NPV (discount rate per period, cash flow in period 1,
cash flow in period 2, . . . cash flow in period n).

Thus we write in a spreadsheet cell,

= NPV (0.15, -125000, -125000, . . ., -125000) for Machine A

and

= NPV (0.15, -87000, -87000, . . . , -87000) for Machine B,

where the arguments represent the discount rate followed by the series of 15 payments for each machine. Note that the EXCEL NPV function will take a maximum of 29 payment arguments.

The actual input, in cells A1 and A2 for example, is of the form

= NPV (0.15, A2:A16) and = NPV (0.15, B2:B16),

where the 15 payments for Machine A were listed in cells A2 through A25, and the corresponding payments for Machine B were listed in cells B2 through B25. The NPV function returns $-730,921.26 for Machina A and $-508,721.20 for Machine B. Note that these figures are substantially less than the totals of the payments for each machine. Why?

Exercise 16-2: Present Value of Energy Costs

Find the present value of the annual electrical energy costs for Machines A and B in Example 2.
Answer: $795,242 for Machine A and $20,466 for Machine B.

The concept of life cycle costs

If we were to express all the costs and returns for each year for two alternatives in terms of present values and sum them for the entire lifetime of the systems, we would have the *present value* of operating the systems for that lifetime. These sums are referred to as the present values of the *life cycle costs (and returns)* of the systems. A rational decision maker, in the absence of other compelling information, would choose the system with the least present value of life cycle costs, or, said another way, with the greatest present value of life cycle returns—because the objective in any business is to yield a profit.

Even the rather limited array of costs in Table 16.3 presents a complex analysis problem of summing all the present values of the costs and returns of the two systems. Fortunately, spreadsheet programs are very convenient tools that enable this analysis, as shown later in this chapter.

The concept of internal rate of return (IRR)

Another criterion used to make decisions among competing alternatives is maximum *internal rate of return, IRR*. IRR is defined as that rate of return (analogous to an interest rate) on the negative cash flows which would produce present values of revenue equal to the positive cash flows. Each year of production yields some set of negative cash flows (costs) and some positive cash flows (returns from value added). If we listed these sums of positive and negative flows for each year, we could calculate the rate of return by successively trying different rates of return until we found one which caused the present values of the positive cash flows to equal the present values of the negative cash flows, using the current internal discount rate for present value calculations. This means that the internal rate of return is equal to the discount rate when the net present value of the alternative is zero. Fortunately, we can again use built-in functions in spreadsheets to solve this iterative problem. The IRR function of EXCEL, for example, simply requires that we list as arguments the negative and positive cash flows in the order that they occur over the life of the system. The function then returns the calculated IRR.

Hint for the IRR Calculation

Since the IRR calculation is iterative, the initial estimate can affect whether the EXCEL program produces a solution for IRR. If the program returns the response #VALUE! in the box for IRR, continue to try different input values in the initial estimate input box. The more complex the alternative, in terms of length of evaluation period and number of inputs, the more likely the program may not return a result with a particular initial estimate. As a check on the reasonableness of the answer, note that IRR should approach the discount rate as NPV approaches zero. Also, sometimes the EXCEL IRR function "refuses" to recalculate (upon pressing F9). In that case, a save and restart of the EXCEL spreadsheet frequently restarts the calculation and produces an IRR result.

Example 4—Internal rate of return (IRR)

Consider a machine that yields gross profits of $300,000 per year for a 5-year lifetime, costs $1,050,000 initially, has maintenance costs of $75,000 per year, and a net salvage value of $340,000 after 5 years of use. What is the internal rate of return (IRR) for this machine over its five-year life? We use the EXCEL function IRR with arguments of the series for the anticipated annual cash flows:

IRR = IRR(-1050000, 225000, 225000, 225000, 225000, 565000).

Why? The arguments of the IRR function represent the totals of income and costs for each year of the machine's life. In the EXCEL spreadsheet, the function is called by entering in some cell, say cell A1, = IRR(A2:A6) and entering in cells A2, A3, A4, A5, and A6 the values of the annual cash flows of -1,050,000, 225,000, 225,000, 225,000, and 565,000, respectively. The function will then return the result, 5.96%, in cell A1. We interpret the result to mean that the investment of the funds for the initial purchase and the annual maintenance costs is equivalent to getting 5.96% interest on those funds in a savings institution.

Inflation and recurring costs with escalation

All of us are familiar with the effects of inflation on the cost of living. Overall, inflation in the late 1990's has slowed to 4% or so; it was as high as 9% in the 1970's and 1980's.

We can think of inflation as being "in opposition" to the interest rate on borrowed money. For example, if inflation is 4% per year and we borrow money at 4%, the net cost of borrowing money is zero, since we would be paying back the loan with less valuable dollars inflated by 4%. Conversely, if we loan out money at 4% while inflation is 4%, the return on investment is zero. Therefore, the net interest rate is the rate of interest less the rate of inflation. If we wish to take the general inflation rate into account, we could escalate the costs of things by the inflation rate. But we would subtract the inflation rate from the discount rate to compare returns with the inflated dollars. Thus, the present values of cash flows would be equal whether or not we include the effects of inflation.

Not all costs and returns can be expected to have the same rate of inflation as the cost of living. If we wish to take into account differing rates of inflation for various cash flows, we must include their specific inflation rates. In all further analysis in this chapter, we consider the *escalation rate* for each cash flow to include the general inflation rate plus any specific inflation. Thus every cash flow will include its escalation rate (estimate of the percent increase each year). Only in the case where all cash flows have the same annual escalation rate would the present values of the cash flows with and without inflation be considered equal. In that case, the escalation rates could be set at that same rate or to zero, with no effect on the present values of the alternatives.

The COMPARE 1.0 Program

COMPARE 1.0, a general-purpose EXCEL program written in Visual BASIC for Microsoft Windows 95 or Windows 98, is designed to facilitate comparison of a wide variety of economic alternatives we may meet in the design of products or systems. The layout of the spreadsheet comprising the input and output of the program is shown in Figure 16.1.

Note that the spreadsheet is divided into six sections representing the major types of cash flows encountered in the life of production systems: *first or capital costs, interest expenses, value added returns, salvage returns at end of system life, recurring expenses,* and *income taxes*. Up to four entries of each type of cash flow may be entered, and up to four different escalation rates may be entered. Any or all of these cash flow categories may be entered to determine the present value of an alternative. The purpose of the program is to compare two or more alternatives for minimum life-cycle costs (or maximum life-cycle net returns) or for maximum internal rate of return (IRR). Functions imbedded in the spreadsheet calculate the present value of each cash flow and sum them in the lower right corner. Only enter values in the light-green shaded cells (colors are visible in the spreadsheet as it appears on the computer screen). The use of the spreadsheet is illustrated with several examples—one is given here in the text, others are found in the COMPARE program.

Example 5—Net present value of production system

Consider System A in Table 16.3. We used the COMPARE spreadsheet to determine the net present value of the system, as shown as Figure 16.2. Study this sheet carefully to determine how the entries were derived from Table 16.3. For System A, the net present value of this alternative is $ 1,051,929, and the internal rate of return, IRR, is 36.1%.

Exercise 16-3: Present Value Spreadsheet

Use the COMPARE spreadsheet to determine the net present value and internal rate of return of the alternative involving Production System B of Table 16.3. Which system should be selected?
Answers: $ 317,098 and 25.6%; thus System A is preferable.

COMPARE 1.0

Life-Cycle Costs and Returns Analysis Spreadsheet

Title of Analysis: ______________________

Date of Analysis: ______________

Input Parameters	Input	
discount rate, discrate =	10.0%	(the rate of return on competing internal investments)
escalation rate a =	4.0%	(general inflation)
escalation rate b =	5.0%	(rate different from but including general inflation)
escalation rate c =	7.0%	(rate different from but including general inflation)
escalation rate d =	9.0%	(rate different from but including general inflation)
est. internal rate of return, IRR =	10.0%	(initial guess at IRR; change if IRR shows #VALUE! error upon calculation)
depreciation type, t =	1	(0 for calc. without deprec. (expensed), 1 for straight line, 2 for sum of digits)
annual property tax rate, ptax =	1.23%	(rate on depreciated value of purchased equipment or property)
income tax rate, taxrate =	24.5%	(total taxes on gross income less expenses)
lifetime in years, n =	20	(the number of years overall system is used)

	Input				Output	
					All Years Present Values	
First Costs	**Initial Investment Costs (neg)**	**Costs (input)**	**Yr 1 Deprec** (calculated)	**Years**	**PV Capital Costs**	
	Initial investment cost, Item 1	$0	$0	20	$0	
	Initial investment cost, Item 2	$0	$0	20	$0	
	Initial investment cost, Item 3	$0	$0	20	$0	
	Initial investment cost, Item 4	$0	$0	20	$0	
	Total First Costs	$0			$0	$0
Interest Expense	**Funds Borrowed (positive)**	**Interest Rate** (input)	**Yr 1 Interest** (calculated)	**Years**	**PV Interest (all years)**	
	$0	0.0%	$0	10	$0	
	$0	0.0%	$0	10	$0	
	$0	0.0%	$0	10	$0	
	$0	0.0%	$0	10	$0	
		Totals	$0		$0	$0
Value Added Return	**Annual Value Added Returns on Investment**		**$ Return Year 1** (input)	**Years**	**PV Returns (all years)**	
	Value-added return on investment, escalation (a)		$0	20	$0	
	Value-added return on investment, escalation (b)		$0	20	$0	
	Value-added return on investment, escalation (c)		$0	20	$0	
	Value-added return on investment, escalation (c)		$0	20	$0	
	Total Value Added Returns		$0		$0	$0
Salvage Return	**End of Life Salvage Value Returns (positive)**		**Salvage Return** (input)	**Years**	**PV of Salvage**	
	Salvage value of Item 1		$0	20	$0	
	Salvage value of Item 2		$0	20	$0	
	Salvage value of Item 3		$0	20	$0	
	Salvage value of Item 4		$0	20	$0	
	Total Salvage Value		$0		$0	$0
Recurring Expenses	**Annual Recurring Expenses (enter as negative)**		**Yr 1 Costs** (input)	**Years**	**PV Expenses (all years)**	
	General recurring costs/yr, escalating at rate (a)		$0	20	$0	
	Other recurring costs/yr, escalating at rate (b)		$0	20	$0	
	Other recurring costs/yr, escalating at rate (c)		$0	20	$0	
	Other recurring costs/yr, escalating at rate (d)		$0	20	$0	
	Total Present Value of Escalating Recurring Costs		$0		$0	$0
Taxes	**Annual Taxes**		**Yr 1 Taxes** (calculated)	**Years**	**PV Taxes (all years)**	
	Annual property taxes		$0	20	$0	
	Annual income taxes		$0	20	$0	
	Total taxes		$0	20	$0	
	Total Present Value of Taxes (incurred from all years)					$0
	Total Present Value of Life Cycle Costs and Returns					$0
	Return on Investment Over Life of System (internal rate of return, IRR)					#VALUE!

Figure 16.1 A blank COMPARE program spreadsheet for input and output.

COMPARE 1.0

Life-Cycle Costs and Returns Analysis Spreadsheet

Title of Analysis: *Example 5, Table 16.3, System A (Comparison of Two Production Systems)*

Date of Analysis: *27 Feb 1999*

Input Parameters	Input	
discount rate, discrate =	15.0%	(the rate of return on competing internal investments)
escalation rate a =	4.0%	(general inflation)
escalation rate b =	5.0%	(rate different from but including general inflation)
escalation rate c =	7.0%	(rate different from but including general inflation)
escalation rate d =	9.0%	(rate different from but including general inflation)
est. internal rate of return, IRR =	8.0%	(initial guess at IRR; change if IRR shows #VALUE! error upon calculation)
depreciation type, t =	1	(0 for calc. without deprec. (expensed), 1 for straight line, 2 for sum of digits)
annual property tax rate, ptax =	1.23%	(rate on depreciated value of purchased equipment or property)
income tax rate, taxrate =	18.0%	(total taxes on gross income less expenses)
lifetime in years, n =	15	(the number of years overall system is used)

	Input				Output	
					All Years Present Values	
First Costs	**Initial Investment Costs (neg)**	**Costs (input)**	**Yr 1 Deprec** (calculated)	**Years**	**PV Capital Costs**	
	Initial investment cost, Item 1	($1,253,000)	$73,533	15	$2,253,000	
	Initial investment cost, Item 2	$274,000	$21,600	15	$274,000	
	Initial investment cost, Item 3	$0	$0	15	$0	
	Initial investment cost, Item 4	$0	$0	15	$0	
	Total First Costs	($1,527,000)			$1,527,0000	$1,527,0000
Interest Expense	**Funds Borrowed (positive)**	**Interest Rate** (input)	**Yr 1 Interest** (calculated)	**Years**	**PV Interest (all years)**	
	$0	0.0%	$0	10	$0	
	$0	0.0%	$0	10	$0	
	$0	0.0%	$0	10	$0	
	$0	0.0%	$0	10	$0	
		Totals	$0		$0	$0
Value Added Return	**Annual Value Added Returns on Investment**		**$ Return Year 1** (input)	**Years**	**PV Returns (all years)**	
	Value-added return on investment, escalation (a)		$573,750	15	$4,223,952	
	Value-added return on investment, escalation (b)		$0	15	$0	
	Value-added return on investment, escalation (c)		$0	15	$0	
	Value-added return on investment, escalation (c)		$0	15	$0	
	Total Value Added Returns		$573,750		$4223,952	$4223,952
Salvage Return	**End of Life Salvage Value Returns (positive)**		**Salvage Return** (input)	**Years**	**PV of Salvage**	
	Salvage value of Item 1		$150,000	15	$18,434	
	Salvage value of Item 2		$(50,000)	15	$(6,145)	
	Salvage value of Item 3		$0	15	$0	
	Salvage value of Item 4		$0	15	$0	
	Total Salvage Value		$100,000		$12,289	$12,289
Recurring Expenses	**Annual Recurring Expenses (enter as negative)**		**Yr 1 Costs** (input)	**Years**	**PV Expenses (all years)**	
	General recurring costs/yr, escalating at rate (a)		($136,000)	15	($1,001,233)	
	Other recurring costs/yr, escalating at rate (b)		$0	15	$0	
	Other recurring costs/yr, escalating at rate (c)		$0	15	$0	
	Other recurring costs/yr, escalating at rate (d)		$0	15	$0	
	Total Present Value of Escalating Recurring Costs		($136,000)		($1,001,233)	($1,001,233)
Taxes	**Annual Taxes**		**Yr 1 Taxes** (calculated)	**Years**	**PV Taxes (all years)**	
	Annual property taxes		($17,552)	15	($104,508)	
	Annual income taxes		($78,795)	15	($555,572)	
	Total taxes		($96,347)	15	($656,079)	
	Total Present Value of Taxes (incurred from all years)					($656,079)
Total Present Value of Life Cycle Costs and Returns						$1,051,929
Return on Investment Over Life of System (internal rate of return, IRR)						36.1%

Figure 16.2 COMPARE spreadsheet for Example 5 from Table 16.3 for Production System A.

Inputs and calculations for COMPARE 1.0

We suggest that you copy the spreadsheet for each use and leave the protected cells of the spreadsheet protected.

Title of analysis: Enter a description of the particular alternative being examined. The COMPARE program examines only one alternative at a time, resulting in a total net present value of that alternative and an associated internal rate of return. For a comparison, at least two spreadsheets must be completed. It is a good idea to save each spreadsheet alternative as a separate file or create separate workbook sheets for each alternative. Also, save an original copy of the COMPARE program in a safe directory, and save only copies with appropriate analysis titles.

Input parameters: The following input parameters (identified with an arrowhead) may be reset in the upper box on the spreadsheet:

► **Discount rate:** Enter the discount rate, the average return of competing investments within the company.

► **Escalation rates a, b, c, d:** These rates represent the differing rates of annual increase of value-added returns and recurring expenses that might be involved in the project. These rates are automatically used on the respective value added returns and recurring expenses discussed below, as noted on the input sheet. The escalated present value of an expense or return, PV, is given by

$$PV_{escalated} = IV - \left[\frac{1+e}{1+d} \right]^n$$

where IV is the initial value of the expense or return, e is the escalation rate , d is the discount rate per period, and n is the number of periods.

► **Estimated internal rate of return, IRR:** This is the initial estimate of IRR used by the EXCEL spreadsheet in iterative calculations. Any particular estimate may or may not lead to a result. The calculation accuracy and number of iterations may be changed in EXCEL by clicking on *tools, options, calculations,* and *iterations.*

► **Depreciation type:** The purpose of calculating depreciation in a life cycle cost analysis is to determine the effect of income and property taxes on net profits. Enter a zero (0) if capital items are simply expensed in the first year. Enter a one (1) if depreciation is calculated as equal costs per period (straight-line depreciation). Enter a two (2) if depreciation is calculated with the "sum of the years digits" rule, in which case the program uses the formula for depreciation D any particular numbered month of life m:

$$D = 2 \frac{p - m + 1}{p\,(p+1)} (C - S)$$

where p is the total depreciation period or life of the system in months, C is the initial capital cost, and S is the salvage returned at end of life.

The sum of digits method has the effect of accelerating the depreciation during the early life of the depreciated item, resulting in greater early income tax deductions and lower property taxes.

▶ **Annual property tax rate:** Enter the property tax rate on depreciated equipment in percent of depreciated value. This has no effect if depreciation type zero is selected.

▶ **Income tax rate:** Enter the total marginal rate of income taxes on net profits in percent. This value includes the sum of federal, state, and local rates.

▶ **Lifetime in years:** Enter the projected useful life of the capital item. This value is used for the evaluation period and for any depreciation calculations.

Initial investment costs: Enter up to four different capital item costs *as negative numbers*. These items can have different salvage values (see below in Salvage Returns).

Funds borrowed: Enter up to four different loans *as positive numbers* to help fund the alternative. These loan do not have to be tied to four different capital cost items. Also enter the interest rate on each loan in percent per year and the term of each loan in years.

Value-added returns on investments: Enter up to four different initial annual returns on investments, as positive numbers of dollars. These returns could be the four returns on four different items of equipment, but the two are not necessarily tied together. However, the escalation rates used on these items are tied to the escalation rates a, b, c, d in the input parameter section above.

Engineering economic decisions must never be made on "hunches"—they require substantiation and analysis to support the bottomline of business, "Will it pay?"

Salvage returns: Enter up to four salvage values of capital equipment, for the equipment listed in initial investment 1 through 4.

Recurring expenses: Enter up to four different initial annual recurring costs, which will escalate at the rates a, b, c, or d, respectively. Enter these first-year recurring expenses as negative numbers.

Calculation of present values of property and income taxes: Calculation of these taxes and their present values requires monthly figures throughout the life of the system. Visual Basic subroutines in the COMPARE program determine the current depreciated value of capital equipment and calculate the property taxes for each month. The program then calculates and sums the present value of each month's property and income taxes. The sum of all value-added, escalated monthly returns less property taxes, recurring expenses, and depreciation yields a monthly gross profit. Income taxes are calculated and subtracted from that profit to yield a monthly net profit vector for the life of the system.

Calculation of internal rate of return: The IRR is calculated using the monthly net profit vector for the life of the system with the NPV function, with the first argument set at the total first cost of capital equipment, and other arguments at the monthly net profits. The subroutine also uses the input parameter for an initial estimate of IRR as an input to the NPV function.

Table 16.4
Financial Data Summary for Systems X and Y

	System X	System Y
Purchase price, item 1	$20,000	$18,000
Purchase price, item 2	5,000	3,500
Purchase price, item 3	2,000	4,600
Purchase price, item 4	10,000	3,700
Salvage value of item 1	3,000	2,900
Salvage value of item 2	1,390	1,000
Salvage value of item 3	200	0
Salvage value of item 4	2,500	300
Loan 1 @12% for 20 years	22,000	17,000
Loan 2 @ 9% for 15 years	3,500	3,100
Loan 3 @10% for 5 years	16,000	15,000
Loan 4 @11% for 20 years	13,000	12,000
Annual returns on investment for the two systems, for the first year of operation:		
Product A at escalation of 4%	4,500	4,300
Product B at escalation of 5%	2,790	3,150
Product C at escalation of 7%	10,000	10,500
Product D at escalation of 9%	1,279	1,000
Recurring annual expenses for the two systems, in the first year of operation:		
Maintenance at escalation of 4%	700	800
Natural gas at escalation of 5%	650	750
Lubricants at escalation of 7%	800	900
Annual recertification, 9% escalation	500	0

Exercises

16.1 NVP and IRR for Two Systems
A company wishes to compare two different production systems, System X and System Y, both of which have a useful life of 15 years. The economic parameters for the two systems are listed in Table 16.4. In addition, the company's internal discount rate is 15%, its property tax rate is 1.23%, and the total rate of all income taxes is 24.5%. The company uses straight-line depreciation. Find the present value of life cycle costs and the internal rate of return for both systems. Make a recommendation on which system should be acquired based on your analysis.

Answers:
For System X, NPV= $58,374, IRR = 42.7%.
For System Y, NPV= $73,783, IRR = 55.8%.
Recommend System Y as the better option.

16.2 Apply COMPARE to Your Project
Use the program to evaluate at least two alternatives in your design project. Make PowerPoint slides of the COMPARE pages to present in your final oral presentation.

References

18.1 *ASHRAE Handbook, 1984 Systems*, especially Chapter 42, "Life Cycle Costing," American Society of Heating, Refrigeration, and Air Conditioning Engineers, Atlanta, GA, 1984.

18.2 Ted G. Eschenbach, *Engineering Economy: Applying Theory to Practice,* Irwin, Chicago, 1995. This is a basic introduction to evaluating projects, recommended for students who want to know more.

18.3 Jan F. Kreider and Frank Kreith, *Solar Energy Handbook,* especially Chapter 28, "The Microeconomics of Solar Energy," McGraw-Hill, 1981.

17

Design Documentation

The 24 example formats in Chapter 17 form a comprehensive model of the 12-step design process described in Chapter 14. Refer to the second half of Chapter 5 for a narrative description of these documents as part of design communication. The formats constitute a resource for:

- Set of standard documents in an engineering design office.
- Set of assignments for a capstone course in engineering design.
- Context for selected assignments in freshman conceptual design.

Table 17.1 Summary of Design Communication Formats

Format numbers are associated with the twelve-step design process of Table 14.1

Figure	Format	Page	Document	Communication
F.1	DP-1	380	Project Concept Statement	Short written report
F.2	DP-2	380	Table of Design Constraints	Table
F.3	DP-3	381	Survey of User (Customer) Needs	Short written report
F.4	DP-4	382	Table of Design Objectives	Table
F.5	DP-5	383	Design Problem Analysis (Briefing Document)	Short written report
F.6	DP-6	384	Design Project Plan	Chart
F.7	DP-6A	385	Design Project Proposal	Formal written report
F.8	DP-6B	386	Executive Summary	One-page written report
F.9	DP-6C	387	Design Project Proposal Presentation	Verbal presentation
F.10	DP-7	388	Modified Pugh Matrix Format	Pugh matrix
F.11	DP-7B	388	Description of Design Concepts	Brief summaries (list)
F.12	DP-7C	389	Concept Drawing	"Formal" sketch
F.13	DP-8	390	Design Project Progress Report	Formal written report
F.14	DP-8A	391	Design Decisions	Written summary report
F.15	DP-8B	392	Assembly Drawings	Formal drawing
F.16	DP-8C	392	Bill of Materials	Table
F.17	DP-8D	393	Design Project Progress Presentation	Verbal presentation
F.18	DP-9	394	Detail Drawings	Formal drawing
F.19	DP-10	395	Test Plan	Short written report
F.20	DP-11	396	Evaluation Results (Report on Design Review)	Short written report
F.21	DP-12	397	Final Design Project Report	Formal written report
F.22	DP-12A	398	Final Design Project Presentation	Verbal presentation
F.23	DP-12B	399	Sales Drawing	Artistic rendering
F.24	DP-12C	400	Final Project Evaluation by Design Team	Form

Project Concept Statement

Project Title

(as descriptive as possible in one line, or one line and a subtitle)

The purpose of this project is to design a *(design product)* which would be used by *(users)* for *(application)*. The design project is sponsored by *(customers, people bringing design request to designer)* who will coordinate design requirements and specifications. Other stakeholders who will be interested in or impacted by the design include *(list and describe stakeholders)*. Major goals of the design include *(list and describe the driving forces behind the design, including new customer needs, opportunities opened by emerging technology, or marketing opportunities)*. A team of *(#)* engineers and *(specify if multidisciplinary)*, *(team members' names)* designated *(team name)*, will conduct the project, with *(leader name)* serving as project leader.

Figure 17.1 Example format for a project concept statement (DP-1).

Design Constraints — Color Inkjet Printer XJ5			
Constraint	**Method of Measurement**	**Target**	**Acceptable Limits**
Small footprint	Width x length of rest pad area	W = 16, L = 14	W < 18, L < 15
Requires standard power without adapter	Total of all heating, cooling, and mechanical power requirements, operating at max. printing speed	V = 115 v, F = 60 Hz, I = 1.5 amperes	105v < V < 120 v 58 Hz < F < 62 Hz I < 1.8 amperes
Accepts standard paper sizes	Width of paper accepted x length (in direction of feed) of paper accepted	Max. width =11" Max. length =14"	W > 11 L > 14
Uses available ink cartridges	Compatible with at least one major cartridge supplier's replacement cartridges.	Epson S020034 (black) and S020097 (color)	Any current Epson, Canon, or HP cartridge

Figure 17.2 Example of a table of design constraints (DP-2).
This example is for the design of a color ink-jet printer for home use.

Survey of User Needs

This is an example format. Each survey must reflect the context of the particular design project and the potential users being surveyed. Adapt your survey format from this example as needed. Also consider that other survey forms such as focus groups, telephone interviews, and simple observation of buying preferences may be more suited to a particular application.

Address the respondent. Very briefly, thank the respondent for taking the time to fill out the survey. Mention any incentives for doing so (premiums, entry into drawings for prizes, or return of results). Explain how the survey results will be used and whether they will be anonymous. If you say the results will be anonymous, do not ask for any information that will identify the respondent. Do not promise anything that you do not intend to fulfill (such as returning results).

Configure for easy scoring. Format the questions and answers for the type of scoring that will be used. For large surveys with machine scoring, you can ask for much detail on each question, but do not exceed the number of answer spaces on the scoring machine. If possible, place the answer blocks on the question sheet to avoid cumbersome separate question and answer sheets.

Obtain needed demographics. If you intend to correlate or stratify the responses with respect to characteristics of the respondents, you must ask for the specific demographics you need, such as sex, zip code, age range, family income, number of family automobiles, owned or rented housing, level of education, etc. Such demographic information can alert you to a non-representative sample group or confirm that it is representative. Be aware, however, that some questions, such as age range or family income, may be very sensitive, depending on the context of the questionnaire. Ask only what you really need to know, and then only if you know exactly how you will use the information to help make decisions in design or in survey validation. Avoid a data swamp.

Obtain needed user opinions or preferences. Ask questions that will help you make design decisions. Ask them in a way that maximizes the information obtained with each question and minimizes the chance of misunderstandings. See Reference 5.9 for more suggestions. Ask as few questions as possible to get the needed information. Try to keep the survey under one page; at the most, use the front and back of one page.

Example. Rather than asking,

1. Do you usually buy electronic Christmas gifts?
2. Do you personally prefer CD-equipped clock radios over cassette-equipped clock radios?

ask instead,

1. Last Christmas, approximately how much did you spend (total) for electronic gifts? $ ____
2. About how much would you be willing to pay for each of the following radios for your own use, if all have equal high-quality sound and tuning capability. Check only one in each row.

	Under $20	$20-39	$40-79	$80-119	Over $120
Simple Am/FM clock radio	_____	_____	_____	_____	_____
AM/FM clock radio with CD player	_____	_____	_____	_____	_____
AM/FM clock radio with cassette player	_____	_____	_____	_____	_____
AM/FM clock radio with both CD and cassette player	_____	_____	_____	_____	_____

Figure 17.3 Example format for a survey of user needs (DP-3).

Table of Design Objectives

for high-efficiency, environmentally friendly, split system, residential heat pump

Objective or Criterion	**Weight**	**Method of Measurement or Estimation**	**Target**
Reliability	20%	Reliability estimate of mean time between failures (MTBF) for system in heating or cooling operating modes, based on life-tested individual reliabilities of all components.	MTBF > 60,000 hrs.
Long life	10%	Estimated lifetime of system requiring no repair cost greater than 1/3 the inflation-adjusted original cost of the system.	Lifetime > 15 years.
High efficiency	20%	Seasonal energy efficiency ratio (SEER)	SEER > 17.
Robust performance	20%	Estimated degradation in peak efficiency and cooling and heating output for all typical weather and environmental conditions with minimal maintenance.	Degradation in efficency < 20%, peak output < 15%.
Low maintenance	10%	Recommended minimum level of homeowner (hrs/yr) and service technician maintenance (hrs/2 years) to maintain robust performance.	Recommended maintenance < 0.3 hrs/yr by homeowner, < 1.0 hrs/yr by technician.
Low noise	10%	Predicted noise level at 20 ft distance from outdoor unit; noise level from indoor unit eight feet from unit with typical ducting and insulation.	<40 dB outdoor unit, <20 dB indoor unit.
Easy installation	10%	Predicted installation time (hours) for average qualified installation technician and assistant.	< 12 hours installation time.
	100%		

Figure 17.4 Example of a table of design objectives (DP-4).
This example is for a residential heat pump. The accompanying table of design constraints would include entries for required ozone-friendly refrigerant (R-410A), standard electrical power requirements (208/230 volts, single phase, 60 Hz), and minimum cooling capacity (5 tons).

Design Problem Analysis

Format

[approximately 1-2 pages plus tables and charts]

Introduction—Design Project Concept: Incorporate the information from the design project concept statement (DP-1) as an introduction to this document.

Design Constraints: Discuss the constraints imposed on the design by the sponsor, by engineering convention, by industry code, or competitive factors. Include and discuss the table of design constraints (DP-2).

User Profiles and Needs: Summarize the results of research, surveys, focus groups, or interviws of potential users of the design product. Incorporate any surveys of user needs (DP-3). Distill the essence of information gathered into tables or plots. Use bar graphs as appropriate to document the preferences of the various groups of potential users. Present prioritized user preferences or user problems in the form of Pareto charts (see Appendix C and the curling iron case study on page 196).

Design Objectives: Present and discuss the design performance objectives synthesized from user profiles and needs. Incorporate the table of design objectives (DP-4). If the user needs surveys or profiles justify it, weight the design objectives according to the users' preferences. Discuss the method(s) selected to measure the performance of the design against these objectives and the quantitative goal of performance established for each.

Design Problem Statement: Summarize the concepts, constraints, user needs, and design objectives from the above analysis into one concise directive sentence to guide the design work. Do not try to say everything, just distill the essentials.

Example:

> *Design an inexpensive (less than $350 retail) automobile navigation system, using latest GPS satellite technology for simple (1 hour) installation and use by car owners, which will locate the vehicle anywhere in the continental USA within 30 feet, present the location on an electronic street map of the area, and provide user-friendly input/output, plot and track trip progress between any two US locations reliably for a useful life of at least 10 years, with mean time between failures of more than 4000 hours.*

Figure 17.5 Example format for a design problem analysis (DP-5).
See Chapter 7 for a discussion of design problem statements.

Design Project Plan

ID	Task Name	Days	Start	Finish	Prior Tasks	Doer
1	**Step : Identify Forces Driving Design(s)**	**7**	**8/24/98**	**9/1/98**		
2	Select design team and ground rules	4	8/24/98	8/7/98		Team
3	Brainstorm possible project topics	3	8/24/98	8/6/98		Team
4	Identify driving forces for each	1	8/27/98	8/27/98		Team
5	Write 3 project concepts (DP-1)	2	8/28/98	8/31/98	4	Jones,Smith,Burke
6	Select design project concept	1	9/1/98	9/1/98	5	Team
7	Select tentative project plan (use this one)	3	8/28/98	9/1/98	2	Adams
8	**Step 2: Identify Constraints**	**3**	**9/2/98**	**9/4/98**	**6**	
9	Identify sponsor-imposed constraints	1	9/2/98	9/2/98		Team
10	Identify any compatibility constraints	1	9/2/98	9/2/98		Jones
11	Identify any restraints from codes/regulations	2	9/2/98	9/3/98		Smith
12	Write table of constrains (DP-2)	1	9/4/98	9/4/98	9,10,11	Burke
13	**Step 3: Identify User Needs**	**11**	**9/4/98**	**9/18/98**	**9,10,11**	
14	Brainstorm hypothetical needs	1	9/4/98	9/4/98		Team
15	Step 3A: Analyze present methods	5	9/4/98	9/10/98		Jones
16	Research new technology	10	9/4/98	9/17/98		Smith
17	Survey potential users (DP-3)	10	9/4/98	9/17/98		Team
18	Summarize user quality characteristics	1	9/18/98	9/18/98	17	Adams
19	**Step 4: Define Objectives or Specifications**	**6**	**9/18/98**	**9/25/98**		
20	Translate user needs into objectives/specifications	2	9/18/98	9/21/98	17	Jones
21	Define targets and measures for each	2	9/22/98	9/23/98	20	Smith
22	Write objectives table (DP-4)	2	9/24/98	9/24/98	21	Burke
23	**Step 5: Design Problem Statement** (DP-5)	**2**	**9/28/98**	**9/29/98**	**19**	**Jones**
24	**Step 6: Plan Design Process to Stay on Track**	**77**	**8/24/98**	**12/8/98**		**Adams**
25	List all tasks, predecessors, and doers (DP-6)	2	8/24/98	8/25/98		Adams
26	Update project monthly	70	9/2/98	12/8/98	7	Adams
27	**Step 6A: Develop Design Project Proposal**	**8**	**9/30/98**	**10/9/98**		
28	Develop design evaluation approach	2	9/30/98	10/1/98	23	Jones
29	Write formal project proposal (DP-6A)	5	9/30/98	10/6/98	23	Smith
30	Write executive summary (DP-6B)	1	10/7/98	10/7/98	29	Burke
31	Develop/deliver oral presentation (DP-6C)	3	10/7/98	10/9/98	29	Adams
32	Develop detailed early concept sketch	3	9/30/98	10/2/98	23	Smith
33	**Step 7: Develop Alternative Concepts**	**46**	**9/1/98**	**11/3/98**		
34	Develop weighted list of user criteria	2	9/21/98	9/22/98	18	Adams
35	Brainstorm alternative concepts	20	9/1/98	9/28/98	5	Team
36	Write up each concept, minimum of 3 (DP-7B)	3	9/29/98	10/1/98	35,38	Jones,Smith,Burke
37	Develop concept sketch for each (DP-7A)	3	10/2/98	10/6/98	36	Jones,Smith,Burke
38	Perform Pugh method analysis, Round 1 (DP-7)	15	10/7/98	10/27/98	37,18,34	Team
39	Develop improved concept sketches (DP-7C, rev DP-7B)	3	10/25/98	10/28/98	38	Jones,Smith,Burke
40	Perform Pugh method analysis, Round 2 (DP-7)	5	10/28/98	11/3/98	39	Team
41	**Step 8: Complete System Level Design**	**41**	**9/30/98**	**11/25/98**		
42	List each major decision in the design (DP-8A)	2	9/30/98	10/1/98	23	Adams
43	Identify alternatives for each decision (minimum of 2)	3	11/4/98	11/6/98	33	Jones,Smith,Burke
44	Analyze alternatives against all criteria	10	11/9/98	11/20/98	43	Jones,Smith,Burke
45	Perform trade-off analyses	10	11/9/98	11/20/98	43	Adams
46	Perform any physical/simulated concept testing (DP-10)	10	11/9/98	11/20/98	43	Team
47	Document analyses and rationale for decisions	10	11/9/98	11/20/98	43	Adams
48	Complete detailed assembly drawing (DP-8B,C)	10	11/9/98	11/20/98	43	Jones
49	Write design progress report (DP-8)	3	11/23/98	11/25/98	48	Smith
50	Prepare and deliver oral progress report (DP-8D)	3	11/23/98	11/25/98	48	Burke
51	**Steps 11/12: Evaluate/Improve/Communicate Design**	**33**	**11/4/98**	**12/1/98**		
52	Evaluate effectiveness of design concept	1	11/4/98	11/4/98	40	Adams
53	Implement improvements to design concept	10	11/5/98	11/18/98	52	Team
54	Evaluate effectiveness of system design	1	11/26/98	11/26/98	41	Smith
55	Implement improvement to system design	2	11/27/98	11/30/98	54	Team
56	Evaluate results of any tests	2	12/1/98	12/2/98	55	Burke
57	Implement improvements to final design	5	12/3/98	12/9/98	56	Team
58	Report on design effectiveness and evaluations (DP-11)	2	12/10/98	12/11/98	52,54,56	Adams
59	Final design project evaluation (DP-12C)	1	12/12	12/12	58	Team, Adams
60	Write final design project report (DP-12)	5	12/14/98	12/18/98	58	Burke
61	Prepare and deliver final oral report (DP-12A)	5	12/14/98	12/18/98	58	Smith

Figure 17.6 Example of a design project plan (DP-6) for a hypothetical one-semester student project.

Design Project Proposal

Format for Formal Report

[less than 10 pages, typed double spaced, with each major heading starting on a separate page]

Title Page *(centered)*

Proposal for Design of a *(name of product)*

Submitted to *(organization)*

by *(designer's name)*

(add organization, address, phone number on separate lines)

Date of Proposal

Executive Summary: *(Begin new page, place project title at the top.)*
Incorporate document DP-6B here to summarize the proposed project. Typically about one page, this section summarizes the purpose, goals, approach, and plans for completing the project. It is ***not*** *a listing of the headings in the report. It could begin as:*

Team Synergy proposes to design a small industrial robot to rotate small parts under a vision system which will read the barcode label on each part. The robot design is constrained by the small size required (less than 2 cubic feet to fit within typical work cells), a weight capacity of at least 5 pounds, power requirements of operating on 115 volts AC, and a cost target of less than $5000. The finished robot must opimize design objectives of low weight, rapid response time, adaptable to personal computer control, reliability, maintainability, precision movement, and safety. We will meet these design objectives by selecting components and designing assemblies which ...

For the body of the proposal, integrate the design problem analysis (DP-5) with a discussion of the design project plan (DP-6) and a new section of how you will evaluate your design, resulting in the following major report sections, each to begin on a separate page.

I **Introduction** *(Repeat the title at the top of the first page.)*
II **Design Problem Statement** and discussion.
III **Design Constraints**
IV **User Profiles and Needs**
V **Design Objectives** *(These objectives may be modified or refined at later stages of the project.)*
VI **Design Evaluation**
Describe how you will check the performance of your design against your design objectives. Typically this will involve one or more of
(a) building and testing a prototype design,
(b) a computer simulation of the design in one or more of its design parameters,
(c) expert evaluation of the design by external consultants, and/or
(d) re-evaluation of the design performance by an independent analysis technique, different from that used to produce the design.
VII **Project Plan** Discuss the project plan (DP-6), assignment of duties of team members, and contingencies for which you have made provisions.
VIII **References** List books, reports, and journal articles in a standard professional format.

Figure 17.7 Example format for a design project proposal (DP-6A).

Executive Summary

Proposal to Design a Pedestrian Bridge over Freedom Boulevard at Morningdale Avenue Pumpkin Center, North Carolina

The firm of Acme Engineers and Designers proposes to design the subject pedestrian bridge at the request of the design sponsors, the City of Pumpkin Center and the North Carolina Department of Transportation. The need for this structure comes from three sources:

- Numerous citizen requests to the city council for a pedestrian route which shortens the average 1-mile walk to the nearest overpass across Freedom Boulevard.
- Several fatalities in the past two years occurred when pedestrians attempted to cross Freedom Boulevard between intersections.
- An independent random survey of 250 citizens of Pumpkin Center by ACME engineers shows that 78% would support a modest tax levy to assist in building the bridge.

Acme Engineers' initial concept for this bridge addresses the constraints imposed by the City of Pumpkin Center and the NC Department of Transportation, including

- Accommodation of pedestrian traffic only.
- Chain-link (or alternative) enclosed pathway on bridge.
- Design meets all applicable codes and statutes.

Acme's survey of 250 Pumpkin Center citizens and two focus groups of potential users have identified the following design objectives listed in order of importance to these users.

- Safety in entrance and exit areas as well as on bridge pathway.
- Clear visibility of users (also a safety concern).
- Accessibility of entrance and exit areas to connecting sidewalks.
- Attractiveness of structure.
- Low cost.
- Low maintenance costs to maintain safe and attractive structure.
- Low susceptibility to vandalism and graffiti.

Our initial design concept, an attractive suspension-style structure, addresses each of these objectives while meeting the constraints imposed. Our proposal is to explore this and other similar concepts to develop a design that optimally satisfies all of these objectives. The design will be evaluated with computer simulations of 100-year wind loads and earthquake levels. Models of the alternative concepts will be constructed for evaluation by focus groups for attractiveness and accessibility prior to selection of final concept. The design will be conveyed to sponsors by interim and final reports, including the scale models.

Figure 17.8 Format for an executive summary (DP-6B)—hypothetical example.
Here, the executive summary is for a proposal, but the same general format can be used for other documents, including interim or progress reports and final design reports.

Design Project Proposal

Format Guidelines for Oral Presentation

General: This is an oral presentation for briefing design project sponsors (and perhaps representatives from marketing, sales, and manufacturing) on the design approach, constraints, weighted objectives based on user needs, and tentative project plans for the purpose of obtaining approval to proceed with the project.

Media: Overhead transparencies or projected computer image.

Time Constraints: Typically 10 to 30 minutes depending on the scope of the project and audience needs, plus up to 30 minutes for questions and discussion (for student presentations, 15 minutes with 5 minutes for questions).

Elements of Presentation: Each of the following major elements should be presented with a separate slide and discussed extemporaneously.

1. **Title:** Title of project, speaker, design team, design sponsor, affiliations, occasion, and date.
   - Introduce speaker and give affiliation.
   - Introduce or acknowledge design team and affiliations, and design sponsor.
   - State the purpose of the presentation in audience terms.
   - Acknowledge the needs and expectations of the audience.
2. **Introduction:**
   - Preview the major points of the project or features envisioned in the product that will be elaborated in the presentation. (Note: this is not simply ticking off the elements of the presentation as with a table of contents—it previews the main selling points of the design or project to be covered.)
   - Summarize the constraints imposed on the design by the sponsors.
   - Present background on the design team's experience, qualifications, and interests.
   - Preview the organization of the remainder of the presentation.
3. **Design Problem Analysis:** Summarize the main points of document DP-5. This may require several slides.
   - Design project concept (include tentative concept sketches).
   - Design constraints.
   - User profiles.
   - Design objectives.
   - Design problem statement.
4. **Design Evaluation:** Describe how you will check the performance of your design against your design objectives.
5. **Project Plan:** Present and discuss a Gantt chart or similar graphic plan.
6. **Summary of Main Points:** Summarize the main selling points of the project or product. (Again, do not give a mere listing of presentation topics.)

Figure 17.9 Format for an oral design project proposal (DP-6C).

Modified Pugh Method Comparison Matrix of Cenceptual Design Alternatives for Lightweight, Quiet, Low-Cost Window Fan

Objectives or Performance Criteria	Percentage (Criteria) Importance Weight	Best Competition ACME ACE Ref.	Alternative Conceptual Designs		
			Whirlwind A	TropicBreeze B	FeatherFan C
Low noise	12	0	0	1	1
Low cost	12	0	1	0	-1
Lightweight	12	0	0	1	1
Reliable	9	0	1	0	1
Portable	9	0	0	1	1
Attractive	9	0	0	1	1
Easy to install	8	0	0	1	1
High flow rate(s)	8	0	0	1	1
Low maintenance	7	0	1	0	1
Easy to clean	6	0	0	1	1
Variable flow	5	0	0	1	1
Efficient	3	0	1	0	-1
Totals	**100**	**0**	**31**	**69**	**70**

Figure 17.10 Example of a modified Pugh matrix (DP-7).

Key to Design Concepts
A — Whirlwind: Stamped aluminum blades, steel housing, high efficiency, sealed motor; cut-to-fit plastic side seals with rubber strip; enamel finish; two-speed control, 16.2 lbs, estimated retail price: $24.50.
B — TropicBreeze: Molded plastic blades; plastic housing; standard motor, slide-to-fit side seals; molded-in finish; snap-out cleaning grill; three-speed control, 12.5 lbs, estimated retail price, $28.50.
C — FeatherFan: Molded lightweight, low-noise plastic blades; lighweight plastic housing; slide-to-fit side seals; molded-in finish; snap-out cleaning grill; lightweight, low-noise, sealed motor; sound dampener in housing; three-speed control; 8.7 lbs; estimated retail price: $34.25.
Ref. ACME Ace: Stamped aluminum blades; standard motor with oiling cup, steel housing; cut-to-fit plastic side seals with rubber strip; enamel finish; two-speed control; 18.2 lbs; retail price: $29.95.

Figure 17.11 Example of brief design concept descriptions (DP-7B); these are always presented with DP-7).

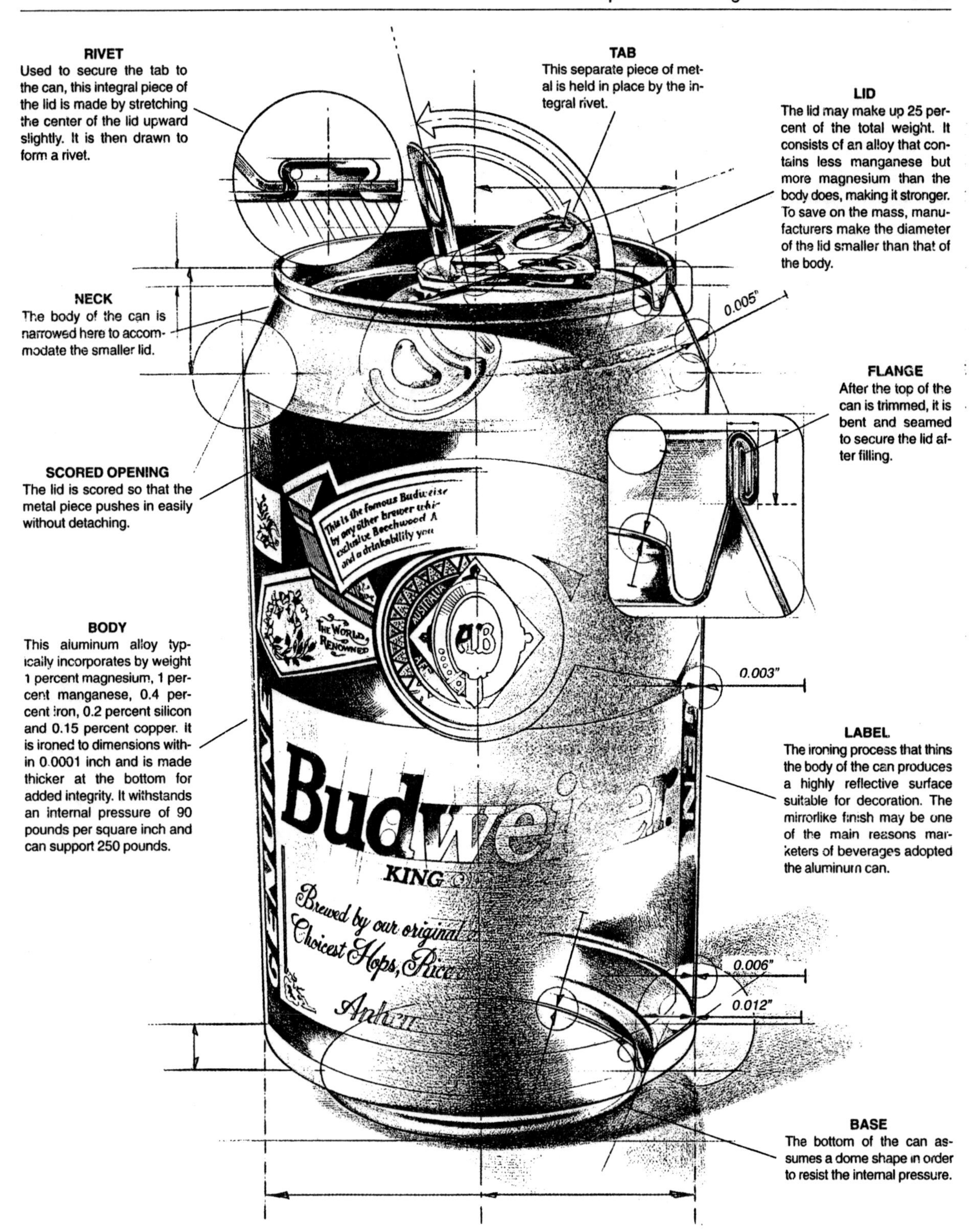

Figure 17.12 Example of a design concept drawing (DP-7C) showing the features of a modern beverage can (used with permission of C. Bruce Morser and reprinted from *Scientific American,* September 1994, page 49).

Design Progress Report
Format

General: This is a report supplementing and updating the material from the design project proposal (DP-6A) with a discussion of the design concepts considered and selected, a list and discussion of the design decisions in progress, and an updated Gantt chart showing how the project will be finished in the time remaining.

Layout: Each of the following sections should begin on a separate page:

Title Page Title of project, designer name(s), sponsor, date, organization.
Executive Summary Update from design project proposal (DP-6A).

I. **Introduction** Update from design project proposal (DP-6A).
II. **Design Constraints** Discuss and update from design project proposal (DP-6A).
III. **Design Objectives** Discuss and update from design project proposal (DP-6A).
IV. **Progress on Design Decisions** Update from document DP-8A.

This section should list *each of the major decisions of the design* and describe how the design activities are moving closer to making each decision. You may include the table of decisions made in a previous document (DP-8A), but the section here is a *discussion* of that table. The table does not substitute for the discussion. ***Note that all decisions are evaluated against all design objectives.*** For example:

***1. Overall design concept(s)*:** Discuss how the major design concept was selected; show alternatives from which the one or two concepts were selected for further work, and show how the concept decision was made (perhaps by including the Round 1 and Round 2 Pugh design concept selection charts). Fully elaborate on all the alternatives which were considered, and comment on the advantages of the chosen concept(s).

2. Size/shape of beams (Example 1): *Calculations for loads have been made in all orientations; shock loads have been estimated; several shapes and sizes are being analyzed to meet design objectives (lightest weight, minimum vibrations, required safety factor, lowest cost, ease of installation).*

3. Motor selection (power and torque requirements, Example 2): *The maximum power required has been calculated, and the torque needed to bring the system up to speed within the specified 10 seconds has been calculated. Three possible motor selections will be evaluated against the design objectives (light weight, minimum vibrations, required safety factor, low cost, easy installation).*

4. Etc.

V **Design Evaluation:** Update from design project proposal (DP-6A).
VI **Project Plan:** Discuss updated project plan (DP-6), assignment of duties of team members, and provisions for contingencies. *Gantt charts should always reflect how the project will be finished in the time remaining.*
VII **References:** List books, reports, and journal articles in a professional format, updated from the design project proposal (DP-6A).
VIII **Appendices:** Include supporting documents (such as survey data), each keyed to an appendix index, updated from DP-6A.

Figure 17.13 Example format for a design progress report (DP-8).

Design Decisions for Portable Window Fan

Design decision	How alternatives will be identified and decisions made
1. Basic concept, shape, and form	Review competing designs; brainstorm to produce alternatives, make detailed sketch of each; evaluate based on most likely to satisfy objectives by using the Pugh matrix.
2. Fan size	Typical window sizes to be fitted.
3. Blade type	Flow analysis for three typical blade types.
4. Blade material	Comparison of performance of three most commonly used blade materials (aluminum, plastic, and steel).
5. Blade thickness and structural shape	Analysis of strength and reliability of four possible shapes; choose on basis of performance against design objectives, especially safety, vibration, and noise.
6. Motor type and capacity	Comparison of performance of two most commonly used motor types against design objectives, sized according to flow power calculations.
7. Shape and style of housing	Have industrial designer sketch 3 basic shapes and select most attractive based on poll of typical consumers.
8. Etc.	*(List all important product characteristics and decision points.)*

Figure 17.14 Example format for the design decisions document (DP-8A).

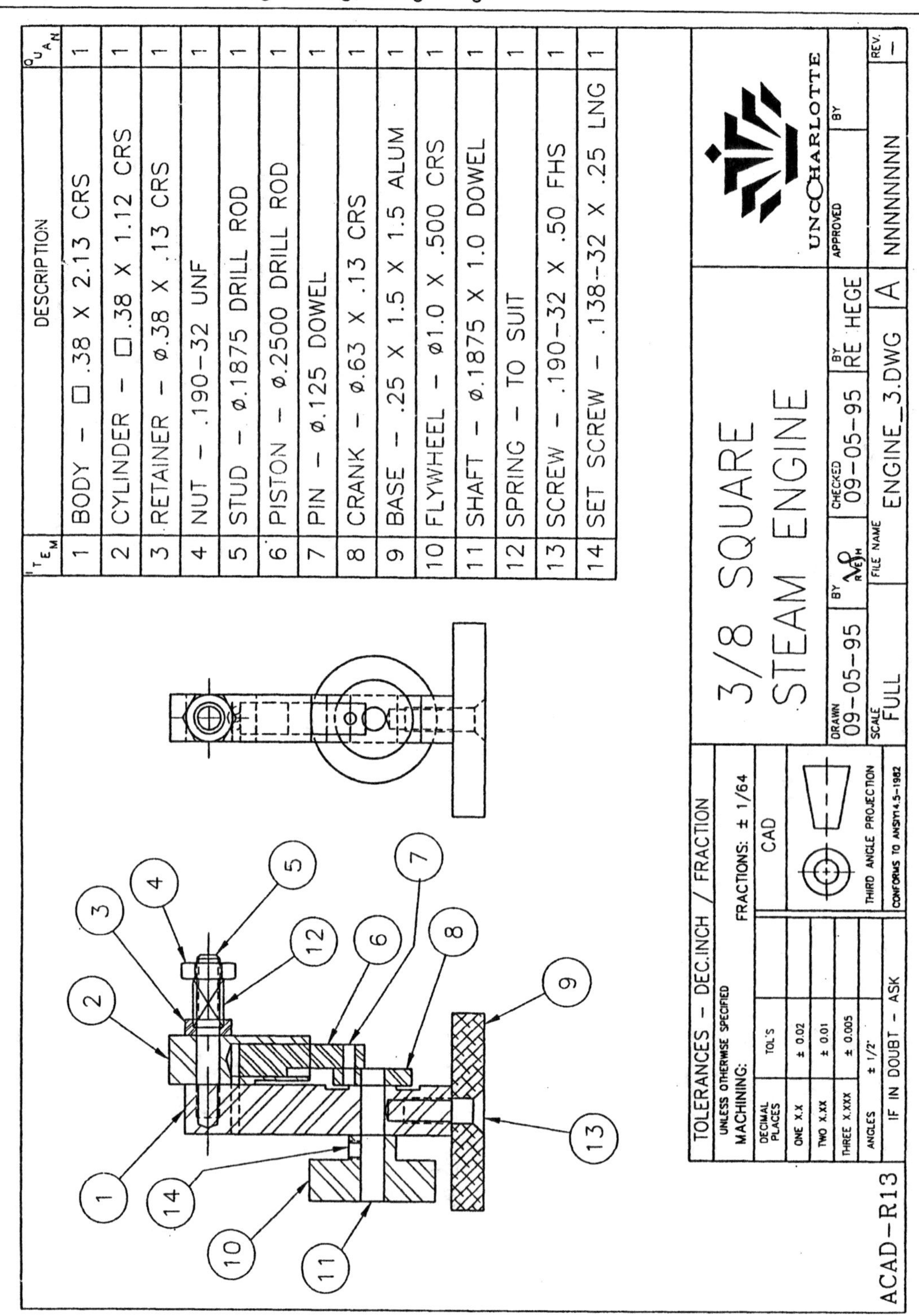

ITEM	DESCRIPTION	QUAN
1	BODY - □ .38 X 2.13 CRS	1
2	CYLINDER - □ .38 X 1.12 CRS	1
3	RETAINER - Ø.38 X .13 CRS	1
4	NUT - .190-32 UNF	1
5	STUD - Ø.1875 DRILL ROD	1
6	PISTON - Ø.2500 DRILL ROD	1
7	PIN - Ø.125 DOWEL	1
8	CRANK - Ø.63 X .13 CRS	1
9	BASE - .25 X 1.5 X 1.5 ALUM	1
10	FLYWHEEL - Ø1.0 X .500 CRS	1
11	SHAFT - Ø.1875 X 1.0 DOWEL	1
12	SPRING - TO SUIT	1
13	SCREW - .190-32 X .50 FHS	1
14	SET SCREW - .138-32 X .25 LNG	1

Figure 17.15 Example of an assembly drawing (DP-8B).

Figure 17.16 Bill of material (DP-8C); this list is always part of the assembly drawing.

Students at the University of North Carolina at Charlotte built this steam engine in a precision manufacturing class and ran it on compressed air. Courtesy of Professor Robert Hocken. Drawn by Roland Hege.

Design Project Progress Report

Format for an Oral Presentation

General: This is a relatively brief presentation summarizing the status of the project and accomplishments to date to the design sponsors along with representatives from marketing and manufacturing (and others as appropriate).

Media: Overhead transparencies or projected computer images.

Time Constraints: Typically 10 to 30 minutes depending on the scope of the project and audience needs, plus up to 30 minutes for questions and discussion (for student presentations, 15 minutes with 5 minutes for questions).

Elements of Presentation: Each of the following major elements should be summarized with a separate slide or slides and discussed extemporaneously.

1. **Title:** Title of project, speaker, design team, affiliations, design sponsor, occasion, date.
   - Introduce speaker and give affiliation.
   - Introduce or acknowledge design team and affiliations, and design sponsor.
   - State the purpose of the presentation in audience terms.
   - Acknowledge the needs and expectations of the audience.
2. **Introduction:**
   - Preview the major accomplishments of the project to date that will be elaborated in the presentation. (Note: This is not simply ticking off the elements of the presentation as with a table of contents—it previews the main points of progress of the design or project.)
   - Briefly summarize the constraints imposed on the design by the sponsors.
   - Preview the organization of the remainder of the presentation.
3. **Design Problem Summary:** Briefly summarize the design project approach, including
   - Initial design project Concept (include initial concept sketches)
   - Design constraints
   - User profiles (very briefly)
   - Design objectives
   - Design problem statement
4. **Progress on Design Decisions:** List the major decisions of the design on a slide (beginning with the selection of a concept) and discuss for each decision.
   - Alternatives considered.
   - Progress toward making the decision.
   - Analytical, graphic, statistical, or other tools used. (Use separate slides to show how the decision was made).
   - Any testing or evaluation to date.
5. **Updated Project Plan:** Present and discuss a Gantt chart or similar graphic plan. (Note: Never present an outdated plan—always show how the project will be completed in the time remaining).
6. **Summary of Main Points:** Again, this is not a listing of presentation topics, but a summary of the main accomplishment of the project or design product to date.

Figure 17.17 Format for the oral design project progress presentation (DP-8D).

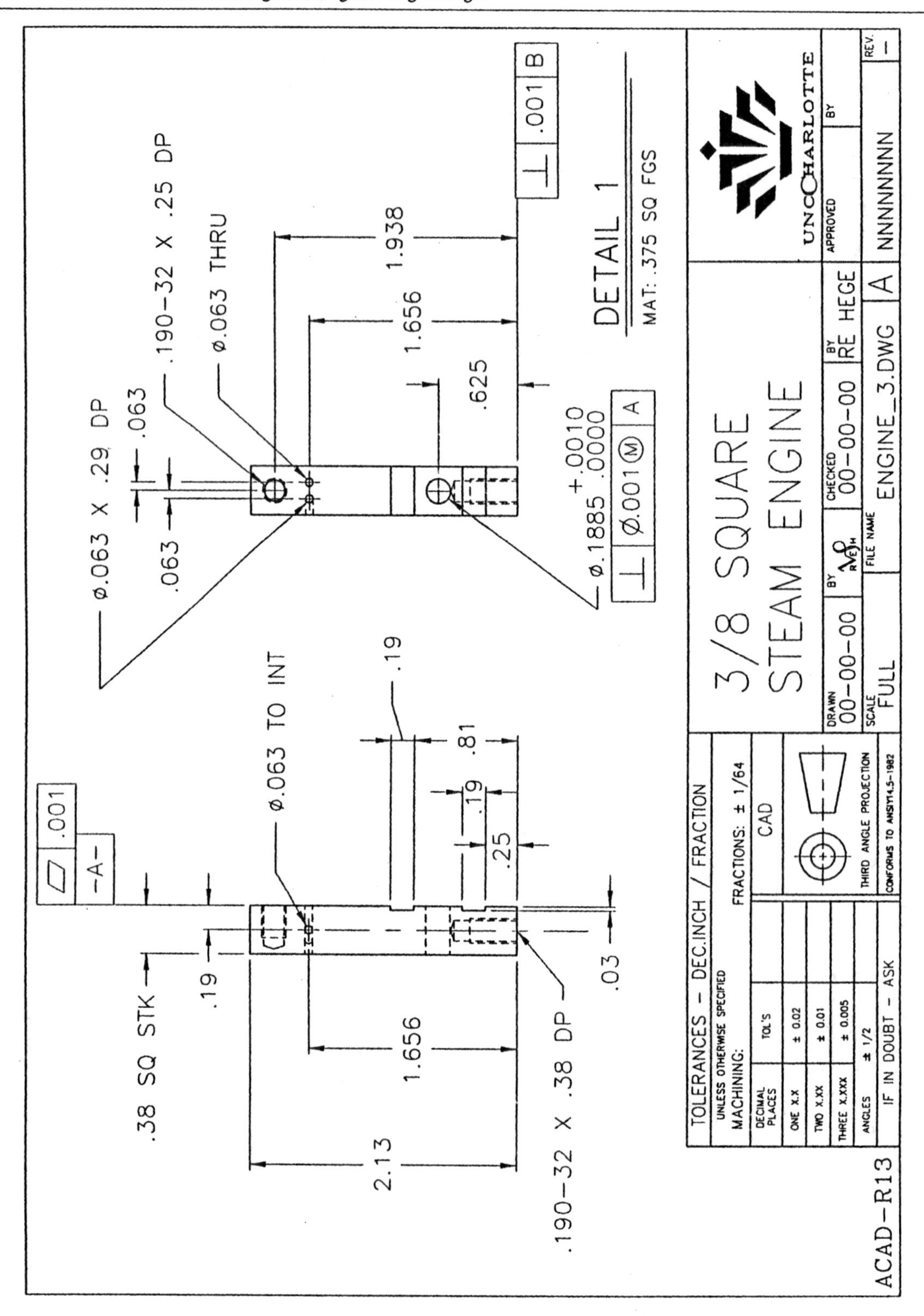

Figure 17.18 Example of a detail drawing with production specifications and tolerances (DP-9). The drawing uses conventions conforming to Standard ANSI Y14.5. This drawing is a detail of the steam engine of Format DP-8B. Drawn by Roland Hege.

Prototype, Component, or Production Model Test Plan
Format

General: This is a brief report detailing the purpose, objectives, procedure, test regimen, and expected results for tests of product components, prototypes and production models.

Purpose of Tests: Explain briefly the overall purpose of the tests, including

- The product prototype model, production number, or other identifier of test object.
- Any desired certification or sanction which is desired from successful tests.
- Specific limitations on the scope of the tests. (e.g., overall concept only, packaging durability only, susceptibility to voltage surges only, etc.).

Test Objectives: Detail exactly what is desired from the tests, for example:

- The output current of the power supply as a function of load resistance to an accuracy of 0.01 milliamperes and 0.01 milliwatts.
- Heat pump output and electrical power input in heating mode (both within 0.1 kilowatts) as a function of outdoor temperature ranging from –20°C to 60°C (in increments of 5°C or less, measured to within 0.2°C).
- Deflection (within 0.01 mm) of beam support B-12°C as a function of load ranging from 1000 N to 10,000 N in increments of 1000 N (measured to within 1 N).

Test Procedure: Specify step-by-step instructions for carrying out the tests or refer to a standard testing protocol published by an engineering society such as IEEE, ASME, ASCE, ASTM, or ASHRAE. Include specific equipment required by model number. Also specify the exact sequence in which test will be run or how test runs will be ordered (random selection, determined by test performance, or other methods). Any required environmental conditions should be noted (such as temperature, humidity, pressure, etc.). Include specifications for any specialized data collection equipment.

Expected Results: Summarize the expected outcomes from the tests. This projection is necessary to plan data collection equipment ranges, set up data sheets or computer data bases, and anticipate safety needs (for tests to destruction in particular).

Format of Output Required: Specify whether handwritten data sheets, computer spreadsheets, computer generated plots, or other specific formats will be used.

Figure 17.19 Format for prototype, component, or production model test plan (DP-10).

Design Evaluation Summary Report

Design of a Modern Thinline Residential Ceiling Fan

Design stage: Evaluation completed at conclusion of system design stage: 8 April 1999.

Evaluation of success of each design criterion (objective or specification) ranked in order of potential user ratings:

Reliability — Target: 20 years without failure): Projected MTBF of entire system (mechanical and electrical components combined) > 50,000 hours (or 25 years typical operation) based on individual MTBF of all components from manufacturers data.
Conclusion: Meets target; however, MTBF could be raised 30% with higher quality, sealed on/off switch costing an additional $2.08 retail.

Attractiveness — Target: 85% approval in potential user poll: 91% of potential users polled found fan to be "attractive" or "highly attractive" based on viewing mock-up and sales drawings.
Conclusion: Exceeds target.

Low noise — Target: indiscernible at low speed, < 20 dB at high speed: Sound chamber tests of prototype showed noise levels < 2dB at low speed, < 10 dB at medium speed, and < 18 dB at high speed.
Conclusion: Exceeds target for all ordinary residential rooms.

Reasonable cost — Target: < $ 89 retail cost: Estimated retail cost at standard configuration totals $91.57, using component prices at volume of 280,000 units per year, $24.35/hr manufacturing labor, 87% indirect manufacturing costs, 10% manufacturing profit, and 55% distribution and retail markup.
Conclusion: Fails to meet target. Could re-examine manufacturing method for housing, possibly replacing dyed polymer housing with painted steel for lower die and machinery costs for projected volume, but this risks reduced life due to possible corrosion.

Effectiveness of air flow — Target: comfortable flow for 80% of users at 8-12 foot radius for high and low speeds respectively: User poll with prototype test reported comfortable or highly comfortable air flow at all speeds and radii for 87% of users.
Conclusion: Exceeds target.

Ease of installation — Target: installation time < 3 hrs by qualified electrician or < 6 hrs by informed do-it-yourself layman: 10 trial installations of mock-up prototypes under various ceiling conditions took an average of 2.8 hours by electricians and 7.2 hours by do-it-yourselfers.
Conclusion: Fails to meet target. Problems with layman installation could be addressed with pictorial instructions, list of required tools, and simplification of balancing procedures.

Adaptability to various lighting modules — Target: compatibility with lighting modules from top 4 fan manufacturers: Fan design is compatible only with Hunter fans lighting modules.
Conclusion: Fails to meet target. However, redesign for compatibility with other manufacturers would require many specialized parts, increasing costs and complicating ease of installation.

Figure 17.20 Example of a design evaluation report on design review (DP-11).

Final Formal Report

Format

General: This is a formal report summarizing the entire design project, emphasizing the design rationale, decisions, overall concept, and evaluation of the design. Begin each of the following major sections on a separate page. *Bind the report with GBC type binding (the one with the rectangular punched holes and a plastic retainer, along with clear plastic covers front and back).*

Layout: Each of the following sections should begin on a separate page:

i **Title Page**: Title of project, designer name(s), sponsor, date, organization.

ii **Table of Contents:** List of report headings and page numbers, including appendix pages.

iii **Executive Summary**: Update from design project progress report (DP-8).

iv **Acknowledgements:** Briefly mention and thank those who have contributed to the design project.

I **Introduction:** Update from design project progress report (DP-8).

II **Design Constraints:** Discuss and update from design project progress report (DP-8).

III **Design Objectives** : Discuss and update from design project progress report (DP-8).

IV **Design Summary:** This section should be a brief, straightforward *description* of the design in its final form, perhaps within one page. It is essential that the section include an overall concept drawing of the design. (Design details are covered in Section V, with complete working drawings in the Appendix, including assembly and associated details.)

V **Discussion of Design Decisions:** Update from design project progress report (DP-8). Section V should discuss and refer to working design drawings which are put in the Appendix. Include drawings illustrating the decision-making process in this section, such as alternative concepts, immediately following the respective discussions.

VI **Design Evaluation:** Update from design project progress report (DP-8). This section should present the results of design evaluation based on any prototype tests, simulation, expert evaluation, or analysis by an independent method from that used to derive the original design decisions, as summarized in Document DP-11. Summarize extensive results in tables, charts, or graphs.

VII **Conclusions and Recommendations:** This section should be a paragraph or two discussing the overall performance of the design in meeting its objectives. If additional work would be worthwhile to refine the design in some area, point out exactly what should be done and what improvement would be anticipated.

VIII **References:** List books, reports, and journal articles in a professional format, updated from the design project progress report (DP-8).

IX **Appendices:** Include in separate appendices, with numbered pages, supporting documents such as

- A. Survey data (if applicable).
- B. Complete working drawings, with each drawing carefully folded and inserted so that it unfolds into reading position for the reviewer.
- C. Final Gantt chart (updated to reflect actual work schedule).

Figure 17.21 Format for the final design project report (DP-12).

Design Project Final Report
Format for Oral Presentation

General: This is a formal presentation to project sponsors and other stakeholders, summarizing the results of the design project and stressing the set of decisions constituting the design and their rationale.

Media: Overhead transparencies or projected computer images.

Time Constraints: Typically 20 to 30 minutes depending on scope of project and audience needs, plus up to 30 minutes for questions and discussion (for student presentations, 20 minutes with 5 minutes for questions).

Elements of Presentation: Each of the following major elements should be presented with a separate slide or slides and discussed extemporaneously.

1. **Title:** Title of project, speaker, design team, affiliations, design sponsor, occasion, date.
   - Introduce speaker and give affiliation.
   - Introduce or acknowledge design team and affiliations, and design sponsor.
   - State the purpose of the presentation in audience terms.
   - Acknowledge the needs and expectations of the audience.
2. **Introduction:**
   - Preview the major accomplishments of the project and the major selling features of the final design that will be elaborated on in the presentation. (Note: this is not simply ticking off the elements of the presentation as with a table of contents—it previews the main selling points of the design.)
   - Briefly summarize the constraints imposed on the design by the sponsors.
   - Preview the organization of the remainder of the presentation.
3. **Design Problem Summary:** Briefly summarize the design project approach, including
   - Initial design project concept (include initial concept sketches).
   - Design constraints.
   - User profiles (briefly).
   - Design objectives.
   - Design problem statement.
4. **Description of Final Design:** Describe the final design, pointing out the major decisions and how they were made—use drawings and sketches liberally. Include:
   - Alternatives considered, especially overall concepts.
   - Analytical, graphic, statistical, or other tools used. (Use separate slides to show how each major decision was made to satisfy the set of user objectives optimally.)
5. **Design Evaluation Summary:** This section gives the results of design testing and evaluation as summarized in Document DP-11. Present extensive results in tables, charts, or graphs.
6. **Conclusions and Recommendations:** Based on the design evaluation, present any conclusions or recommendations with respect to improvement of the design or decision to proceed to production stage. If additional work would be worthwhile to refine the design in some area, point out exactly what should be done and what improvement would be anticipated.
7. **Summary of Main Points:** (Again, this is not a listing of presentation topics, but a summary of the main accomplishments of the project or design product.)

Figure 17.22 Format for the final oral presentation of the design project (DP-12A).

MX Series Two Wheel Drive Tractor with Cab http://www.casecorp.com/agricultural/newequip/tractors/mxseries/dimensions.html

Case IH Agricultural Equipment

Home | Site Map | Search | Dealer Locator

MX Series MAXXUM® Tractors

Introduction
Engines
Cab/Controls
Transmission
Hitch
Hydraulics

More Features:
Powertrain
Row Crop
Hay & Forage
High-Clearance
Open Deck
L300 Loader
Attachments
Serviceability

Specifications:
Std. Equip.
MX100
MX110
MX120
MX135
Loader Specs

Dimensions

MAXXUM® TRACTOR DIMENSIONS

Two Wheel Drive Tractor with Cab

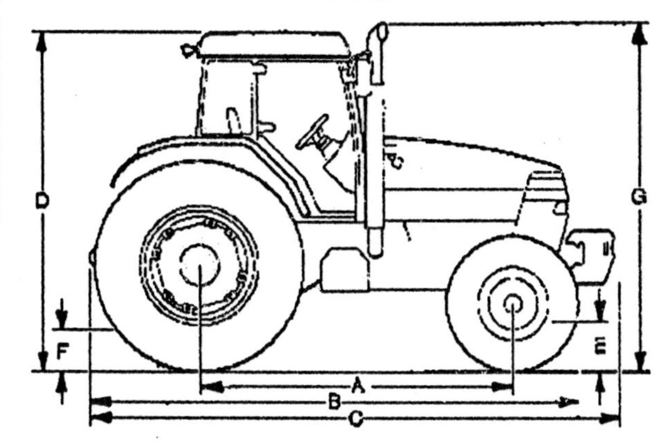

MODEL	*A*	*B*	*C*	*D*	*E*	*F*	*G*
MX100 MX110 & MX120	2812 mm (110.7 in.)	4465 mm (175.8 in.)	5034 mm (198.2 in.)	2871 mm (113 in.)	570 mm (22.5 in.)	368 mm (14.5 in.)	2885 mm (113.6 in.)
MX135	2812 mm (110.7 in.)	4465 mm (175.8 in.)	5034 mm (198.2 in.)	2914 mm (114.7 in.)	570 mm (22.5 in.)	419 mm (16.5 in.)	2928 mm (115.3 in.)

A - Wheelbase
B - Overall length less front weights
C - Overall length including front weights
D - Height to top of Cab
E - Ground Clearance under the axle
F - Ground Clearance under the drawbar
G - Height to top of muffler

Figure 17.23 Example of a sales drawing for a new product (DP-12B).
This drawing was taken from the web site for Case Corporation's new MAXXUM Tractors (used by permission).

Final Design Evaluation by Design Team

Design Project Title: ______________________________

Date of Rating: ____________________ **Rater:** ____________________

Rate the success of this design project by writing a number from 0 to 10 corresponding to the given scale in response to each of the questions below. For each rating less than 4 or above 6, explain reasons for your rating in the space below each question.

unsuccessful		somewhat successful			successful	very successful			completely successful	
0	1	2	3	4	5	6	7	8	9	10

____ 1. How successfully did the final design meet all of the original design constraints?

____ 2. How successfully did the final design meet all of the original design objectives?

____ 3. How successful was the final design in terms of creativity and innovation?

____ 4. How successful was the design project in meeting deadlines and plan dates?

____ 5. How successful was the design team in maximizing synergy in the design process?

____ 6. How successful (clear, concise, credible, and complete) was the final design report in documenting the design decisions?

____ 7. How successful was the final design presentation (clear, concise, credible, and memorable) in presenting the design process and validity of the design decisions?

Figure 17.24 Form for the final project evaluation by the design team (DP-12C). To be filled out by each design team member and any expert evaluators.

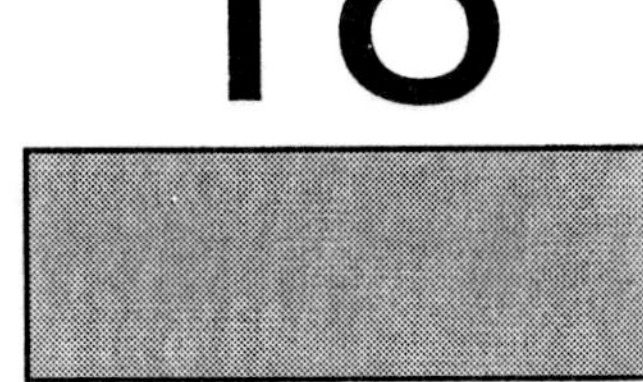

18 Innovation in the Workplace

What you can learn from this chapter:

- Prerequisites for creativity and innovation in organizations.
- Who are the innovators and what are their motives?
- Four pillars for sustained innovation in an organization: 1. Education—by individuals. 2. Application—by teams. 3. Climate—by the organization. 4. Communication—linking all three areas.
- How to identify the commitment of an organization to creativity and innovation. Concluding remarks.
- Further learning: reading list and action checklist.

At the beginning of the book, we gave a working definition of creativity. We also looked at innovation as applied creativity. Creativity is a necessary but not sufficient condition for innovation. Innovation is more than invention. Innovation involves a change and ultimately results in a useful product or process. Innovation usually requires several rounds of creative problem solving. Thus we can expand the definition and say that innovation is the *practical* application of creativity in a group or organization. Creativity originates in an individual mind (possibly enhanced by group interaction). Innovation requires the involvement of a team and of the wider organization. Good communication is central to innovation, as we will see later in this chapter.

Innovative ideas can be destructive, because they affect the value of investments in capital and equipment, as well as people's careers vested in an old paradigm. On the other hand, they also represent opportunities for entirely new careers and investments, if we are flexible and see these changes coming.

Why do people innovate? What are the prerequisites for innovation in industry? Who are the innovators and where do they come from? How do we recognize a creative organization or innovative group? What is the difference between innovation and continuous improvement? How can companies and groups improve their innovative potential? And the bottom line—how do we personally respond to innovative ideas? This chapter does not give a recipe for building a creative company, but it will provide some guidelines that will improve the chances for innovation to take place. You will be able to recognize an innovative organization or workplace. With everything that you have learned, you will be empowered to work for change in whatever environment you will find yourself.

Prerequisites for organizational innovation

Invention, innovation, and continuous improvement often become blurred, in people's minds as well as in the way innovation happens in organizations. Let's explore how the three processes are related.

Conventional wisdom says that if you want to invent something highly technical, you must have extensive technical knowledge in the subject area. Yet many inventors are outsiders; they learn on their own and create knowledge as they go along. Think of the Wright Brothers. How much did these bicycle makers know about aerodynamics or the manufacturing of airplanes? Igor Sikorsky, an outstanding pioneer in aircraft designs, inventor, and "father" of the helicopter, left the Kiev Polytechnic Institute after only one academic year because he concluded that the abstract sciences and higher mathematics at that time had little relationship to the solution of practical problems and innovation. He left school to work on his own in his shop and laboratory and to study Leonardo Da Vinci. Chester Carlson was working in the patent office of P.R. Mallory Company when he invented the xerox process. Henry Ford had little formal education, yet he came up with a new way of manufacturing automobiles for a mass market.

Creativity is thinking up new things, innovation is doing new things.

3 M Company

Almost as the exception to the rule, we can find some cases of invention by a highly educated group inside a corporate environment, such as the invention of the transistor by Shockley, Brattain and Bardeen at Bell Labs. But many important and well-known devices, such as the MRI scanner, personal computer, and bar coding, were invented outside the corporate environment. Alan Robinson and Sam Stern, in their book *Corporate Creativity,* state that "a company cannot expect creative acts in a particular area to come from experts in that area."

Developing inventions into more useful and more sophisticated products is another story. This happens through many successive improvements and innovative steps by people with expertise in these areas of science and engineering. Think of going from the Wright Brother's plane at Kitty Hawk to a Boeing 747 capable of flying nonstop across the Pacific. Think of the years of development work on the xerox process at Battelle to produce a practical copier. When Alistair Pilkington thought of the idea of float glass (while helping his wife wash dishes), it then took a team of six people working in secret for seven years (and a huge investment) to develop the process into a technical and commercial success. A key ingredient was the optimism maintained by the team while struggling to overcome problem after problem. Today, the process is completely computerized. Over 115 float lines are operating around the world, producing a ribbon of glass that could circle the Earth thirty-four times each year. Jet propulsion was invented independently by Frank Whittle in England (who wrote his graduate thesis on the idea and got the patent while working as a pilot) and Hans von Ohain (an engineer) in Germany in the 1930s. It took continuous improvement to go from the turbojet to the more efficient high-bypass ratio turbofans that are necessary to develop high thrust without excessive noise. Turbofans are used exclusively today to power commercial jet aircraft.

Innovation requires an organization specifically designed for that purpose. The* innovating *organization's components are completely different from and often contrary to those of existing organizations, which are generally* operating *organizations.

Jay R. Galbraith

Thus we could say that continuous improvement of an invention is the vehicle used in corporations and organizations to achieve innovation. One of the tools used in this process is benchmarking (see Appendix B). Continuous improvement has one advantage in that it diffuses the risk involved with innovation. Continuous improvement can be thought of as focused, limited, or incremental innovation and is thus much more acceptable, manageable, and likely to occur in the corporate environment which would feel threatened by unpredictable and unleashed creativity. An attitude of continuous improvement dispersed throughout a company over the long run prepares the way for creativity. Breakthrough ideas are more likely to occur, *be recognized,* supported, and implemented in such an environment. A company that is organized to efficiently produce a successful product is unlikely to welcome the chaos and uncertainty that comes with innovative ideas, because "they are not good at doing something for the first time" (Ref. 18.5).

Although creative acts seem random, it is possible to increase their frequency by optimizing the four conditions that can sustain creativity and innovation in an organization. In all companies, the creative potential represented by its people far exceeds the performance. That potential can be tapped and developed. Who are the innovators? We will look at some examples to see if we can discern some of their characteristics.

Who are the innovators?

The first question we might want to ask is, "What motivates innovators?" The Bayer quote may seem at first a bit humorous, but money is a key ingredient driving innovation—both making money and saving money. When Dr. Paul MacCready, one of the major innovators of this century, was asked why he took on the challenge of making the first human-powered flight across the English channel (in the *Gossamer Albatross*), he replied, "for the money." He went on to explain that someone was unable to repay him a large sum of money. When he heard on the radio that the Kramer Prize for crossing the English channel had remained unclaimed for decades despite numerous efforts by famous aerodynamicists, the sum mentioned was exactly what he needed. So he took on the challenge and won the prize. Table 18.1 lists some common reasons and motivations for inventing new products.

Research turns money into knowledge. Innovation turns knowledge into money.

Bayer Corporation, Germany

Table 18.1 Reasons and Motivations for Inventions

- As a response to threat ➛ radar, weapons.
- As a response to existing need ➛ can opener (invented 50 years after cans).
- As a response to an imagined future need ➛ high-temperature ceramics.
- For the fun of it, as an expression of creativity.
- To satisfy intellectual curiosity.
- As a response to an emergency ➛ Band-Aid.
- To increase the chance of survival and security.
- To increase comfort and luxury in life-style.
- Through better problem solving ➛ hydraulic propulsion system.
- Through turning a failure into a success ➛ Post-it notes.
- To overcome flaws ➛ magic tape.
- Accidentally, on the way to researching something else ➛ polyethylene.
- As a deliberate synthesis ➛ carbon brakes for aircraft.
- Through brainstorming with experts or outsiders ➛ myriads of ideas.
- From studying trends, demographic data, and customer surveys.
- Through cost reduction and quality improvement efforts ➛ float glass.
- Through finding new uses for waste products ➛ aluminum flakes in roofing.
- Through continuous improvement of work done by others.
- Through having new process technology ➛ proteins from hydrocarbons.
- By finding new applications for existing technology.
- Through having new materials available.
- To win a prize or recognition ➛ human-powered aircraft.
- Meeting tougher legal and legislated requirements ➛ catalytic converter.
- By having research funds available to solve a specific problem ➛ Kevlar.
- By being a dissatisfied user of a product ➛ typewriter correction fluid.
- To satisfy a personal need ➛ the Topsy Tail hair styling (pony-tail inverter) tool.
- Responding to a challenge, opportunity, or assignment.
- "It's my job—I do it for a living."
- To improve the organization's competitive position.
- To get around someone else's patent.

Some of the items mentioned in Table 18.1 were strong personal incentives for the individual inventors. Some of the items depend on a supportive environment provided by an employer. But even there, inventions frequently come from people outside the area of expertise, and technological change originates outside an industry. The invention of the transistor is a typical example. Although we usually think of inventions and innovation as products, processes are less frequently and sometimes more difficult to recognize as important innovations. Henry Ford combined the idea of Carl Benz (who designed and built the world's first practical automobile powered by an internal combustion engine) with the invention of Eli Whitney (who mass-produced firearms using interchangeable parts) to come up with the assembly line way of manufacturing automobiles for the mass market.

Jay R. Galbraith (Ref. 18.5) describes the characteristics of an innovator: Hard to get along with; strong ego; persistent; disruptive to an "operating" organization; driven to pursue and succeed at innovation; attitude of contempt for the status quo; comfortable with major change.

Creative acts are not planned for and come from where they are least expected. Nobody can predict who will be involved in them, what they will be, when they will occur, or how they will happen.

A. Robinson and S. Stern

Innovators occur at random

To illustrate this universal phenomenon—that innovation can come from anyone, anywhere, at any time—here are two stories from Japan (see Ref. 18.14 for more details). The first involves a project by Japan Railways East. While boring through Mount Tanigawa during construction of a new line north of Tokyo, engineers had a particularly troublesome problem with water leaking into the tunnel. They drew up elaborate plans on how this water could be drained away. In the meantime, construction and maintenance crews inside the tunnel used some of this water for drinking. One worker noticed that this water had an exceptionally fine taste; he suggested to the company that they should bottle and market it as premier mineral water instead of pumping it into runoff channels. The brand name for this water is Oshimizu, and in 1994 the sales were $47 million.

Hedy Lamarr, the movie star, and George Antheil, an avant-garde composer, invented and patented a device that controls torpedoes by radio. Their concept is used today in anti-jamming systems in defense communications satellites.

Tomoshige Hori was a young researcher at Snow Brand Milk, a Japanese dairy company. Hori's job was to find ways to improve the taste and nutrition of milk. In April 1980, he decided to attend a symposium on thermophysical properties of materials where he happened to attend a lecture on a new way of measuring thermal conductivity of a liquid using a "hot-wire." When he returned to his lab he decided to build an experimental apparatus to measure the thermal conductivity of milk. It should be noticed at this point that in many corporate environments, Hori would have had a hard time justifying attending a conference outside his field and building a device not directly related to his work assignment. One afternoon, Hori forgot to turn off the hot-wire. After several hours, he found that the milk had curdled. This observation made him curious. He went to talk with people in cheese production at Snow Brand and learned that being able to monitor the amount of curdling in milk is crucial to making good cheese. Timing was everything; if the process is terminated too soon, the yield is poor, if too late, the taste is poor. The timing, as it had for centuries all over the world, relied on the judgment of skilled workers and was not infallible. Hori realized that his discovery with the hot-wire could lead to a very accurate and automated process for making cheese. But because of the reactions from his boss and colleagues, he stopped spending time on what they called "weeds" research for a year and a half.

Because Hori felt that his discovery was important, he published an article in a scientific journal and applied for patents. When the article generated much interest from foreign researchers, the management of Snow Brand Milk at last decided to support his project. It took two years of working with technical personnel to develop the hot-wire probes. Within four years, the probes were installed in all milk-curdling vats in the company's cheese plants in Japan, and in 1990, Hori received a national award from the Japan Institute for Invention and Innovation. In

The original meaning of the word **serendipity** *is "a fortunate accident that meets with sagacity (keenness of insight)."*

A. Robinson and S. Stern

the U.S., Land O'Lakes was one of the first companies to adopt the new technology. The biggest hurdle to acceptance was not technological difficulty but the reluctance of the skilled cheese workers whose influential position in their companies was being threatened.

Hori's example illustrates key characteristics of innovators: broad interests, curiosity, vision, intrinsic motivation, ability to make connections between unrelated fields or ideas, persistence despite initial opposition, patience, and resourcefulness in finding ways to get their ideas known. Basically, he was a creative problem solver. Compare this list with the traits of creative thinking discussed in Chapter 6.

Someone might say, "Oh, it was just a lucky accident." In many inventions, we can see the action of chance. However, it takes a prepared mind—the mind of an "explorer"—to recognize when something is unusual and unexpected *and* to make use of the discovery. If invention and innovation appear at random or by serendipity, is it possible to increase the probability of its happening and succeeding in an organization? The following section will examine this question.

Four pillars for sustaining innovation in an organization

There are four conditions—if met—that will increase the likelihood of innovation in a company: education, application, climate, and communication. They are not a prescription that will guarantee that innovation will happen—they are guidelines (or pillars) that will increase the probability of success.

(1) Education — individual level

Since invention and innovation come from people who can think creatively, it seems that educating and training people in creative thinking and creative problem solving would be an obvious first step. Although many companies require their technical staff to periodically take courses, a course in creativity and innovation is rarely on the list. Creativity is presently in the same position as the quality movement in the 1970s. Now, TQM pervades the business world. Quality today is not the finish line but just the starting gate. This quality of "exceeding customer expectations" is essential if a product is to survive, and engineers have adopted many of the analytical quality tools such as SPC. Continuous improvement demands that businesses use a whole-brain approach and foster the creative and innovative potential of their employees. Adoption of the whole-brain knowledge creation cycle and a creative vision will prove more difficult but is necessary. As described by a former General Electric executive, companies must "innovate or evaporate."

A firm that wishes to innovate needs both an operating and an innovating organization.

Jay R. Galbraith

Everything I did
in my life
that was worthwhile
I caught hell for.

Earl Warren

What kind of education are we talking about? Universities (with rare exceptions) don't provide it, most industries don't have it, and only employees in a few organizations have received special training (mostly through consultants). With little accountability, the training seems to be largely hit-or-miss. Perhaps this is one reason why innovation is so random. Education in creativity and innovation must be of the same magnitude as that in the quality movement, if we are to increase the probability of success—it must become part of the organization's culture. This education is much more than a matter of taking creativity courses. Ideally, instructors working with teams must go to their work sites to understand the working environment and provide for just-in-time and on-the-job learning. As a first step, those that have not taken courses in college on creative problem solving should have at least one workshop on the subject. This is very important because a group or team must have a common understanding and shared values about creativity and innovation. Sustainable change does not happen when individuals work alone, are ridiculed for being different, and cannot communicate the excitement of discovery. But education—learning and applying what has been learned—is above all still the responsibility of each individual.

Activity 18-1: Personal Application

How will you use what you are learning from this book—personally, for developing habits of creative thinking and a mindset of "explorer"? In what way can creativity and innovation be part of your continuing education program?

Activity 18-2: Team Application

How is your team developing its collective creative thinking and problem solving skills beyond the study of this book?

Activity 18-3: Organizational Application

How can you and your team help your organization improve the climate for creativity, for change, and for innovation?

In the quality movement, industry was the driver and universities and colleges responded by offering courses focusing on quality issues. Will there be a similar development for creativity and innovation? Nanyang University and the University of Singapore (both ranked in the top five universities in Asia) are now required by their government to teach courses in creativity starting in the freshman year. An American who spent his entire career at DuPont in innovation and holds numerous patents has been hired as a lecturer in innovation. The new Innovation Center at Nanyang University is truly impressive and rivals many U.S. college campuses in size. Who is leap-frogging the competition?

(2) Application —team level

Innovation is fostered when teams are trained and are routinely using the creative problem solving process. Teams are cross-functional, and innovation is a shared process. Teams understand that its members have different thinking preferences; they have learned to work together and value the contributions each person brings to the team. They do not segregate the process and assign different steps to "only" the quadrant D thinkers, for example. No, everyone participates in the entire innovation process. This shared understanding enables the team to properly define problems, develop out-of-the-box solutions, and optimize. It also enables them to develop effective "selling" strategies to get their ideas accepted and implemented.

Successful innovators were seen to have a much better understanding of user needs. They get this superiority in different ways. Some may collaborate intimately with potential customers to acquire the necessary tacit knowledge of user requirements. Others learn about these explicitly through market studies.

Project SAPPHO Study of Innovation, University of Sussex (1970-72)

Patience is an important virtue for innovators and teams when selling innovative ideas. The motto, "If at first you don't succeed, try, try again" may sound trite, but it is essential. A book by Michael Geshman entitled *Getting It Right the Second Time* discusses forty-nine products that are now established household items but had serious trouble from day one in the market. It took a Herculean effort to bring them to being fully accepted by consumers. An example is Kimberly-Clark, the maker of such products as Kleenex and Kotex. Both of these products were unsuccessful initially—both were invented to use an enormous surplus of Cellucotton at the end of World War II. Kleenex was first marketed as a new way for ladies to remove cold cream from their faces. Kotex could not be advertised for its intended use at all, as the subject was unmentionable at that time. When Kleenex was creatively marketed for blowing one's nose and Kotex in a white, plain package in vending machines (where women could "anonymously" buy the product), sales took off, to where the brand name now is synonymous with the product.

Just because an innovative idea, product, or process does not succeed immediately (and it frequently won't), this does not mean that it will continue to fail. Hundreds of products that are successful today can attest to that, including the Post-it notes, the Xerox machine, bar coding, Jell-O, and Pepsi-Cola (which at one time was offered to the Coca-Cola Company for sale at a ridiculous rock-bottom price and still was refused). An organization can have training in creative thinking and a supportive climate, but if the problem-solving tools that are used routinely are strictly analytical, the likelihood of innovation will be significantly reduced. Creative problem solving must become the paradigm in the organization, not just for use in targeted projects, such as product development, but in solving many different problems (as summarized in Table 7.8).

When creativity is seen as messy and chaotic (which it is) and feared instead of understood and appreciated, an organization can pilot a small-scale implementation on the model of a "skunk works" where a creative team can operate outside the regular bureaucratic channels and control. Since nothing is more powerful than success when selling an idea, this may be an approach that can make innovation more acceptable in a very conservative, cautious organization.

The first principle of empowerment is the promotion of entrepreneurship —an insistence upon freedom in the workplace to pursue innovative ideas.

3M

The discussion so far has taken the view of supporting creativity from the top down, through managerial leadership and providing the right kind of training in creative thinking and problem solving. But organizations can also grow creative—at least in the initial two stages—from the bottom up. Most organizations are at the first stage, where a few creative individuals work quietly in isolation so they will not be noticed and get into trouble. In the second stage, two or more of these people discover each other; they begin to collaborate and mutually enhance each other's creativity. This process may start a chain reaction and grow to involve teams—creativity has become contagious and acceptable. Finally, in the third stage, management will come to support these efforts with creative challenges and assignments; it will evaluate, appreciate, and implement good solutions, with the result that the organization has become a creative community because creative individuals have started the process. Three or more knowledge creation cycles are needed to spiral creativity up to the management and organizational level.

(3) Climate—organizational level

One important factor in the organizational level that influences the acceptance of new ideas is the current paradigm of the business. If someone came to 3M (which is a paragon of an innovative company) with an idea for a new zipper, that would be a much more difficult "selling" job than finding acceptance for a brand-new class of sand paper. Chester Carlson's invention was rejected by 20 companies over 8 years. A major reason was that the photography business environment was not ready for this new idea of "electrostatic photography"—now known as the xerox process. Thus the environment includes not only a nurturing atmosphere within a particular company, but includes the broader context. As we have seen in Hori's case, it was the interest of outsiders that kept his invention alive in the early stages.

To illustrate the influence of environment, we want to begin with the story of the bar code (from Ref. 18.14). In 1948, Sam Friedland, the president of Food Fair, the largest supermarket chain in Philadelphia, made a visit to the dean of engineering at Drexel University, a well-known private university that prides itself on innovative and cooperative programs with industry. Mr. Friedland wanted the dean to start a research project to automatically add up the prices of groceries at the check-out counter. His stores were losing huge amounts of money due to clerical errors and inefficiency. The dean was not interested; he said that it was not Drexel's mission to be involved in this type of applied research (as humorously illustrated in Figure 18.1).

You can provide an environment in which creativity will flourish.

Allen F. Jacobson, CEO, 3M Company

Figure 18.1 Why innovation rarely happens in universities. ©1999 Don Kilpela, Jr.

Now it happened that Bob Silver, an electrical engineering instructor at Drexel, was outside the dean's door waiting for his appointment and overheard the conversation between the dean and Sam Friedland. Silver later that day talked to Joseph Woodland, a colleague in mechanical engineering, and they decided to work on this problem. Through a set of interesting circumstances, Woodland found himself on a Florida beach, when the idea of the vertical bars for the code came to him as he happened to notice the pattern of streaks his fingers had formed in the sand. Friedland worked on developing a code reader, and by October 1949 they filed for the patent which they received in 1952. But it took more than 20 frustrating years before the bar code would be commercialized.

It is easy to blame the dean for being short-sighted. Most likely, he was only reflecting the view of the general university environment which valued fundamental research and disdained practical, applications-oriented projects. Fortunately, these attitudes are now changing.

Risk is like staying behind a slow truck on a two-lane road. If you don't overtake the truck, it will be a long time before you get to your destination. You might also get rammed from behind by a faster vehicle. If you decide to overtake the truck, you must use good judgment to evaluate the risks and pick a moment when you can see that the road ahead is clear.

Empowerment is much talked about these days, especially by upper management. What we see here frequently is a mismatch between the vision (which may emphasize quadrant C and D values) and the actual day-to-day behavior of the organization. At the level of middle managers whose main objective is to cut costs and do more with less, the only empowerment of employees is the right to quit—a typical response of a quadrant A and B culture. Charles Prather and Lisa Guntry (Ref. 18.12) came up with nine dimensions or questions for a manager who wants to examine the organizational climate for innovation, as summarized in Table 18.2. Figure 8.2 conveys the same message with a bit of humor.

Table 18.2 Nine Dimensions of an Innovative Organizational Climate

1. **Challenge and involvement.** Is the climate dynamic and inspiring? Are people engaged and motivated? As an example, one of 3M's "stretch" goals (not quotas)—publicized throughout the company—is to have thirty percent of its sales from products that are less than four years old, and ten percent from products less than one year old!
2. **Freedom.** Do the people have independence? Are individual employees and teams given autonomy to define and do their work?
3. **Idea time.** Do people have time to elaborate on new ideas or consider alternate solutions to a problem? The famous "15% of your time to pursue new ideas" policy at 3M is a good example.
4. **Idea support.** Are new ideas greeted with silence or an attitude that "we have done this before"? Or are they received in a supportive and attentive way?
5. **Conflict.** Do people behave in a mature way or is there a proliferation of interpersonal warfare where gossip, slander, plots, and traps are part of the daily routine? It is important in a creative organization that people control their negative impulses. One negative person can pollute the entire environment. If these people are not removed (or do not change), it is impossible to grow a creative environment.
6. **Debate.** Are diverse opinions in a group allowed and accepted?
7. **Playfulness and humor.** Are good-natured jokes and a degree of playfulness accepted and valued as part of a relaxed environment.
8. **Trust and openness.** With a few exceptions, are the doors to the managers' offices open. Do people have a sincere respect for each other? Or are things discussed behind people's backs?
9. **Risk taking.** Do people feel that whenever they are dealing with things with an unknown outcome they are "going out on a limb"? Or are they encouraged to take a calculated risk. Do they feel protected when they are working on a new initiative on behalf of the company?

Risk taking

The issue of risk needs to be addressed. A good environment is one that encourages the employees to explore, network, experiment, question and think of more than one solution to a problem. A creative organization is a learning organization that is prepared to take calculated risks. The initial application of technology involves high risk and high failure. This is why these projects are frequently undertaken by quadrant D individuals (in their garages and a chaotic environment). When technologies are modified in a process of continuous improvement by a team in an organization, the risk is moderate. When the technology is developed and has matured for mass production and wide marketing to customers, the risk and failure rate are low, and the environment has become a quadrant B culture: ordered, disciplined, scheduled, and tightly managed. Unfortunately, this risk-averse behavior not only discouraging the discovery of new paradigms and further invention, it leads to morale problems.

Figure 18.2 What response do new ideas receive in your organization? ©1999 Don Kilpela, Jr.

When employees note the discrepancy between the vision of upper management and their daily work climate, they lose respect for their organization. Management must evaluate innovation from two different points of view: short-term payback must be balanced with the long-term benefits. Unfortunately, many managers (especially at the middle level) are simply not going to put their job on the line for something that may have a long-term payback. Although managers often want the fruits and benefits of creativity (and pay lip service to the concept), they do not fully understand the risks, challenges, and consequences that this would entail and the fundamental changes that would be required in their management style and procedures.

Rewards

We haven't said much about rewarding creativity. Rewards can be effective, if used judiciously, if they are truly merited, and if they are meaningful. Otherwise, they can backfire or be counterproductive. An example is the system of extrinsic rewards and quotas for creative ideas developed in Russia under Lenin. What happened there is that people divided creative ideas into minuscule variations to claim rewards. This did not lead to an increase in innovation—it only led to the proliferation of spoof ideas and the implementation of simply "stupid" ideas to meet quotas. Only with the collapse of the USSR did this policy stop. Fostering intrinsic motivation is a much superior approach in the long run. The best reward is for people to see their ideas implemented. The organizational culture for innovation should be such that there will be a social climate against not being creative. Managers who model and reinforce these values by their own daily actions are the most effective. They help develop rituals (such as newsletters and special celebrations) to reinforce the values of creativity and innovation. Most importantly, the emphasis must be on the quality of the innovation or idea, not on quantity.

The example I set is important, so it is critical that I encourage ideas.

Bill Gates, president of Microsoft

Another factor to consider is to give rewards to the right person. It is almost the norm to reward the developers and to ignore the originators of creative ideas. Both are needed—both should be acknowledged.

(4) Communication

If you want to succeed, you need to double your failure rate.

Thomas Watson

As we have seen, communication is central to engineering design—it is also central to innovation! There has to be communication between individuals, teams, and management. You need to develop a plan to network and get information through to other people in your organization. Individuals or groups simply going about their business and doing their own thing will rarely innovate. Communication is the vital link between the pillars of education, application, and organizational climate.

Let's say you are in a group of designers and you serendipitously discover an innovative approach to design using voice commands rather than pushing icons on your workstation. You do not know enough about voice activation, so you need to communicate with people who do have such knowledge. You need to communicate with your managers to convince them that this work is part of their business plan (unless the company already has a procedure for developing innovative ideas). You need to draft up an analysis of different scenarios if the idea succeeds and for the implications if it fails. You will need cost analyses. Some companies will have committees or technical forums that can evaluate your idea; others will not. At this point you might think that you can patent the idea and let the company take it from there. But ninety-five percent of patents never get off the shelves. On the other hand, many products which became very successful never could be patented, or their inventors never bothered with getting a patent. If you have failed before, the learning experience should increase your chance for success, but committees may only see you "coming again with those crackpot ideas." Thus it is essential that the lines of communication be open at all levels so you and your team can build understanding and support for your idea.

The failure of large organizations in America to innovate is primarily the result of a communication gap, not a decline in ingenuity.

Robert Rosenfeld and Jenny Servo,
The Futurist, *August 1981*

The 3M Company probably would have never developed *Scotchgard,* one of its most successful products, if it had not been for communication between a few key people. A young research chemist, JoAn Mullin, happen to spill three drops of a chemical onto her deck shoes. When she washed her shoes, she noticed that those spots simply refused to get wet. She knew that work was going on elsewhere in the company on oil- and water-repellent treatments, so she was excited about her discovery. However, no one paid attention to her discovery, except her supervisor who eventually handed swatches of treated fabric for testing to the research group. However, nothing was done because of the "failure" of earlier tests. JoAn became so discouraged that she switched to another job. But her supervisor followed up. Fortunately, the timing was better this time around, as a young researcher had just finished a project and was ready to do something else. The results completely changed the direction of the research program. As it turned out, the spilled substance did not work out, but it was a stepping stone to the ultimate product. JoAn Mullin never received credit for her original discovery (see Ref. 18.14).

Open communication still goes against the grain of many people who are used to keeping good ideas to themselves. Some have learned through hard experiences that ideas get "stolen" by unethical colleagues. But in any organization, whether small or large, both formal and informal communications are necessary to present and "grow" innovative ideas. It is precisely the informal, unanticipated exchange between employees who do not normally communicate on a regular basis that often moves a project forward. Therefore, companies must foster formal and informal means for communicating innovative ideas. This is not the traditional "suggestion box." Also, people should be responsive to requests for information or help from others in the company (and should be allowed the time to do so). Indifference can squelch ideas just as effectively as ridicule.

In the formal arena, companies should have committees, technical groups and management groups that regularly evaluate ideas. They should have a formal channel for employees to communicate ideas, and employees must know how to tap into the company's expertise. Informal channels should be developed as well, where employees who do not normally interact are given opportunities to do so. Successful networking at 3M happens through a number of creativity "clubs." This is similar to the user groups at Ford Motor Company who meet regularly to discuss their experiences with the solid modeler I-DEAS.

⌛ Activity 18-4: Identifying a Creative Organization

Table 18.3 is a checklist for identifying traits of a creative organization—for different users. No organization will have a perfect score on this "exam," but collecting this information and evaluating the answers objectively can give you a feel for the climate for creativity in the organization.

- If you are a student soon to enter the job market (as a new engineer or as an intern in a coop program), the list may give you some hints— if you can talk candidly with some of the employees to collect the data.
- If you are part of an innovation team and want to work for change, the list helps identify areas that may need improvement.
- If you are a manager, the list can gauge the "state of your organization" and your own alignment; it will help you set priorities if you want to pioneer a paradigm shift.

Also think about the personal application: Does the organization match your own creativity and innovation level? Table 18.3 does not mention one case: what if neither you nor your organization are creative? Unfortunately, this is still a prevalent situation. At present, it may not (yet) be a problem for you or the organization, but in a global environment and a rapidly changing world, things could change quickly. Also, if you are applying what you are learning from this book, you will not remain in this category long—you will have to evaluate the third option in Table 18.3.

Table 18.3 Checklist for Identifying a Creative Organization

____ Do all employees know the company's vision in the area of creativity and innovation? Do they support the vision wholeheartedly?
____ Is everyone being encouraged to always look for improvement?
____ Are managers aware that creative ideas can come from anyone, anywhere, at any time?
____ Is technology recognized as a competitive tool that can make difficult products achievable?
____ Is calculated risk acceptable and bureaucracy at a minimum?
____ Are creative thinkers and innovators rewarded? How?
____ Do you encourage your employees to maintain a portfolio of creative accomplishments?
____ Are people from different departments involved in creative problem-solving teams and in the evaluation and implementation process?
____ Does the organization have policies and procedures that encourage a process of never-ending innovation and improvement?
____ Are innovators allowed to learn and work by playing?
____ Is management walking the talk about empowerment for creativity and innovation?
____ Is management for creativity more than paying lip service by providing a "suggestion box"? Do definite mechanisms exist for developing and funding creative ideas? Does this system respond in a timely and fair manner? Are successes publicized?
____ Is everybody in the organization able to recognize a potentially useful idea?
____ Is it easy for an employee in one part to find out about ideas in a different part?
____ What types of formal and informal networking opportunities and communication channels exist in the organization?
____ How would the organization respond to an employee working "unofficially" on a project?
____ Is there a process of "appeal" for independent review of an idea that has been rejected?
____ Are people encouraged to look for the unusual and unexpected as a potential opportunity for creative ideas? Do they have "permission" to be creative and take independent action?
____ Is there a policy of job rotation and developing new skills? Is the organization making full use of the skills their employees have? How can you find out what skills and expertise is available? If contacted for help, how readily are people responding?
____ Do employees have an opportunity for mini-sabbaticals to gain new stimuli?
____ How easy (or difficult) is it in the organization to interact directly with the customers?
____ Are success stories being circulated about recent innovations and inventions?
____ What training programs in creativity, whole-brain thinking, team development, etc., have been or are being used in the organization?

Go/no go checkpoint at the personal level (for evaluating your job situation):

1. If both you and the organization are creative, go for it—take the job.
2. If the organization is creative and you are not, you might get your creativity unleashed—if you get or keep the job. The assessment should give you confidence to take some risks.
3. If you are creative and the organization is not, you will have to use your judgment. If you need the job to survive, take it but be on the lookout for opportunities (within and outside the company)—you might be able to make a difference, but realize that it will not be easy. How good are your communication skills, and your persistence? If these two characteristics are not your strong points, try to find a more favorable work environment. You will have to evaluate the risks for yourself against the potential opportunities.

Concluding remarks

It is better to light one candle than curse the darkness.

Adlai E. Stevenson, in an eulogy to Eleanor Roosevelt, 1962

We want to encourage you to develop your own creativity, your teamwork, and your communication skills within the framework of your engineering or technology profession. This is what this book is all about. We also want you to consider the wider context, not just the organizational environment, but what is happening on a global perspective.

Our involvement with creative problem solving has certainly changed us. First of all, we deeply regret that we did not know—that we were never taught—these thinking skills. We were, in various ways, driven through ridicule to hide and lock away our creativity by the time we were 15, until we no longer even remembered that we once had this ability. As an engineer, Ed used his analytical skills to solve problems and come up with some improvements to products. But when he had a creative idea for an invention, it was set aside because he did not know how to appreciate it, much less how to develop it into a usable product. Even raising our children would have been easier had we known then what we know now about creativity and creative problem solving.

We believe that the last few years have been the most productive and enjoyable in our lives, because we are solving problems creatively. We have created this book through a multistage process of continuous improvement and teamwork; the addition of a third author has been a fascinating and rewarding process of synergy. We are glad there are no limits to what we can yet discover and learn. We are encouraged to see how our example and our efforts are helping make a change in how students are taught in schools and universities.

If you are creative (or want to become more creative), you should be in an organization where being innovative is the norm not the exception.

The dreams of space travel notwithstanding, we only have one world, and we must take care of it. We must learn to get along with each other. No one group or nation is superior—we all have things to share; we all can learn from each other; we all have unique abilities and achievements that can be appreciated and further developed. Our world requires the service of all our energies, our talents, our minds as well as our hearts—then we will have the power to make change for the better. Learn to make wise choices—spend your time and your resources on what has lasting value. Be a unique individual as well as a team member and a citizen of the global community. Be involved in teaching, because it is your responsibility to pass on the values of your culture past and present in a framework understandable to the young. You must build the bridges and make the connections—computers and television will not do this for you. Model whole-brain critical thinking by the choices you make in your life and the responsibilities you assume for yourself and others. Thus you will create and build a precious legacy of values and accomplishments, whatever surprises the twenty-first century will bring.

Activity 18-5: Inventions and Inventors

After surveying this chapter, select a specific invention as your topic (see Table 18.4). Do an in-depth study of the subject from library materials as well as by interviewing people inventing, developing, manufacturing, or using the invention or product.

Alternatively, select an inventor as the subject of your investigation. You can select a special focus: thinking style, environment, motivation, as well as the process, effort, and final outcome. What were the costs (and benefits) to the inventor?

Activity 18-6: Invent!

Alone or in a small group, create an invention! Brainstorm ideas, then focus on one area. Use the creative problem solving process (or the engineering conceptual design process) to develop your ideas. Then build a prototype to demonstrate that the invention actually works. You will need to have one extra piece of documentation that has not yet been mentioned—an inventor's log or journal. This has two purposes: to help think through and develop ideas, and then to protect the completed invention.

Guidelines for the Inventor's Log
• Use a bound notebook and make notes each day about the things you do and learn while working on your invention. • Record your idea and how you got it. • Write about problems you have and how you solve them. • Write in ink and do not erase. • Add sketches and drawings to make things clear. • List all parts, sources, and costs of materials. • Sign and date all entries at the time they are made and have them witnessed.

Activity 18-7: Culture, Technology, and Creativity

With a team, brainstorm and then discuss the values in your culture that support creativity and inventiveness. How can these values be strengthened? How do they support the integration of technology with the culture? Focus your discussion on one or more of the following: your family culture, your organizational culture, the culture of your local community (political unit, educational system, etc.), or the culture of your ethnic community.

Activity 18-8: Lifelong Applied Creativity

We have given you a match, now light a candle. Determine to expand and nurture your thinking skills and creativity. Apply whole-brain creative and critical thinking to solve problems, to help others, and to preserve the environment for a future generation. Be a bridge builder. Strengthen the values that have undergirded your culture. The state of the world at the threshold of a new century could give cause for pessimism; yet we believe that with positive thinking and creative problem solving we can make a difference right where we are. So can you!

Table 18.4 A Sample List of Inventions and Inventors

1. Earmuffs were invented in 1873 by 13-year -old Chester Greenwood.
2. Band-Aids were invented by an employee of Johnson & Johnson.
3. LifeSavers® were invented in 1913 by Clarence Crane.
4. Stephanie Kwolek, a chemist with the DuPont company, invented the miracle fiber, Kevlar. What is the connection to Paul MacCready's Gossamer Albatross?
5. The *Bobcat* loader was invented by Louis Keller, a farmer from North Dakota who had dropped out of school after the eighth grade.
6. The pop-up toaster was patented by Charles Strite in 1918.
7. The ballpoint pen was patented in 1938 by two Hungarian chemists, George and Ladislao Biro.
8. It took Guideon Sundback 30 years to perfect the slide fastener—the zipper—patented in 1913.
9. The mousetrap was invented by Charles F. Nelson of Galesbury, Illinois (Patent Number 661068; you might want to look it up to see the original model).
10. Charles Babbage and Ada, Lady Lovelace (Byron's daughter) were the first to understand and invent machine computation—the first computer. However, their designs could not be built with the technology available in 1833.
11. Alexander Graham Bell made his first invention at the age of 14. What was it?
12. Robert Fulton—how many different things did he do with boats?
13. Margaret Knight holds 26 patents in diverse fields. What did she invent?
14. Bette Graham, a housewife and typist, invented "liquid paper"—her company was sold in 1980 for over $47 million.
15. Mary Anderson received a patent for windshield wipers in 1903.
16. Josephine Cochrane patented a dishwasher in 1914.
17. Marion Donovan received a patent for the first disposable diaper in 1951.
18. Rose Totino patented a dough for frozen pizza in 1979.
19. Elijah McCoy had over 50 patents for his inventions; he was the "real McCoy."
20. Keith D. Elwick of Vinton, Iowa, invented a manure spreader that became so popular in England that he received a silver medal and trophy from Queen Elizabeth for his creativity.
21. Robert F. LeTourneau of Longview, Texas, received 187 patents and founded a college, even though he only had a seventh-grade education. He wrote an autobiography, *Mover of Men and Mountains.*
22. Henrietta Bradberry invented an underwater torpedo "cannon" during World War II.
23. Patent $821,393 for a "flying machine" was granted to Orville and Wilbur Wright in 1906.
24. Patent #121,992 for "An Improvement in Adjustable and Detachable Straps for Garments," was granted to Samuel L. Clemens, of Hartford, Connecticut in 1871 (Mark Twain's first patent).
25. Abraham Lincoln received patent #6,469 for a "Manner of Buoying Vessels" in 1849.

Reading List

18.1 James L. Adams, *Flying Buttresses, Entropy, and O-Rings: The World of an Engineer,* Harvard University Press, Cambridge, Massachusetts, 1991. This contextual overview presents a brief history of technology, diversity engineering, the design and invention process, failure, and many other topics.

18.2 McKinley Burt, Jr., *Black Inventors of America,* National Book Company, Portland, Oregon, 1989. This book contains copies of successful patent applications by many African-American inventors.

Even if you are on the right track, you will be run over if you just sit there.

Will Rogers

18.3 Rick Crandall, editor, *Thriving on Change in Organizations,* Select Press, Corte Madera, California, 1997. Written for the Institute of Organizational Change, this book contains practical insight from 19 authors.

18.4 David A. Fussell, *The Secret to Making Your Invention a Reality—The Workbook,* Ventur-Training L.P., 3794 Meeting Street, Duluth, Georgia. The book and tape series were published in 1994. The author holds 14 patents. His own company helps small companies and inventors achieve success through his unique method of product licensing.

18.5 Jay R. Galbraith, "Designing the Innovative Organization," Chapter 9 in *How Organizations Learn,* Ken Starkey, editor, International Thompson Business Press, London, 1996. This is just one of the chapters in this book which focuses on organizational learning, innovation, and management.

18.6 Michael Gershman, *Geting it Right the Second Time,* Addison-Wesley, Reading, Massachusetts, 1990. Written for people in business, it is fun for anyone interested in the history of consumer products.

18.7 Mike Gray, *Angle of Attack: Harrison Storms and the Race to the Moon,* Norton, New York, 1992. This book about Project Apollo centers on North American Rockwell engineer Harrison Storms, Jr., and how he inspired his team to surmount every obstacle toward winning the NASA contract. Written in beautiful non-technical language.

18.8 Gordon D. Griffin, *How to Be a Successful Inventor: Turn Your Ideas into Profit,* Wiley, New York, 1991. This book is filled with good advice on how to recognize the economic potential of ideas and how to develop a mind for the business side of invention.

18.9 Jerry Hirshberg, *The Creative Priority: Driving Innovative Business in the Real World,* HarperBusiness, New York, 1998. The author is the founder and president of Nissan Design International and reveals his strategy for designing an organization around creativity.

The learning requirements for major innovation are high for everyone affected by the innovation. This is one factor why change is difficult.

Teamwork and communciation are key in this organizational learning—this is the direct connection between knowledge creation and innovation in the context of problem solving.

18.10 Tracy Kidder, *The Soul of a New Machine,* Avon Books, New York, 1981. This book tells the fascinating story of the development of a new computer by Data General Corporation.

18.11 James Mattson and Merrill Simon, *The Story of MRI: The Pioneers of NMR and Magnetic Resonance in Medicine,* Dean Books Company, Jericho, New York, 1996. This 800 plus page book details the fascinating story of a medical breakthrough and the different researchers involved; it is a fascinating history as well as a good biography.

18.12 Charles W. Prather and Lisa K. Gundry, *Blueprints for Innovation: How Creative Processes Can Make You and Your Company More Competitive,* American Management Association, New York, 1995. Concise, useful!

18.13 Struan Reid, *Handbook of Invention and Discovery,* Usborne, London, 1986. This softcover reference book with colorful illustrations presents much condensed information about the most important inventions and discoveries, beginning with the Stone Age. The format used clearly brings out the development over time of a particular invention and related products or processes.

18.14 Alan G. Robinson & Sam Stern, *Corporate Creativity: How Innovation and Improvement Actually Happen,* Berrett-Koehler Publishers, Inc., San Francisco, 1997. This is a practical book for managers, with many detailed examples to help unleash the creative potential in companies.

18.15 Autumn Stanley, *Mothers and Daughters of Invention,* Scarecrow Press, Metuchen, New Jersey, 1992. This book brings out the difficulties most women have had in getting their talents and inventions recognized.

18.16 Neil Steinberg, *Complete & Utter Failure: A Celebration of Also-Rans, Runners-Up, Never-Weres and Total Flops,* Doubleday, New York, 1994. Very entertainingly written by a reporter of the Chicago Sun-Times.

18.17 David Tanner, *Total Creativity in Business & Industry,* Advanced Practical Thinking Training, Inc., Des Moines, Iowa, 1997. This book has lots of valuable examples and a focus on developing thinking skills.

18.18 Robert Temple, *The Genius of China: 3000 Years of Science, Discovery and Invention,* Simon & Schuster, New York, 1986. This beautifully illustrated volume summarizes the discoveries of Joseph Needham about the technological achievements of China between 1300 B.C. and A.D. 1500.

18.19 Charles Thompson, *What a Great Idea! The Key Steps Creative People Take,* Harper Perennial, New York, 1992. Exercises in this workbook help overcome barriers to creativity. It includes an interview with Yoshira NakaMats, inventor of the floppy disk, digital watch, and compact dis player and holder of thousands of patents.

Innovation occurs when organizations function as effective learning systems, and learning comes through experimentation and failure. Truly innovative organizations are those where people can take risks, reap the rewards of success, and survive constructive failures.

Michael Tushman and David Nadler (Ref. 18.5)

Action checklist

☐ Write your own summary of the key concepts in this chapter.

☐ What are the most important things you learned from this book? Periodically flip through some of the chapters and review. Revisit the action checklists for ideas that might be suitable to implement at this time in your life. Keep the book—you will find it a valuable resource worth many times the price of what you could get if you sold it.

☐ Travel and spend some time in another part of the world. Live in another culture; learn the language. Find out how different peoples solve problems; find out what values you have in common.

☐ Practice creative problem solving, alone, in your team, with your friends and family, and on the job, until it becomes second nature.

☐ Innovation is a never-ending story in the knowledge creation cycle. We, the authors of this book, would like to hear about your experiences as you implement what you have learned. E-mail to: lumsdain@mtu.edu or shelnutt@uncc.edu. Or network with other instructors and learners through our new website at **www.engineering-creativity.com**. This is our own 1999 learning goal—how to set up and manage the site.

Appendix

The purpose of this appendix is to introduce the topics with an overview and vocabulary. We highly recommend that engineering and technology students (seniors and graduate students), as well as engineers and managers in industry, take courses or workshops to learn these techniques used in industry for problem solving and analysis. Many companies offer such training in-house. Professional societies and conferences are another source of short courses. Also be on the lookout for courses available on-line from "virtual" universities—any information we could give here would be outdated within a few month, as this educational field is developing very rapidly. Keep your eyes out for new techniques, and keep on learning!

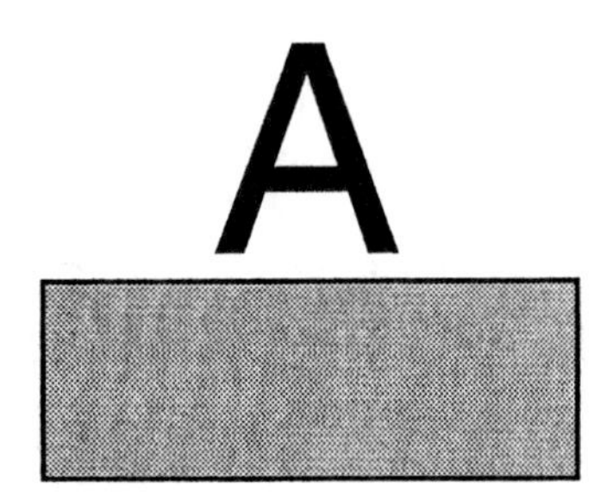

QFD (Quality Function Deployment)

Chapter 11 shows how the Pugh method of creative design concept evaluation can help us develop a superior product. Creative thinking and problem solving are also needed at various stages throughout a company's manufacturing, sales, and service activities. To achieve companywide or total quality control, organizations have devised various procedures. Among these is quality function deployment (QFD). QFD is a very structured team approach to quality control that offers many opportunities for creative thinking and brainstorming. As seen in Figure 4.2 (page 91), QFD is an important tool in concurrent engineering.

Scope of QFD

Basically, QFD is a mechanism to ensure that customer needs drive the entire product design and production process in a company, including market planning, product design and engineering, process development and prototype evaluation, as well as production, sales, and service.

Key Principle

The context is never irrelevant. When we develop a product or a service, the customer's needs and wishes contribute as much to the context as do cost, production or implementation, innovation, and environmental effects. To do the job well, we cannot work in isolation.

QFD is an extremely complicated procedure that originated in Japan, where it has been very successful in companies such as Toyota and many of its suppliers. Many American company are learning to use parts of the process, notably the House of Quality. A variety of QFD workshops are available through the American Supplier Institute of Dearborn, Michigan; they provide a thorough introduction and customized training in this technique of total quality control. Why are Japanese industries sharing this technique with their competitors? One reason is that Japanese companies are increasingly becoming purchasers of U.S.-made components for their American manufacturing plants and products; another reason is that the Japanese have progressed far beyond QFD in their quality control efforts.

QFD is based on the Japanese philosophy of quality, which considers the loss or cost to society if something does not have perfect quality. Efforts that will reduce these costs result in an increase in quality . In engineering terms, hitting the target is quality in Japan, whereas in the United States, we still think of quality as falling within a range or band

Table A.1 Customer-Driven Quality Control

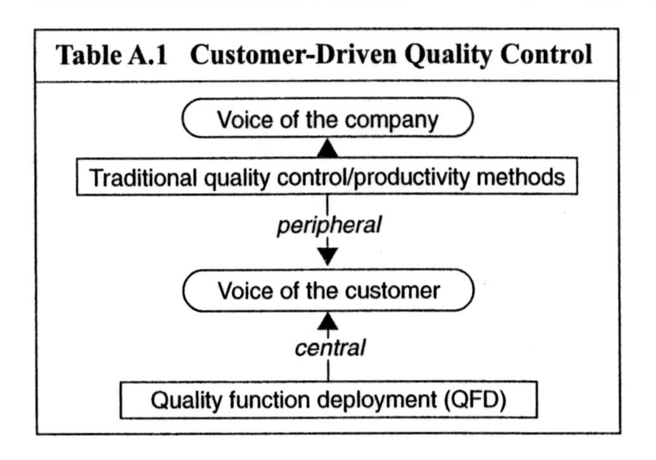

of tolerances around a specified value. Many of us operate on the principle that a certain number of defects are acceptable. In Japan, no defect is acceptable. QFD differs from quality control in most U.S. companies, as shown in Table A.1. In the U.S., the "voice of the company" drives quality efforts and involves mostly manufacturing process quality control. In QFD, it is the voice of the customer; it is centered on quality in product development. Quality is designed into the product; it is not something that enters the picture only during the manufacturing phase.

QFD has four distinct phases: **Phase 1—product planning**—translates the customer's wants into design requirements through an analysis matrix called the House of Quality. **Phase 2—part-deployment** or component planning—takes the critically important design requirements down to the level of part characteristics in a scaled-down House of Quality. **Phase 3—process planning**—identifies key process operations related to the important part characteristics. **Phase 4—production planning**—relates the key process operations to production requirements through the operating and control charts. This results in prototype construction and production start-up. Operational factors such as process monitoring functions and worker training requirements are also identified.

A primary benefit of QFD is its function as a comprehensive data base; all important information about a product is collected and stored in one place. This enables new staff to be informed quickly about all aspects of a product, from design objectives to production, sales, and service. Restructuring a company according to QFD requires a tremendous initial effort; however, once this data base and the procedure have been established, further development and changes are easy to make. QFD can identify targets for improvement and areas where technological development is required, with both directly linked to consumer needs and the market. QFD disperses all essential information about the product and product quality horizontally as well as vertically throughout the company—there is a common understanding of the quality issues.

The QFD House of Quality product planning chart

QFD information flow is shown schematically in Table A.2. Many steps are involved in building the House of Quality planning matrix (and the subsequent QFD charts). Three major activities are associated with each step and identified with the thinking modes shown in the sidebar and in Tables A.3, A.5, A.6, and A.7.

ACTIVITIES CODE

a = data analysis
b = brainstorming
c = data collection

Table A.2 QFD Information Flow (Schematic)

PHASE 1

Product Planning Chart

1 Customer wants

2 Product characteristics

⊙ Critically important relationships

PHASE 2

Component Deployment Chart

1.1 Detailed critical customer wants

2.1 Detailed product quality features

3 Critical components

2.1.1 Critical control characteristics

3.1 Critical parts specifications

PHASE 3

Process Planning Charts

4 Process

5 Control points

6 Process monitoring checkpoints

PHASE 4

Production Planning Charts

7 Operating charts

8 Quality control charts

9 Other charts (installation, training, etc.)

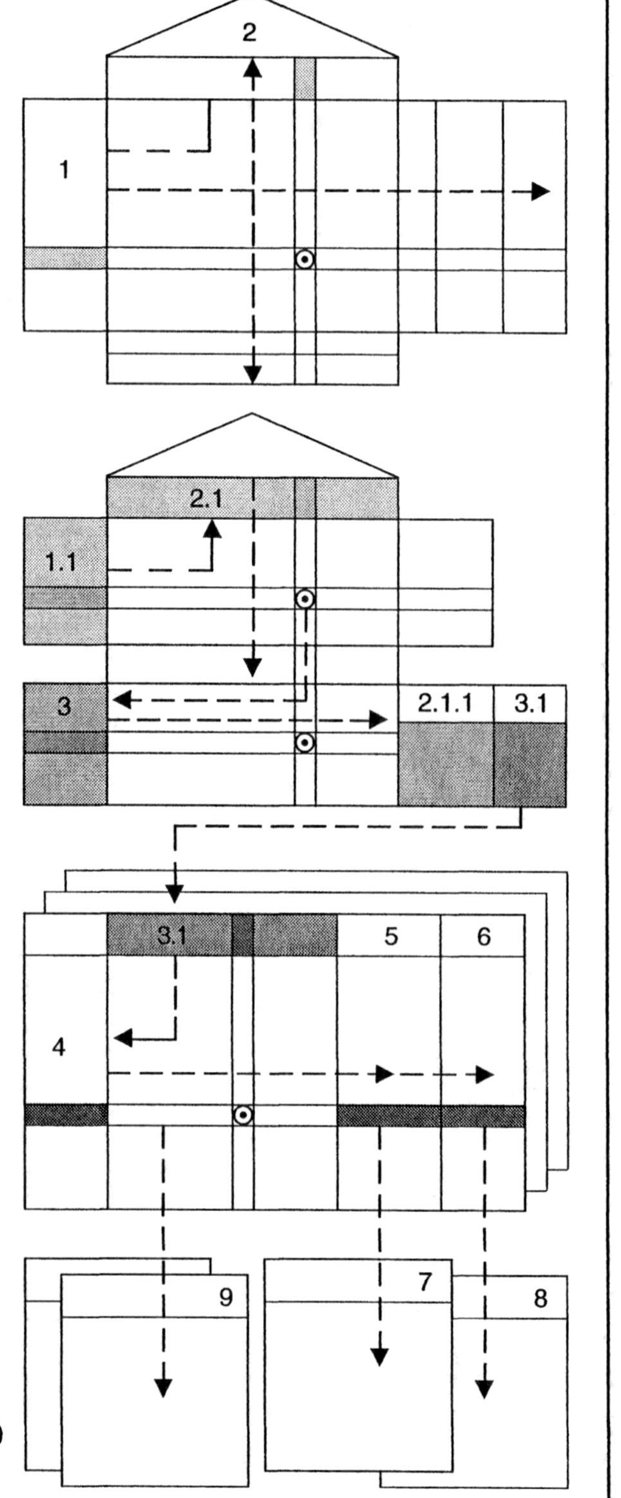

Table A.3 Initial Steps in the House of Quality Planning Chart

bac	1. Data on the "voice of the customer" are collected, sorted into tiers of related categories and entered as customer wants in the left-hand column. **Example:** 4—long life, 4.2—strong car, 4.2.1—many years durability.
bc	2. The customer needs are translated into product characteristics, features, or requirements and entered as headings across the matrix. **Example:** 7—rust prevention, 7.1—initial-stage rust.
ab	3. The relationship matrix between customer wants and product features is done. **Example:** Durability and rust prevention are strongly and directly related.
ab	4. The crucially important interaction matrix between the product features is done. **Example:** Preventing dust and rust during manufacturing will result in increased painting surface quality ➤ positive interaction. Washing the car (to reduce dust) during manufacturing can result in initial-stage rust ➤ negative interaction.

We will briefly survey items in the QFD charts that are identified with creative thinking. In Step 1 of the House of Quality (Table A.3), customer data are collected through brainstorming with the sales department and through interviews, questionnaires, and other surveys. Problem definition considers such questions as:

- Who will be the users of the planned product?
- Who will be the purchasers?
- What is the product expected to accomplish?
- How should the product perform?
- Are there any warranty claims against a similar "old" product?
- Why do users use this product?
- Why are others not using it?
- What features would turn the non-users into users?
- How does the product compare to the competition?

The customer "wants" data are sorted into related categories in a process similar to the one we use during the idea evaluation phase in creative problem solving. The Japanese call this process the affinity diagram or the KJ method. Table A.4 shows an example for a car door. A complicated product such as a car may have as many as nine levels of categories. What is important here, though, is that none of the customer requirements are either left out inadvertently or ignored on purpose. These groupings, as well as all customer wants, are listed in the left-hand column of the evaluation matrix. In the House of Quality, the first three tiers of customer wants are listed; in the subsequent component characteristics deployment matrix, the relevant customer wants are listed from the third tier on down to the last item in each subcategory to ensure that not a single identified customer want that is related to a critical part or component is left out.

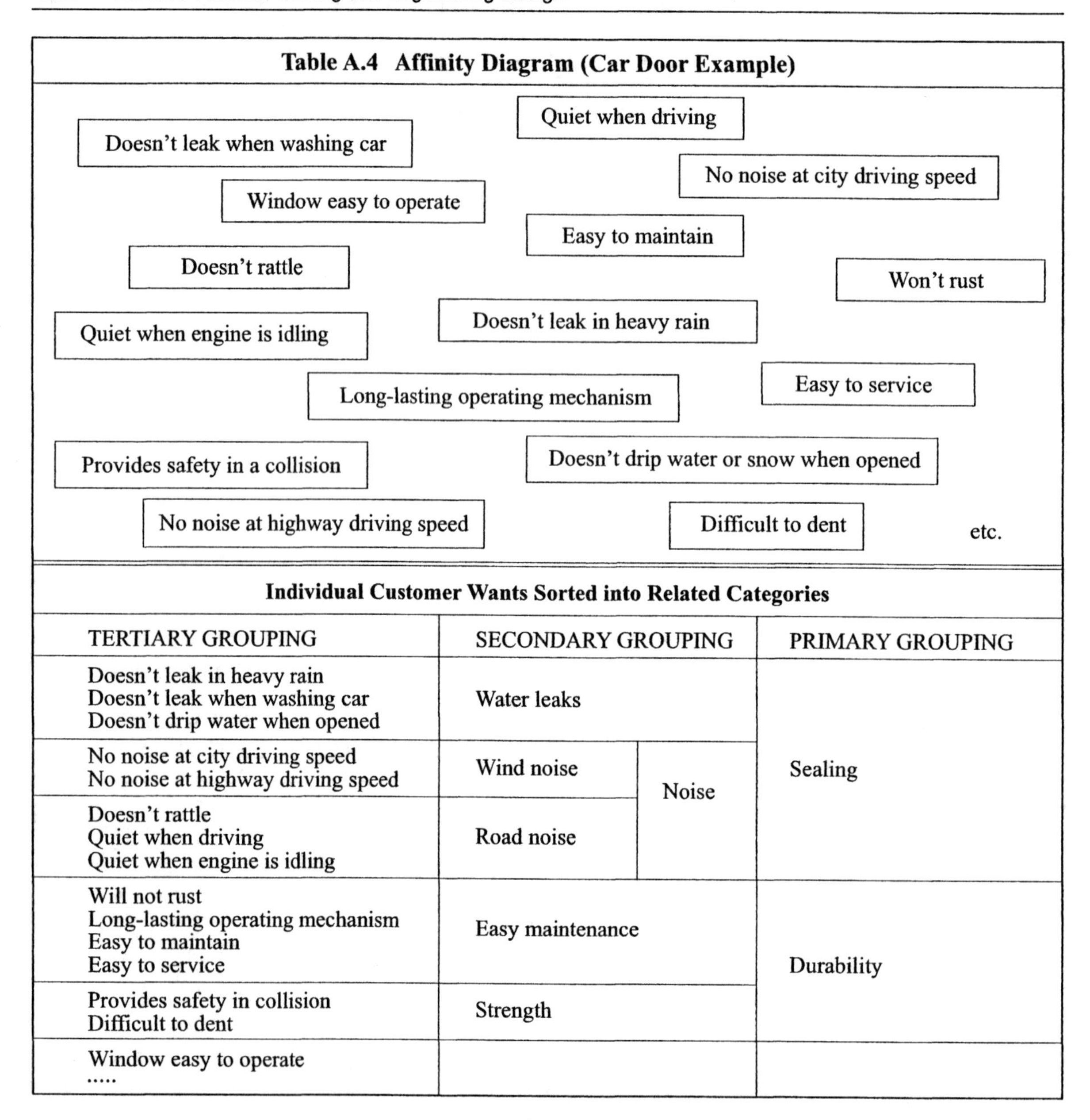

Table A.4 Affinity Diagram (Car Door Example)

Quiet when driving
Doesn't leak when washing car
No noise at city driving speed
Window easy to operate
Easy to maintain
Doesn't rattle
Won't rust
Quiet when engine is idling
Doesn't leak in heavy rain
Long-lasting operating mechanism
Easy to service
Provides safety in a collision
Doesn't drip water or snow when opened
No noise at highway driving speed
Difficult to dent
etc.

Individual Customer Wants Sorted into Related Categories

TERTIARY GROUPING	SECONDARY GROUPING		PRIMARY GROUPING
Doesn't leak in heavy rain Doesn't leak when washing car Doesn't drip water when opened	Water leaks		Sealing
No noise at city driving speed No noise at highway driving speed	Wind noise	Noise	
Doesn't rattle Quiet when driving Quiet when engine is idling	Road noise		
Will not rust Long-lasting operating mechanism Easy to maintain Easy to service	Easy maintenance		Durability
Provides safety in collision Difficult to dent	Strength		
Window easy to operate			

In Step 2 of Table A.3, engineers translate the customer wants into product features or characteristics through brainstorming. These product characteristics should be expressed in measurable engineering terms. Identifying those characteristics that actually cover the customer wants or needs is difficult and time-consuming but gets easier with practice. These product features are listed across the top of the House of Quality evaluation matrix.

In Step 3, the matrix is evaluated by the team by indicating whether there is a strong relationship, a weak relationship, or no relationship between each product characteristic and each customer need. If only a few strong relationships are identified, this means the product is not doing a good job of meeting the customer's needs. Another important task is to identify conflicting requirements. This is done in Step 4 with a second matrix above the product characteristics, making up the "roof" of the House of Quality. Here, each product feature is compared with all the others to see if there is a positive (reinforcing) relationship, no relationship, or a negative (conflicting) relationship. For areas of conflicting relationships, creative problem solving is needed because the design must be optimized to meet target values without traditional compromise and trade-off between cost and quality.

Identified negative relationships are starting points for creative thinking that can lead to innovation in a company's efforts to achieve highest quality at lowest cost.

When it is impossible to resolve the conflicting requirements for an important product feature, the development or application of new technology is indicated. With creative problem solving, promising ideas for investigation and development can be generated. For example, increasing the thickness of a metal panel could increase the strength of a weak component, but this would cause the hemming process to be more difficult—a negative interaction with present technology. Are there other ways of increasing strength—other materials or metal-working processes that could increase panel strength or would make hemming unnecessary in this part? Is this part necessary—could its function be taken over by some other component?

The House of Quality is continued with several additional analyses (see Table A.5). In Step 5 (or the "back porch"), the marketing analysis is listed with a customer importance rating and an analysis of competing products (related to customer wants). Strong selling points are noted in a separate column (Step 10). In the "basement," competing products are evaluated against the product characteristics. The targets or goals that the planned product has to satisfy are established; constraints and specifications that must be met are listed. These targets can form the basis of the list of criteria for the Pugh method (Steps 7, 8, 9, 11). See also Appendix B for a description of the benchmarking process. The House of Quality analysis is important because it determines the product characteristics that are critical for meeting the customer wants, those that are strong selling points, and those that have difficult targets (Step 12). Only these critically important product characteristics are carried forward for further deployment at the component level. Each model year, four or five items are typically selected for further deployment. When this process is repeated year after year, it becomes a process of continuous improvement that results in quality (and innovation) that is difficult to beat.

The House of Quality is only the initial document in QFD. Additional documents are required to assure that the voice of the customer via the identified critical product characteristics is carried through every step of product and process development and continued into production, marketing, and sales. We will briefly survey of each of these steps.

Table A.5 Additional Steps in the House of Quality

cab	5. The market analysis against customer wants is completed (including warranty data, customer importance rating, and competitive evaluation). **Example:** Customers have lodged many complaints about rusting, and three competing vehicles have outperformed the company's best product.
c	6. Technical standards in quantitative terms are listed as horizontal "basement" headings. **Example:** Initial-stage rust prevention, minimum 30 cycles.
ca	7. Competitive benchmarks (test data) are listed and evaluated against product features and technical standards (this listing is also known as the product evaluation chart). **Example:** The company's best product has the worst showing for rust.
c	8. Regulatory, warranty, liability, and other constraints and control items are listed in the lower left-hand column. **Example:** Important warranty component: rust-resistant paint job.
a	9. The relationship matrix between the constraints and technical standards is completed. **Example:** Initial-stage rust prevention and a rust-resistant paint job are strongly related.
ba	10. The key selling points are developed (from a comparison study of market analysis, the product evaluation chart, and the technical matrix) and listed to the right of the "back porch." **Example:** In this particular model, no selling points related to rust were listed; however, new features were a sunroof and a streamlined backdoor damper.
ba	11. Targets are developed from the market analysis, the product evaluation chart, the technical matrix, and the key selling points; these targets are entered on the product evaluation chart. **Example:** The rust performance target is set to substantially exceed that of the best competitor.
ab	12. The critical product features are selected from the importance indications in the relationship matrix, in the market analysis, and in the targets; these critical targets are "flagged" in the bottom line of the House of Quality. **Example:** Rust prevention is selected as one of the critical consumer wants because of high warranty claims and low performance when compared to competitive benchmarks.

QFD Phase 2—part deployment (component planning chart)

What parts or components are instrumental for achieving the selected critical product characteristics? Creative design concepts must be developed and evaluated in those important areas identified in the House of Quality. If the first idea that appears to work is taken without developing better options, the result is a rush to experimental hardware and wasted, unfocused effort. As shown in Chapter 11, the Pugh method can creatively generate and evaluate a variety of design concepts until a superior design emerges. This design is then processed further and culminates in the prototype to confirm the design.

QFD Phase 2 resembles a magnifying glass in that the QFD analysis zeroes in on detailed customer needs and the related detailed quality characteristics of the critical targets and product features identified in Phase 1. Another House of Quality analysis is conducted for the selected items, resulting in further clarification of critically important quality characteristics. The subsystems, components, and parts involved in these characteristics are then identified and the final, critical control characteristics are selected and parts specifications developed. Particular attention is focused on those parts that are critical to safety and to the

Table A.6 QFD Phase 2—Steps in the Component Deployment Chart

c	1.	Consumer wants are listed in detail for the critical items selected in the last step of the main House of Quality in the left-hand column of the component's House of Quality. **Example:** For rust prevention, 53 original customer wants are listed.
cb	2.	More detailed product quality requirements for the critical items above are listed as matrix headings; sometimes, weighting factors are assigned to these. **Example:** The initial three rust characteristics are expanded to six; resistance against spot rusting is targeted as most important.
ab	3.	The relationship matrix and the interaction matrix are completed.
cb	4.	The market quality evaluation (related to the customer wants) is graphically noted on the "back porch," together with the target values.
acb	5.	The competitive evaluation from test data and targets are entered.
ab	6.	The technical difficulty in achieving the important quality requirements is assessed and noted at the bottom of the chart. Also, areas where R&D is needed are pinpointed.
ab	7.	Testing specifications for problem clarification are developed (if applicable) and entered below the technical difficulty assessment.
acb	8.	The critical components, subassemblies, and parts related to the important quality characteris tics are identified and entered in the left-hand column in the bottom section of the chart.
a	9.	The relationship matrix between the critical components and the important quality character istics is completed. Only those components or parts that show critical relationships are deployed further. In particular, critical safety parts and critical function parts are identified. Value engineering analyses, FMEAs, and cost analyses are performed as appropriate on parts that have high warranty claims, high potential for cost reduction, and high variance.
acb	10.	Detailed final quality (or control) characteristics are entered for the critical parts.
bac	11.	Parts specifications are entered for each critical part (including numerical values and sketches), addressing the important quality requirements identified in the matrix. This activity demands creative problem-solving skills from the group because here the group specifies "how" the problem is to be solved in order to meet the customer needs and quality requirements.

quality performance of the product. Testing needs for the selected critical areas are established (if any). Process capability and R&D requirements are assessed. These steps are listed in more detail in Table A.6. It is interesting to note that the Japanese do not use the Pugh method—thus this is an area where creative thinking can give a competitive advantage.

QFD Phase 3—process planning and QFD Phase 4—production planning

Process planning and production planning occur concurrently with product planning and part deployment (product design). These steps are described in Table A.7. The process planning and quality control charts

Table A.7 Steps for the Production and Process Planning Charts

	Process Planning Charts (QFD Phase 3)
b	1. The final critical quality (or control) characteristics for each critical part are entered from the component deployment chart as headings in a new matrix. Each part has a separate chart.
abc	2. The processes needed to accomplish the control characteristics in the matrix heading are listed in the left-hand column.
a	3. The relationship matrix is completed.
abc	4. The control points are identified for each process and control characteristic.
acb	5. The process checkpoints are identified from the results of the matrix.
acb	6. Specifications for process monitoring are established for the process checkpoints (method, schedule, etc.).
	Quality Control Charts (QFD Phase 4)
c	7. The control points are listed (in the second column) from the process planning chart. For each process, a separate quality control chart is developed.
abc	8. The process flow schematic diagram is developed and sketched in the left-hand column of the quality control chart.
ac	9. The control method, sample size/frequency, and check method are determined and indicated for each control point.
	Additional Charts
acb	10. To complete the QFD procedure, operating charts, equipment installation charts, personnel training charts, etc., are developed as needed with information from the quality control charts, the process planning charts, and the component deployment chart.

that are developed in Phase 3 identify critical product and process parameters for each critical part, together with their control points and checkpoints. This information is used to set up and prove out the production lines. Optimum processes and sequencing can be developed with brainstorming. Process planning is usually done concurrently with product evaluation; creative ideas are exchanged between the teams working on these two activities.

The importance of this method cannot be overemphasized, regardless of whether the project involves the design of a whole system or of a single component. Its effectiveness dictates that it be followed from the beginning of all design projects.

It is important to first worry about* what *needs to be designed and only after that is fully understood to worry about* how *the design will look and work.

David G. Ullman, The Mechanical Design Process *(Ref. 11.4)*

In QFD Phase 4, the control and operating charts for each process identified in Phase 3 are developed; these charts specify operations and checks to be done by plant personnel to assure that all important customer needs are achieved, including SPC activities to monitor and solve problems of product variance. The necessary plans for equipment installation and personnel training are also developed. These large QFD charts, including the House of Quality, are posted in the factories where the shop floor workers study and understand them. They can see the direct relationship between their job and achieving the quality that the customer wants and buys. Production planning—although still conducted as a team activity—is primarily an analytical process supplemented with creative thinking to avoid arbitrary specifications. Table A.7 also summarizes the steps involved in the Phase 4 charts.

References

Reference 11.4 includes a discussion of QFD, with details on how to develop the list of customer requirements in terms of firm engineering targets in design applications (see also the quote in the sidebar). The following resource material on QFD is available from the American Supplier Institute, Dearborn, Michigan. In addition to several years of proceedings from QFD symposia, they have the following books and helpful software:

A.1 William E. Eureka and Nancy E. Ryan, *The Customer-Driven Company: Managerial Perspectives on QFD,* second edition, ASI Press, 1994.

A.2 QFD DESIGNER. Interactive software by Qualisoft. A demonstration package can be ordered from ASI.

A.3 Nancy E. Ryan, editor, *Taguchi Methods and QFD: Hows and Whys for Management,* ASI Press, 1988.

The design or planning and development phase of a product or service is extremely important to quality and offers many opportunities for the productive use of team creative problem solving, even if you are not following the entire QFD process.

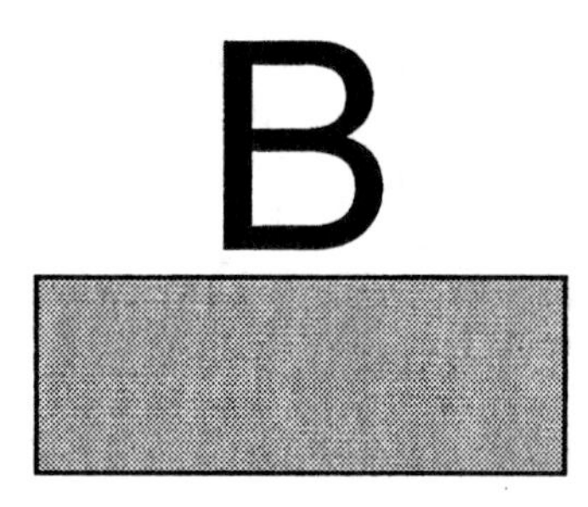

Benchmarking

Applications for benchmark studies

Benchmarking is part of a process of continuous improvement, not a one-time effort.

1. In developing the QFD House of Quality (see Appendix A), one of the steps involves analyzing the quality characteristics of competing products. Such benchmark data helps set target values for a new product that will exceed the competition in meeting customer requirements.

2. Benchmarking provides benefits in the context of engineering design, because it leads to a better understanding of the design problem. Although a benchmarking study concentrates on comparing customer requirements that are measurable, some factors can be more subjective and have to be evaluated with customer opinion surveys.

3. The Pugh method of design optimization uses the best existing product as a benchmark or datum for evaluating new concepts. The new designs, in an iterative process of optimization, must perform significantly better for all criteria than the datum (see Chapter 11).

4. Benchmarking is a way of learning from others. However, when this outside knowledge is integrated into an organization's culture and vision to create something new in a process of knowledge creation, that process can ultimately lead to true innovation, which lets the business "leap-frog" the competition (Ref. 3.8).

5. High product quality, low unit cost, fast development cycle, reliable service—these areas receive simultaneous attention and efforts for improvement in companies that want to be "best in class"—they are not satisfied with having to make trade-offs between the different areas. These companies use benchmarking to identify global leaders in their business and to measure their own progress—see Table B.1 for examples.

Lesson #1
You're probably not as good as you think you are.

Lesson #2
No company or plant is good at everything.

Lesson #3
Use value added per employee for a quick, meaningful measure of performance.

Edith A. Wiarda and Daniel D. Luria, Industrial Technology Institute, Ann Arbor

What is benchmarking?

Large companies develop and customize their own benchmarking procedures and manuals which tend to include proprietary information. Here is a simplified outline of the benchmarking process which brings out its major features. Even in a limited study, it is a complicated and lengthy

Table B.1 Examples of Benchmarking Results

Xerox undertook its first benchmark study in 1979, in an effort to turn the company around. Benchmarking analysis showed that the Japanese copiers had superior performance in several key areas when compared to Xerox: (a) production costs were 50% less; (b) cycle time to develop a new product and bring it to market was 50% shorter; (c) it involved 50% fewer people on the design and development team while (d) achieving quality —in terms of defective parts—that was an astounding 10 to 30 times better. The results of the study were so remarkable that, from 1989 to 1992, Xerox conducted 200 benchmarking studies in its manufacturing division alone, and many of its best practices have since been adopted by other companies. In one well-publicized study of a partner outside the industry, Xerox benchmarked the best-in-class shipping process at L.L. Bean—a key activity that not only helped improve warehousing but many other operations at Xerox.

Ford Motor Company (which in 1978 recalled more cars than it produced) surveyed the practices of other firms all over the world during the development of the Taurus, with the goal of meeting or exceeding performance specifications for over 400 parts. The Taurus became one of its most successful products. The lessons learned were: quality comes first; customers are the focus of everything the company does; improvement must be continuous; employees, dealers, and suppliers must be involved; global integrity is vital; and process is as important as the product. The use of many quality tools (Deming's principles, Taguchi methods, SPC, and QFD) in cross-functional teams led to new approaches in manufacturing.

Hewlett-Packard, as a result of a benchmarking study on "best scheduling practices," adopted QFD to gain a better understanding of customer requirements and improve its scheduling. The concept or measure of break-even time was adopted. Under old practices, the DeskWriter jet printer would have taken 4.5 years to develop; it took just 22 months under the new approach. Examples of other results are: (1) reduction in the manufacturing cycle from 4 weeks to 2 days on one line alone; (2) drop in order turnaround time from 6 weeks to 3 days in another operation.

process requiring teamwork, cooperation, and good communication with many people at all stages, from working for commitment, planning the study, acquiring the data, finding benchmarking partners, establishing trust, and implementing the changes.

Objective

The objective of benchmarking is to make changes that lead to significant, measurable improvements in products, processes, and services to gain a sustainable competitive advantage. The focus is on achieving *total customer satisfaction* with timely, cost-effective, high-quality products and services. Benchmarking leads directly to action plans that are integrated with the organization's mission and strategic plans at all levels, such as investment, marketing, quality, and technology development.

American firms must be willing to learn from the best performers in their industry, whether those best performers be domestic or foreign.

MIT Commission on Industrial Productivity (Ref. F.3).

Definition

Benchmarking is a commitment to excellence in a continuous process of defining *the critical success factors*—these are factors the customers think are important to superior performance. Benchmarking is a comparison study, either with the toughest competitors or the "best in class" in different fields. Ways are found and implemented to close the identified gap in the targeted area, resulting in fast change and the substantial and verifiable performance gains required to survive and be successful.

Benchmarking is an agent for change. It becomes as natural a part of work as answering a telephone.

Fred Bowers DEC, 1992

Prerequisites

Among the important prerequisites for successful benchmarking are: commitment from the top; willingness to learn from others; ood communication throughout the organization; employee understanding, involvement, and support; teamwork, and a project "owner" willing to change and committed to implement the process and results.

Scope

Large benchmarking projects involve upper-level management and a team of about a dozen people. They can take a year or longer, with change and implementation extending over three or more years. Medium-level projects involve mid-level management and experts, a team of five to eight people, last about half a year, and changes take up to two years to implement. Departmental projects involve mostly departmental employees, a team of two or three, a time frame of up to three months, with implementation as fast as possible. The costs can range from $5,000 for a modest study to $100,000 or more for larger projects—thus it is crucial to focus on the "20 percent of activities that affect 80 percent of the results" (paraphrasing the Pareto principle).

Procedure for benchmarking

The benchmarking process used by companies such as DEC, Motorola, Boeing, Xerox, and Ford is an approach that has five major steps for the team to accomplish—it looks at the "what" and "how" of a critical area in the organization and then compares them to best-in-class performers, followed by an analysis of results and an implementation plan.

A benchmark study can increase the awareness that quality must be designed into a product early in the design process to minimize manufacturing costs.

An attitude that refuses to adopt better ways because they are "not invented here" can hurt the ability of a business to keep up with the competition.

1. What do we want to benchmark?

- Analyze the organization's strategic or functional plan.
- Seek out the voice of the customer.
- Find what is considered to be *best in class* from the user's viewpoint.
- Identify those "what" factors that are critical to the success of your organization. Find, "Why do customers want to buy from us?" Factors that customer consider high priority but are seen as weak in your organization are a potential area to benchmark. For a benchmark study, these factors must be quantifiable and measurable.
- Select a benchmark topic from the high-priority items that directly relates to the organization's strategic plan.

2. How do we perform in this particular area?

- Identify "how" factors (processes and practices) that need to operate well for superior performance in the identified area.
- Collect data from key people on current performance; include flowcharts and information on the processes involved. This data collection takes time and must be carefully done and verified. Clearly

define the processes, data, and terms used, so the benchmarking partners can easily understand and compare equivalent items.

- Use the collected information to develop a questionnaire for the benchmarking partners, as well as for comparing your own future performance with your current status.
- As potential ideas for improvements surface during the study, jot them down for future discussion.

3. Who is best in the particular area of study?

- Develop a list of characteristics a partner must have to be considered best in class. Example: defect-free performance (six sigma or 3.4 defects per million).
- Select a list of potential candidates (from other fields as well as direct competitors).
- Obtain preliminary data on the gap of "where they are" and your current performance.
- Evaluate against the list of criteria and select three or four compatible organizations as benchmarking partners.

There are benefits of staying close to product users, not merely to better understand their rquirements, but because they are often sources of incremental product innovation.

Eric von Hippel, MIT

4. How do the partners do it?

- Finalize the questionnaire for collecting information from your partners, based on your own performance analysis. Focus on " What factors enable the partners to maintain the competitive gap?"
- First seek answers to your specific questions using information in the public domain.
- Contact the partners to obtain additional data and exchange information to determine the key factors and practices that support superior performance. The team must be willing to share information discovered during the benchmark study with the partners, and they may have to be sold on the benefits to them for participating in the study.
- Be efficient; don't waste your partners' time; carefully plan what you need to know.
- Determine if a site visit is needed for an accurate comparison or to confirm findings.

5. Analysis and implementation of change

- Prepare a summary of best practices and the resulting gap analysis.
- Set goals for improvement and if possible make recommendations for innovation targets (to gain a competitive advantage).
- Develop action plans for implementing the required changes. These action plans specify what is to be done by whom and when.
- Obtain approval for the action plans; then implement the plans!
- Monitor the implementation and measure the results.
- Celebrate the effort and success of the team; give credit for all help.
- Decide on the next step for continuous improvement.
- Strive for balance—optimizing one process may negatively affect the larger system; thus look to benefit the entire organization.

Table B.2 Useful Sources for Benchmarking Information

- ❑ Determine and use sources within your company, including lists in the proprietary benchmark manual.
- ❑ Do a library search on general references and industry survey publications, such as *Moody's Industrial Manual* or Dun & Bradstreet's *Principal International Businesses.* Librarians in the business section of large libraries are very knowledgeable about data base information that can help in your research.
- ❑ Check business periodicals, Wall Street analysts, market research reports, trade associations.
- ❑ Check U.S. government publications (such as *U.S. Industrial Outlook).*
- ❑ Also check colleges and universities, as well as on-line data bases.
- ❑ Contact customers, suppliers, distributors, and other industry insiders. Although these people can be helpful, do not expect them to share proprietary information obtained from their benchmarking partners.
- ❑ Take a plant tour in the area being benchmarked (not necessarily at the benchmarking partner's).
- ❑ Contact the benchmarking partners with specific questions—don't go fishing.

Where to get benchmarking information

Benchmarking requires a lot of data, and this data is not always easy to find. It is helpful to have a structured approach to avoid duplication. It may not always be necessary to conduct a site visit; the needed information may be in the public domain, or it can be collected by telephone through interviews with key personnel in your own organization and at your partners. Table 8.2 outlines possible steps—not all sources are equally helpful or reliable. However, information found in one area can lead to other valuable sources through cited references. It is often frustrating to find that the data one has been searching for is only available as an estimate, but it may still be very useful for the benchmarking study.

Benchmarking etiquette

The success of the benchmarking process depends on having good interactions with the people involved. This requires that everyone is familiar with the expected proper conduct, as outlined in Table 8.3. A protocol or set of conventions prescribing correct etiquette and procedures was developed jointly by the American Productivity & Quality Center's International Benchmarking Clearinghouse (APQC IBC) and the Strategic Planning Institute (SPI) Council on Benchmarking and includes the following important principles:

1. LEGALITY: Although benchmarking relationships are expected to be based on trust and openness, the interaction between partners who are close competitors requires ground rules about the areas where information can be shared and those that will be restricted. Do not ask for sensitive data. Also, stay away from discussions or actions that could imply a restraint of trade, price fixing, disclosure of proprietary information, or similar questionable practices. If there is a doubt, seek legal counsel.

Under no circumstances should improper, unethical means be used to uncover business secrets.

2. FAIR EXCHANGE: Make sure that each partner gets something of value from the investment of time in the study. Never ask for any information that your own company would not want to share with another.

Table B.3 Proper Conduct for "Benchmarkers"

PREPARATION CHECKLIST:
___ Be knowledgeable about the benchmarking process.
___ Before contacting benchmarking partners, know what you want and how it will benefit "them."
___ Have the first two steps of the process already completed, as well as the list of interview questions.
___ Be thoroughly prepared to conduct the visit in an efficient manner.
___ Make sure you have the authority to share information with the benchmarking partners.
___ Go through the proper channels. Agree on meeting schedules and logistics (including the agenda).
___ Check with your superiors or company policy if confidentiality agreements will be required.
___ Operate on a firm commitment to go through with the study and implement the results.

SITE VISIT GUIDELINES:
___ Follow the agreed-on process without demanding changes or additions.
___ Your conduct must be above reproach: honest, courteous, on-time, straightforward, professional.
___ Introduce all the people in your party and explain why they are included.
___ Keep to the agenda and the schedule; honor the benchmarking protocol for all interactions.
___ Observe good communication skills; avoid jargon.
___ Share information from your internal process analysis.
___ Offer to host a reciprocal visit and thank your hosts for their assistance.

Complete, open, and honest communication in the initial stages of establishing the benchmarking relationship can help each partner understand the value of the exchange to be made.

3. CONFIDENTIALITY: Treat all information gained from the partners as proprietary and confidential—it should not be revealed to third parties or in a public forum. Even the names of the partners should not be revealed to others without special permission.

4. USE OF INFORMATION: Have a common understanding of how information will be handled. In general, the information gained is for the sole purpose of improving operational processes within a company, not for advertising, marketing, or selling.

The Benchmarking Cycle

Explicit Knowledge:

1. **Here is where you are.**
2. **Here is where you want to be.**

Tacit Knowledge:

3. **Here is an example of how you can get there.**
4. **NOW GO and DO IT!**

References

B.1 Robert C. Camp, *Business Process Benchmarking: Finding and Implementing Best Practices,* ASQC Quality Press, Milwaukee WI, 1995.

B.2 Peter Merrill, *Do It Right the Second Time: Benchmarking Best Practices in the Quality Change Process,* Productivity Press, Portland OR, 1997.

B.3 Edith A. Wiarda and Daniel D. Luria, "The Best-Practice Company and Other Benchmarking Myths," *ASQ Quality Progress,* Vol. 31, No. 2, pp. 91-94, Feb. 1998. The authors are part of the Industrial Technology Institute, Ann Arbor, Michigan—a resource on benchmarking for smaller businesses. The article cites a number of helpful web sites for benchmarkers.

B.4 Gregory H. Watson, *Strategic Benchmarking: How to Rate Your Company's Performance against the World's Best,* John Wiley and Sons, New York, 1993. This is a basic overview which includes interesting case studies.

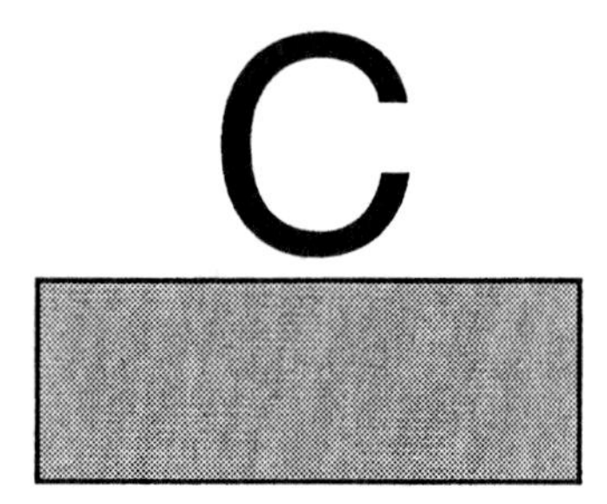

SPC (Statistical Process Control)

SPC is a tool that uses statistical data and comparisons to monitor processes; its primary objective is to prevent problems, similarly to what a doctor is doing during your annual physical exam. A process is simply a sequence of actions or activities which result in a product or service. One of the primary goals of quality control is to assure that the results of a process remain uniform—or in other words, that variations and the resulting decrease in quality are kept to a minimum. When a process is controlled, this means that it is monitored to assure the product meets the quality targets. SPC is also a valuable tool for problem analysis: by making graphs of the data and then analyzing the results, the causes of problems can be identified. The originators of SPC were two American quality experts: Dr. Walter A. Shewhart of Bell Laboratories in the 1920s and Dr. W. Edwards Deming after World War II.

The first step in using SPC to analyze a problem is to collect data. This data must be relevant; it must be usable; and it must be accurate and reliable. Always check the calibration of the instruments that are used for measurements. Carefully think about data collection to avoid creating a data swamp with too much information. SPC has seven tools: check sheet, histogram, fishbone diagram, Pareto diagram, scatter diagram, control charts, and additional documentation.

THE TOOLS OF SPC

1. **Check sheets**
2. **Histogram**
3. **Fishbone diagram**
4. **Scatter diagram**
5. **Pareto diagram**
6. **Control charts**
7. **Other graphs**

1. Check sheets

Basically, there are three different types of check sheets. The *checklist* is a memory aid; it helps us to follow procedures correctly. A simple example is the shopping list you take to the grocery store. Following a checklist scrupulously can be very important—airline pilots use checklists before takeoff. In a manufacturing plant, checklists are frequently used for maintenance procedures and schedules of various sorts. Your car dealer or service station may use a check sheet for your vehicle's 20,000-mile maintenance work. *Recording check sheets* are for collecting data on frequency. Instructors may use a recording check sheet for class evaluation at the end of a term to get a statistical look for identifying areas for improvement. A *location check sheet* is a drawing of a product part or map; the location of defects is marked so that the investigation can zero in on the most critical areas. Another example is a floor plan of the plant, where all locations of accidents are noted that occurred within a specified time period. Or a state's highway department marks accidents and fatalities on roads and highways on a large map to identify danger spots and areas for improvement. Data visualized this way can be very helpful in pinpointing critical areas for an investigation and corrective action; then this baseline data simplifies comparison once the improvements have been made. Check sheets are important in problem solving because they collect data in easily used form. If set up carefully, they make data analysis and troubleshooting easier.

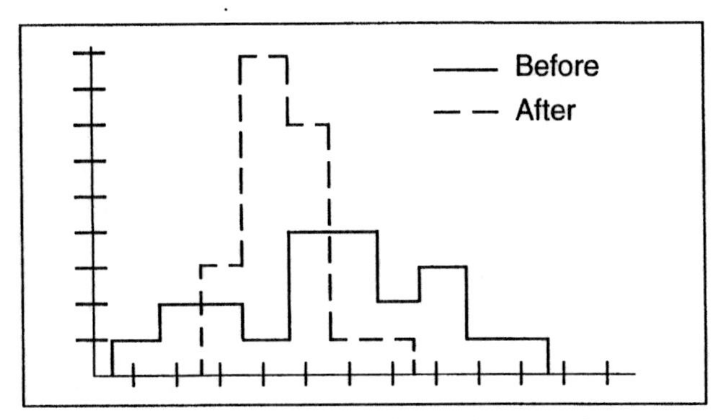

Figure C.1 Generic histograms showing reduction in variability after corrective action was implemented.

2. Histogram

A histogram is a bar graph depicting frequency versus variation of some product parameter. It is often used in charting the precision of machines or in process capability studies to determine the relationship between target values and actual production values. Its aim is to eliminate defects and improve yield and product quality; it is a key tool in the factory to chart and monitor "continuous improvement." Figure C.1 shows a comparison of two histograms, with the dotted line showing improved performance.

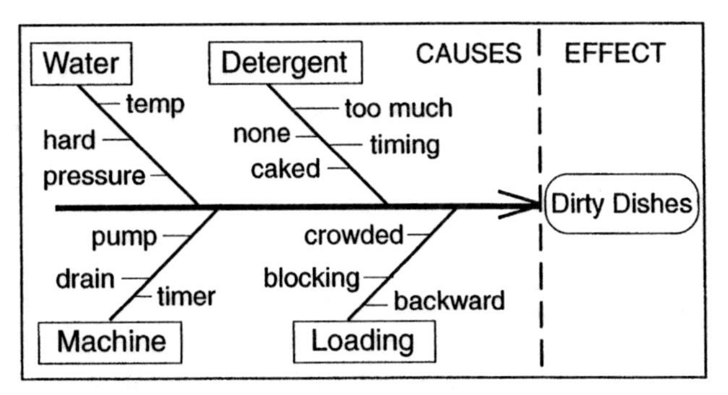

Figure C.2 Example of a fishbone diagram (excerpt) for dishwasher not getting dishes clean.

3. Cause-and-effect (fishbone) diagram

Raw materials, work methods, equipment, and measurement can be the cause of quality variation. Through brainstorming, a cause-and-effect diagram can be built-up in detail by asking the question: "Why does this variation occur?" Such diagrams clearly illustrate the various causes affecting product quality, and they can aid record-keeping of production deviations. The cause-and-effects diagram was invented by Kaoru Ishikawa in 1943. The main characteristic that needs improvement is placed along the main horizontal arrow (or "fish vertebra") as illustrated in Figure C.2. The main factors that may be causing dirt on dishes after washing are determined and entered as side branches in the diagram. The diagram can show various levels of complexity.

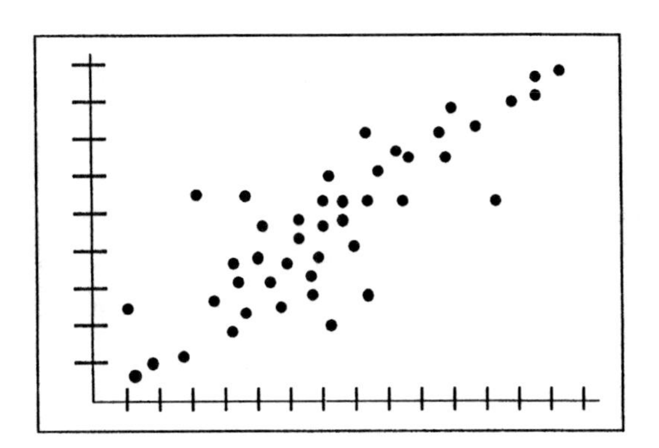

Figure C.3 Generic scatter diagram showing a positive correlation.

4. Scatter diagram

Scatter diagrams are frequently used to study the relationship between a cause and its effect, with cause values commonly plotted on the horizontal axis and the effect on the vertical axis (as shown in Figure C.3). The resulting scatter patterns will indicate positive correlation if both axes show an increasing trend, negative correlation

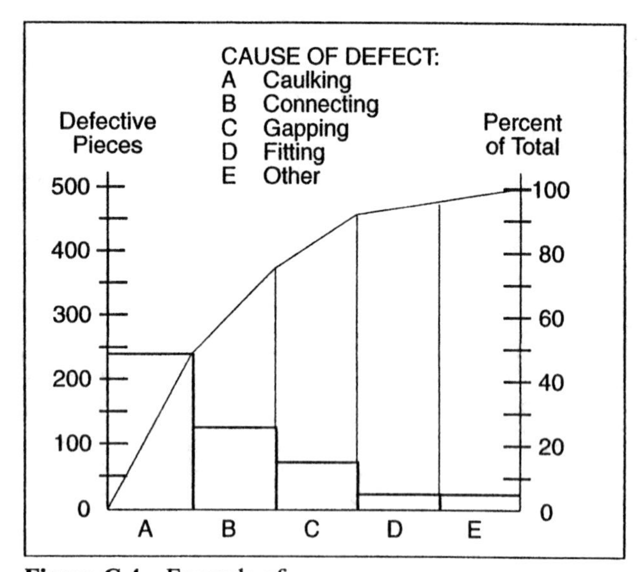

Figure C.4 Example of a Pareto diagram (adapted from Ref. C.3).

if the effect decreases for increasing cause values, or no correlation if the plotted dots are randomly scattered. If the pattern is crescent-shaped, more than one cause may be involved, and further analysis should be done to clarify these effects.

5. Pareto diagram

The Pareto diagram is a specialized bar graph used to identify and separate the vital, the most important, causes of trouble from the more trivial items. The Pareto diagram arranges classification items by order of importance and thus points out which factors or causes need to be addressed first for improvement. The vertical axis can be expressed in numbers or percentage of cases, but the most useful parameter is money lost by the defect. This diagram was "invented" by Vilfredo Pareto, an Italian economist who was struck by the fact that 20 percent of the population in a country control 80 percent of the wealth This "80/20 principle" makes it possible to concentrate resources on removing the top 20 percent of the causes and thus cure 80 percent of the problems. The Pareto diagram is useful for assigning priorities for continuous improvement; an example is given in Figure C.4. Among the operations causing defects, caulking should be the initial target to improve quality.

6. Control charts

Control charts present data plotted in a chronological sequence to show how the influence of various factors in the production process (materials, workers, methods, equipment) changes over a period of time (see Figure C.5). Data points falling outside the limit lines signal an abnormal situation that needs appropriate action. The control limits are calculated mathematically—they are not specifications. The daily data, for example, are averaged to obtain an average (mean) value for that day. Each of these values then becomes a point on the control chart which represents the characteristics of that day. Both the mean value $\underline{x}$ and the range R are plotted, where R is the difference between the largest and smalles value in each subgroup. The $\underline{x}$ portion of the chart shows changes in the mean value of the process, while the R portion shows

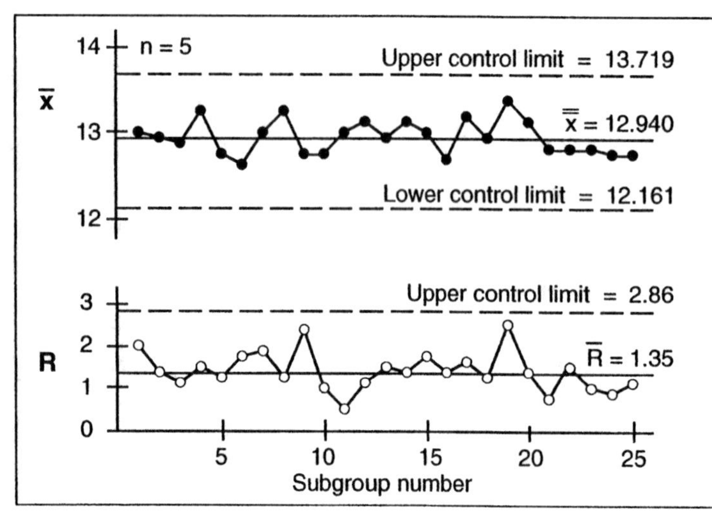

Figure C.5 Process control chart (adapted from Ref. C.3).

changes in the dispersion of the process, using a minimum of 100 samples. Control charts are powerful tools for preventing the fabrication of defective products; they are also used to track quality improvements and identify when a process has achieved a decrease in variation. Quality circles or teams use this "problem-definition" data to brainstorm ways of improving processes and quality.

7. Additional graphs and documentation

Other types of graphs can be used as "detective" tools to document defects or visualize identified causes, such as flow diagrams, stem-and-leaf plots, line graphs, circle diagrams, spaghetti charts, and matrices.

Quality control circles are a unique feature in Japan where workers study and analyze the quality control process on their own initiative.

References

C.1 Don P. Clausing, *Total Quality Development: A Step-by-Step Guide to World-Class Concurrent Engineering,* ASME Press, Fairfield, New Jersey, 1994. This book addresses the problem of quality in engineering and is written for highly technical readers.

C.2 W. Edwards Deming, *Out of the Crisis,* MIT Center for Advanced Engineering Study, Cambridge, Massachusetts, 1982. Quality in manufacturing is discussed by one of the early leaders of the quality movement in Japan.

C.3 Kaoru Ishikawa, *Guide to Quality Control,* Asian Productivity Organization,(available from Unipub, White Plains, NY, 1976). This book, written in Japanese in 1968 and later translated into English, gives a nice overview of the tools of statistical process control.

C.4 Kaoru Ishikawa, *Introduction to Quality Control,* 3A Corporation, Tokyo, 1990. This is an updated version of Dr. Ishikawa's quality principles and quality control techniques.

C.5 Kaoru Ishikawa (translator: D.J. Lu), *What Is Total Quality Control?* Prentice-Hall, Englewood Cliffs, NJ, 1985. This is a key tool for TQM.

Biographical note on Dr. Kaoru Ishikawa

Dr. Kaoru Ishikawa was born in 1915 and graduated in 1939 from the Engineering Department of the University of Tokyo with a major in applied chemistry. In 1947 he was appointed assistant professor at Tokyo University. After earning the doctorate in engineering, he was promoted to the position of full professor in 1960. He won the Deming prize and the Nihon Keizai Press Prize, the Industrial Standardization Prize for his writings on quality control, and the Grand Award (1971) from the American Society for Quality Control for his education programs in quality control. He has written many books on quality control and statistics, and his methods have contributed greatly to his country's economic and industrial development. The *Guide to Quality Control* was written specifically for the "Quality control circle members—let's study!" campaign.

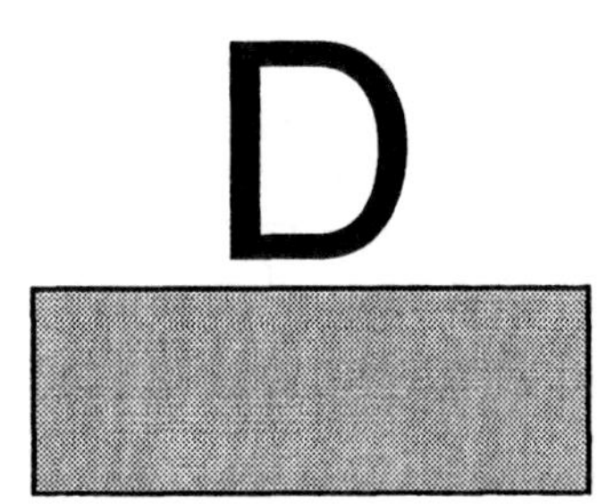

FMEA

(Failure Mode and Effects Analysis)

Definition and objectives

A failure mode and effects analysis (FMEA) is a systematic, analytical technique performed by an experienced design, manufacturing, or quality control engineer on a planned product, manufacturing process, or quality control system to assure that the product characteristics will meet customer needs. The FMEA allows the engineer to assess the probability as well as the effect of a failure. It identifies significant process variables to be controlled. The FMEA should be performed as early as possible in an engineering program (or product development) and be updated as later information becomes available. By identifying potential problem areas, an early FMEA will aid engineers in *directing timely design actions to prevent defects* and in planning appropriate test programs. The FMEA also documents the rationale for the manufacturing or assembly process being developed—it represents a valuable data base to track future actions and improvements and to train new personnel. The objectives of an FMEA are (1) identification of potential and known failure modes; (2 identification of the causes and effects of each failure mode; (3) ranking in priority of the identified failure modes according to frequency of occurrence, severity, and detection, and (4) specific directions for problem follow-up and corrective action.

Description of steps

An FMEA is a living document and should always reflect the latest product design and manufacturing process actions.

Each organization can devise its own FMEA worksheet format. The following sequence of steps may be followed to record the analysis data.

1. General identification: These items are usually entered in the space above the worksheet headings and include:

a. Subsystem or process name and numerical designation.
b. Model year, lines that will utilize the product or process.
c. Division or office that has the design responsibility.
d. Other departments involved, including manufacturing and design.
e. Outside suppliers of a major component within the subsystem.

f. Engineers: Name and phone number of the subsystem, design, or process engineer responsible for the FMEA; system supervisor.
g. Scheduled date for release of complete subsystem.
h. Date of first FMEA and dates of later revisions.

The following items are entered as separate columns in the worksheet.

2. Part or process identification: Specify the assembly, process, or component being analyzed exactly as shown on the design drawings and indicate all identifying information, including design level.

3. Part or process function: Concisely describe the function of the analyzed part or process, to help identify the consequence of failure.

Typical failure modes:

Bent	Cracked
Leaking	Blistered
Damaged	Loose
Porous	Bound
Deformed	Melted
Rough	Brittle
Discolored	Misaligned
Shorted	Broken
Distorted	Tight
Corroded	Grounded
Omitted	Wrinkled
Misassembled	

4. Failure mode: Describe each possible failure mode by considering the question, "What could possibly go wrong with this part, system, or process?" No judgment is to be made on whether or not it will fail, only on *how it might possibly fail.* As a starting point, review the problems in past design FMEAs or with the quality, warranty, durability, and reliability of comparable components.

Effect of failure:

Designs: Excessive operating effort. Engine will not start. Fuel fumes. Oil leakage. Insufficient A/C cooling. Luggage compartment water leaks.

Processes: Stops the line. Generates loud noise. Damages parts. Impairs safety.

5. Effect of failure: Describe, "What does the customer experience as a result of the failure mode just listed?" In the case of a process FMEA, control items are flagged with a special symbol, such as ♦.

6. Causes of failure: Analyze what conditions can bring about the failure mode; list all potential causes for each failure mode. For example, would poor wire insulation cause a short? Would a sharp sheet metal edge cut through the insulation and cause the short? Other examples of causes are listed in the sidebar.

Causes of failure:

Broken wire
Inadequate venting
Damaged part
Incorrect speeds, feeds
Handling damage
Material failure
Heat treat shrinkage
Missing operation
Improper surface prep
No lubrication
Inaccurate gauging
Out-of-tolerance
Inadequate clamping
Packaging damage
Inadequate control system
Worn tooling

7. Current controls: This step is applicable to process FMEAs only. List all current controls of process variables that are intended to prevent the causes of failure from occurring or to detect the causes of failure or the resultant failure mode.

8. Estimate of occurrence frequency: Estimate the probability that the given failure mode will occur, using an evaluation scale of 1 to 10, with the low number indicating a low probability of occurrence and a 10 indicating near certainty of occurrence as shown in Table D.1. Probability in Table D.1 means the statistical proportion outside the specification limits. Each organization can develop its own ranking tables; the only requirement is that the tables will be used consistently throughout the company's FMEAs.

9. Severity: Evaluate the severity or estimated consequence of the failure on a scale of 1 to 10, with the low number indicating a minor nuisance and a 10 indicating a severe, total failure, as shown in Table D.2.

Table D.1 Occurrence Ranking Criteria

Criteria	Ranking	Probability
Remote probability of occurrence. Process capability shows at least $x \pm 4\,\sigma$ within specifications.	1	1/10,000
Low probability of occurrence, with process in statistical control. Capability shows at least $x \pm 3\,\sigma$ within specifications.	2 3 4 5	1/5,000 1/2,000 1/1,000 1/500
Moderate probability of occurrence, for processes experiencing occasional failures. Process is in statistical control, with $x \pm 2.5\,\sigma$ within specifications.	6	1/200
High probability of occurrence, with frequent failures. Process is in statistical control, but capability shows $x \pm 2.5\,\sigma$ or less within specs.	7 8	1/100 1/50
Very high probability of occurrence. Failure is almost certain to occur sooner or later.	9 10	1/20 1–1/10

Table D.2 Severity Ranking Criteria

Criteria	Ranking
Minor nature of failure, no noticeable effect on performance, undetectable by customer.	1
Low severity, causing only slight customer annoyance due to very minor subsystem performance degradation.	2 – 3
Moderate failure causing some customer discomfort, dissatisfaction, and annoyance due to subsystem or total performance degradation.	4 – 6
High degree of customer dissatisfaction due to nature of the failure (inoperable subsystem or total system).	7 – 8
Very high severity ranking for failure mode involving potential safety problems and/or nonconformance to federal regulations. Nonregulated components with a 9 or 10 severity ranking and occurrence rankings > 1 should be designated as control items (♦).	9 – 10

Table D.3 Detection Ranking Criteria

Criteria	Ranking	Probability
Remote likelihood that product would be shipped containing such an obvious defect, since it is detected by subsequent factory operations.	1	1/10,000
Low likelihood for shipment with defect which is visually obvious or has 100% automatic checking.	2 3 4 5	1/5,000 1/2,000 1/1,000 1/500
Moderate likelihood for shipment with defect, since the defect is easily identifiable through automatic inspection or functional checking.	6 7 8	1/200 1/100 1/50
High likelihood of shipping with subtle defect.	9	1/20
Very high likelihood that defect will not be detected prior to shipping or sale (checks are impossible or defect is latent).	10	1–1/10

This summary is based on two Ford Motor Company document sources:
• Reliability Methods: Failure Mode and Effects Analysis—Module XIV (January 1972).
• Potential Failure Mode and Effects Analysis for Manufacturing and Assembly Processes (Process FMEA) Instruction Manual (December 1983).

10. Failure detection: Estimate the probability that the problem will be detected before it reaches the customer. A low number indicates high probability that the failure would be detected; a 10 means that it would be very difficult to detect the failure before the product is shipped or sold (see Table D.3). For example, an electrical connection left open and thus preventing engine start might be assigned a detection of 1; a loose connection causing intermittent no-start might be a 6; a connection that corrodes causing no-start after a period of time might be assigned a 10.

11. Calculation of risk priority number (RPN): When the numbers of (h), (i), and (j) are multiplied, the RPN index is obtained, providing an indication of the relative priority of the failure mode. Regardless of the RPN value, components or systems receiving a high occurrence ranking should be given special attention and corrective action should be taken.

12. Recommended corrective action: Follow-up is critical to the success of the FMEA. Responsible parties and timing for completion should be designated for all corrective actions. Give a brief description of the corrective action or actions recommended to prevent the failure mode, including the person responsible for implementing the solution and the status of the corrective action (transmittal numbers, promise date, etc.). Based on the analysis, actions can be specified for the following:

- To reduce the probability of occurrence, process or design revisions are required. An action-oriented study of a process via SPC (see Appendix C) should be implemented with an ongoing feedback of information for never-ending improvement and defect prevention.
- To reduce the severity of product failure modes, part design actions are required. As a means of highlighting significant process effects, the severity ranking may be increased, depending on the effect of the failure on subsequent process operations in the plant.
- To increase the probability of detection, process revisions are required. Generally, improving detection controls is costly and ineffective for quality improvement, since inspection is not a positive corrective action. In some cases, a design change to a specific part may be required to assist in detection. Changes to the current control system may be implemented to increase the probability of detection. Emphasis must be placed on preventing defects (that is, reducing the occurrence) rather than on detecting them.

EXAMPLE

4. **Plastic lever breaks.**
5. No drive, locked in park.
6. Overload on lever, inferior plastic material, brittle when cold, damaged in handling.
7. None.
8. Occurrence ranking = 3
9. Severity ranking = 10
10. Detection ranking = 9
11. RPN = 270
12. **Corrective action:** Redesign lever with thicker material and strengthen ribs to carry 100% overload.

13. Actions taken: The need for the best possible FMEA has been reinforced by studies of safety and emission recalls. These studies found that in a number of cases a fully implemented FMEA program with follow-up on critical concerns would have made the recall unnecessary. An effective FMEA is a living document and should always reflect current status of processes and product design. Therefore, a section on actions taken is added to the worksheet periodically, together with a new listing of occurrence, severity, and detection and the revised PRN calculation, and with identification of the individual responsible for the action.

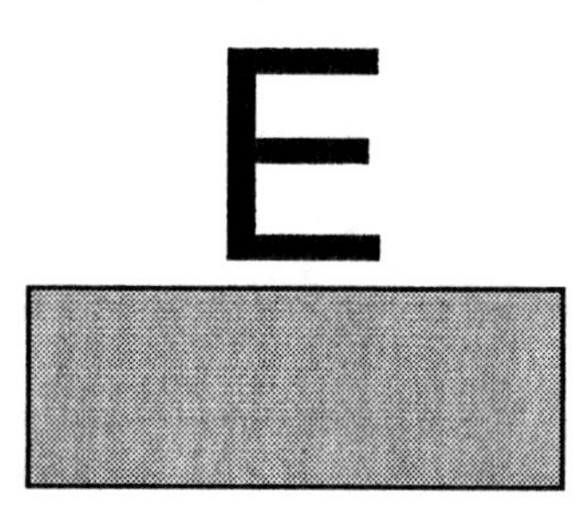

FTA (Fault Tree Analysis)

Definition and objectives

Fault tree analysis (FTA) is a method of system reliability/safety analysis. The fault tree analysis provides an objective basis for analyzing system design, justifying system changes, performing trade- off studies, analyzing common failure modes, and demonstrating compliance with safety requirements. The concept of fault tree analysis was originated in 1961 by H. A. Watson of Bell Telephone Laboratories to evaluate the safety of the Minuteman launch control system. Fault tree analysis is a deductive analysis that requires considerable information about the system; it is a graphical representation of Boolean logic associated with the development of a particular system failure.

Benefits of fault tree analysis

- It provides options for qualitative and quantitative reliability analysis.
- It helps analysts to understand system failures deductively.
- It points out the aspects of a system that are important with respect to the failure of interest.
- It provides insight into system behavior.
- It is restricted to the consideration of one undesirable event.
- It graphically communicates the sequence of causes leading to the undesirable event.
- It directs the analysis toward elimination of the undesirable event.
- It is concerned with ensuring that all critical aspects of a system are identified and controlled.

The FTA is restricted to the analysis of a particular failure and the sequence of causes leading to the failure.

Definitions

- Fault tree—model that graphically and logically represents combinations of possible normal and fault events leading to the top event.
- Top event—system failure under investigation.
- Primary events—basic failures (causes) leading to the top event.
- Fault event—abnormal system state.
- Normal event—event that is expected to occur.
- Event—dynamic change of state occurring in a system element.
- System element—hardware, software, human, and environmental factors.
- Logic gate symbols are defined in Table E.1, event symbols in Table E.2.

This material has been condensed from 1986 lecture notes prepared by Dr. K. C. Kapur of the American Supplier Institute, Dearborn, MI.

Table E.1 Definition of Logic Gate Symbols

Symbol and Gate Name	Causal Relationship
"And" gate	Output event occurs if all input events occur simultaneously.
"Or" gate	Output event occurs if any one of the input events occurs.
Inhibit gate	Input produces output only when a conditional event occurs.
Priority gate	Output event occurs if input events occur in order from left to right.
Exclusive gate	Output event occurs if one (but not both) of the input events occurs.
Sample gate (m, n)	Output event occurs if m out of n input events occur.

- Gate symbols connect events according to their causal relations.
- A gate may have one or more input events but only one output event.

Table E.2 Definition of Event Symbols

Event Symbol	Meaning
Rectangle	The rectangle defines an event that is the output of a logic gate; it is dependent on the type of gate and the inputs to the gate.
Circle	The circle defines a basic inherent failure of a system when operated under specified conditions → primary failure. Sufficient data.
Diamond	The diamond represents a failure other than a primary failure that is purposely not developed further. Insufficient data.
Switch or house	The switch or house represents an event that is expected to occur (ON) or to never occur (OFF) because of design and normal operating conditions. It is thus used to examine special cases by forcing some events to occur and other events not to occur.
Oval	The oval represents a conditional event used with the inhibit gate.
IN OUT	Arrowheads are transfer symbols to and from the fault tree to simplify the representation.

- The fault tree sequence leads down from system failure through a sequence of events to its basic causes.
- Elimination of the causes eliminates the top event.

Description of steps

Whereas FMEAs consider all possible failure modes for a product or process, fault tree analysis is restricted to the identification of the system elements and events that lead to a single, particular system failure. The steps for fault tree analysis are: (1) Define the top event. (2) Establish boundaries. (3) Understand the system. (4) Construct the fault tree. (5) Analyze the fault tree. (6) Recommend and take corrective action.

Fault Tree Construction

The fault tree is structured so that the sequence of events that leads to the failure is shown below the top event; these events are related to the failure by the logic gates. The input events to each logic gate are shown as rectangles; they are usually the output of lower-level logic gates. These events are developed to lower levels still until the sequence of events leads to basic causes of interest (represented in circles). Diamonds at the lower edge of the fault tree indicate the limit of resolution of the analysis. Three types of causes can contribute to failure:

Table E.3 Example System

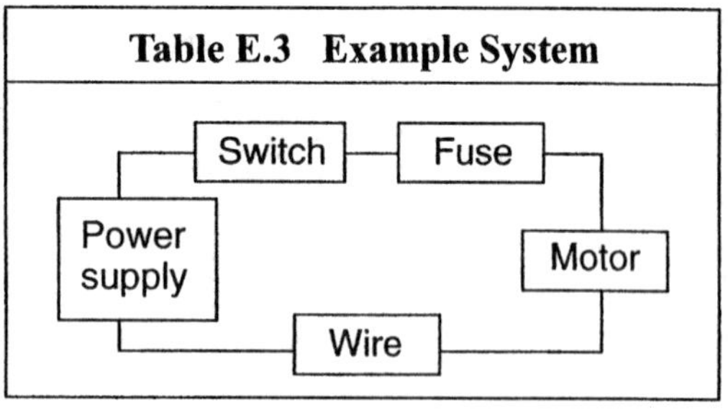

1. A primary failure can be due to the internal characteristics of the system under consideration.
2. A secondary failure can be due to excessive environmental or operational stresses.
3. A command fault can result from the inadvertent operation or nonoperation of a system element due to failure of elements that can control or limit the flow of energy to respond to system conditions.

Table E.4
First Fault Tree for the Example System

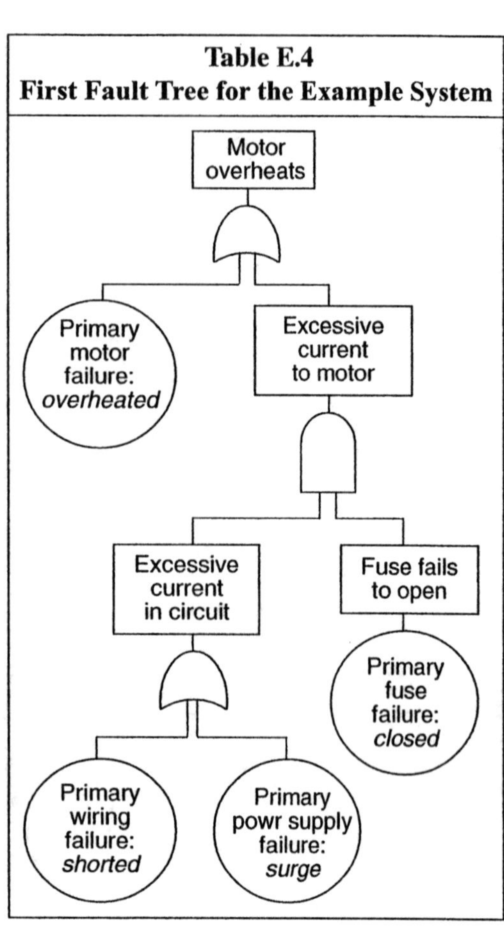

Example of fault tree analysis

Table E.3 shows a diagram for the operation of an electric motor. The system boundary conditions are:

Top event = Motor overheats.
Not-allowed events = Failures due to effects external to the system.
Existing events = Switch closed.

Table E.4 shows by inductive reasoning that the motor overheats if an electrical overload is supplied to the motor or if a primary failure within the motor causes the overheating (through bearings losing their lubrication or through a short within the motor). From a knowledge of the components, the fault tree construction is continued to two other causes, which in turn are caused by primary failures (in circles).

Table E.5 shows a fault tree for the same system that was given in Table E.3, but for different boundary conditions:

Top event = Motor does not operate.
Initial condition = Switch closed.
Not-allowed events = Failures due to effects external to the system.
Existing events = None.

Here the diamond symbol has been chosen to show that the "open switch" is a failure external to the system boundaries and that insufficient data are available to develop the event further. Secondary fuse failure can occur if an overload in the circuit occurs, because an overload can cause the fuse to open. However, the fuse does not open every time an overload is present in the circuit because not all conditions of an overload result in sufficient overcurrent to open the fuse. Thus the oval off the diamond inhibit gate gives the condition that would cause the secondary fuse failure. Even though the development and analysis of the fault tree are nominally separate tasks, there is much interaction between the two activities. During the course of analysis, additions and changes are made to the tree as new insight is gained into the failure paths. The analysis must be followed by recommendations for action that will lead to the elimination of the main event failure.

Table E.5 Second Fault Tree for the Example System

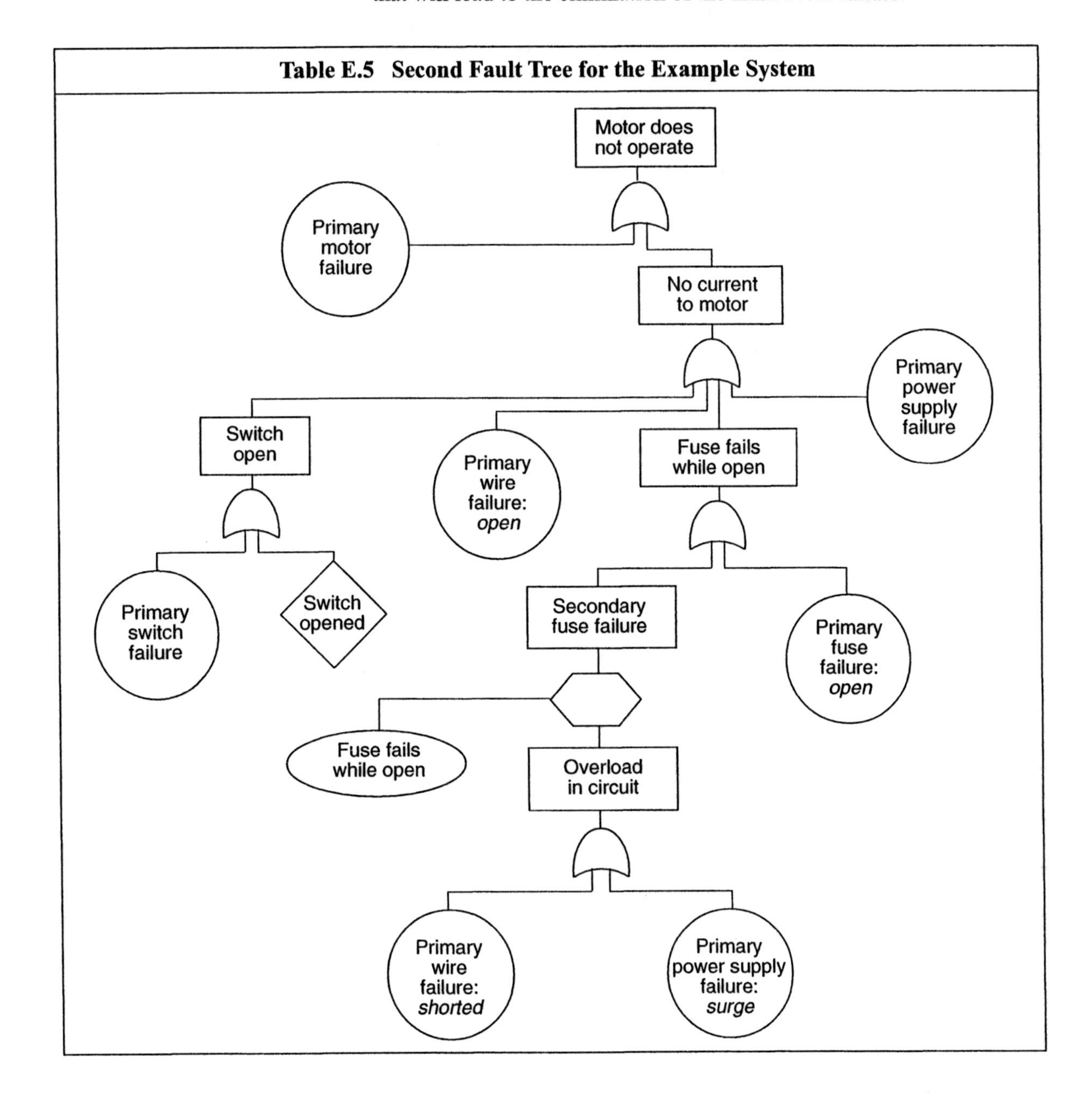

F TQM (Total Quality Management)

What is TQM (total quality management)? Basically, TQM is an attitude and a way of life in an organization, not merely a technology or procedure. With TQM, quality is designed into the entire organization and its products—it is not achieved by inspection. It involves learning and continuing education by everyone. It is different for each organization because it must be custom-fitted to its purposes and culture, to its social and economic context. Creative problem solving can be used for planning and implementing TQM in an organization. Although quality improvements are measured and monitored in a variety of ways, the ultimate success is "voted on" with the customer's pocketbook or market share. We must know who our customers are and be in constant contact with them because customer needs and expectations change frequently. change. Appendix A on quality function deployment includes a discussion of quality as it relates to the voice of the customer when manufacturing a product or providing a service. The voice of the customer is central to total quality management. Who our customers are is not always obvious at first glance. Anyone who is a user or who is impacted by our business activities is a customer. If our manufacturing activities impact the environment, our community's citizens are our customers, not just the people buying our product.

Challenge your people to think about the future, to pay attention to marketplace and technological developments and how they will impact the organization. Seeing the challenges and the threats mobilizes action.

James A. Belasco, Teaching the Elephant to Dance

Why should TQM be used? After all, isn't it a lot of trouble for organizations to change the way they are used to doing things? If the world were stable, with conditions much as they were ten, twenty, or more years ago, then doing business the usual way might be adequate. But we live in a rapidly changing world, a world with a global marketplace and increased competition from industrialized as well as developing countries. Sophisticated communications and data-processing technology are increasingly being used as tools for identifying and meeting customer needs. Consumers everywhere now demand quality—it is no longer a luxury reserved for the well-to-do. TQM has become essential to the long-term survival of companies. With TQM, they can learn how to manage change, anticipate future markets, and develop possibilities into innovative products and services. If your organization does not provide improved products and services, someone else will. In TQM, employees are seen not as subordinates but as associates who are given decision-making responsibilities and power.

If better is possible,
then good is not enough.

Ron Meiss

Three people have been crucial in the quality movement: W. Edwards Deming, Joseph Juran, and Philip Crosby. Although they differ somewhat in the specifics on how to achieve quality, it is important to know the main principles that each man has developed to give you a better understanding of TQM. W. Edwards Deming's fourteen points of quality management are probably best known; they are listed in Table F.1. Deming developed his philosophy while working with teams on the improvement of U.S. military materials during World War II. His ideas were received with open arms by the Japanese who were trying to reconstruct their economy after the war, and his ideas of teamwork were well suited to their culture. He did not become well known in the United States until the early 1980s.

Table F.1 Deming's Fourteen Points of Total Quality Management

1. Create a common goal and constant purpose throughout your organization, beginning with management.
2. Adopt the TQM philosophy.
3. Stop depending on mass inspection to achieve quality.
4. Do not award business on price tag alone; minimize total cost.
5. Constantly improve production processes and service systems.
6. Institute on-the-job training.
7. Institute leadership at all levels.
8. Drive out fear (of change and of risk taking).
9. Break down barriers between departments.
10. Eliminate slogans, exhortations, and numerical targets.
11. Eliminate work standards (quotas) and management by objective.
12. Encourage people to take pride in their workmanship.
13. Institute a vigorous program of education and self-improvement.
14. Everyone in the organization works to accomplish the transformation.

Joseph Juran was a quality expert who worked with Deming and Kaoru Ishikawa (see Appendix C) in Japan as part of General MacArthur's program to rebuild Japan's industry. Together they developed the concepts of employee involvement, just-in-time delivery, and SPC. Juran's emphasis is on building awareness of the need and opportunity for improvement, setting goals, organizing and training teams for problem solving, putting the teams to work on projects, reporting and recognizing progress and results, and making annual improvement a part of all processes and systems of an organization.

Philip Crosby, formerly with ITT, developed his zero-defect movement while at Martin Marietta. He expanded and modified Deming's fourteen points. Many companies have adopted or built on his ideas. The 8-D approach developed at Ford Motor Company depends heavily on his approach which features committed management; quality improvement teams; measurements that allow objective evaluation; a systematic method for identifying problems and permanently resolving them through corrective action; employee education, especially in defining the type of training needed to carry out the quality improvement process; improved communications; recognition of team results, and continuous improvement. Education and training are stressed by all three quality leaders.

DEFINITION

TQM is a never-ending quest to satisfy the needs of the customer better than anyone else through extra value, no variation in performance, and no defects, and through change and continuous improvement in response to changing conditions and expectations—through all the people in an organization working together.

Every product or service produced is a creation of the whole organization.

Benefits

What benefits can a company expect from its efforts to institute TQM? First, we must ask—what are the costs of not using TQM? It ultimately comes down to economic survival, because in today's and tomorrow's business climate, no organization can afford to stand still and rest on the accomplishments of the past. U.S. companies, even such giants as General Motors and Xerox, have been slow to recognize this.

Here are some examples of success stories. Through TQM, Corning doubled its profits in seven years. Ford, Xerox, Motorola, Procter and Gamble, and many others have significantly increased market share. Xerox established a 10-year program to achieve 100 percent customer satisfaction. It spent $125 million on employee education and trained 480 suppliers in quality as well. In 5 years, it achieved a 78 percent decrease in defects of machinery, 40 percent in emergency maintenance, and 27 percent in the response time of its service calls. In 1989, it won the coveted Malcolm Baldrige National Quality Award.

In 1981, someone at Motorola had the courage to tell the board of directors the shocking news: "The quality of your products stinks." In response, the company set a 5-year goal to improve quality by a factor of 10. This was achieved. But when in 1986 its executives went to Japan, they discovered that companies there were 1500 times better in lowering the number of defects per unit of work than Motorola (as reported in the cover story of *Template: The Magazine of Engineering Systems and Solutions*, Volume 5, published by Xerox). With this benchmark, Motorola then set a goal to improve its performance by 68 percent every year to achieve six sigma quality (this means only 3.4 defects per million). Thus the company reduced manufacturing costs by $1.5 billion over 4 years! Although some investment may be needed in the beginning to initiate quality improvements—around 1 percent of revenues—most companies find that this investment in quality will quickly pay back since—as shown by Genichi Taguchi—an increase in quality leads to cost reduction.

Ten steps for implementing TQM

Ron Meiss, in his workbook *Total Quality in the Real World (from Ideas to Action),* discusses ten steps that must be taken by an organization that wants to adopt TQM. Each of the steps is outlined in this section. The general requirements are that everyone is involved at all levels of the organization. It is a continuous process, with ever-expanding goals as the demand for quality by internal and external customers expands.

1. Awareness: As a first step, the organization's awareness about the need for quality improvement must be increased through effective communications. A supportive climate for creativity, risk taking, teamwork, and lifelong education must be nurtured. Paradigm shifters throughout the organization must be recognized and rewarded.

2. Commitment: Fostering commitment to TQM by everyone in the organization is crucial. All must participate and contribute to decision making. This can be accomplished through creative problem-solving teams and will lead to a basic change in the organization's culture. Communication is the key, and change must be championed. It is managed through the whole-brain approach shown in Figure F.1.

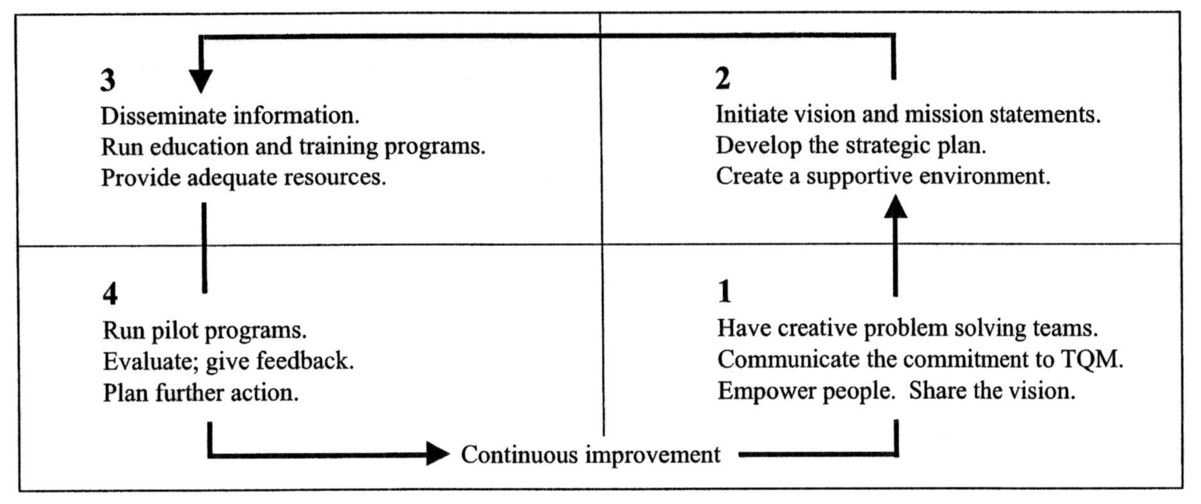

Figure F.1 Implementing TQM in an organization using the steps in the knowledge-creation cycle.

3. Involvement: Learning to use and implement TQM requires a change in the mindset of management from control to facilitation. Teamwork, flexibility, and creative problem solving are encouraged, not blind adherence to bureaucracy and pedantic procedures.

4. Change: Change is seen as positive, with concrete benefits to the survival of the organization. A climate for acceptance of new ideas and innovation is fostered. The organization encourages creative problem solving as it implements the philosophy of quality improvement through concrete team projects and processes. People do not fear change; they learn to act as change masters.

5. Improvement: Improvement is achieved in many small steps. Failure is acceptable if it serves as a learning experience and stepping stone to improvement. People learn from each other. Creative thinking is needed to identify areas and ideas for improvement. Success will usually come easily at first, since obvious quality problems are tackled first. Persistence and hard work are needed to keep the process going and effect permanent change in the organization.

6. Components: Many tools, techniques, and procedures are part of TQM, as it permeates all areas of an organization. The overriding maxim is: "The customer is always right!" This attitude involves all areas of communication: the information handled within the organization as well as to the outside; the financial resources and billing practices; supplier partnerships; manufacturing and inventory procedures; the delivery of the product; service, warranty claims, and response to mistakes through apology, restitution, and adding extra value; and the prevention of problems. Each person is responsible—there is no "passing the buck." Each person is qualified to come up with creative ideas on how to improve his or her own job. All good ideas are important, recognized and rewarded.

7. Measurement: Success is gauged through quantitative measurements, such as the seven tools of SPC. Reports are made on the actions taken and the results: How many errors have been detected? What is the percentage decrease in errors? How much time has been saved? What were the savings in expenditures? What were the increases in sales or decreases in complaints, etc.? Long-term trends are documented also.

Quality isn't an event, a decision, or a program with starting and finishing dates. It is ongoing performance, integrating excellence into marketing, manufacturing, planning, research and development, in our interrelationships with our customers and ourselves.

Edgar S. Wollard, Jr., chairman, DuPont

8. Leadership: The essence of TQM leadership is to get people to do what is needed because they want to, not because they have to. Managers should help them set goals and succeed in reaching the goals. This takes a whole-brain approach of maintaining a balance between caring and expecting top work. A unique feature is to have a mindset that catches people doing something right! Superior performance is easily recognized and rewarded when well-publicized standards and criteria exist.

9. Training: Training in the following TQM skills is especially needed:

- Interpersonal—knowing how to work with others.
- Working as part of creative problem-solving teams.
- Identifying the customers and being sensitive to their needs.
- Doing quality work with SPC techniques and other procedures.

Each person receives cross-training and continuing education to expand job skills and understand the context of the job as related to total quality.

10. Implementation: Everyone becomes involved in implementing TQM. In this ongoing process, each person is 100 percent responsible for quality—including setting goals, achieving the customer requirements, and maintaining a support network. It is difficult to change an attitude of fault-finding to a mindset that finds solutions when requirements are not met, but with the commitment of the entire team, it can be done. Continuous innovation can keep people excited about maintaining their efforts in meeting the customer requirements and increasing quality.

What are the customer requirements? On what criteria do customers judge a product? The absence of actual defects is just the starting point. Performance, reliability, durability, conformance, serviceability, extra

Quality is the product of everything we do. It is what sets us apart in the way our products perform, in the way our people respond. It is a philosophy and business practice that keeps our customers satisfied and willing to do business with us again and again.

Bill Pittman, CEO, Xerox

features, aesthetics, and reputation are all contributing factors. Receiving extra value for the cost is especially important for services. Because the requirements keep changing, communication with the customer is vital. Tools—such as the QFD House of Quality, the Taguchi loss function analysis, and the Taguchi robust design with its emphasis on uniformity around a target—can help determine quality goals (see Refs. F.12 and F.13 for more information).

Keeping variability to a minimum by manufacturing products as close to set target values as possible is central to the Japanese concept of quality. This is in contrast to manufacturing parts within specification ranges. Process variation around a target is measured using the C_{pk} (process capability) index. A C_{pk} of 1.00 means that the process average $\pm$ 3 σ falls just within the specification limits. In Japan, minimum acceptable performance has a C_{pk} of 1.33 (8 σ); for important quality characteristics, this increases to a C_{pk} of 1.66 (10 σ). But frequently, C_{pk} values exceeding 3.00, 5.00, and even 8.00 are encountered. Traditional managers are satisfied when they find parts meeting specifications with a C_{pk} = 1.00 and thus would not spend money to improve a process that exceeds this performance, not realizing that inspection, scrap, and rework will be eliminated for a process that shows a high process capability index and thus would result in overall cost savings.

To implement TQM requires creative problem solving and the best thinking skills of everyone! With TQM, people take pride in their work—they love their jobs and enjoy their working environment. They have a clear purpose and a commitment to that purpose. Everyone's mission is the same! People take the responsibility for shifting from outdated problem-solving paradigms to innovation as they continuously try to perfect what they do. They are supportive, they listen, they communicate positive feedback, they work well in teams. Where frequent communication between different departments is important, the key people's offices are located in the same area (not more than 75 feet apart). Negative attitudes and criticism are recognized as detrimental to a quality environment; instead, positive reinforcement and rewards are used.

Anyone who contributes to the creation of a product or service and does his or her best is a "producer" in the highest sense of the word. To do the best requires creative problem solving by the individual and the teams within the TQM framework of the organization.

TQM is a never-ending activity because every process and service can be improved; cost can be lowered and productivity increased. It is also a never-ending process of formal education and informal continuous learning. Be on the lookout for new books and journal articles about TQM implementation case studies—on successes and on failures. This approach to companywide quality is by no means universally accepted in the United States, and even companies that made easy progress at first now find that they are not reaping the expected profits, cooperation, and acceptance. We believe that the implementation of fundamental changes in an organization requires careful preparation—most of all in the area of teaching everyone creative thinking and problem-solving skills.

References

F.1 James A. Belasco, *Teaching the Elephant to Dance: Empowering Change in Your Organization,* Crown, New York, 1990.

F.2 Philip Crosby, *Quality Is Free: The Art of Making Quality Certain,* McGraw-Hill, New York, 1979.

F.3 Michael L. Dertouzos, Richard K. Lester, and Robert M. Solow, *Made in America: Regaining the Productive Edge,* MIT Press, Cambridge, Massachusetts, 1989.

F.4 Peter F. Drucker, *Innovation and Entrepreneurship: Practice and Principles,* Harper and Row, New York, 1985.

F.5 Andrea Gabor, *The Man Who Discovered Quality: How W. Edwards Deming Brought the Quality Revolution to America—The Story of Ford, Xerox, and GM,* Random House, New York, 1990.

F.6 David Haberstam, *The Reckoning,* Morrow, New York, 1986.

F.7 J.M. Juran and Frank M. Gryna, *Quality Planning and Analysis: From Product Development through Use,* third edition, McGraw-Hill, New York, 1993.

F.8 Ron Meiss, *Total Quality in the Real World: From Ideas to Action,* Ron Meiss and Associates, Kansas City, Missouri, 1991.

F.9 Rosabeth Moss Kanter, *Change Masters: Innovation for Productivity in the American Corporation,* Simon & Schuster, New York, 1985.

F.10 Tom Peters and Robert H. Waterman, Jr., *In Search of Excellence: Lessons from America's Best-Run Companies,* Warner Books, New Yok, 1982.

F.11 Peter M. Senge, *The Fifth Discipline: The Art and Practice of the Learning Organization,* Doubleday, Garden City, New York, 1990.

F.12 Genichi Taguchi, *Taguchi on Robust Technology Development: Bringing Quality Engineering Upstream,* ASME Press, Fairfield, New Jersey, 1993.

F.13 Genichi Taguchi, *Introduction to Quality Engineering: Designing Quality into Products and Processes,* Kraus International, UNIPUB (Asian Productivity Organization), White Plains, NY, 1986.

F.14 Robert H. Waterman, Jr., *The Renewal Factor: How the Best Get and Keep the Competitive Edge,* Bantam Books, New York, 1987.

F.15 James P. Womack, David T. Jones, and Daniel Roos, *The Machine That Changed the World (The Story of Lean Production),* Harper, New York, 1991.

Any institution has to be organized so as to bring out the talent and capabilities within the organization; to encourage people to take initiative, give them a chance to show what they can do, and a scope within which to grow.

Peter Drucker

The ability to learn faster than your competition may be the only sustainable advantage.

Arie de Geus, Head of Planning for Royal Dutch/Shell

INDEX